LANDMARK PAPERS: METAMORPHIC PETROLOGY

Selected by B. W. Evans
University of Washington, USA

2007
Published by the Mineralogical Society of Great Britain & Ireland

THE MINERALOGICAL SOCIETY

The Mineralogical Society of Great Britain & Ireland was instituted in 1876. The prime aim of the Society is to advance the knowledge not only of Mineralogy but also of Crystallography, Geochemistry and Petrology, together with kindred subjects. This is done principally through the publication of scientific journals, books and monographs, and through arranging or supporting scientific meetings. The Society speaks for Mineralogy in Great Britain, linking with British science through the Royal Society and cooperating closely with the Geological Society. It maintains close liaison with other European mineralogists as a member society of the European Mineralogical Union, and is the body that nominates British representatives to the International Mineralogical Association. There are some 700 members of the Society, of whom approximately half reside abroad.

The Society's contact details are:
Address: 12 Baylis Mews, Amyand Park Road, Twickenham TW1 3HQ, UK
Tel. +44 (0)20 8891 6600
Fax +44 (0)20 8891 6599
E-mail: info@minersoc.org
Website: www.minersoc.org

Copies of this book are available from the Mineralogical Society office.

First edition 2007.

Typeset in 10.5/12 pt Times by Westfield Typesetting, West Kirby, Wirral, UK
Printed and bound by Lightning Source, UK

ISBN: 978-0-903056-24-3

© 2007 The Mineralogical Society of Great Britain and Ireland (chapters written by B.W. Evans). Copyright for all other material resides with the original publishers and/or the authors. The latter material is reproduced here with permission for which the Society is grateful.

No reproduction, copy or transmission of this publication may be made without written permission, except within the provisions of the Copyright Licensing Agency, 90 Tottenham Court Road, London W1P 9HE, UK, or the Copyright Clearance Center, 27 Congress Street, Salem, MA 01979, USA. Enquiries concerning reproduction outside these terms should be directed to the Mineralogical Society at the London address printed on this page or to the appropriate publisher on the list given below.

A catalogue record for this book is available from the British Library.

The Publishers make no representation express or implied, with regard to the accuracy of the information contained in this book, and cannot accept any legal responsibility for any errors or omissions.

Publishers of original papers, reproduced with permission in this volume, together with links to their websites or publications:

Geologists' Association (http://www.geologist.demon.co.uk/)
Geological Survey of Norway (http://www.ngu.no/)
University of Chicago Press (http://www.press.uchicago.edu/)
Mineralogical Society of America (http://www.minsocam.org/)
Carnegie Institution of Washington (http://www.carnegieinstitution.org/yearbooks.html)
Yale University (http://www.geology.yale.edu/journals/Ajs.html)
Springer (http://www.springer.com)
Geological Society of London (www.geolsoc.org.uk)
Oxford University Press (www.oup.co.uk)
Blackwell Publishing (http://www.blackwellpublishing.com/)
American Association for the Advancement of Science (http://www.aaas.org/)

Preface

Much of Earth's crust and arguably parts of its mantle are composed of rock that has undergone partial to complete textural and mineralogical reconstitution as a result of changes in conditions imposed on it. Metamorphic rocks carry a record of surface, shallow and deep geological events and processes going back to 4 Ga. Early in the last century, the descriptive science of metamorphic petrography began a gradual evolution into metamorphic petrology and petrogenesis, much as we know it today. Researchers came to depend more and more on related sciences, such as thermodynamics, materials science, mineralogy, tectonophysics, and isotope geochemistry, to provide a fuller understanding of the facts coming from the field and the laboratory. Fundamental principles and procedures from these borrowed sciences helped keep metamorphic petrology moving and contributed to its endless fascination.

For me personally, it was exciting to have been active during what Touret and Nijland (2002) have called the "golden years" of metamorphic petrology that commenced with university expansion after World War II. Many of the brightest and scientifically best-qualified workers were attracted into our branch of the geosciences, and over a period of some 30 years the growth in documented observations, laboratory data, and conceptual understanding of metamorphic systems was enormous. It soon became clear that metamorphic petrology had much to contribute to the dynamics of major crustal processes resulting from movements of Earth's plates. It also helped our feeling of satisfaction that the investigation of mineral compositions and assemblages, reaction textures, and phase equilibria were amenable to quantitative theoretical analysis. While the heady days may be gone – notwithstanding continuing developments of the most exciting kind (e.g. UHPM research) – the solid accomplishments of the 20th century are of course amply recorded. The monograph by Spear (1993) on *Metamorphic Phase Equilibria and Pressure-Temperature-Time Paths* is, in my opinion, outstanding testimony to the scientific maturity of metamorphic petrology near the close of the 20th century.

I have found it an absorbing and in many respects eye-opening experience to go back a hundred years and read what was then being written; I hope readers will react in the same way. I have chosen to start in 1912 with two very different papers by Victor Moritz Goldschmidt and George Barrow. As you will see, the rest of the papers reproduced here seem to flow from these early writings. Brief historical accounts of the development of metamorphic petrology are commonly given in our textbooks, but I would like to draw readers' attention to the recent comprehensive reviews of metamorphic petrology in general by Miyashiro (1994) and Touret and Nijland (2002), and of thermodynamics and metamorphism by Fritscher (2002). These accounts follow the flow of ideas, revealing at the same time something of the problems that, in the early days, hindered orderly linear progress: communication difficulties, the influence of persuasive and influential players, the inertia of certain schools of thought, etc. The purpose of this Landmark series volume is to let students read for themselves in the original how some of the giants of the field set down their ideas. Their papers convey something that is not necessarily obvious in the summaries found in our textbooks, namely a feeling and respect for the environment of intellectual discourse in which the early thinkers worked. Many things that we consider self-evident today were not at that time part of the general scientific understanding, yet they wrote with admirable clarity and logic and made the best of what information was available. I should also mention here Yoder's (1993) 'Timetable of Petrology', a comprehensive listing of first discoveries in igneous and metamorphic petrology, an eye-opener for most of us.

Many metamorphic rocks are attractive in appearance, in hand specimen and in thin section under the microscope. Their aesthetic appeal touches senses beyond the merely scientific, a pleasure that most geochemists forego. Added to this is the satisfaction we derive as they reveal each new instance of nature's fidelity to scientific laws. But the aesthetic appeal comes at a cost. Metamorphic petrologists need to be adept in very many things: crystal chemistry, physical chemistry, phase equilibria, microscopy, textural analysis, regional geology, structural geology, geodynamics and geochronology. We recognize that curriculum revision must proceed, but the total elimination of instruction in polarized-light microscopy can only be viewed as a huge mistake.

Nowadays the international literature of science is almost exclusively written in the English language. In the first half of the 20th century this was not the case, and metamorphic petrology was no exception. Important papers were written in French, German, Swedish, Japanese and other languages. In selecting only one of these in translation, I am left with the uneasy feeling that I might have searched further among them for overlooked treasures.

My task has been simplified by the decision of the editor of this series to devote one or more later volumes to landmark papers in experimental petrology. High-PT phase equilibrium and other kinds of laboratory experiments (e.g. in kinetics, calorimetry) have played such an important role in the 20th century development of metamorphic petrology that, absent these other volumes, I would have been grossly remiss had I failed to include some of these papers in this volume. The advances in metamorphic petrology made in the second half of the 20th century would not have been possible without the creative energies of those who developed and exploited the cold-seal hydrothermal pressure vessel, the piston-cylinder apparatus, and more recently the multi-anvil press. Experimental petrology went hand-in-hand with growth in the thermodynamic database of metamorphic minerals, and in the thermodynamics skills of the average petrologist.

The electron microprobe introduced in the 1960s supplanted the tedious separation of minerals and bulk analysis that necessarily accompanied descriptive petrology of new field areas. Its micron-size resolution opened up new areas of study and understanding of metamorphic systems, and it became a valuable tool in the phase-characterization of experimental run-products.

With one exception, each chapter in this book is devoted to one selected paper, even though in many cases this meant I had to select one from a series of equally important contributions on closely related subject matter by the same author or authors. Some of them were, or still are, controversial. In the introduction to each chapter, I have tried to set the scene prior to the appearance of the paper, comment – helpfully, I hope – on its contents, and talk about ensuing developments. These remarks focus on the impact the paper made on current understanding of the topics covered, and are not a substitute for a more thorough and comprehensive review. Some of the papers were the product of superior observations coupled with a heightened recognition of their significance, some the product of unusually creative thinking – to which we tend to react like Thomas Huxley on reading Darwin's *Origins*: "How incredibly stupid not to have thought of that!" While preparing the various chapters in this volume, I was struck by how interconnected the topics were as ideas evolved through the twentieth century; each topic is not a neat separate package. Difficult choices had to be made, and doubtless my own biases show. If my selection of landmark papers induces indignant reaction among my readers, I will judge it to have succeeded in at least generating the desired interest. It is unfortunately true that selecting the work of just a few players leaves out the contributions of the majority who have helped shape our science.

Volkmar Trommsdorff passed away suddenly in the summer of 2005 while I was preparing this volume, a great shock to family and friends. He and I had collaborated in writing a good many papers on the petrology of metamorphosed ophiolitic rocks in the Alps. I recall with great pleasure our fieldwork in some of the world's most beautiful places. Volkmar's field interests extended from Spain to the Himalaya, and his investigations of mixed volatile systems, phase equilibria, ophiolites and Alpine geology and tectonics won him wide recognition and awards. I am pleased that the Mineralogical Society has allowed me to dedicate this volume to his memory.

I would like to acknowledge miscellaneous kinds of help in the preparation of this volume from J.J. Ague, R.G. Berman, C. Chopin, M.J. Holdaway, B. Mason, E. Mullen, P. O'Brien, D.M. Pattison, R. Powell, S.S. Sorensen, D.K. Tinkham and V. Trommsdorff. Mineral abbreviations are those recommended by Kretz (1983).

References

Fritscher, B. (2002) Metamorphism and thermodynamics; the formative years. Pp. 143–166 in: *The Earth Inside and Out: Some Major Contributions to Geology in the Twentieth Century* (D.R. Oldroyd, editor). Special Publication, **192**, Geological Society, London.
Kretz, R. (1983) *American Mineral*ogist, **68**, 277–279.
Miyashiro, A. (1994) *Metamorphic Petrology*. Oxford University Press, New York, 404 pp.
Spear, F.S. (1993) *Metamorphic Phase Equilibria and Pressure-Temperature-Time Paths*. Mineralogical Society of America Monograph, 799 pp.
Touret, J.R.L. and Nijland, T.G. (2002) Metamorphism today: new science, old problems. Pp. 113–142 in: in: *The Earth Inside and Out: Some Major Contributions to Geology in the Twentieth Century* (D.R. Oldroyd, editor). Special Publication, **192**, Geological Society, London.
Yoder, H.S. (1993) Timetable of Petrology. *Journal of Geological Education*, **41**, 447–489.

Bernard W. Evans
University of Washington
May, 2007

Volume 3 Editor: Bernard W. Evans

Bernard Evans took a B.Sc. at King's College London in 1955, and D.Phil at Oxford in 1959 with a thesis on Dalradian rocks in Connemara. A postdoctoral position with W.S. Fyfe at the University of California Berkeley in 1961–62 morphed into an assistant research position with responsibility for the new microprobe. He joined the faculty there in 1964, teaching mineralogy, petrology and geochemistry. In 1969 he moved to the University of Washington Seattle, and became emeritus in 2001. His research interests have been primarily in the phase petrology and petrogenesis of crystalline rocks of many kinds, and the thermodynamic properties of the amphiboles. He has worked in Hawaii, the Alps, Greece, and North America.

LANDMARK PAPERS: METAMORPHIC PETROLOGY

Foreword

Metamorphic Petrology is a field that was barely defined at the beginning of the 20th century, but blossomed rapidly to the point where many of its first order problems had been resolved by the century's end, opening up a whole range of new research opportunities. The core of the subject is the recognition of the interplay of rock composition and metamorphic environment in giving rise to mineral assemblages that have recurred through space and time (and are doubtless still developing today), and the determination of the conditions implied by specific assemblages. This collection of papers documents the recognition of these systematic patterns of mineral assemblages, the realisation of their significance and the value of physical chemistry for their interpretation, and then the progressive refinement and especially calibration of the approach to yield information about depths and temperatures of metamorphism. Little more than 40 years ago, I am told that a Professor of Geology from Imperial College was able to stand in front of the Geological Society of London waving an experimentalist's platinum capsule and declare that nothing that came out of it was going to convince him how much cover had been eroded from the rocks of north east Scotland; here we see that not only has the metamorphic project been vindicated in its central tenets, it has gone on to identify new geodynamic regimes leading to the return of ultra high pressure rocks from far greater depths than had seemed possible.

Bernard Evans is especially well qualified to draw together its landmark papers, because his own contributions span the 3 core elements that have come together to make metamorphic petrology a coherent discipline and an essential part of studies of the workings of the deep earth: fieldwork and the study of natural samples, experimentation, and the application of equilibrium thermodynamics. His determination of the upper stability limit of muscovite plus quartz was itself a landmark contribution in experimental petrology, but he has since made major contributions to our understanding of the metamorphism of basic and ultrabasic rocks, based on fieldwork in both the USA and Europe, and to the thermodynamics of their major minerals. This has given him a uniquely balanced perspective on what has made the subject tick, and of course he has known many of the main participants personally. His choice of papers embraces the major phases of the development of the subject, from the early recognition of metamorphic patterns and their significance by the pioneers Barrow, Goldschmidt and Eskola, through the blossoming of approaches to quantify the relative and absolute conditions under which metamorphism takes place, and on to recent work that highlights the profound statements that can be made about the cycling of rocks to depth and their return to the surface thanks to the robust basis of the discipline. While the main emphasis is on the development and validation of an approach to metamorphic rocks that is rooted in equilibrium thermodynamics, improved understanding of metamorphic processes and rates are also represented.

The papers reproduced here are drawn from a motley collection of sources, based only on the influence that they had on the subject; a number are from publications that would not figure in the Science Citation Index. What makes this collection especially valuable is the commentary that Bernard Evans has written to go with each chapter, placing it in the context of what had gone before and what happened next. It will be something of a surprise to many readers to learn that the work of figures such as Barrow and Eskola was not always well appreciated at the time, running contrary as it did to the outdated ideas of senior professors whose names are now largely forgotten. In the commentaries he has also been able to highlight other influential contributions by authors whose work may never have been encapsulated in a single paper suitable for inclusion, but was nonetheless of seminal importance for the development of the field.

Within these pages the practicing metamorphic petrologist will find the context and rationale of much that has been taken for granted, while the general reader may perhaps gain a slightly more sympathetic understanding of why it is that metamorphic petrologists set such store by ever more complex phase diagrams. The papers document the major achievements of the field, but also the new challenges that it has set itself. Here are the definitive statements of the subject alongside the papers that defined current areas of endeavour, including some that are still controversial. That is how it should be: the study of metamorphic rocks is not a dead or a static topic, but one that continues to move on and evolve.

Bruce W. Yardley

LANDMARK PAPERS: METAMORPHIC PETROLOGY

by B. W. Evans

Table of Contents

Preface	iii
Editor	iv
Foreword	v

1: The laws of metamorphism
    Goldschmidt, V.M. (1912) Die Gesetze der Gesteinsmetamorphose. *Norsk Videnskapsselskaps Skrifter I. Matematisk-Naturvidenskapelig Klasse*, **22**, 16 pp.   L1

2: Barrovian metamorphism
    Barrow, G. (1912) On the geology of the lower Dee-side and the southern Highland border. *Proceedings of the Geologists Association*, **23**, 274–290.   L9

3: The metamorphic facies
    Eskola, P. (1920) The mineral facies of rocks. *Norsk Geologisk Tidsskrift*, **6**, 143–194.   L25

4: Metamorphism of limestone and the petrogenetic grid
    Bowen, N. L. (1940) Progressive metamorphism of siliceous limestone and dolomite. *Journal of Geology*, **48**, 225–274.   L63

5: Composition projections
    Thompson, J. B. (1957) The graphical analysis of mineral assemblages in pelitic schists. *The American Mineralogist*, **42**, 842–858.   L109

6: Mixed volatiles
    Greenwood, H.J. (1962) Metamorphic reactions involving two volatile components. *Carnegie Institution of Washington, Yearbook*, **61**, 82–85.   L131

7: Blueschists
    Ernst, W.G. (1971) Do mineral parageneses reflect unusually high-pressure conditions of Franciscan metamorphism? *American Journal of Science*, **270**, 81–108.   L139

8: Pseudosections
    Hensen, B.J. (1971) Theoretical phase relations involving cordierite and garnet in the system $MgO\text{-}FeO\text{-}Al_2O_3\text{-}SiO_2$. *Contributions to Mineralogy and Petrology*, **33**, 191–214.   L171

9: Thermal models of collision belts
    England, P.C. and Richardson, S.W. (1977) The influence of erosion upon the mineral facies of rocks from different metamorphic environments. *Journal of the Geological Society of London*, **134**, 201–213.   L199

10: Mineral thermometry
    Ferry, J.M. and Spear, F.S. (1978) Experimental calibration of the partitioning of Fe and Mg between biotite and garnet. *Contributions to Mineralogy and Petrology*, **66**, 113–117.   L219

11: Thermodynamic database of minerals

(a) Helgeson, H.C., Delaney, J.M., Nesbitt, H.W. and Bird, D.K. (1978) Summary and critique of the thermodynamic properties of rock-forming minerals. *American Journal of Science*, **278-A**, 1–229.

(b) Berman, R.G. (1988) Internally-consistent thermodynamic data for minerals in the system Na_2O-K_2O-CaO-MgO-FeO-Fe_2O_3-Al_2O_3-SiO_2-TiO_2-H_2O-CO_2. *Journal of Petrology*, **29**, 445–522.

(c) Holland, T.J.B and Powell, R. (1990) An enlarged and updated internally consistent thermodynamic dataset with uncertainties and correlations: the system K_2O-Na_2O-CaO-MgO-MnO-FeO-Fe_2O_3-Al_2O_3-TiO_2-SiO_2-C-H_2-O_2. *Journal of Metamorphic Geology*, **8**, 89–124.   L229

12: Fluid driven metamorphism

Ferry, J.M. (1983) Regional metamorphism of the Vassalboro Formation, south-central Maine, U.S.A.: a case study of the role of fluid in metamorphic petrogenesis. *Journal of the Geological Society of London*, **140**, 551–576.   L255

13: Kinetics: experiment and theory

Wood, B.J. and Walther, J.V. (1983) Rates of hydrothermal reactions. *Science*, **222**, 413–415.   L287

14: Ultra-high pressure metamorphism

Chopin, C. (1984) Coesite and pure pyrope in high-grade blueschists of the Western Alps: a first record and some consequences. *Contributions to Mineralogy and Petrology*, **86**, 107–118.   L297

15: Kinetics: field

Austrheim, H. (1987) Eclogitization of lower crustal granulites by fluid migration through shear zones. *Earth and Planetary Science Letters*, **81**, 221–232.   L317

CHAPTER 1: THE LAWS OF METAMORPHISM

Goldschmidt, V.M. (1912) Die Gesetze der Gesteinsmetamorphose. *Norsk Videnskapsselskaps Skrifter I. Matematisk-Naturvidenskapelig Klasse*, **22**, 16 pp.

Our first two landmark papers were published in the same year, a little short of a century ago. Both have had an enduring influence on the entire framework of later studies in metamorphic petrology. They appeared at the time when the study of metamorphic rocks began to have its own identity within the field of 'petrography', thanks to the publication of influential texts by Becke (1903), Van Hise (1904), and Grubenmann (1904).

Readers will notice immediately the markedly different manner in which the subject matter is handled in the two papers. Together, they show that writing which attempts to quantify, theorize, and generalize, on the one hand, and that which largely documents a natural history, on the other, can play equally important roles in the development of a science. That is not to say that the Goldschmidt and Barrow papers fall neatly into either category; both were supported by extensive fieldwork. There is ample justification even today for field and theoretical studies to go hand-in-hand.

I might have chosen any one of three or four of Goldschmidt's classic papers written during the metamorphic part of his "petrographic years" (Mason, 1992) from 1907 to 1921. (Read Brian Mason's biography for fascinating scientific and personal information about this legendary scientist.) The 1912 paper is a concise synthesis of Goldschmidt's ideas on metamorphism at the time, incorporating discoveries developed in the course of his doctoral work on hornfelses in the Oslo region. Some of his contemporaries thought that the title of his paper was a shade presumptuous for a 24 year old. But when nature meets the expectations of theory to the extent that Goldschmidt found in the inner-zone hornfelses, we can perhaps forgive the title as simply an expression of youthful exuberance.

Goldschmidt's father was a colleague of Van't Hoff and later a professor of chemistry, and so the emphasis on physical chemistry in his work comes as no surprise.

Goldschmidt's doctoral thesis (1911) recognized that Gibbs' phase rule ($P + F = C + 2$), previously applied outside the laboratory by Van't Hoff to the mineral parageneses of salt deposits, applied equally well to metamorphic systems in nature. With degrees of freedom, F, represented by pressure and temperature, a practical "mineralogical phase rule" is simply $P = C$, or in words: "the number of minerals present can equal but will not exceed the number of components". This explained why, in the products of contact metamorphism of shale-marl-limestone sediments around alkali gabbro, diorite, syenite and granite intrusions in the Oslo (Christiania) region, Goldschmidt was able to recognize ten compositional classes of inner-zone hornfelses represented by simple 2- and 3-mineral associations in the (SiO_2-saturated) system Al_2O_3-CaO-(Mg,Fe)O. Chemical analyses of representative rocks showed consistent correlations between bulk-chemical and mineralogical composition. The classes (associations) occur repeatedly, whereas other possible mineral pairs (that we would call incompatible) never occur. These observations were the basis for Goldschmidt's main argument that thermodynamic equilibrium had prevailed. Shortly after, Eskola (1915) showed the correlation between bulk-chemical and mineralogical compositions in an ACF diagram, and the Oslo region became the type locality for the hornfels facies (see Chapter 3). The hornfelses in the Comrie aureole (Tilley, 1924) were found to show the same equilibrium features as those in Norway. We may feel that these simple mineral–chemical correlations are self-evident, but early on this was not the case.

In the 1912 paper, Goldschmidt asserted that whole-rock composition is not significantly changed during "normal" metamorphism. This view excluded other classes of rocks such as skarns, ore deposits and pneumatolytic rocks, of whose existence he was fully aware. By no means all of Goldschmidt's contemporaries (see Harker, 1918) accepted the simple notion that the mineralogical compositions of metamorphic rocks can be completely determined by the composition of the original sediment and the temperature and pressure of recrystallization; the formation of new minerals in any rock could, for example, be driven by the introduction of new elements.

Goldschmidt admitted the tentative nature of his calculated curve for the calcite-quartz-wollastonite equilibrium, but it was nevertheless a petrological *tour de force* for that time. Experiments performed much later (Harker and Tuttle, 1956; Haselton et al., 1978), and calculations at different times (Danielsson, 1950; Ellis and Fyfe, 1956; Mäder and Berman, 1991), show that Goldschmidt's curve is some 100°C high in the low-pressure region and 100–200°C low in the high-pressure region. Overall, he was commendably close! In his calculation formula, Q is the heat of reaction, 4.571 is 2.303R, the second term is a heat capacity term, and the constant C is an entropy term. Nernst's heat theorem is a statement of the Third Law. CO_2 was taken to be an ideal gas, which is partly why his curve was at too low a temperature at high pressure. He would not have thought that this and similar reactions in thick limestone beds could ever have relevance to Earth's climate (Kerrick and Caldeira, 1993)! Goldschmidt also recognized that the (now familiar) logarithmic equilibrium P-T curve must also apply to dehydration reactions, and is in contrast with the straight dP/dT slopes of solid-solid reactions. He hinted at the eventual development of what we now call a petrogenetic grid (see Chapter 4). In identifying a curve for the reaction en + an = ky + di, he was probably thinking of its application to eclogites.

For generalizations about the pressure and temperature of metamorphism, the wollastonite equilibrium was not of course a good choice for the first devolatilization reaction on our (future) petrogenetic P-T grid. In most lithological types, the pressure of H_2O is much more likely than CO_2 to approach that of lithostatic pressure during prograde metamorphism.

Goldschmidt thought, as we do today, that the limit of metamorphism is reached when the rock is largely molten. This meant that ideas of magma temperatures current at the time determined the temperature range of metamorphism. The lower limit of 1000°C for the inner zone hornfelses in the Oslo region (Goldschmidt 1911, p. 110) was taken from the eutectic point for alkali-feldspar-quartz mixtures (Vogt, 1908, p. 117), after allowing a small lowering for magmatic H_2O. This was long before Tuttle and Bowen's classic granite memoir (1958). The 1200°C upper limit of temperature (Fig. 1) was provided by the absence in the hornfelses of pseudowollastonite (1180°C, Goldschmidt 1911, revised to 1120°C by Buckner et al., 1960), and the lack of signs of partial melting. Our ideas of conductive heat flow in contact aureoles have also evolved (Lovering, 1935; Jaeger, 1959). These considerations would dramatically change the various fields of metamorphism on his P-T diagram, and solve the problem of the scarcity of wollastonite in the regionally metamorphosed rocks of the high mountains of central Norway. Conversely, a down-temperature correction in the wollastonite curve, as noted above, would provide more reason for the development of wollastonite in the low-pressure (400–1500 atm.) hornfelses of the Oslo region. Later (Chapter 4), we shall see that Bowen advocated the use of the final disappearance of calcite + quartz, rather than the growth of wollastonite, as indicative of upgrade passage across the wollastonite curve; this approach minimizes uncertainty due to the influence of components other than CO_2 in the fluid, which is something Goldschmidt did not discuss.

Although the "depth zone" model of regional metamorphism favoured by Grubenmann and others obviously influenced Goldschmidt's Fig. 1, he departed from the "static" idea of Grubenmann (that T is dependent on P), as Eskola and Harker did later, by discussing the possible roles of magmatic heat and tectonic overpressure in perturbing the relation. In the last sentence of his paper, Goldschmidt admitted that we have not yet completely "mastered the laws of regional metamorphism".

Any petrological paper that is nearly a century old is likely to have shortcomings, particularly when the emphasis is on petrogenesis. Rather, the paper is to be enjoyed for its unequivocal emphasis on the application of thermodynamics and physical chemistry to what is observed in the field. In that respect the paper is sound; obviously things have changed because we have many new thermochemical data and enormous volumes of field information at our disposal. Yet I think his general views on the controls of metamorphism have in fact survived very well. Sadly, it was quite some time before mainstream metamorphic petrologists began to view their rocks as chemical systems.

Goldschmidt went on to develop the emerging field of geochemistry and justifiably earned the title of Father of Modern Geochemistry (Mason, 1992). It is tantalizing to contemplate where we might now be in our field had he stayed with his first-love, *die Gesteinsmetamorphose*.

In my translation I have tried as much as possible to retain the writing style and terminology that was in use at the time. "Gesteinsmetamorphose" has been simplified to "metamorphism" except in the title. Note that "thermometamorphism" or thermal metamorphism is not the same as "contact metamorphism", whereas in modern usage we tend to equate thermal with contact metamorphism.

References

Becke, F. (1903) Über Mineralbestand und Struktur der kristallinischen Schiefer. *Denkschriften der kaiserliche Akademie der Wissenschaften, Mathematisch-Naturwissenschafte Klasse*, **75**, 1–53, 1913 (first published 1903).

Buckner, D.A., Roy, D.M. and Roy, R. (1960) Studies in the system CaO-Al_2O_3-SiO_2-H_2O. II: The system $CaSiO_3$-H_2O. *American Journal of Science*, **258**, 132–147.

Danielsson, A. (1950) Das Calcit-Wollastonitgleichgewicht. *Geochimica et Cosmochimica Acta*, **1**, 55–69.

Ellis, A.J. and Fyfe, W.S. (1956) A note on the calcite-wollastonite equilibrium. *American Mineralogist*, **41**, 805–807.

Eskola, P. (1915) On the relations between the chemical and mineralogical composition in the metamorphic rocks of the Orijärvi region. *Bulletin de la Commission Géologique de Finlande*, **44** (English summary pp. 109–145).

Goldschmidt, V.M. (1911) Die Kontaktmetamorphose im Kristianiagebiet. *Videnskapsselskapets Skrifter I. Matematisk-Naturvidenskapelig Klasse*, No. 1, 483 pp.

Grubenmann, U. (1904–1906) *Die Kristallinen Schiefer*. Gebrüder Borntråger, Berlin.

Harker, A. (1918) The present position and outlook of the study of metamorphism in rock masses (presidential address). *Geological Society of London Quarterly Journal*, **74**, li–lxxx.

Harker, R.I. and Tuttle, O.F. (1956) Experimental data on the P_{CO_2}-T curve for the reaction: calcite + quartz = wollastonite + carbon dioxide. *American Journal of Science*, **254**, 239–256.

Haselton, H.T., Jr., Sharp, W.E. and Newton, R.C. (1978) CO_2 fugacity at high temperatures and pressures from experimental decarbonation reactions. *Geophysical Research Letters*, **5**, 753–756.

Jaeger, J.C. (1959) Temperatures outside a cooling intrusive sheet. *American Journal of Science*, **257**, 44–54.

Kerrick, D.M. and Caldeira, K. (1993) Paleoatmospheric consequences of CO_2 released during early Cenozoic regional metamorphism in the Tethyan orogen. *Chemical Geology*, **145**, 213–232.

Lovering, T.S. (1935) Theory of heat conduction applied to geological problems. *Geological Society of America Bulletin*, **46**, 69–93.

Mäder, U.K. and Berman, R.G. (1991) An equation of state for carbon dioxide to high pressure and temperature. *American Mineralogist*, **76**, 1547–1559.

Mason, B. (1992) *Victor Moritz Goldschmidt: Father of Modern Geochemistry*. The Geochemical Society, Special Publication No. **4**, 184 pp.

Tilley, C.E. (1924) Contact metamorphism in the Comrie area of the Perthshire Highlands. *Quarterly Journal of the Geological Society of London*, **80**, 22–71.

Tuttle, O.F. and Bowen, N.L. (1958) Origin of granite in the light of experimental studies in the system $NaAlSi_3O_8$-$KAlSi_3O_8$-SiO_2-H_2O. *The Geological Society of America, Memoir* **74**, 153 pp.

Van Hise, C.R. (1904) *A Treatise on Metamorphism*. United States Geological Survey, Monograph XLVII, 1286 pp.

Vogt, J.H.L. (1908) Physikalisch-chemische Gesetze der Krystallisationsfolge in Eruptivgesteinen. *Tschermaks mineralogische und petrographische Mitteilungen*, **27**, 105–180.

THE LAWS OF ROCK-METAMORPHISM, WITH EXAMPLES FROM THE GEOLOGY OF SOUTHERN NORWAY

V.M. Goldschmidt: *Videnskapsselskapets Skrifter I. Matematisk-Naturvidenskabelig Klasse*, 2 Bind, No. 22, 1–16.

Translated from German by B.W. Evans

One of the most interesting fields of geology and the related sciences of petrography and mineralogy is the metamorphism of rocks. The phenomenon whereby limestone is changed into marble, shale into hornfels or a coarsely crystallized garnet-micaschist, and granite into gneiss, has long been observed in numerous places, perhaps nowhere with such great diversity as in Norway, where the greatest part of the bedrock consists of metamorphic rocks.

Investigation into the occurrence and properties of metamorphic rocks has provided the material for numerous valuable studies that have given us a rich database. However, we are still poorly informed about the causes of metamorphism, and about the conditions that are necessary for the alteration of rocks; interpretations in this field still remain unclear and inconsistent.

Of course, as with all natural phenomena, general laws must exist which regulate with quantitative accuracy the occurrence of individual types among the diversity of metamorphic rocks. The known laws of nature are in part based directly on experience, and in part derived from logic and mathematics. An especially clear position is derived from those laws that can be based purely on thermodynamics, which require no assumptions other than the three laws of thermodynamics. The science that applies thermodynamic ways of thinking about the changes of observable bodies is physical chemistry; this science must furnish us with the key to understanding metamorphic rocks.

In order to be able to formulate the use of thermodynamic laws in metamorphism, we must first determine as exactly as possible what the rocks were before their transformation, and what they have become due to the transformation. The study of thin sections by optical means offers us a working method. The procedures which serve this work have been perfected significantly in the last decade; here I need only refer to the advances associated with the name of F. Becke.

The general result of such investigations with regard to the nature of the alteration process indicates that normal metamorphism, be it contact or regional, takes place in the solid state. A shale, which is transformed into a micaschist, is neither melted nor completely in solution during the metamorphism; during each individual stage of the metamorphism the principal mass of the rock remains solid; only a small fraction is in a state other than solid, or – to express it differently – the mass of the crystal phases is always much greater than that of its associated solution.

This can be perceived from the fact that the layered structure of metamorphic sediments is preserved in the metamorphic product; even more convincing are the traces of fossil organisms that are found in metamorphosed sediments. Thus, W.C. Brögger described well preserved fossils in high-grade contact metamorphic rocks, and H. Reuch in regional metamorphic rocks.

Furthermore, chemical investigation has shown that the quantitative whole-rock composition is not significantly changed during normal metamorphism (only certain volatile substances, water and carbon dioxide can be driven out). As a rule, no addition of material to the rock takes place (exceptions are various kinds of pneumatolytic metamorphism).

Alteration during metamorphism involves only exchanges among the individual components within the rock-mass itself; during the transformation certain combinations are newly formed, while others are simultaneously eliminated. Metamorphism primarily concerns the mineral content of the rocks.

Whereas for example a shale comprises the following constituents: the smallest quartz grains, flakes of muscovite, chlorite, kaolinite and amorphous Fe hydroxide, the metamorphic product of the same rock in one case consists of quartz, andalusite, cordierite, biotite and feldspar, and, in another case, quartz, muscovite, biotite and garnet.

The following question then remained to be answered: why does metamorphism cause exchanges among the individual components of a rock?

Thermodynamics shows us that every mixture of a given composition at given values of temperature and pressure can produce only one single stable combination of compounds in the solid state. If in some way a mineral assemblage has formed that differs from that corresponding to the state of equilibrium, then transformations must occur until finally only the stable state is reached.

The real advantage of this thermodynamic approach lies in the fact that we need no special hypothesis as to how this transformation takes place, whether in gas, liquid or solid state, with or without the help of foreign solvents; the final result is clearly determined by the type and amount of materials as well as P and T.

Thus the question becomes the following: how does one find the equilibrium conditions for given materials at given values of pressure and temperature?

In this initial form the problem appears complicated, since the substances here under consideration, silica, alumina, alkalis, etc., can yield a great variety chemical combinations that we must then locate among the minerals of the rock. However, the solution becomes easier when we use a simplified form of the phase rule which refers to processes in condensed systems as they exist in metamorphism (both contact and regional metamorphism); from it we get an upper limit to the number of coexisting minerals in stable combinations. The law that I have named the 'mineralogical phase rule' says that in a stable combination the maximum number of individual minerals that can occur is the number of independently variable components determined by the bulk composition. In a mixture of silica, magnesia and alumina, no more than three minerals can occur simultaneously at any time.

I have carried out a detailed application of this law for the products of contact metamorphism. Those stratified rocks that consist of mixtures of limestone and shale give only 10 classes of metamorphic products when they are subjected to the heat-effect of immediately neighbouring molten plutonic rocks. The classification of these metamorphic rocks, the so-called hornfelses, is naturally not valid only for the Christiania region (Oslo) region where it was first established, but for wherever sediments of the shale-limestone series are transformed under similar conditions of temperature and pressure. To name some examples, we can cite the works of O.H. Erdmannsdörffer in the Harz and P. Sustschinsky in Finland.

The application of the phase rule to the crystalline schists of regional metamorphism was recently treated by P. Niggli, who with its help demonstrated the complicated conditions of formation of chloritoid schists.

In the following, we will not limit ourselves to contact metamorphism, but discuss the more general rules that are valid for metamorphism under all temperature and pressure values that can occur in the earth's crust.

The physical conditions of metamorphism are given by two variables: pressure and temperature. We choose the two variables as coordinate axes in a graphical representation (Fig. 1), with pressure in atmospheres as the horizontal axis and temperature in Celsius as the vertical axis. For the origin of the diagram we choose the physical conditions which reign on the earth's surface, namely 0°C and one atmosphere.

In such a diagram every kind of rock formation or alteration corresponds to a definite field that is given by corresponding values of temperature and pressure.

At the origin of the coordinate system, we have the stability field of those rocks that form on the surface of the earth at ordinary temperatures; these are the products of weathering and the sediments that form from them. A little further up and to the right lies the principal region of diagenesis.

Accordingly, we find in sedimentary rocks (except for possible relics) those minerals that can exist at low temperatures and pressures, such as kaolinite, muscovite and chlorite.

The case that corresponds to a vertical band above the origin, that is, metamorphism of rocks at high temperatures and low pressures, is also found in nature. This is the region of those transformations that xenoliths undergo in the lavas of volcanoes. During metamorphism an intense heating takes place; however, the pressure is of the order of one atmosphere. Thus, in a shale are formed the following minerals which are collectively stable under the new temperature-pressure conditions: cristobalite (or tridymite), sillimanite, hypersthene, cordierite and others. Such cases of alteration have been particularly thoroughly described by A. Lacroix and recently also by R. Brauns; I myself was able to study them in the inclusions in the volcanic rocks of Spitzbergen. Lower down, the conditions for this metamorphism are limited by temperatures at which a noticeable effect no longer occurs; those reactions of importance which proceed at the lowest temperatures appear to be the formation of specular hematite from iron hydroxides. To be sure, the field is also limited above by the melting temperatures of the minerals. As soon as the entire rock is melted under the influence of added heat, further increase in temperature cannot influence the metamorphism; we change only the transient temperature of the melt, without causing any modification of the product of new crystallization during subsequent cooling. This upper limit is taken to be ~1550°C; here the most important minerals under consideration are melted. The kind of metamorphism discussed here, which is caused purely through the effect of heat, can be termed thermo-metamorphsim after the proposal of F. Rinne.

To the right of the band of thermo-metamorphism, and joined to it without a break, lies the field of rock alteration that we call contact metamorphism. Contact metamorphism

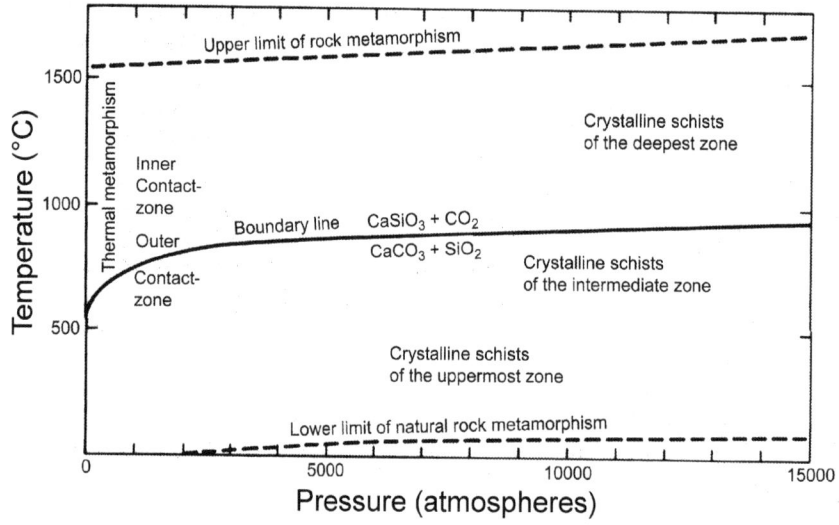

Fig. 1.

can be defined as a thermo-metamorphism at greater depth, that is, under higher pressure. It is especially intensive in the vicinity of large magma bodies solidified at depth, where the alteration can be detected up to two, rarely four, kilometers distance from the immediate contact.

The temperature during contact metamorphism is obviously highest at the immediate contact surface against the igneous rock, and is lower with increasing distance from this boundary. Correspondingly, we can distinguish between an outer and an inner contact zone. The maximum temperature during the metamorphism can be inferred at least approximately from the minerals present and their conditions of formation; in this way I obtained numbers between 1000 and 1200°C for the inner contact-zone of the deep-seated rocks in the Christiania region. Relatively seldom the temperature of the innermost contact-zone rises to higher values, which can be recognized by the occurrence of melting phenomena; here too we reach as the uppermost limit for metamorphism the temperature range of melting, which might possibly lie a little higher than in the case of thermo-metamorphism, since melting points are increased by pressure.

The pressure during contact metamorphism is in many cases likewise accessible by numerical calculation. If the geological relations during the metamorphism are precisely known, then the thickness and nature of the overlying rock-masses are given, hence we know also the weight exerted by the overburden pressure. When such a calculation is done for various places in the contact zones in the Christiania region, one gets, as I have shown, pressures from 400 to about 1500 atmospheres.

The minerals of contact metamorphism are not quite the same as in thermo-metamorphism, since here we encounter some minerals which require higher than atmospheric pressure for their stability at high temperatures; let us name only grossular as an example; also some minerals with a content of volatile substances, biotite, vesuvianite and others can exist in stable form in the inner contact-zone since decomposition is here avoided due to the pressure, although they release their volatiles when heated at atmospheric pressure.

With grossular, the stability field is increased with rising pressure, since it takes up less volume than its breakdown products. This is only one special case of a quite general conformity to the law that favours at high pressure those reactions which lead to a volume reduction of the reacting phases. This principle, long known already to petrographers was, of course, contained in the law of Van't Hoff and Le Chatelier. The volume law is already in force when one goes from thermo- to contact-metamorphism; its influence becomes greater still in the transition from contact to regional metamorphism.

Regional metamorphism got its name from the fact that it occurs consistently over greater regions. It is not, as is the case in contact metamorphism, linked to individual contacts with younger igneous rocks. Whereas the width of a contact zone very rarely reaches 4 kilometers, the dimensions of regional metamorphic terrains must be measured in hundreds of kilometres.

It is now mostly assumed that regional metamorphism occurs in such a way that larger parts of the earth's surface are depressed to significant depths, whether by simple superimposition of younger rocks, or by geotectonic movements. At depth, the interior heat of the earth combined with a high overburden pressure is said to effect the transformation.

Regional metamorphism, experience shows, is specially connected to those parts of the earth's crust in which great folding deformation can be demonstrated. Regional metamorphism occurred more-or-less simultaneously with these movements. Many geologists and petrographers, therefore, see in the folding-movements the real cause of regional metamorphism, which must therefore be designated as dynamic metamorphism. For many cases of purely mechanical rock-deformation, this may prove correct; in regional metamorphism, also, a significant participation in the metamorphism is attributable to movements in the earth's crust. This contribution can occur in the following ways: (1) through movements in the earth's crust the pressure can locally rise strongly over that value which corresponds purely to the overburden pressure; (2) differences in pressure can appear with direction, giving rise to stress effects (schistosity, stretching); (3) individual parts of the crust can be depressed to especially great depths due to folding; (4) the metamorphic products of great depths, through folding, can be returned to such elevations that with minimal erosion they are accessible to observation; (5) as F. Becke first called to our attention, stress can have an accelerating effect on transformations.

The rocks that form through regional metamorphism are called crystalline schists, that is, for the large part they possess pronounced schistose structures. The schistosity is occasioned by oriented pressure or stress, in which flaky parts of the assemblage take on a parallel arrangement, while such minerals as originally had no flaky habit are often likewise rolled out into parallel layers by the stress. At this point we will not go into these stress effects more closely, but limit ourselves to the influence of medium pressure and temperature. These are the two ultimate factors that are determinative for the stability fields of minerals; stress as a rule alters only the form of the individual mineral grains and their mutual arrangement.

To discuss the formation of crystalline schists, we can undertake the following thought experiment.

Let a rock-mass sink deeper and deeper in the earth's crust. Its temperature will then increase, according to the available data ~3°C for every 100 metres depth; simultaneously the overburden pressure to which the rock is subjected rises in proportion to the height of the overlying rock-column. If it were permitted to extrapolate the temperature rise to very great depths, one could calculate the depth at which melting would occur, that is, the maximum depth of metamorphism. Such an extrapolation will obviously yield low values of depth, because the temperature rise with increasing depth is slower than in the horizons accessible to us. If, with all reservations, we carry through the extrapolation, then we get for a temperature of 1700°C a depth of almost 60 kilometres, which corresponds to an overburden pressure of approximately 15,000 atmospheres.

If our assumptions about the mode of formation of crystalline schists and about the temperature gradients are correct, the field of formation of the crystalline schists in the deepest zone of the diagram (Fig. 1) would lie to the left and below this point. Those crystalline schists that are formed at lower values of pressure and temperature should show a region of origin still further to the lower left on the diagram. I have indicated on the diagram such a zonal division of the crystalline schists as has been presented by F. Becke, C.R. van Hise and U. Grubenmann.

According to the calculation method employed above, two values of pressure and temperature should always correlate in regional metamorphsim, because pressure and temperature increase regularly towards the deeper levels. This relationship is disturbed by two influences that, moreover, counteract one another. First, the temperature of regional metamorphism, in comparison to a normal temperature gradient, will as a rule increase faster with depth, because folding and regional metamorphism are mostly linked with the contemporaneous uprise of magmatic rocks; these magmatic rocks naturally increase the temperature of their surroundings (many geologists and petrographers would like to attribute the principal role in regional metamorphism to these very rocks). The final result of rock transformation is thus placed intermediate between ideal regional metamorphism and contact metamorphism. The second influence is the fact that during the folding, and as a cause of the folding, internal stresses must occur that can increase the pressure in the location concerned far above the value that corresponds to the load pressure of the overlying rocks. This circumstance shifts the field of regional metamorphism to the right in the diagram. These two influences may well operate simultaneously, and so must partly compensate one another. The magnitude of the resultant departure from the expected is indeterminate; it probably goes in the direction of higher pressure, that is, moves the field to the right in the diagram.

The diagram for rock-metamorphism is unconstrained in the direction of higher pressure; in all other directions fixed limits are found. Also, below to the right we see a boundary line against a band that is not realized in nature. In those parts of the earth's crust where high pressures rule, the temperature cannot simultaneously be very low. High overburden pressures can only begin to occur at a certain depth; high pressures due to previous movement are accompanied by frictional heat. Metamorphism in this band can of course be realized experimentally, for example, in the interesting experiments of Frank D. Adams.

A diagram like the one discussed above gives us in many respects a convenient overview of the regularities of mineral formation in metamorphic rocks.

Individual minerals and mineral combinations possess stability regions that are characterized by defined intervals of temperature and pressure. Related types of metamorphism that border each other on the diagram must therefore show common traits in the mineral composition of the rocks.

Thus, for example, hypersthene, spinel and cordierite are common for thermo-metamorphism and contact metamorphism of the inner zone. Muscovite is a characteristic mineral in the uppermost and in the middle zone of crystalline schists.

On the other hand, the diagram is intersected by important division lines. The most important of these boundary lines is the one shown in Fig. 1.

The curve refers to the reaction:

$$CaCO_3 + SiO_2 = CaSiO_3 + CO_2,$$

wollastonite formation from calcite and quartz. In common sediments calcium is found only in the form of calcium carbonate; pure shale always contains less than 1% CaO, higher contents are mixtures with lime carbonate. Also, coarse-grained sediments are almost always free of lime silicates, the only exception being tuff-sediments.

For lime-bearing silicates in sediments to result from metamorphism, the carbonate of the calcite must be expelled by the silica. The course of this reaction depends on both temperature and pressure; rising temperature favours the formation of the silicate, rising pressure that of the carbonate. For every temperature that a mixture of calcite and quartz can reach, there is a definite pressure of CO_2. If the external pressure is less than this pressure, the silicate is formed; if, on the other hand, it is greater, then calcite remains along with quartz. I have already had the opportunity to make these qualitative observations during my investigations of contact metamorphism.

So it would be of the greatest geological interest to discover the CO_2 pressure of the system for different temperatures, for we would then be able to draw conclusions about the temperature-pressure-values from the presence or absence of wollastonite.

A solution to this exercise is offered by thermodynamic means, in which we can use the third fundamental principle of heat theory, the theorem of Nernst. This theorem gives us the size of the integration constant that is left undetermined in the second principle. In our case for numerical computation we can use only the approximation formula given by Nernst (*Theor. Chemie*, 1909, p. 711), as follows:

$$\text{Log } p = -Q/4.571T + 1.75 \log T + C$$

Here, p is the desired pressure in atmospheres at the absolute temperature T, Q is the heat of reaction in calories per gm-molecule, and C is the so-called chemical constant. For Q we have an older determination by Le Chatelier and a new, slightly different one published by D. Tschernobajew and L. Wologdine (*Comp. rend. Paris*, **154**, 1912, p. 206). For the calculation I chose the newest figure, 25,300 cal.[1] The size of C is given as 3.2, and E.H. Riesenfeld (*Journ. Chim. Phys.*, **7**, 1909, p. 561) finds the same number.

To produce a complete calculation by using the exact equation instead of this approximation, we would also have to know the correction for the variation of heat of reaction with temperature (and pressure). To evaluate this, an experimental investigation of the system is greatly desired.

The curve calculated according to the approximate formula above should nevertheless in its principal features come close to the true location. It shows the characteristic feature of vapour-pressure curves, first a rapid, then a

---

[1] Unfortunately, it is not indicated which modification of $CaSiO_3$ the figure refers to.

progressively slower rise in temperature with increase in pressure. Solid phases along the curve in the range represented on our diagram are α- and β-quartz[2], wollastonite (β-$CaSiO_3$) and β-calcite.

Quartz is stable beside calcite below the curve, and wollastonite occurs above the curve. So the occurrence of wollastonite in metamorphosed sediments gives us a geological thermometer; our boundary curve must be overstepped for the formation of wollastonite. The trajectory of the boundary curve agrees well with geological-petrographical experience. Wollastonite is a characteristic mineral in altered sediments of the inner contact-zone and the deepest zone of crystalline schists; in contrast, it is lacking in the outermost contact-zone and in all crystalline schists of the two upper zones.

Analogous curves can be constructed for the formation of lime-alumina-silicates in metamorphosed sediments, as long as we know the heat of reaction of their formation. Since Q here is plainly smaller than in the formation of wollastonite, these curves will lie a little lower than the wollastonite curve; however, they must exhibit a similar shape, first a steep, then a flatter rise in the higher pressure region. The curve for clinozoisite should lie the lowest (among the lime-magnesia-silicates tremolite appears to possess the lowest curve).

Each of these (apparently closely neighbouring) curves divides the diagram into two parts, an upper left part in which lime-bearing silicate occurs, and a lower right part in which the original calcite remains present along with alumina-bearing silicate and quartz.

Thus we get two different types of metamorphism; in the upper left field there occur rocks of the *wollastonite type*, and in the lower right the *quartz-calcite* type. This separation into two different types of metamorphism[3] has great importance for geology.

It must be noted however that the importance of the lime silicates applies only to metamorphosed *sediments*, whose calcium content was originally present as carbonate. In metamorphosed igneous rocks that contained primary calcium in silicate form, secondary lime-silicates can form at all temperatures and pressures.

One can also draw up curves analogous to the wollastonite curve for the decomposition of water-bearing silicates with increasing temperature. Here, Q becomes the heat of reaction for dehydration, and C the chemical constant for steam. These curves must also run from the lower left towards the upper right; the occurrence of water-bearing silicates in the outermost contact-zone and the crystalline schists in the upper zones corresponds to this.

Boundary curves of a different form result from reactions in condensed systems, possibly quartz:cristobalite, or anorthite + enstatite = diopside + kyanite.

The significance of the physical-chemical discussion of metamorphism lies not only in our obtaining an insight into the mechanism of the transformation. Of at least equal importance is the fact that in this way geology gets values for those temperatures and pressures at which specific rock-types must have formed.

We find examples of metamorphism of the wollastonite type everywhere in the contact-zones of the Christiania region. We see wollastonite in this tract of land wherever the limestone-sandstone sequence is cut by younger intrusives; and in the hornfelses that I have described is found the complete sequence of lime-aluminium and lime-magnesian silicates, from anorthite and amphibole all the way to vesuvianite.

The uprise of magmatic rocks can naturally lead to local compression, and we get from this a condition called stress; such signs of stress are detectable here and there in Langesundfjord.

Through the combination of contact metamorphism with simultaneous local stress action (a condition that W.C. Brögger called 'contact-pressure-metamorphism' as early as 1890), contact rocks with the structure of crystalline schists can be produced, where all sheet-like mineral particles become oriented parallel. Such a development of schistosity is most clearly shown in the contact-metamorphic 'Ergussgesteinen', which W.C. Brögger has described from Langesundfjord. The primary schistose rocks of Langesundfjord also belong to the wollastonite type.

The high mountain chain of southern Norway is admirably suited for the study of regional metamorphism. The Cambro-Silurian shales are for the most part developed as phyllite ('Glanzschiefer'), and in some cases already as true micaschists. We are dealing then with crystalline schists of the uppermost zone, and in part the middle zone. One never encounters wollastonite or other lime silicates in the interlayered metamorphosed limestone-sandstone; the calcite did not react with quartz during the metamorphism. The metamorphism of this region thus belongs to the quartz-calcite type.

It is a known fact that metamorphism of a similar type is found in the fold-mountains of other countries.

How does metamorphism manifest itself in those places where younger magmatic rocks are in contact with phyllite? From the outset, one would absolutely expect that at the direct contact so high a temperature prevailed that the boundary curve of the wollastonite type would be overstepped despite the high pressure. The high mountains of Norway provide a good opportunity to test this question. Intrusive rocks, which on good geological grounds are considered to be younger than the Cambro-Silurian sediments, are found in many places. Well known are the large basic plutonic rock-masses of Jotunheimen, which mostly belong to the norite group. Last summer I spent a great deal of time trying to detect wollastonitic metamorphism of the phyllite formation at their contact, but up to now it has not been located with certainty.

Previously, one sought to avoid this kind of problem by assuming that wollastonite-formation is prevented by high pressure. The curve of CO_2-pressure for the system calcite-quartz does not support this hypothesis, as long as one is unwilling to assume pressures of hundreds of thousands of atmospheres; the temperature of 1200°C should correspond

---

[2] The heat of inversion α-quartz:β-quartz is very small (180 calories per mol.)
[3] They correspond approximately to E. Weinschenk's distinction between 'normal' and the 'alpine' types of metamorphism.

[4] The question of overthrusting will not be dealt with here.

to a CO_2 pressure of 96,000 atmospheres (of course the extrapolation to 1200°C seems uncertain inasmuch as we are here dealing with H.E. Boekes' α-calcite).

Similar difficulties also emerge, according to the available data, in the case of many intrusive rocks in the Alps (E. Weinschenk has already alluded to the quartz-calcitic schist envelopes of the alpine granites).

The physicochemical approach will certainly succeed ultimately in explaining these apparent contradictions. And I believe that specifically the union of petrography and physical chemistry must provide the solution to the currently most important geological problems on the Scandinavian Peninsula, namely the interpretation of the alpine metamorphism and tectonics.

And not until we completely master the laws of regional metamorphism will the time have come to attack with positive success a still larger problem: the geology and petrography of the crystalline basement.

CHAPTER 2: BARROVIAN METAMORPHISM

Barrow, G. (1912) On the geology of the lower Dee-side and the southern Highland border. *Proceedings of the Geologists Association*, **23**, 274–290.

Because it is a more complete account of metamorphic zones in the metapelites of the southeast Scottish Highlands, I have chosen to reprint a part of the second of George Barrow's two classic papers (1893, 1912), even though the first one has been cited more frequently.

There is a certain irony in the fact that one of the most enduring early accounts of metamorphic petrology on a regional scale was provided not by an author who considered himself a petrographer or petrologist, but by a field geologist in the employ of the British Geological Survey. As a skilled observer of rocks in the field, Barrow entered on his map of the southeast Highlands of Scotland whatever was mappable. In rocks that were formerly shales, this information included lines delineating the "outer limits" of the occurrence of the metamorphic "index minerals" sillimanite, kyanite and staurolite (Barrow, 1893) and (in 1912) garnet, biotite and a zone of digested clastic micas. Barrow unambiguously interpreted the outer limit lines as isothermals, and noted that the most coarsely crystallized rocks were in the higher temperature "zones" (so that coarse grain-size could be used as a proxy for the staurolite, kyanite and sillimanite zones). The terms "isograd" and "isogradic surface" were introduced later (Tilley, 1924). Barrow recognized that biotite and garnet (as well as muscovite) were present in the higher-temperature zones, a fact we now fully understand. He also found a narrow belt containing chloritoid coinciding with the biotite and garnet zones.

In the words of Tilley (1925), Barrow's papers represent the "first attempt in the petrological literature to bring precision to the study of regional metamorphism, by laying upon a map zonal lines indicative of varying grades of metamorphism". Conventional wisdom at the time was for a threefold division of regional metamorphism into epi-, meso-, and kata-zones of alteration (Grubenmann, 1910), implying a correlated downward increase in temperature, pressure, and plastic deformation. Barrow was the first to provide convincing field evidence of a sensitive thermal structure for regional metamorphism.

In his earlier paper (Barrow, 1893), the formation of sillimanite, kyanite and staurolite was attributed to "thermometamorphism" or "contact-metamorphism". He found that the distribution of index minerals was related to their nearness to the older granitic masses, which were believed to continue to the southeast below the ground surface. At the time, sillimanite was considered to be a mineral of contact metamorphism. Based on laboratory work by Vernadsky, the temperature of the kyanite to sillimanite transition was quoted as being between 1320 and 1380°C. The later discovery of mullite, which is optically identical to sillimanite, showed that Vernadsky had almost certainly produced mullite. The area mapped by Barrow in the 1912 paper was about five times larger than covered in the 1893 paper. The metamorphism was described as "regional thermo-metamorphism", representing an aureole around an intrusion of "older granites".

It was many years before Barrow's work became widely known internationally. Goldschmidt (1915) described a similar zonation in regionally metamorphosed rocks in the Trondheim region, apparently unaware of Barrow's papers. Tilley (1949) believed that the title of Barrow's 1893 paper was partly responsible for the lack of attention it received. Over time, the term Barrovian metamorphism has become strongly fixed in the lexicon of geology. It signifies a type of regional metamorphism of worldwide distribution, which, through its sequence of index minerals, corresponds to a specific gradient in metamorphic pressure and temperature conditions; metamorphism of the "kyanite-sillimanite facies series" is nowadays often used synonymously. Other baric types of metamorphism are contrasted with the widely occurring Barrovian reference. If metamorphic petrology is mentioned at all in beginning geology courses, students are likely to remember Barrow's zones when much else has faded from memory. While a student at King's College London, I heard frequent proud mention of Charles Lyell (who was the first to discuss the meaning of the term "metamorphism") as its first Professor of Geology, but, curiously, I have no recollection of any reference to George Barrow as a former science student there.

Alfred Harker, one of Britain's earliest metamorphic petrologists, was certainly impressed by Barrow's maps, but he regarded "the lines laid down as at once isothermals and isodynamics" (Harker, 1918); the latter term was meant to imply the dominance of shearing stress over hydrostatic pressure, a view he held consistently (Harker, 1932, 1939, 1950).

Tilley (1925) extended the mapping of Barrow's three lowest-grade zones about 100 miles to the southwest, as far as the Isle of Arran on Scotland's west coast. Tilley combined Barrow's zones of "clastic mica" and "digested clastic mica" into a single "chlorite" zone, not because these kinds of mica were hard to distinguish under the microscope but because the division was not readily determined in the field, and chlorite was the most characteristic mineral of this lowest grade. The entry of biotite ended the chlorite zone, but Tilley noted that this change needed confirmation with the petrographic microscope, especially for distinguishing brown biotite from the green or brownish-green biotite found with epidote in the Green Beds outside the biotite zone for pelites. Tilley changed the name garnet to almandine and claimed that the surface of this isograd was inverted. He asserted that the older granites were an "incident" in the metamorphism rather than a cause; the metamorphic zones go far beyond the area of the older granites. In Elles and Tilley (1930), the isograds were extended into the central and southwest Highlands, and further field evidence was given for the inversion of isogradic surfaces as a result of Caledonian recumbent folding. Later, the rocks were correlated with the inverted limb of the Iltay nappe (Read and MacGregor, 1948). Kennedy (1948) continued the map of zones across the Great Glen fault into the Moine Series, and showed that as a whole the isogradic surfaces, when restored to the pre-fault configuration, appeared to define a large-scale thermal anticline plunging to the southwest parallel to the main Caledonian folding, a view later regarded as oversimplified. With its beautiful collection of thin section drawings, Harker's textbook (1950) provides a delightful introduction to the petrography of the pelites of the Barrovian type-area. He, too, thought in terms of a dome-like uprise of isothermal surfaces. Fieldwork over the years by H.H. Read made important contributions to Grampian regional geology (e.g. Read, 1952); his observations on the older granites and migmatites were the source in part of his strong transformist views in the granite controversy of the 1950s.

In the second half of the 20th century, regional, structural, petrological and geochemical studies of Barrovian terranes worldwide were pursued intensively, using techniques and ideas as they evolved over time. There is hardly a subject more central to metamorphic petrology than the phase relations, conditions, and driving forces of Barrovian-style metamorphism. A Google search for "Barrovian + metamorphism" elicited 4660 entries! The type locality in the Scottish Highlands itself naturally became an attractive target for study, despite evidence for great structural and metamorphic complexity. In what follows I shall also comment on some of the studies elsewhere that have had an important influence on our understanding of Barrovian metamorphism and the significance of the Barrovian zones.

Chapter 5 of this volume recognizes the introduction in 1957 of the AFM-projection by J.B. Thompson as a landmark event in metamorphic petrology. It was unquestionably a shot in the arm for studies of Barrovian metamorphism. By its graphical representation of the Mg/Fe ratios and Al-contents of minerals in metapelites, it enabled us to understand clearly the Barrovian isograds in terms of a sequence of univariant (discontinuous) and divariant (continuous) reactions that (ideally) take place in the progressive metamorphism of pelites under "Barrovian" P-T conditions. Applications of the AFM-projection in terranes of Barrovian metamorphism, for example in Nova Scotia (Phinney, 1963), Vermont (Albee, 1965), and Ontario (Carmichael, 1970), soon appeared. Carmichael advocated the use of the product mineral assemblage of each isogradic reaction as a more rigorous way of naming the isograd than the index mineral, for example, kyanite + biotite, rather than simply the kyanite isograd; this designation recognizes either kyanite or biotite as minerals possibly occurring singly below the isograd. Carmichael (1969) also found a way to reconcile thin-section textures observed at Barrovian isograds that were seemingly inconsistent with the inferred reactions. In a proposal that has resonated widely in the metamorphic community, he argued for local metasomatic ion-exchange reactions in a rock that, when added up, correspond to the balanced metamorphic reaction.

Chinner (1965) began the assignment of P-T conditions and metamorphic histories for the various parts of the Dalradian in NE Scotland with a petrogenetic grid based on the Al_2SiO_5 polymorphs.

With the newly introduced electron microprobe, Harte and Henley (1966) discovered cryptic compositional zoning of Mn, Ca, Mg and Fe in garnet porphyroblasts in a sillimanite-zone schist from Glen Esk, correctly attributing the Mn-profile to depletion during growth. Atherton (1968) discussed the phase-equilibrium consequences of the bell-shaped compositional growth zoning of garnet that appeared to be common in Dalradian pelites, pointing out that unreactive garnet cores yield a residual chemically active rock matrix richer than before in Mg, and in some cases Al, signifying motion of "bulk-rock" compositions to the right across the AFM projection. Harte and Johnson (1969) found structural and textural evidence to be consistent with the synchronous growth of garnet, staurolite and kyanite in the Barrovian type area, and possibly later growth of sillimanite in association with migmatite development, as proposed by Chinner (1961). Thus, despite questions raised by some, the general idea that Barrovian metamorphism in the Highlands represented progressive metamorphism climaxing in a single, maximum temperature "event" gained significant support.

Many writers struggled with reactions that might be responsible for the biotite isograd in prograde metamorphism. Mather (1970) showed that in Barrovian metamorphism the reaction of chlorite with K-feldspar produced biotite in metagreywackes and in some metapelites (see summary in Spear, 1993), but in aluminous metapelites it required a gradually less celadonitic muscovite (less Si) to open up the composition field to biotite (as seen in the AKF projection); many biotite-zone pelites do not contain biotite. Ferry (1984) showed that a reaction of muscovite with carbonate and quartz was an important but overlooked source of biotite in biotite-zone metapelites.

Albee (1972) drew a pseudobinary T–X_{Mg} section (a pseudosection, see Chapter 8), which greatly helped to explain the specific sequence of possible mineral assemblages observed with increase in grade during Barrovian metamorphism.

The classic Barrovian zones of the type-area did not include the two highest-grade zones now recognized in the kyanite-sillimanite facies series. In his textbook, Harker (1950, p. 228) mentioned the dissociation of muscovite to aluminosilicate and potash feldspar, but he did not cite a precise location for this observation. An "orthoclase" or "second sillimanite isograd", corresponding to the formation of orthoclase + sillimanite from the reaction of muscovite with quartz, was first mapped by Heald (1950) in New Hampshire. The P-T conditions of this important dehydration reaction were not known at the time, but several experimental investigations between 1955 and 1975 indicated a melt-forming reaction at ~700°C and 6 kbar (Thompson, 1976). Subsequently, Baker and Droop (1983) and Baker (1985) discovered a metamorphic high along a

SW–NE axis going through Glen Muick where K-feldspar occurs with sillimanite in pelites, and they were able to map the isograd. At the beginning of granulite-facies conditions, reaction of biotite with sillimanite plus quartz leads to a stable join between garnet and cordierite (+ orthoclase). Baker did not report this pair, although the conditions he inferred from thermobarometry (820°C and 8 kbar) were appropriate. Both Sil + Or and Grt + Crd regional zones are developed in central Massachusetts (Tracy et al., 1976), where conditions reached at least 750°C at 8 kbar at the termination of a kyanite-sillimanite facies series. In the Caledonides, Grt + Crd metapelites are located only in the vicinity of synorogenic intrusions (Ashworth and Chinner, 1978).

Richardson and Powell (1976) modelled possible heat flow in Dalradian metamorphism consequent upon tectonic thickening of crust to ~50 km. They calculated a steady-state geotherm of 28°C/km, consistent with the kyanite-sillimanite facies series, based on radioactive decay in the Dalradian and underlying Moinian rocks, and normal mantle heat flow. A transient model involving erosion above a warming crust produced a T_{max} at ~40 my after the cessation of thickening. An advective supply of heat was not needed. Shortly after this paper appeared, conductive heat-flow modelling of tectonically thickened crust began to play an influential role in metamorphic studies, as a result of the publication of the landmark paper by England and Richardson (1977; see Chapter 9).

Atherton (1977) superimposed whole-rock compositions of Barrovian pelites (which are not particularly aluminous, except perhaps for the chloritoid-bearing schists near Stonehaven: Chinner, 1967) onto Albee's pseudobinary T–X_{Mg} section. This threw further light on the occurrence of specific mineral assemblages observed with increase in grade in the Barrovian type-area. Atherton was not prepared to rule out magmas as a source of heat for the metamorphism. Later, Baker (1987) noted that the extra-high T area in Glen Muick was associated with the newer gabbros.

Harte and Hudson (1979) derived a primitive PT "assemblage diagram" for the eastern Dalradian (and lower-pressure facies-series in the Buchan area to the north) from a petrogenetic PT grid established for the system KFMASH. Citing England and Richardson (1977), they interpreted the shallow dP/dT metamorphic field gradients as indicative of a conduction situation followed by convection of heat by magma injection.

The suspected inversion of the garnet isograd (Tilley, 1925; Elles and Tilley, 1930) was confirmed by Watkins (1985). Chinner (1978, 1980) suggested that the kyanite isograd might also be inverted along the SE edge of the Barrovian metamorphic zones (the Flat Belt). Structural evidence argues against post-metamorphic folding. Instead, kyanite-grade metamorphism was believed to have climaxed in a negative thermal gradient that was initiated by emplacement of the Tay nappe. Then we have to ask why the temperature inversion was preserved during thermal relaxation of a thickened crustal pile (see Chapter 9). Jamieson et al. (1996) pointed out that an inversion of metamorphic grade need not represent an inverted geothermal gradient, but can result from syn-metamorphic differential exhumation across crustal-scale ductile shear zones. It is commonly the case that metamorphic grade increases structurally up-section in Barrovian terranes (Jamieson et al., 1998, p. 23).

Thermobarometry of high-grade rocks from the type area (Baker, 1985; McLellan, 1985; Dempster, 1985; Ague, 2001) has been generally consistent with PT conditions passing across the experimental kyanite-sillimanite curve, namely 600–700°C and ~ 6 kbar for the sillimanite zone. Baker (1985) believed that an overthrust-erosional, England-Richardson-Thompson model (England and Thompson, 1984) was probably appropriate for the Dalradian of the central Highlands, in contrast to the general consensus that the lower-pressure Buchan area involved addition of magmatic heat (Harte and Hudson, 1979; Hudson, 1980). Given the structural evidence for large-scale tectonic thickening of the Grampian pile, and the suspected late growth of sillimanite, Baker (1987) figured a thermal relaxation + decompression P-T-time path for the Barrovian metamorphism and for 10 kbar rocks in the Central Highlands to the west. In the Dalradian of western Ireland (Yardley et al., 1987), crossite-bearing blueschist/greenschist transitional rocks were overprinted by Barrovian metamorphism. Dempster (1985) described margins of Grt, Sta and Bt crystals in the Ky and Sil zones that were affected by retrograde cation exchange, and correlated them with variable uplift rates based on isotopic closure ages.

McLellan (1985) showed in the Barrovian type-area that a progressive contraction in the compositional field of staurolite, in the presence of muscovite and quartz, was what produced regional kyanite and additional biotite (as proposed by Chinner, 1965), and, at higher-grade, "regional" sillimanite and more biotite. "Overprint" fibrolitic sillimanite formed structurally slightly later during erosion of the Tay nappe and isothermal decompression. Detailed PTt-paths based on thermobarometry in individual samples of metapelites, that might perhaps be oblique to the metamorphic field gradient and suggestive of thermal relaxation (Chapter 9), have apparently not been obtained from the type-area. One wonders whether McLellan's (1985, Table 13) estimates of aH_2O in Ky and Sil zone pelites (mostly 0.3 to 0.7) are perhaps an artifact of problems with the thermobarometry (Chapter 10). Somewhat higher temperatures (e.g. Baker, 1985) would raise H_2O activities to what might be considered more reasonable values, but probably not require reversal of her conclusion (McLellan 1989) that the Ky-zone migmatites are subsolidus. The release of H_2O during the progressive metamorphism of pelites suggests that at least the main stage of Barrovian metamorphism was "wet", in the sense of producing fluid-filled pores at least episodically. For the migmatites in the Sil zone, McLellan (1989) suggested that melting was aided by the introduction of aqueous fluid. Interestingly, Symmes and Ferry (1995) proposed that partial melting in the innermost zone of Onawa aureole, Maine, was aided by up-T transport of metamorphic fluid. What happens to the released fluid in metamorphism is one of the most important but undeniably controversial issues in metamorphic petrology. We take up this subject in Chapter 12, but it is appropriate to mention here the work by Ferry and collaborators in the early 1990s, who presented the case, based on reaction-progress and isotopic data, for Barrovian metamorphism of carbonate-bearing rocks in Vermont as being in part driven by substantial layer-parallel infiltration of hydrous fluid on a regional scale. This scenario has been accepted as applying elsewhere, and seems likely to be appropriate also for metamorphism in the Barrovian type-area.

Modelling of prograde reactions in an average pelite composition by Symmes and Ferry (1991) showed that, at least during the early stages, a minimum fluid flux of $\sim 10^4$ m^{3}m^{-2} was necessary to replicate the sequence of isogradic reactions in Barrovian metamorphism; otherwise, carbonate and chlorite would survive to higher grades than observed and K-feldspar would appear below the Sil isograd. This conclusion would not be justified if the pelites were totally free of carbonate at the outset, however.

Several local studies in the Scottish Dalradian, starting with Graham et al. (1983), have shown that infiltration of H_2O-rich fluid derived from the devolatilization of pelitic, psammitic and metabasic rocks has driven metamorphic reactions. In the SW Highlands, Fein et al. (1994) documented regionally extensive dolomitization of calcite marble by infiltrating fluid. Skelton et al. (1995, 1997) showed that substantial fluid migration through a phyllite unit induced metamorphic reactions in metabasite sills. Ague (1991) showed that the conventional assumption of fundamentally isochemical metamorphism of pelites based on many studies, but notably that of Shaw (1965) on metapelites of the Littleton formation, New Hampshire, is not entirely correct. Substantial loss of Si takes places with increase in grade, presumably in solution in evolved H_2O, much of it getting deposited in veins and the necks of boudins. Yardley and Bottrell (1992) showed in the Connemara schists that the oxygen isotope compositions of vein quartz reflect those of nearby matrix rocks (which are quite variable), and thus the veins reflect a local segregation process rather than large-scale fluid migration. A comprehensive study of Barrovian metamorphism of pelites in the Wepawaug Schist by Ague (1994a) revealed 20–30% volume losses with increasing metamorphic grade due to the loss of Si, P and Na, although in some zones various enrichments took place. Ague (1994b) calculated an average regional fluid flux of $\sim 6 \times 10^4$ m^{3}m^{-2} in metapelites in the amphibolite-facies portions of the Wepawaug terrane. This fluid was derived from devolatilization reactions at greater depths and was focused through a regional fracture network. It created quartz veins surrounded by cm to dm thick, highly aluminous, silica- and alkali-depleted selvages, in which index minerals such as staurolite, kyanite, and garnet were concentrated. Further discussion of the possible non-isochemical nature of metamorphism of the Littleton pelites surfaced in 1995 (Moss et al., 1995; Moss et al., 1996), with a response from Ague (1997a). If there was any agreement, it was that statistically rigorous sampling of compositional variation with grade is difficult, given the nature of local-scale variation in protolith compositions (see review in Vernon, 1998). Back in Scotland, in garnet-zone rocks north of Stonehaven, Ague (1997b) described vein quartz of local origin next to cm-wide alteration selvages representing bulk-rock non-volatile element enrichments and depletions. Based on the growth of plagioclase, he argued for major up-T infiltration of Cl-bearing aqueous fluid. Seemingly contradictory conclusions of down-T infiltration were made by Dempster et al. (2000) in a study nearby in the biotite zone, where they found replacement of plagioclase by muscovite resulting from an alkali-exchange reaction. These results are perhaps reconcilable if the events were separated in time.

Zenk and Schultz (2004) showed that compositional zoning of Ca-amphibole in metabasites from the Barrovian type-area, unlike the associated metapelites, preserve evidence of both prograde and retrograde PTt-paths. No evidence for a thermal relaxation path emerged; their prograde and retrograde paths with positive dP/dT slopes are almost identical. No indications of early, high-P/T blueschist conditions were found.

In apparent support of the ideas of Eugster (1959) regarding the persistence of Fe redox states through metamorphism, a much-cited paper by Chinner (1960) on high-grade pelitic gneisses from Glen Clova reported the presence of metapelites with widely varying proportions of ferrous and ferric iron, some with ilmenite and some with hematite. Surprisingly, Ague et al. (2001) found that the large variations in bulk-rock oxidation ratio reported by Chinner did not correlate with results of calculations of log f_{O_2} (2 log units f_{O_2} above the FMQ oxygen buffer), based on the present mineral assemblage. If order-of-magnitude gradients in f_{O_2} existed between layers prior to metamorphism, Ague et al. (2001) suggested that they were erased by fluid interaction during metamorphism. A local influx of reducing fluid was invoked to explain the presence of a small percentage of more reduced layers (0.1 log units below FMQ).

Marginal growth and recrystallization of detrital Archaean zircons was shown by Breeding et al. (2004) to have resulted from fluid flux contemporaneous with Barrovian metamorphism in Glen Clova at 462 ± 8 Ma. Thus, overall, there seems to be abundant evidence of active fluids attending Barrovian metamorphism, but definitive answers to the questions of fluid sources, flow paths, and volumes remain elusive, and will, I am sure, be the object of much future work. A regional study of bulk-rock oxygen isotopes in the Barrovian pelites of the type-area would be useful. In other terranes, declines of ~5‰ or more in oxygen isotope ratios from low to high grade have been attributed to an up-T regional flow of aqueous fluid (Chapter 12).

Recent work has shown that the geological time-span for the Grampian Orogeny, during which the Barrovian metamorphism of the Dalradian took place, was brief, on the order of 15–20 m.y. (Soper et al., 1999; Oliver et al., 2000; Oliver, 2001). Ages of metamorphism based on U/Pb zircon and Sm/Nd garnet/matrix geochronology fall in the range 478–460 Ma (Oliver et al., 2000; Baxter et al., 2002). In conjunction with microstructural evidence (Harte et al., 1984; McLellan, 1985, 1989), Baxter et al. (2002) interpreted these results as indicative of the near contemporaneity of peak temperatures for the garnet, kyanite and sillimanite zones. This conclusion is seemingly inconsistent with what has become the "traditional" orogenic model of thermal relaxation of overthickened crust (Chapter 9), and was interpreted to support the proposition that the Barrovian metamorphism was driven by heat from local intrusions (Baxter et al., 2002), as has generally been accepted for the Buchan area. The geological and chronological evidence for synmetamorphic intrusions of gabbros and granites, both in Scotland (Rogers et al., 1994; Oliver et al., 2000) and in the Dalradian extension in Connemara, western Ireland (Friedrich et al., 1999a,b), further supports the idea of magmas as a regional source of heat. Thus, the original ideas of Barrow (1893, 1912) remain alive to the present day, although I dare say we have not heard the last of this discussion. Burg and Gerga (2005), for example, argue that the supply of magmatic heat is augmented by a contribution from deformation-generated viscous dissipation in the upper mantle

References

Ague, J.J. (1991) Evidence for major mass transfer and volume strain during regional metamorphism of pelites. *Geology*, **19**, 855–858.

Ague, J.J. (1994a) Mass transfer during Barrovian metamorphism of pelites, south-central Connecticut, I: Evidence for changes in composition and volume. *American Journal of Science*, **294**, 989–1057.

Ague, J.J. (1994b) Mass transfer during Barrovian metamorphism of pelites, south-central Connecticut, II: Channelized fluid flow and the growth of staurolite and garnet. *American Journal of Science*, **294**, 1061–1134.

Ague, J.J. (1997a) Compositional variations in metamorphosed sediments of the Littleton Formation – discussion. *American Journal of Science*, **297**, 440–449.

Ague, J.J. (1997a) Crustal mass transfer and index mineral growth in Barrow's garnet zone, northeast Scotland. *Geology*, **25**, 73–76.

Ague, J.J., Baxter, E.F. and Eckert, J.O. (2001) High fO$_2$ during sillimanite zone metamorphism of part of the Barrovian type locality, Glen Clova, Scotland. *Journal of Petrology*, **42**, 1301–1320.

Albee (1965) Phase equilibria in three assemblages of kyanite-zone pelitic schists, Lincoln Mountain quadrangle, central Vermont. *Journal of Petrology*, **6**, 246–301.

Albee, A.L. (1972) Metamorphism of pelitic schists: reaction relations of chloritoid and staurolite. *Geological Society of America Bulletin*, **83**, 3249–3268.

Ashworth, J.R. and Chinner, G.A. (1978) Coexisting garnet and cordierite in migmatitic rocks from the Scottish Caledonides. *Contributions to Mineralogy and Petrology*, **65**, 379–394.

Atherton, M.P. (1968) The variation in garnet, biotite and chlorite composition in medium grade pelitic rocks from the Dalradian, Scotland, with particular reference to the zonation in garnet. *Contributions to Mineralogy and Petrology*, **18**, 347–371.

Atherton, M.P. (1977) The metamorphism of the Dalradian rocks of Scotland. *Scottish Journal of Geology*, **13**, 331–370.

Baker, A.J. (1985) Pressure and temperature of metamorphism in the eastern Dalradian. *Journal of the Geological Society of London*, **142**, 137–148.

Baker, A.J. (1987) Models for the tectonothermal evolution of the eastern Dalradian of Scotland. *Journal of Metamorphic Geology*, **5**, 110–118.

Baker, A.J. and Droop, G.T.R. (1983) Grampian metamorphic conditions deduced from mafic granulites and sillimanite-K-feldspar gneisses in the Dalradian of Glen Muick, Scotland. *Journal of the Geological Society of London*, **140**, 489–497.

Barrow, G. (1893) On an intrusion of muscovite-biotite gneiss in the south-eastern Highlands of Scotland, and its accompanying metamorphism. *Geological Society of London Quarterly Journal*, **49**, 330–358.

Barrow, G. (1898) Chloritoid schists from Kincardineshire. *Quarterly Journal of the Geological Society of London*, **54**, 149–155.

Barrow, G. (1912) On the geology of the lower Dee-side and southern Highland border. *Geologists Association Proceedings*, **23**, 274–290.

Baxter, E.F., Ague, J.J. and DePaolo, D.J. (2002) Prograde temperature-time evolution in the Barrovian type-locality constrained by Sm/Nd garnet ages from Glen Clova, Scotland. *Journal of the Geological Society, London*, **159**, 71–82.

Breeding, C.M., Ague, J.J., Grove, M. and Rupke, A.L. (2004) Isotopic and chemical alteration of zircon by metamorphic fluids: U-Pb age depth-profiling of zircon crystals from Barrow's garnet zone, northeast Scotland. *American Mineralogist*, **89**, 1067–1077.

Burg, J.-P. and Gerga, T.V. (2005) The role of viscous heating in Barrovian metamorphism of collisional orogens: thermomechanical models and application to the Lepontine Dome in the Central Alps. *Journal of Metamorphic Geology*, **23**, 75–95.

Carmichael, D.M. (1969) On the mechanism of prograde metamorphic reactions in quartz-bearing pelitic rocks. *Contributions to Mineralogy and Petrology*, **20**, 244–267.

Carmichael, D.M. (1970) Intersecting isograds in the Whetstone Lake area, Ontario. *Journal of Petrology*, **11**, 147–181.

Chinner, G.A. (1960) Pelitic gneisses with varying ferrous/ferric ratios from Glen Clova, Angus, Scotland. *Journal of Petrology*, **1**, 178–217.

Chinner, G.A. (1961) The origin of sillimanite in Glen Clova, Angus. *Journal of Petrology*, **2**, 312–323.

Chinner, G.A. (1965) The distribution of pressure and temperature during Dalradian metamorphism. *Quarterly Journal of the Geological Society of London*, **122**, 159–186.

Chinner, G.A. (1965) The kyanite isograd in Glen Clova, Angus, Scotland. *Mineralogical Magazine*, **34**, 132–143.

Chinner, G.A. (1967) Chloritoid, and the isochemical character of Barrow's zones. *Journal of Petrology*, **8**, 268–282.

Chinner, G.A. (1978) Metamorphic zones and fault displacement in the Scottish Highlands. *Geological Magazine*, **115**, 37–45.

Chinner, G.A. (1980) Kyanite isograds of Grampian metamorphism. *Quarterly Journal of the Geological Society of London*, **137**, 35–39.

Dempster, T.J. (1985) Garnet zoning and metamorphism of the Barrovian type area, Scotland. *Contributions to Mineralogy and Petrology*, **89**, 30–38.

Dempster, T.J., Fallick, A.E. and Whittemore, C.J. (2000) Metamorphic reactions in the biotite zone, eastern Scotland: high thermal gradients, metasomatism and cleavage formation. *Contributions to Mineralogy and Petrology*, **138**, 348–363.

Elles, G.L. and Tilley, C.E. (1930) Metamorphism of the south-western Highlands of Scotland. *Transactions of the Royal Society of Edinburgh*, **56**, 621–646.

England, P.C. and Richardson, S.W. (1977) The influence of erosion upon the mineral facies of rocks from different metamorphic environments. *Journal of the Geological Society of London*, **134**, 201–213.

England, P.C. and Thompson, A.B. (1984) Pressure-temperature-time paths of regional metamorphism I. Heat transfer during the evolution of regions of thickened continental crust. *Journal of Petrology*, **25**, 894–928.

Eugster, H.P. (1959) Reduction and oxidation in metamorphism. Pp. 397–426 in: *Researches in Geochemistry* (P.H. Abelson, editor). John Wiley, New York.

Fein, J.B., Graham, C.M., Holness, M.B., Fallick, A.E. and Skelton, A.D.L. (1994) Controls on the mechanisms of fluid infiltration and front advection during regional metamorphism; a stable isotope and textural study of retrograde Dalradian rocks of the SW Scottish Highlands. *Journal of Metamorphic Geology*, **12**, 249–260.

Ferry, J.M. (1984) A biotite isograd in south-central Maine, U.S.A.: Mineral reactions, fluid transfer, and heat transfer. *Journal of Petrology*, **25**, 871–893.

Friedrich, A.M., Hodges, K.V., Bowring, S.A. and Martin, M.W. (1999a) Geochronological constraints on the magmatic, metamorphic and thermal evolution of the Connemara Caledonides, western Ireland. *Journal of the Geological Society, London*, **156**, 1217–1229.

Friedrich, A.M., Hodges, K.V., Bowring, S.A. and Martin, M.W. (1999a) Short-lived continental magmatic arc at Connemara, western Irish Caledonides: Implications for the age of the Grampian Orogeny. *Geology*, **27**, 27–30.

Goldschmidt, V.M. (1915) Geologisch-petrographische Studien im Hochgebirge des südlichen Norwegens. III. Die Kalksilikatgneise und Kalksilikatglimmerschiefer im Trondhjem-Gebiete. *Videnskapsselskaps Skrifter I. Matematisk-Naturvidenskapelig Klasse*, No. **10**.

Graham, C.M., Greig, K.M., Sheppard, S.M.F. and Turi, B. (1983)

Genesis and mobility of the H₂O-CO₂ fluid phase in greenschist metamorphism: a petrological and stable isotope study in the Scottish Dalradian. *Journal of the Geological Society of London*, **140**, 577–599.

Grubenmann, U. (1904, 1910) *Die Kristallinen Schiefer*. Bornträger, Berlin.

Harker, A. (1918) Anniversary Address of the President. *Quarterly Journal of the Geological Society of London*, **74**, li–lxxx.

Harker, A. (1932, 1939, 1950) *Metamorphism*. Methuen, London.

Harte, B. and Henley, K.J. (1966) Occurrence of compositionally zoned almandine garnets in regionally metamorphosed rocks. *Nature*, **210**, 689–692.

Harte, B. and Hudson, N.F.C. (1979) Pelitic facies series and the temperatures and pressures of Dalradian metamorphism in E. Scotland. Pages 323–337 in: *The Caledonides of the British Isles – reviewed*. Geological Society of London.

Harte, B. and Johnson, M.R.W. (1969) Metamorphic history of Dalradian rocks in Glens Clova, Esk and Lethnot, Angus, Scotland. *Scottish Journal of Geology*, **5**, 54–80.

Harte, B., Booth, J.E., Dempster, T.J., Fettes, D.J., Mendum, J.R. and Watts, D. (1984) Aspects of the post-depositional evolution of Dalradian and Highland Border Complex rocks in the Southern Highlands of Scotland. *Transactions of the Royal Society of Edinburgh: Earth Sciences*, **75**, 151–163.

Heald, M.T. (1950) Structure and petrology of the Lovewell Mountain quadrangle, New Hampshire. *Bulletin of the Geological Society of America*, **61**, 43–89.

Hudson, N.F.C. (1980) Regional metamorphism of some Dalradian pelites in the Buchan area, N.E. Scotland. *Contributions to Mineralogy and Petrology*, **73**, 39–51.

Jamieson, R.A., Beaumont, C., Fullsack, P. and Lee, B. (1996) Tectonic assembly of inverted metamorphic sequences. *Geology*, **24**, 839–842.

Jamieson, R.A., Beaumont, C., Fullsack, P. and Lee, B. (1998) Barrovian regional metamorphism: where's the heat? Pages 23–51 in: *What Drives Metamorphism and Metamorphic Reactions?* (P.J. Treloar and P. O'Brien, editors). Special Publications, **138**, Geological Society, London.

Kennedy, W.Q. (1948) On the significance of thermal structure in the Scottish Highlands. *Geological Magazine*, **85**, 229–234.

Mather, J.D. (1970) The biotite isograd and the lower greenschist facies in the Dalradian rocks of Scotland. *Journal of Petrology*, **11**, 253–275.

McLellan, E. (1985) Metamorphic reactions in the kyanite and sillimanite zones of the Barrovian type area. *Journal of Petrology*, **26**, 789–818.

McLellan, E. (1989) Sequential formation of subsolidus and anatectic migmatites in response to thermal evolution, eastern Scotland. *Journal of Geology*, **97**, 165–182.

Moss, B.E., Haskin, L.A. and Dymek, R.F. (1995) Redetermination and reevaluation of compositional variations in metamorphosed sediments of the Littleton Formation, New Hampshire. *American Journal of Science*, **295**, 988–1019.

Moss, B.E., Haskin, L.A. and Dymek, R.F. (1996) Compositional variations in metamorphosed sediments of the Littleton Formation, New Hampshire, and the Carrabassett Formation, Maine, at sub-hand specimen, outcrop, and regional scales. *American Journal of Science*, **296**, 473–505.

Oliver, G.J.H. (2001) Reconstruction of the Grampian episode in Scotland: its place in the Caledonian Orogeny. *Tectonophysics*, **332**, 23–49.

Oliver, G.J.H., Chen, F., Buchwaldt, R. and Hegner, E. (2000) Fast tectonometamorphism and exhumation in the type area of the Barrovian and Buchan zones. *Geology*, **28**, 459–462.

Phinney, W.C. (1963) Phase equilibria in the metamorphic rocks of St. Paul Island and Cape North, Nova Scotia. *Journal of Petrology*, **4**, 90–130.

Read, H.H. (1952) Metamorphism and migmatization in the Ythan Valley, Aberdeenshire. *Transactions of the Edinburgh Geological Society*, **15**, 265–275.

Read, H.H. and MacGregor, A.G. (1948) *British Regional Geology: The Grampian Highlands*. H.M. Stationery Office, Edinburgh.

Richardson, S.W. and Powell, R. (1976) Thermal causes of the Dalradian metamorphism in the central Highlands of Scotland. *Scottish Journal of Geology*, **12**, 237–268.

Rogers, G., Paterson, B.A., Dempster, T.J. and Redwood, S.D. (1994) U-Pb geochronology of the 'Newer' Gabbros, NE Grampians (abstract). In: *Caledonian Terrane Relationships in Britain, British Geological Survey Programme with Abstracts*, **8**.

Shaw, D.M. (1965) Geochemistry of pelitic rocks. III. Major elements and general geochemistry. *Bulletin of the Geological Society of America*, **67**, 919–934.

Skelton, A.D.L., Graham, C.M. and Bickle, M.J. (1995) Lithological and structural controls on regional 3-D fluid flow patterns during greenschist facies metamorphism of the Dalradian of the SW Scottish Highlands. *Journal of Petrology*, **36**, 563–586.

Skelton, A.D.L., Bickle, M.J. and Graham, C.M. (1997) Fluid flux and reaction rate from advective–diffusive carbonation of mafic sill margins in the Dalradian, southwest Scottish Highlands. *Earth and Planetary Science Letters*, **146**, 527–539.

Soper, N.J., Ryan, P.D. and Dewey, J.F. (1999) Age of the Grampian Orogeny in Scotland and Ireland. *Journal of the Geological Society of London*, **156**, 1231–1236.

Spear, F.S. (1993) *Metamorphic Phase Equilibria and Pressure-Temperature-Time Paths*. Monograph, 799 pp., Mineralogical Society of America, Washington, D.C.

Symmes, G.H. and Ferry, J.M. (1991) Evidence from observed mineral assemblages for infiltration of pelitic schists by aqueous fluids during metamorphism. *Contributions to Mineralogy and Petrology*, **108**, 419–438.

Symmes, G.H. and Ferry, J.M. (1995) Metamorphism, fluid flow and partial melting in pelitic rocks from the Onawa contact aureole, Central Maine. *Journal of Petrology*, **36**, 587–612.

Thompson, A.B. (1976) Mineral reactions in pelitic rocks: II. Calculation of some P-T-X(Fe-Mg) phase relations. *American Journal of Science*, **276**, 425–454.

Thompson, J.B. (1957) The graphical analysis of mineral assemblages in pelitic schists. *The American Mineralogist*, **42**, 842–858.

Tilley, C.E. (1924) The facies classification of rocks. *Geological Magazine*, **61**, 167–171.

Tilley, C.E. (1925) A preliminary survey of metamorphic zones in the southern Highlands of Scotland. *Quarterly Journal of the Geological Society of London*, **81**, 100–112.

Tilley, C.E. (1949) Victor Moritz Goldschmidt (1888–1947). *Obituary Notices of Fellows of the Royal Society*, VI, 1948–1949, 51–66.

Tracy, R.J., Robinson, P. and Thompson, A.B. (1976) Garnet composition and zoning in the determination of temperature and pressure of metamorphism, Central Massachusetts. *American Mineralogist*, **61**, 762–775

Vernon, R.H. (1998) Chemical and volume changes during deformation and prograde metamorphism of sediments. Pages 214–246 in: *What Drives Metamorphism and Metamorphic Reactions?* (x and X, editors). Special Publications, **138**, Geological Society, London.

Watkins, K.P. (1985) Geothermometry and geobarometry of inverted metamorphic zones in the W central Scottish Dalradian. *Journal of the Geological Society of London*, **142**, 157–165.

Yardley, B.W.D. and Bottrell, S.H. (1992) Silica mobility and fluid movement during metamorphism of the Connemara schists, Ireland. *Journal of Metamorphic Geology*, **10**, 453–464.

Yardley, B.W.D., Barber, J.P. and Gray, J.R. (1987) The metamorphism of the Dalradian rocks of western Ireland and its relation to tectonic setting. *Philosophical Transactions of the Royal Society, London*, **A321**, 243–270.

Zenk, M. and Schulz, B. (2004) Zoned Ca-amphiboles and related P-T evolution in metabasites from the classical Barrovian metamorphic zones in Scotland. *Mineralogical Magazine*, **68**, 769–786.

ON THE GEOLOGY OF LOWER DEE-SIDE AND THE SOUTHERN HIGHLAND BORDER.

By GEORGE BARROW, F.G.S.

THE Excursion of the Association to the west of Scotland in September, 1912, was undertaken in order to enable the Members to examine two portions of the great mass of more or less crystalline rocks that cover so large a part of the north of Scotland and of the north-west of Ireland. As it is now well known that the same metamorphosed sediments occur in both countries it will be convenient to call these rocks the Highland series. The object of this examination was to get a true insight into the nature of this great mass of crystalline material; into the cause, in part at least, of its crystallisation, and to study the phenomena occurring along the margins of the mass where it ceases to be crystalline.

I.—THE NATURE OF THE HIGHLAND ROCKS.

Starting from any point along the south-eastern, or outer, margin of the Highlands it is found that as we proceed in a north-westerly direction the rocks steadily become more and more crystalline. But the rate of increase in crystallisation varies; it is most rapid in the area nearer the coast, and least rapid in the area north-west of Dunkeld, which may be conveniently taken as the limit of the south-eastern Highlands. Moreover, in the coastal area we soon reach the most highly crystalline portion of the Highland series. North of Stonehaven, on the "outer margin" of the series, the first rocks met with, when examined in microscopic sections, prove to contain no biotite; but, on the other hand, all clastic micas have been completely digested in the rocks that were originally shales or gritty shales, but now are slates or gritty slates. In this district this "outer margin" of the Highland series is only about 300 yds. broad, but some distance farther inland it gradually widens out, having a breadth of rather more than half a mile near the Bridge of Cally, on the Ericht above Blairgowrie. At this point the "outer margin" of the Highland series, which has so far continued from Stonehaven in a nearly straight line, suddenly projects for nearly a mile in a south-easterly direction, or in the direction of decreasing metamorphism. This brings on the belt of least-altered rocks along the Highland Border*; in this the clastic micas are still present, though the rocks are much

* See the Folding Map.

hardened. This belt with the clastic micas, nearly a mile broad, continues at least as far as Callander, but it must gradually diminish in breadth farther to the south-west, for in Arran it has either disappeared or is only a few yards broad.

So far then we have two belts or zones along the Highland Border; in the least altered the clastic micas are still present; farther within the area, at the Bridge of Cally, we find the clastic micas digested, but no brown mica developed. Zone i.—Proceeding across Zone i we reach rocks in which biotite is developed; this marks the commencement of Zone ii. But no matter how far to the north-west we go, brown mica is usually present in many of the rocks; so that the distance to which it extends has no zonal value; it is the line marking the first oncoming or the "outer limit" of the mineral that gives the zonal line. Farther north-west the common (non-calcareous) garnet is met with; this, too, extends into the highest zones, so that again the line showing its first oncoming or "outer limit" is the zonal line. After this, first staurolite, then cyanite and then sillimanite are met with, and the line marking the oncoming or "outer limit" of each mineral gives a zonal line.

It will be seen that as we pass from Zone i to the highest zone containing sillimanite the rocks become steadily more and more coarsely crystalline. In the highest and most coarsely crystalline, sillimanite is developed on a great scale, and as this mineral requires a specially high temperature for its development it gradually becomes clear that these "outer limit" lines correspond with isothermals; *i.e.*, they indicate the point where the rocks have been raised to a sufficiently high temperature to develop the index-mineral of the zone. These "outer limit" lines have been traced over a large part of the south-eastern Highlands, and exclusively by the writer.*

There are thus seven of these "outer limit" lines, or isothermals, that can be drawn more or less continuously in the south-eastern Highlands, and they are given below in order of decreasing temperature.

6. "Outer limit" of the sillimanite zone.
5. "Outer limit" of the cyanite zone.
4. "Outer limit" of the staurolite zone.
3. "Outer limit" of the garnet zone.
2. "Outer limit" of the biotite zone.
1. "Outer limit" of the zone of digested clastic micas.

Outside this comes the less continuous zone in which the clastic micas are still preserved.

Before proceeding further two specially important points arise:

* See the Map.

Firstly: The existence of a zone in which the clastic micas have been digested, but unaccompanied by the development of brown mica, is believed by the writer to be an exclusively Archæan phenomenon. In all cases where it has been tested by the writer it has been found that this Zone i is never present in any aureole of thermal metamorphism round a post-Torridon granite. In all cases the small clastic micas persist into the zone within which biotite has been developed. The attention of petrologists is invited to this problem in all parts of the world.

Secondly: Cyanite and staurolite are developed in rocks of a special composition, and though by good luck these lines can be traced for great distances in the south-eastern Highlands, there are, even here, areas where there are gaps. But the associated rocks show a special coarseness of crystallisation; if specimens of these rocks be collected for comparison, as standards, there will be little difficulty in drawing approximately accurate lines till the minerals can be picked up again.

There is thus seen to be a close analogy between the phenomena of these areas of crystalline regional metamorphism and the aureoles of thermo-metamorphism round a post-Torridon granite. Many years ago, Dr. Barrois noted this resemblance in his description of the metamorphism produced by the Brittany granites, but he specially drew attention to the fact that the resemblance was fundamentally due to the rocks having been sheared before they were heated; when heated they became foliated crystalline schists. Recently the close resemblance in structure between the crushed and thermally altered rocks in Cornwall and the regionally altered rocks of the Highlands has been insisted on in the description of the aureoles round the Cornish granites. The most important distinction has already been described by the writer in the *Geological Magazine*.* Put briefly, the increase in coarseness of crystallisation is far more rapid as we enter the higher thermal zones in the case of areas of regional thermo-metamorphism than in the case of an aureole round a post-Torridon granite. In addition, the vast development of the non-aluminous garnets and the more local great development of cyanite and staurolite are distinctive features of regional metamorphism; even where these garnets are developed by post-Torridon metamorphism they are far smaller, on the whole, than the garnets of the Highland rocks.

A key to the form of the zones of this area of regional metamorphism is given by the "outer limit" lines of the sillimanite zone (6) and of the biotite zone (2). In the south-eastern Highlands, sillimanite-bearing rocks occur over an area of, roughly, 200 square miles, and within this area every rock that could, from its composition, reasonably be expected to contain sillimanite, always does so. The area is roughly triangular, with its base or shortest

---

* See *Geol. Mag.*, dec. v, vol. v (1908), pp. 423-424.

side to the sea; it is merely a fragment of a much larger area.

But the line marking the outer limit of biotite is almost straight; it has been followed by the writer along the greater part of its course, where not covered by newer deposits or by the sea, as far as Omagh in Ireland, a distance of 250 miles. There is not the least sign of its cessation either at Omagh or Stonehaven, and one is gradually led to the conclusion that the great masses of crystalline rocks in Scotland and Ireland are but a continuation of the similar rocks in Scandinavia. When the intervening "outer boundary" or zonal lines are traced it is seen that they are not straight, but that (3), defining the limit of garnet, slowly diverges from the (2), just described. The divergence is more marked in the case of staurolite (4); its boundary begins to assume a rude parallelism with (6), defining the outer limit of sillimanite. The parallelism is still closer in the case of cyanite (5).

In the south-western Highlands Mr. Clough has traced the southern limit of the garnet zone and has further shown that this is the highest zone reached in that district; we cross it and come again to the biotite zone. It thus gradually becomes apparent that while the line marking the lower zones is almost straight, the intermediate zones surround lenticles of the most highly altered rocks containing sillimanite; further, the whole masses of crystalline rocks are somewhat lenticular in form.

Returning to the south-western Highlands, after entering the biotite zone, we reach the area mapped by Mr. Hill, and after crossing the outer zonal line (1) we actually enter the belt with clastic micas once more, near Loch Awe. A perusal of the various editions of Sir A. Geikie's small scale map of Scotland shows that a belt of little-altered rocks continues up to and along the south side of the Caledonian Canal. If we turn south-east from the latter we approach the area already described and meet with the same zones, but they are in reverse order; the higher temperature zones now come on as we proceed toward the south-east.

It thus appears that the Highland area, south of the Caledonian Canal, is essentially built up on the lines of an aureole of metamorphism round a granite intrusion; but instead of aureoles we have zones or belts which diverge more and more from a lenticular, highly crystalline nucleus, until the lines bordering the lower temperature zones are nearly straight.

The area between the Caledonian Canal and the Moine-thrust plane can be similarly zoned. "The Geological Survey Memoir on the North-West Highlands" and minor publications make it clear that if we start from any point along the outcrop of the Moine-thrust plane and proceed in a south-easterly direction the rocks invariably become more and more crystalline; the phenomena are identical with those seen in the South-Eastern Highlands.

Owing to the low hade of the Moine-thrust plane, denudation has eaten more deeply into the margins of the crystalline mass. So far as is known the outer fringe with clastic micas has gone entirely, and fragments only are left of the zone with no biotite. But there is no difficulty in tracing the outer limit of the garnet-zone. The staurolite zone may not be traceable as the number of infolds of suitable rock is small; but with the cyanite zone there can be little difficulty.*

The extension of the sillimanite area is not yet known, and the difficulty of ascertaining this is increased by the difficulty of finding this mineral in the highly crystallised Moine-gneisses of which much of the area is composed. It requires special experience to know which types to examine, but it may be briefly stated that the mineral has not, so far, been found in those rich in microcline.

If Dr. Horne's introduction to the "Memoir on the North-West Highlands" be read carefully, and also the papers by Dr. Hicks and Canon Bonney, it gradually becomes clear that the Lewisian gneiss† is really a third mass of crystalline rock, of the same age and containing part of the same sediments as the other two already described. The second great mass has been driven along the Moine-thrust plane on to the third. The distance to which it has been driven varies: in the north-east the "outer margin" of the second mass has been driven on to the highest zone of alteration; in the south-west (Gairloch area to Loch Maree) it has been driven on to the cyanite zone only; indeed, in some places, it is doubtful if the rocks of the Lewisian gneiss of this area had been raised to a temperature sufficient for the production of this mineral.

With the exception of the portion of the Southern Highlands mapped zonally by the writer, no attempt has been made to produce a similar map for the rest of the Highlands. Indeed, the zonal lines actually engraved on some of the one-inch maps are omitted from the more recent 4-mile-to-the-inch map. The members of the Association are invited to join in this work, which can be taken up in any part of the area.

The map which accompanies this paper, showing approximately the position of these zones in the South-Eastern Highlands, proves that there can be no connection between them and the Newer Granites, which are also shown on the map. The question then arises—Is there any granitic material that can be connected with the highest phases of metamorphism? The answer is yes; but in place of forming great coherent masses, like the Newer

---

* See "Petrographica Characters of the Inliers of Lewisian Rocks among the Moine-gneisses of the North of Scotland," by J. S. Flett, M.D., D.Sc. *Summary of Progress of the Geological Survey for* 1905, p. 155, etc.

† Twenty years ago the writer sent in a memorandum to this effect.

Granite, it tends to infinite subdivision; *lit-par-lit* intrusion is met with over a great area and pegmatite veining often occurs on a grand scale. The first part of the excursion was devoted to the examination of one of these granitic floodings and the vast development of sillimanite by which it is accompanied.

II.—THE CAIRNSHEE OR INTERMEDIATE GRANITE IN LOWER DEE-SIDE.

In the south-eastern Highlands there are granite intrusions of widely different ages, separated by an epoch of great faulting. The more recent, commonly known as the Newer Granites, tend to occur in large coherent masses, such as the Kincardine Granite, which occupies so large an area in the one-inch sheet 66. The Older Granites rarely form large coherent masses in the area at present exposed by denudation; they tend rather to form a great series of small intrusions, too small in most cases to be shown on the one-inch map. A brief account of these Older Granites and their distribution in the south-eastern Highlands is given in Dr. Hatch's Petrology,[*] where it is shown that there are two groups of these intrusions; one containing the normal proportion of oligoclase to alkali-felspar, the other containing oligoclase in excess, at times almost to the exclusion of alkali-felspar. Each group can be divided into three parts; the first containing biotite but no muscovite; the second, biotite somewhat in excess of muscovite; the third, muscovite usually much in excess of biotite.[†]

For reasons that will be described in the sequel, these intrusions are termed "magmas," and it should be understood that these magmas occur at the surface within definite areas, and the lines forming the outer boundaries of these areas correspond to some extent to the outer boundaries of the newer coherent granite masses, though the subdivision of the older material tends to obscure the fact. It is the opinion of the writer that all these older intrusions are not only of pre-Torridon age, but are actually part of the Lewisian gneiss. The portion that the Association was invited to examine is the normal granite magma of intermediate composition, known as the Cairnshee Granite[‡] or gneiss, which attains its greatest development at the present surface in the lower part of Dee-side.

The Lower Dee-side area shows exceptionally well a feature that is more or less characteristic of much of the Older Granite, but which is specially noticeable in the case of the Cairnshee magma in this area. These intrusions occur as dykes from the

---

[*] "Text Book of Petrology," Dr. F. H. Hatch, Lond., 1909, pp. 289-292.
[†] For a description of the third, see "On an Intrusion of Muscovite-biotite gneiss," by G. Barrow, *Quart. Journ. Geol. Soc.*, vol. xlix, 1893, pp. 330-358.
[‡] It is believed by the writer to be the Laxford Granite of the Lewisian gneiss.

sides of which sills are thrown off at irregular intervals. At one part of the area, near Banchory, the upper portion of the dykes can be easily examined, the sills here are small, and the coherent patches of granite are small. But farther down Dee-side the sills become larger, and it is gradually seen that they really proceeded from a deeper seated and broader portion of a dyke. The largest mass of all occurs at Aberdeen, and much of the city is built on this rock. It was penetrated in a boring at the Bon Accord Distillery to a depth of 450 feet without any sign of the granite ceasing, but it is not possible to say if this was a large sill or entirely in the dyke-feeder of a sill. The rock was a pink foliated granite or augen-gneiss, in which the felspars had no taper end, but were equally rounded at both ends. The examination of the whole area led to the conclusion that the mode of occurrence of the intrusion might be shown by the diagram (Fig. 42). If this view is correct, the proportion of granite at the surface would be much greater had denudation cut deeper. Further, it would offer an explanation of the occurrence, in

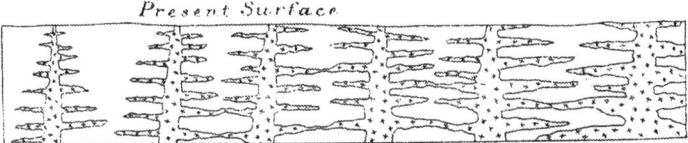

FIG. 42.—TO SHOW DYKE AND SILL MODE OF INTRUSION OF THE CAIRNSHEE GRANITE AND THE INCREASE IN THE AMOUNT OF IGNEOUS MATERIAL FROM BANCHORY TO ABERDEEN.

Archæan areas, where mainly composed of granite, of small patches of intensely altered sedimentary material; as the sills increase in thickness till they almost coalesce, such patches would almost certainly be caught in between them.

It had been intended that the examination of the area should commence at Woodside on the south-east of the Hill of Goach to the south of Banchory; but time did not permit of a visit to this area, and it was found on a trial excursion that the ground was not so clear as when the area was surveyed. Much of it is now thickly wooded. On examining the sides of a dyke it is found that its walls have been burst open, and part of the material that once occupied this vertical fissure has been forced out in the form of small sills. Toward the upper part of the hill the sills become smaller and finally cease; on top of the hill small vertical veins are met with, proceeding from the crest of the main dykes. Toward the base of the hill, the sills become thicker, and can be traced farther. If seen as we go towards the southern margin of the area invaded by this granite magma. On the one-inch sheet 67, in Cowton Burn, just above the high road, a small intrusion of Cairnshee gneiss is shown, in which post-consolidation crushing is well marked. It will be seen that it occurs much south of the main intrusion, in a lower temperature zone. It thus consolidated more quickly and so was crushed after solidifying; in the rest of the area final consolidation took place after the crushing movements ceased.

Age of the Cairnshee Magma.—The published one-inch map sheet 66 and the 4-mile-to-the-inch map both show that the Cairnshee granite and gneiss are cut through by the great mass of the Kincardineshire Newer Granite at two localities, Garrol Hill, and at the north-east point of the great coherent intrusion. When these areas are examined it is seen that the greater mass here has no pegmatite fringe that can be taken seriously. This is partly due to the fact that we are here close to the original top of the great mass. The evidence here is the same as over the whole of the rest of the south-eastern Highlands; there is no appreciable pegmatite veining on the top of the Newer Granite masses.

The phenomena connected with the intrusion of the Cairnshee granite magma have never been met with in Great Britain in connection with any post-Torridon granite; they are characteristic of portions of the Archæan Lewisian Gneiss, and are perfectly reproduced in the Laxford granite and gneiss in the north-west Highlands.

Editor's note: Part III (pp. 282–290) has been omitted.

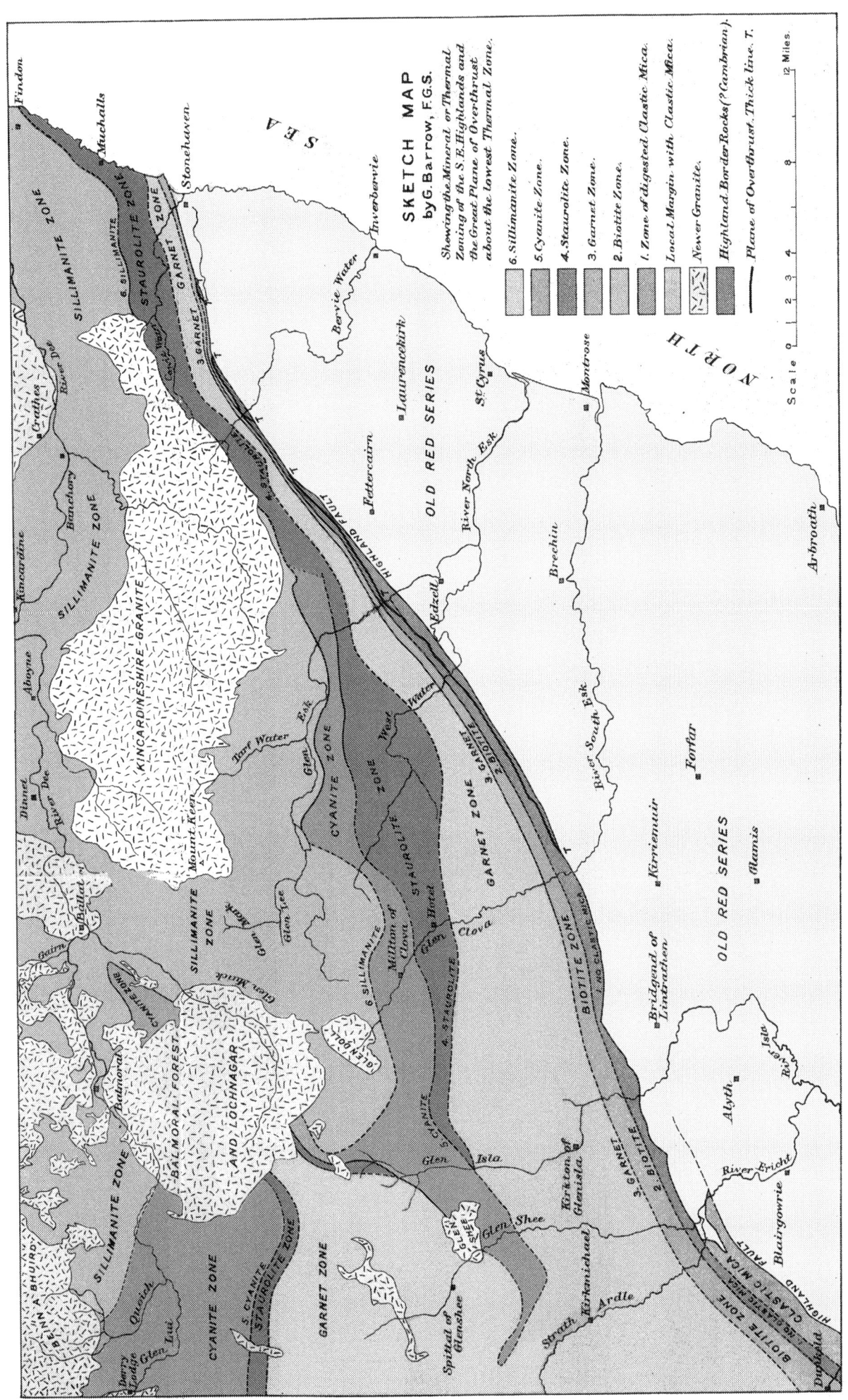

CHAPTER 3: THE METAMORPHIC FACIES

Eskola, P. (1920) The mineral facies of rocks. *Norsk Geologisk Tidsskrift*, **6**, 143–194.

In the summer of 1960, Pentti Eskola, who was then 77 years of age, gave me a personal tour of the Orijärvi district in Finland. Viewing the terrain where the metamorphic facies concept was born, in the company of the individual who conceived it, was as you can imagine an awe-inspiring experience for a freshly minted doctorate in metamorphic petrology. To his obvious pleasure, I confirmed that I had indeed collected some eskolaite (Cr_2O_3) at Outokumpu!

Eskola's (1920) introduction of the concept of mineral facies to the petrological community was a product of his work on the metamorphic rocks of the Orijärvi mining district in southwest Finland (Eskola, 1915) and inspiration provided by Goldschmidt's studies in Norway (especially Goldschmidt, 1911). Eskola's rocks showed the same qualities of chemical equilibrium that Goldschmidt found in the Oslo hornfelses, but the actual assemblages were different, and he attributed this to equilibration under different conditions of temperature and pressure. It was a conceptual breakthrough to take each rock as a chemical system that obeyed the phase rule and possessed a mineral association, rather than a single mineral, as an index of metamorphic grade.

Eskola's definition of a facies (p. 146) bears repeating here: "A mineral facies comprises all the rocks that have originated under temperature and pressure so similar that a definite chemical composition has resulted in the same set of minerals...". Eskola's mineral facies included igneous as well as metamorphic facies, but the igneous facies did not catch on, and so I reproduce here only the first part of his paper. (The second part contains a powerful critique of the CIPW norm.) Eskola recognized five metamorphic facies in 1920 and depicted their different paragenetic relations in the ACF-diagram that he had presented earlier (Eskola, 1915). He hinted at what subsequently became the glaucophane-schist facies, using language that today sounds almost cute: "some little known facies probably related to the eclogite facies" (p. 156) and a "hydrated eclogite facies" (p. 176). He also mentioned a possible epidote-free sub-greenschist facies. Reading Eskola's 1920 paper makes us realise how little field information was available at that time, an enormous contrast to what is now at our disposal.

Given (according to Goldschmidt) that the mineral assemblages of metamorphic rocks depend on the chemical composition of the rock, the pressure and the temperature, that the number of minerals present is constrained by the mineralogical phase rule, and that chemical equilibrium appears in many cases to have prevailed, it was logical to expect that, eventually, someone would attempt to provide a synoptic listing of the diverse sets of mineral assemblages that are repeatedly found in metamorphic rocks in nature. This would not have to be a static document, but one that could be refined from time to time. Eskola's facies represented a stunning affirmation of a school of thought that some of his contemporaries felt unable to embrace.

The metamorphic facies aimed to subdivide the entire accessible PT-plane into portions based on corresponding observable sets of mineral parageneses. Each facies encompassed all possible whole-rock compositions whether or not in nature they exhibited mineral assemblages uniquely distinctive for that facies. As an exercise in systematic petrography, the metamorphic facies were complementary to the equally legitimate subdivision of rocks of fixed chemical composition into zones defined by the first or last occurrence in the field of a mineral or mineral assemblage, as in the Barrovian zones (Chapter 2), and as later developed for carbonate rocks by Bowen (1940, Chapter 4).

Eskola argued that the metamorphic facies offered better potential P-T resolution than the system of depth zones of crystalline schists of Van Hise, Becke and of Grubenmann then in vogue. No certain limits of P or T are defined by the facies, nor a correlation between increasing depth (P) and T. Conforming to classical thermodynamics, the crystallization path followed (via solid, liquid, or vapour states) prior to final (frozen) equilibrium was irrelevant. Rocks of basaltic chemical composition played a central role in the naming and definition of Eskola's facies. He could instead have favoured pelites the mineralogy of which tends to be more sensitive to P and T, but in retrospect I think his choice was a fortunate one. The ACF diagram (p. 157 and following pages) was an important development because it showed mineral compatibilities as well as the relationship between whole-rock chemical composition and the minerals present. The 'other triangle' (p. 168) is the A'KF-diagram, which is more suited to pelitic rocks (see Chapter 5). Possible balanced chemical reactions linking adjacent facies were not spelled out until much later. Nor were the partial pressures of H_2O and CO_2 considered as possible independent variables. Eskola viewed eclogites as having both igneous and metamorphic origins. Note that the Na-free "pseudojadeite" component of eclogitic pyroxene (pp. 155, 173) has 50% vacancy on the $M2$ cation site; nowadays this is known, fittingly, as the Ca-Eskola component, $Ca_{0.5}AlSi_2O_6$, modest amounts of which are found in the omphacitic pyroxene of UHPM rocks (see Chapter 14). The typomorphic minerals in the amphibolite facies at Orijärvi (muscovite, andalusite, cordierite, almandine, anthophyllite/cummingtonite) are, of course, those of a low-pressure variant, variably known as the Buchan or Abukuma type, or a part of the andalusite-sillimanite facies series. The origin of the cordierite-anthophyllite rocks at Orijärvi and similar mineralized localities worldwide was controversial for many years, before we knew about hydrothermal processes at oceanic spreading centres (see Spear, 1993, for a review, and Schneidermann and Tracy, 1991, for phase petrology).

The reception by the petrological community of the idea of metamorphic facies was mixed. Some of Eskola's contemporaries failed to see the novelty of, or the need for, metamorphic facies; they felt there was no need to treat P and T as independent variables. The concept of three depth zones of metamorphism (Epi, Meso- and Kata conditions) continued in use on the European continent for many years (e.g. Grubenmann and Niggli, 1924; Niggli, 1948). Becke (1921), however, accepted the concept of metamorphic facies although he regretted having another "facies" in geology. He expressed the need for an epidote-amphibolite facies, with albite or oligoclase, between the amphibolite and greenschist facies, and questioned whether a state of equilibrium could be assumed for a rock that had undergone metasomatism, for example, by addition of K. He added tremolite, diopside and sillimanite to the list of typomorphic minerals for the amphibolite facies. Tilley (1924, p.170), as usual, was on the mark: "One is led to believe that, provided the limitations imposed by the non-attainment of equilibrium are clearly admitted and recognized, the facies conception in metamorphic petrology will serve as a very useful guide to aid us on the way towards the goal of deeper understanding of metamorphic processes." Harker (1932) did not embrace the metamorphic facies because there appeared to be no place in them for the role of shearing stress in determining the mineral assemblage. Barth (1936) thought that the great advantage of the facies classification was its elasticity, "as observations increase, the system automatically expands." Eskola was fully aware of the requirement for exercising petrographic judgment in distinguishing among the parageneses that may be present in rocks with a complex metamorphic history. I think in fact that we have acquired the experience through the years that enables us now to do this quite well. Challenging petrographic detective work is still required from time to time, however, as we have seen in bringing to light eclogite-facies UHPM terranes (Chapter 14).

With the addition of the epidote-amphibolite facies, the granulite facies (equivalent to his former igneous gabbro facies), and the glaucophane-schist facies, Eskola (1939) increased the number of facies to eight.

Ramberg (1952) feared that if too literal an application were made of Eskola's ideas, there could eventually be an infinite number of facies. Instead, like Turner (1948), he proposed a definition involving physical conditions. This was really unfortunate in that it would have introduced a subjective element into the definition of facies. Yoder (1952) showed that, according to his experimental work in the system $MgO-Al_2O_3-SiO_2-H_2O$, assemblages corresponding to nearly all of Eskola's facies could be obtained at 600°C and 1 kbar, when the H_2O content of the system is allowed to vary. His view was that H_2O behaved as an "initial value" component, and that its amount must be considered along with T and P. From the outset it was known implicitly that the reactions bounding the metamorphic facies are mostly dehydration reactions. Thompson (1955) and Korzhinskii (1959) effectively redefined metamorphic facies to accommodate the chemical potential of H_2O as an external variable in addition to P and T. Miyashiro (1994) discusses this issue very clearly in his textbook. The metamorphic facies are best used when there are consistent paragenetic relations across an area. Outcrop-scale variations in the chemical potential of H_2O (or CO_2) introduce a kind of "noise" that the metamorphic facies cannot always accomodate. Although Eskola and others later talked of "contact-metamorphic facies", and the name pyroxene-hornfels facies lingers still, there is no place for geological setting or the rocks' microstructure in any determination of metamorphic facies.

Coombs et al. (1959) and Coombs (1960, 1961) found that there was need to consider additional facies at the low-T end, with proposals for zeolite, prehnite-pumpellyite, and lawsonite-albite facies.

Miyashiro (1961) linked metamorphic facies into three major baric series, each representing the sequence of zones developed in progressive metamorphism in any one terrane. The idea was to distinguish divergent PT-trajectories that could be related to tectonic environment. It worked effectively for prograde paths, but we now realise that we must contend with individual terrane paths that cross the boundaries of the facies series, as we shall see in Chapter 9.

After the hiatus of World War II, textbook writers seriously embraced the concept of metamorphic facies (Turner, 1948; Turner and Verhoogen, 1951, 1960; Fyfe et al., 1958; Winkler, 1965, 1967), but ultimately they went a little too far by proposing numerous subfacies. These proved problematic to apply on a global scale and, following criticism by Lambert (1965), the subfacies were dropped. Winkler (1974) became sufficiently disenchanted with the metamorphic facies that he proposed they be abolished and replaced by four large divisions of metamorphic grade, with further subdivision of the PT-plane being accomplished with isograds or "isoreaction-grads". This was not a bad idea, but we were apparently not ready to give up on metamorphic facies. Turner (1968, 1981) provided a definition of the facies which gave them a certain robustness and removed any reference to specific external controls: "A metamorphic facies is a set of mineral assemblages, repeatedly associated in space and time, such that there is a constant and therefore predictable relation between mineral composition and chemical composition".

Armed with this kind of general definition, the metamorphic facies have continued to be used by petrologists and the general geological community to the present day. Facies identification requires little more than polarized light microscopy and standard petrographic thin-sections. In terranes that are primarily ones of prograde metamorphism of metasediments, the consistency in the appearance of new assemblages suggests that variations in the chemical potential of H_2O are not so great as to force us to consider it independent of P and T. The same observation argues against sluggish and variable rates of reaction as a major issue in these terranes (Chapter 13). The widespread occurrence of mineral solid solutions (involving notably Mg, Fe^{2+}, Ca, and Al, Fe^{3+}) has meant that facies boundaries as they have conventionally been applied, mostly in metabasites, have a phase-rule variance of more than one, and correspond to bands of pressure and temperature that vary according to the exact composition of the rock. Thus we draw facies boundaries in PT-diagrams as fuzzy lines (e.g. Spear 1993), enclose the facies in non-overlapping loops (Miyashiro 1994), or drop the boundaries altogether (Bucher and Frey, 1994). Specialists desirous of greater precision in extracting values of P and T for their rocks can use appropriate petrogenetic grids (Chapters 4 and 8) for specific bulk compositions, or employ thermobarometric methods (Chapter 10). Furthermore, paragenesis diagrams nowadays make greater use of projection techniques (Chapter 5), which help to overcome some of the problems associated with chemically complex rock-systems. For the non-specialist, the metamorphic facies concept has handicaps that they can live with; it provides a general characterization of metamorphic terranes that is adequate for many purposes. Geophysicists happily use facies names in their sections through subduction zones. In teaching, it is a way of introducing students to the diversity of metamorphic rocks in the context of the entire range of possible temperatures and pressures.

Discussion of facies definitions and boundaries is not over yet. For example, experiments by Green and Ringwood (1967) suggested a substantial PT-region between eclogite and granulite for ordinary metabasalts that was neither one nor the other according to the definitions inherited from Eskola, namely garnet + omphacite, without plagioclase, for the former, and low-Na clinopyroxene + orthopyroxene + plagioclase, without garnet, for the latter. Only more recently have a number of terranes come to light possessing this "high-pressure" granulite-facies mineralogy, consisting of an orthopyroxene-free clinopyroxene + garnet + plagioclase assemblage (e.g. O'Brien and Rötzler, 2003). Pattison (2003) examined the PT-limits of this assemblage with respect to bulk composition and H_2O activity, according to thermobarometry and available experimental constraints.

Are we going to find it useful to add to the list of metamorphic facies, or modify them in any serious fashion? The extraordinary range in P and T encompassed by the eclogite paragenesis omphacite + garnet is problematic. Notwithstanding a call for its dismissal for this reason (Coleman et al., 1965), its use has continued (e.g. Carswell, 1990). Carswell (1990) suggested separation into high-T, medium-T and low-T eclogites. This approach certainly has merits – I confess to having used the similar A, B, C group classification of Coleman et al. (1965) in my teaching for many years – although I cannot help wondering whether it is really necessary to abandon the Eskola tradition of using purely observable mineral criteria for facies boundaries. Recent suggestions along these lines include fields for lawsonite-, epidote-, amphibole-, and dry-eclogite (Liou et al., 1998, Fig. 1; Okamoto and Maruyama, 1999). Subdivisions based largely on experimental phase equilibria can unfortunately be problematic; for a good reason, Turner used the phrase "repeatedly associated in space and time" in his definition of metamorphic facies. In my opinion, for rocks of crustal origin, it might be better to have paragonite-quartz-eclogite, kyanite-quartz-eclogite, and coesite-eclogite.

At the low-temperature end of metamorphism, the substantial overlap, not present in the simple system CMASH, in the calculated PT-fields for complex-system metabasites containing respectively prehnite + pumpellyite, prehnite + actinolite, and pumpellyite + actinolite (Frey et al., 1991) suggests that we might do better for the time being by lumping these facies together as a sub-greenschist facies, cf. Liou et al. (1985). Alternative approaches to metamorphic "rank" at these low temperatures, such as illite crystallinity, seem to be just as useful.

Some of our metamorphic facies are idiosyncratic. Two, the blueschist and eclogite facies, do not blend with uniformitarian ideas (Chapter 7). As pointed out by Zen (1961), diagnostic minerals of the zeolite facies require the attendant fluid to have extremely low CO_2 contents, and for that reason do not occur even when T and P are appropriate. Granulite-facies metamorphism has been attributed to a consistently lower activity of H_2O than in the amphibolite facies, although the explanation, even the justification, for this conclusion remains a subject of debate. These issues may cloud the viability of the metamorphic facies if we identify them too closely simply with P and T. I think we should accept the fact that the facies embody a strong empirical element, and continue to use them as long as it remains practical to do so.

References

Barth, T.F.W. (1936) Structural and petrological studies in Dutchess County, New York, Part II. Petrology and metamorphism of the Paleozoic rocks. *Geological Society of America Bulletin*, **47**, 775–850.

Becke, F. (1921) Zur Facies-Klasssifikation der metamorphen Gesteine. *Tschermaks Mineralogische und Petrographische Mitteilungen*, **35**, 215–230.

Bowen, N.L. (1940) Progressive metamorphism of siliceous limestone and dolomite. *Journal of Geology*, **48**, 225–274.

Bucher, K. and Frey, M. (1994) *Petrogenesis of Metamorphic Rocks*, 6th edition. Springer-Verlag, Berlin, 318 pp.

Carswell, D.A. (1990) Eclogites and the eclogite facies: definitions and classification. Pp. 1–13 in: *Eclogite Facies Rocks* (D.A. Carswell, editor). Blackie, Glasgow, UK, 396 pp.

Coleman, R.G., Lee, D.E., Beatty, L.B. and Brannock, W.W. (1965) Eclogites and eclogites: Their differences and similarities. *Geological Society of America Bulletin*, **76**, 483–508.

Coombs, D.S. (1960) Lower grade mineral facies in New Zealand. *Proceedings of the 21st International Geological Congress, Copenhagen*, **13**, 339–351.

Coombs, D.S. (1961) Some recent work on the lower grades of metamorphism. *Australian Journal of Science*, **24**, 203–215.

Coombs, D.S., Ellis A.J., Fyfe, W.S. and Taylor, A.M. (1959) The zeolite facies with comments on the interpretation of hydrothermal syntheses. *Geochimica et Cosmochimica Acta*, **17**, 53–107.

Eskola, P. (1915) On the relation between the chemical and mineralogical composition in the metamorphic rocks of the Orijärvi region. *Bulletin de la Commission Géologique de Finlande*, **44**, (English summary p. 109–145).

Eskola, P. (1920) The mineral facies of rocks. *Norsk Geologisk Tidsskrift*, **6**, 143–194.

Eskola, P. (1929) Om mineral facies. *Geologiske Forening i Stockholm Förhandlingar*, **51**, 157–173.

Eskola, P. (1939) Die metamorphen Gesteine. Pp. 263–407 in: *Die Entstehung der Gesteine* (T.F.W. Barth, C.W. Correns, and P. Eskola, editors). Julius Springer, Berlin (reprinted in 1960 and 1970), 422 pp.

Frey, M., de Capitani, C. and Liou, J.G. (1991) A new petrogenetic grid for low-grade metabasites. *Journal of Metamorphic Geology*, **9**, 497–509.

Fyfe, W.S., Turner, F.J. and Verhoogen, J. (1958) Metamorphic reactions and metamorphic facies. *Geological Society of America Memoir* **73**, 259 pp.

Goldschmidt, V.M. (1911) Die Kontaktmetamorphose im Kristianiagebiet. *Videnskapsselskaps Skrifter I. Matematisk-*

Naturvidenskapelig Klasse, No. 1, 483 pp.

Green, D.H. and Ringwood, A.E. (1967) An experimental investigation of the gabbro to eclogite transition and its petrological applications. *Geochimica et Cosmochimica Acta*, **31**, 767–833.

Grubenmann, U. and Niggli, P. (1924) *Die Gesteinsmetamorphose, Allgemeiner Teil*, Gebrüder Bornträger, Berlin, 539 pp.

Harker, A. (1932) *Metamorphism: A Study of the Transformation of Rock Masses* (2nd edition, 1939) Methuen, London, 362 pp.

Korzhinskii, D.S. (1959) *Physicochemical Basis of the Analysis of the Paragenesis of Minerals*. Consultants Bureau, New York, 142 pp.

Lambert R.St.J. (1965) The metamorphic facies concept. *Mineralogical Magazine*, **34**, 283–291.

Liou, J.G, Maruyama, S. and Cho, M. (1985) Phase equilibria and mineral parageneses of metabasites in low-grade metamorphism. *Mineralogical Magazine*, **49**, 321–333.

Liou, J.G., Zhang, R.Y., Ernst, W.G., Rumble, D. III, and Maruyama, S. (1998) High-pressure minerals from deeply subducted metamorphic rocks. Pp. 33–96 in: *Ultrahigh-Pressure Mineralogy* (R. Hemley and D. Mao, editors). Reviews in Mineralogy, **37**, Mineralogical Society of America, Washington, D.C.

Miyashiro, A. (1961) Evolution of metamorphic belts. *Journal of Petrology*, **2**, 277–311.

Miyashiro, A. (1994) *Metamorphic Petrology*. Oxford University Press, New York, 404 pp.

Niggli, P. (1948) *Gesteine und Minerallagerstätten*. Verlag Birkhäuser, Basel (English translation by R.L. Parker: *Rocks and Mineral Deposits*, 1954, Freeman and Company, San Francisco, 559 pp.)

O'Brien, P.J. and Rötzler, J. (2003) High-pressure granulites; formation, recovery of peak conditions, and implications for tectonics. *Journal of Metamorphic Geology*, **21**, 3–20.

Okamoto, K. and Maruyama, S. (1999) The high-pressure synthesis of lawsonite in the MORB+H_2O system. *American Mineralogist*, **84**, 362–373.

Pattison, D.R.M. (2003) Petrogenetic significance of orthopyroxene-free garnet + clinopyroxene + plagioclase ± quartz-bearing metabasites with respect to the amphibolite and granulite facies. *Journal of Metamorphic Geology*, **21**, 21–34.

Ramberg, H. (1952) *The Origin of the Metamorphic and Metasomatic Rocks*, University of Chicago Press, 317 pp.

Schneidermann, J.S. and Tracy, R. (1991) Petrology of orthoamphibole-cordierite gneisses from the Orijärvi area, SW Finland. *American Mineralogist*, **76**, 942–955.

Spear, F.S. (1993) *Metamorphic Phase Equilibria and Pressure-Temperature-Time Paths*. Mineralogical Society of America, Monograph, 799 pp.

Thompson, J.B. (1955) The thermodynamic basis for the mineral facies concept. *American Journal of Science*, **253**, 65–103.

Tilley, C.E. (1924) The facies classification of rocks. *Geological Magazine*, **61**, 167–171.

Turner, F.J. (1948) Mineralogical and structural evolution of the metamorphic rocks. *Geological Society of America Memoir* **30**, 342 pp.

Turner, F.J. (1968, 1981) *Metamorphic Petrology* (1st and 2nd editions). McGraw-Hill, New York, 524 pp.

Turner, F.J. and Verhoogen, J. (1951, 1960) *Igneous and Metamorphic Petrology*. McGraw-Hill, New York, 694 pp.

Winkler, H.G.F. (1965, 1967, 1974, 1st, 2nd and 3rd editions) *Petrogenesis of Metamorphic Rocks*, Springer-Verlag, New York.

Yoder, H.S. (1952) The MgO-Al_2O_3-SiO_2-H_2O system and the related metamorphic facies. *American Journal of Science*, Bowen Volume, pp. 569–627.

Zen, E-an (1961) The zeolite facies: An interpretation. *American Journal of Science*, **259**, 401–409.

THE MINERAL FACIES OF ROCKS

BY

PENTTI ESKOLA

Preface.

The present paper is preliminary in a double sense of the word. In the first place, it is a preliminary report of an investigation on the Norwegian Eclogites, which will be treated in detail in another publication. But at the same time it is a report of the present state of a series of investigations carried on since 1914 whith the aim of clearing up the relations between the chemical and mineralogical composition in different kinds of rocks.

The rocks are here dealt with in groups, called Mineral Facies. It follows from the nature of this paper that the various groups will be treated in a rather different manner. The groups called the Hornfels, the Sanidinite and the Amphibolite facies will be here treated shortly, as it is possible to refer to earlier more detailed investigations. In the case of a number of other facies brevity is imperative, because they are still too little studied from the present view-point. In my study of the eclogites I was led to some conclusions that seem to be new and deserving of general interest. It was thought advisable on this occasion to give a summary of these results, without entering into details. The chapters dealing with the eclogite facies are therefore somewhat longer than the other.

This paper is a part of the results of my work done during a visit to Christiania in 1919—1920. My cordial thanks are due to Norwegian geologists who ali showed me the kindest helpfulness. In this connection I wish especially to name

Professor Dr. V. M. GOLDSCHMIDT, Director of the Mineralogical Institute of the University of Christiania, Professor J. SHETELIG, Director of the Mineralogical and Geological Museum of the University of Christiania, Professor Dr. W. C. BRØGGER, Dr. H. H. REUSCH, Director of the Geological Survey of Norway, and Professor Dr. C. F. KOLDERUP, Director of Bergens Museum.

Professor GOLDSCHMIDT has had a very effective share in my work. During frequent colloquies with him most of the problems here treated were discussed, and many of the ideas developed have grown from his suggestions. For all his kindness I wish to express to him my most sincere thanks.

To Mr. OLAF ANDERSEN of the Geological Survey of Norway I am very obliged for much valuable help and especially for his kindness in making grammatical corrections in my manuscript.

These studies have been carried out with assistance from two funds for the advancement of scientific research in my native country, Finland, namely, ALFRED KORDELIN'S General Trust for the Advancement of Progress and Knowledge, and HERMAN ROSENBERG'S Travelling Bursaries Trust of the University of Helsingfors.

The Mineralogical Institute of the
University of Christiania, July 1920.

Pentti Eskola.

Introduction.

F. BECKE, U. GRUBENMANN and C. R. VAN HISE have based the study of metamorphism on the physico-chemical principle of re-adjustment of systems in correspondence with changes in the attendant physical conditions. V. M. GOLDSCHMIDT first proved that metamorphic rocks, by re-crystallization, may have arrived at a very perfect state of chemical equilibrium. This conclusion was the result of a study of the contact metamorphic hornfelses of the Christiania region in Norway.[1] The present writer found, in the rocks of an Archaean metamorphic area in Southwestern Finland,[2] a similar perfect state of equilibrium, though with other minerals as stable phases. The writer proposed to use the term **metamorphic facies** to designate a group of rocks characterized by a definite set of minerals which, under the conditions obtaining during their formation, were at perfect equilibrium whith each other. The quantitative and qualitative mineral composition in the rocks of a given facies varies gradually in correspondence with variations in the chemical bulk composition of the rocks.

A tendency towards a state of equilibrium is, of course, a general feature in all kinds of crystallization, and in fact igneous rocks also may have reached it. Thus we may speak

---

[1] V. M. GOLDSCHMIDT, Die Kontaktmetamorphose im Kristianiagebiete. Vid.-Selsk. Skrifter. I. Math.-Naturv. Klasse, 1911. No. 1. This monograph will in the following be shortly called „Kontaktmetamorphose".

[2] P. ESKOLA, Om sambandet mellan kemisk och mineralogisk sammansättning hos Orijärvitraktens metamorfa bergarter (with an English Summary: On the Relations between the Chemical and Mineralogical Composition in the Metamorphic Rocks of the Orijärvi Region). Bull. Comm. géol. Finl. n:o 44, 1915. In the following called shortly „Sambandet".

of igneous facies, parallel to the metamorphic. Both these conceptions may properly be headed under a more general term, the mineral facies.

A mineral facies comprises all the rocks that have originated under temperature and pressure conditions so similar that a definite chemical composition has resulted in the same set of minerals, quite regardless of their mode of crystallization, whether from magma or aqueous solution or gas, and whether by direct crystallization from solution (primary crystallization) or by gradual change of earlier minerals (metamorphic recrystallization).

By defining for each crystalline rock its place amongst the mineral facies we arrive at a mineralogic-chemical classification, the divisions of which, the largest as well as the smallest, up to the very finest variations, are given by nature itself.

On the first line we have the two parallel branches: that of the igneous and that of the metamorphic rocks. In each of them we have as many facies as there are different sets of minerals in any series of chemically similar rocks. The number of the facies thus being apriori undefined, the system will be so comprehensive and elastic as to be adaptable to every degree of progress and refinement of the science. The definite associations of minerals which appear in each facies in a quantitative correspondence with the variation in the chemical composition represent the smaller divisions of the facies system. An example of this kind of classification was afforded by GOLDSCHMIDT in the case of the hornfels rocks, and the divisions corresponding to definite associations were called classes.

Detailed investigation shows that the mineral associations of various rocks very often represent equilibrium in such a state of perfection that the characteristics of the ideal equilibria may be deduced from them. It is therefore possible to use these associations as a basis of classification. In cases of rocks that do not follow the rules of mineral association in any facies we may at least find, by comparative study, the facies which they have tended to approach during the chief phase of their mineral development.

Certain difficulties arise from the specific kind of action of unequal pressure or stress (cf. Sambandet p. 18 and 116). During stress action no true equilibria are possible. However, the ideal equilibria may even here be used as norms of the classification, as pointed out by J. Johnston and P. Niggli.[1]

We must also be aware of the complication that arises from the fact that pressure influences the origin of minerals in two different manners: (I) by moving the equilibria towards the associations and modifications which have the smallest volume and (II) by preventing volatile components from escaping, whereby the formation of hydrated minerals is favoured and the temperature of crystallization is depressed. The importance of this latter kind of action has been much emphasized by N. L. Bowen in his masterly treatise on the development of the igneous rocks.[2] This fact does not, however, lessen the validity of the facies principle, as this only takes into account the actual mineral development with all components present.

We have before us the task of investigating all varieties of rocks in order to clear up the correlation between their chemical and mineralogical composition. This programme was for the first time expressed by V. M. Goldschmidt in 1914,[3] when he had already begun to explore the Scandinavian Mountain Region with the explicit purpose of determining, by means of the mineral parageneses, the zones of the mountain chain in which conditions of different temperature and pressure had prevailed during the metamorphism. Shortly after I proposed the conception of metamorphic facies and the classification based upon it, without knowing of Goldschmidt's paper. In his first publication on the regional-metamorphic rocks in the Caledo-

---

[1] J. Johnston and P. Niggli, The General Principles Underlying Metamorphic Processes. Journal of Geology 21, 1913, p. 612.
[2] N. L. Bowen, The Later Stages of the Evolution of the Igneous Rocks. Jour. Geol. XXIII, 1915, Suppl.
[3] V. M. Goldschmidt, Om mineralogiens opgaver. (Tiltrædelsesforelæsning holdt den 28. september 1914). „Naturen", novbr. 1914.

nian Chain[1] the latter already used the facies conception proposed by me as a basis of division.

H. VÄYRYNEN recently studied a series of igneous rocks in Southern Österbotten in Finland[2] and dealt with them from view-point of the facies principle. He arrived hereby at many interesting suggestions concerning the mineral development at magmatic crystallization.

Earlier the mineral associations had been subjected to a special study by P. NIGGLI with regard to the chloritoid-schists in the Gotthard region,[3] this paper thus having a direct bearing on the facies-petrology.

A large amount of work, still unpublished, has been done by Professor GOLDSCHMIDT on the rocks of the Scandinavian mountains. I have myself undertaken an investigation of that facies to which the eclogites belong, and a summary of this study will already be given in this paper. Furthermore I am at present working with a series of metamorphic rocks from Eastern Fennoscandia.

From what has been said it will appear that the quantity of work done with the special aim of clearing up the facies relations of rocks is not quite inconsiderable. Much more, however, remains to be done.

Petrographical descriptions in the literature may of course also be utilized to deduce rules of mineral association; in the following will be given a summary of such a literature research on Archaean metamorphic rocks in Finland and Sweden.

---

[1] V. M. GOLDSCHMIDT, Die Kalksilikatgneise und Kalksilikatglimmerschiefer des Trondhjem-gebietes. Vid.-Selsk. Skrifter. I. Mat.-Naturv. Klasse. 1915. N:o 10.

[2] HEIKKI VÄYRYNEN, Etelä-Pohjanmaan graniitti-dioriittisten vuorilajien petrologiaa. Helsinki 1920. (Dissertation. Will be published in German language in Bull. Comm. géol. Finl.).

[3] P. NIGGLI, Die Cloritoidschiefer und sedimentäre Zone am Nordrande des Gotthardmassivs. Beitr. zur geol. Karte der Schweiz, 36, 1912.

Stable and Unstable Relics.

Here it would seem useful to repeat some general statements concerning the relict phenomena in crystalline rocks (Sambandet p. 24 and 118).

Relics may appear whenever a rock has been successively subjected to various physical conditions, each set of which tend to produce a definite facies. If a mineral, stable in the earlier facies, is also stable in the later facies, it will, unless stress action sets in, be preserved in its original form and present a stable relic. This is a very common case, e. g. feldspar in amphibolites etc., quartz in very many various rocks. But when a mineral, or a definite association of minerals, becomes unstable, it may yet escape alteration and appear as an unstable relic.

The discrimination between the stable and unstable relics, so far as I know, never made clear before, throws light on many common and important petrographical phenomena. We may here mention only uralitization, as an example. It has been tacitly assumed that the pyroxenes by metamorphism most likely would be converted into amphiboles. From the view of the facies-conception this statement must be much restricted, and we may say, that uralitization takes place when pyroxene-bearing rocks are brought under the conditions of the amphibolite-facies (which is the name proposed later for a certain facies). But even in the rocks of amphibolite facies the diopside may be a stable constituent, viz. in those rocks in which femic lime is present in excess over the hornblende ratio (in the simplest case $CaO : (Mg, Fe)O = 1 : 3$). We may, in such a rock, find diopside seemingly in process of alteration into hornblende, and yet the diopside is not an unstable mineral. It is only unstable in the presence of excess of magnesia, and in the case supposed the alteration had gone so far as possible at the composition given, and the rock had arrived at an equilibrium.

We may now suppose a gabbro brought under the conditions of the amphibolite facies. Then two cases are possible: —

Either the gabbro has a composition at which the diopside is stable in the amphibolite facies (a rather rare instance), or there are, in the rock, ferrous oxide and magnesia in excess of the ratio named above. In the former case the original diopside of the gabbro will preserve itself as a stable relic, in the latter case it will be uralitized. If, nevertheless, we find diopside in such a rock, then it is an unstable relic. Such gabbroid rocks are not rare e. g. in the Fennoscandian Arhaean, where the largest part of gabbros have been completely uralitized or had crystallized already primarily in the same facies.

A common phenomenon, fairly well illustrative of the tendency towards equilibria, is the formation of **armours** around such minerals which have become unstable in their association, but have not been brought beyond their fields of existence in general (the **armoured relics**).[1] Thereby the associations of minerals in actual contact with one another become really stable. If, however, the constituent minerals of a rock containing armoured relics are named without noting this phenomenon, it may be taken as an unstable association.

Many phenomena at the contacts of two minerals, such as so-called coronas, reaction rims etc., by J. J. SEDERHOLM classified under the collective term synantetic minerals,[2] are such armours.

---

[1] When H. BACKLUND (Geol. Fören. i Stockholm Förh. XXXX, 1918, p. 257) called my definition of armoured relics a „helphypothesis", this proves only that he had failed to understand the phenomenon. There is nothing of a hypothesis in the conception of armoured relics, but a statement of the simple fact that minerals occurring in an association against the phase-rule are commonly isolated from the other minerals by the armour which is formed by the reaction product of the non-congressible minerals. And the occurrence of such phenomena does by no means need any helping hypothesis, as it is itself one of the fairest proofs of the equilibrium theory, the armour-mineral being stable with the associations on both sides and isolating the non-congressible neighbours from each other.

[2] J. J. SEDERHOLM, On Synantetic Minerals. Bull. Comm. géol. Finl. n:o 48, 1916.

Metamorphic Facies.

So far as metamorphic rocks are concerned, the main lines of the facies-classification will be conformable to Becke's and Grubenmann's and Van Hise's well-known divisions in depth-zones. Thus the facies principle, so far from bringing any revolution in the science, may directly continue upon the work founded by those classicists in the metamorphic petrology, with the aim of ascertaining with ever greater exactitude the temperature and pressure conditions under which rocks have originated.

We may actually start from Grubenmann's three depth-zones of crystalline schists, and will find the most typical rocks of each zone to compose a facies in the sense defined. These three facies, however, are not identical with the depth-zone groups (Grubenmann's orders), these latter being more comprehensive and less sharply defined. In each depth-zone we shall find more than one facies.

Even the difference just pointed out would defend the proposition of a new nomenclature for the facies system, in spite of the actual agreement with the depth-zone system. But there is still another reason. The prefixes kata, meso and epi refer to a supposed origin at a certain depth, corresponding to temperatures and pressures within definite limits. The facies-principle, on the other hand, is based directly upon the mineral associations and is thought to serve as a means of clearing up these limits. Therefore it must not be forged into a finished scheme.

The most natural nomenclature in the facies system would, I think, be one where every rock-name would be composed of two parts, the former expressing the typical mineral constituents and the latter the facies. Goldschmidt already practiced this method in the systematics of the hornfelses, using hornfels as a facies-name, though it is not quite correctly a facies name in our sense, as each facies should comprise all possible chemical variations, while e. g. a limestone is not called a hornfels. Terms immediately adaptable to the facies systematics do not exist, they must be fixed by arbitrary definitions.

A mineralogical nomenclature has also already been in use in salt-petrography, a branch of science on the firmest physico-chemical basis.

The facies-classification being still very little elaborated it seems premature to propose any detailed nomenclature. In the following will be given at first a short synopsis of five central metamorphic facies, corresponding to the most essential portions of kata, meso and epi rocks respectively, the contact metamorphic rocks of the Christiania type and the so-called pyrometamorphic rocks. In the present paper I will quite provisionally call the three first-named facies by the names of the rocks having a gabbroid bulk composition, in which the mineral composition is most sensitively variable. Thus we have the greenschist facies, amphibolite facies and eclogite facies. The contact metamorphic facies named will simply be called the hornfels facies, and for the pyrometamorphic rocks we use the name sanidinite facies after their best known representative.

We shall at first see which mineral assosiations are characteristic for these five facies in rocks with a gabbroid bulk composition. I have, in the following table, selected analyses for which the corresponding mineral composition, or the mode, is known, and have considered the exact agreement of the analyses of less importance, as presently no one denies that representatives of all these rocks may be quite identical in composition, disregarding the percentages of water.

As pyrometamorphic rocks (of the sanidinite facies) having a gabbroid bulk composition do not seem to have been analyzed, I have instead of such a rock put an analysis of a hypersthene-augite bearing diabase, belonging to the igneous parallel facies which will be termed the diabase facies.

	Diabase facies: Diabase. Shtsheliki, Olonetz Carelia. W. WAHL, Fennia 24, 3, 1908, p. 20.	Hornfels facies: Essexite-hornfels. Aarvold, Christiania. V. M. GOLDSCHMIDT Kontaktmet. Kristiageb. p. 176.	Greenschist facies: Epidote-chlorite-schist. Val de Bagne, Wallis. U. GRUBENMANN, Die Kristallinen Schiefer, 1910, p. 211.	Amphibolite facies: Amphibolite. Kisko, Finland. P. ESKOLA, Bull. Comm. géol. Finl. 44, 1915, p. 51.	Eclogite facies: Eclogite. Burgstein, Tirol. L. HEZNER, T. M. P. M. 22, p. 466.
SiO_2	49.15	49.19	44.82	49.73	46.26
Al_2O_3	11.48	14.32	20.18	16.05	14.45
Fe_2O_3	3.97	6.00	3.47	2.44	4.41
FeO	13.22 NiO 0.07	8.28	4.04	7.96	5.82
MnO	0.44	0.09	n. d.	0.20	n. d.
MgO	5.39	5.70	7.84	7.84	11.99
CaO	8.63 BaO 0.04	8.55	10.82	10.22	11.66
Na_2O	2.64	3.48	2.03	2.99	2.45
K_2O	1.36	0.79	1.30	0.61	1.51
TiO_2	2.41	2.98	2.38	0.56	0.28
P_2O_5	0.32 FeS_2 0.22	n. d.	n. d.	0.12	n. d.
H_2O	0.57	0.51	3.61	1.03	1.10
Sum	99.91	99.71	100.49	99.75	99.93
Sp. G.	3.090	3.02[1]	3.05	2.99	3.45
Mode	Plagioclase .... 48.4 Hypersthene-augite 37.4 Hornblende, mica, iron ore etc. ... 14.2	Plagioclase . 48 % Hypersthene 17 " Diopside .. 18 " Biotite, iron ore etc. ... 17 "	Epidote ...... 43 %[2] Chlorite ...... 24 " Albite .... 18 " Mica, iron ore etc. 15 "	Plagioclase .. 26.5 % Hornblende .. 71.5 " Quartz ...... 2.0 "	Omphacite 48,5 % Garnet ... 50.5 " Iron ore and rutile ... 1.0 "

[1] Determined by the writer. [2] Calculated by the writer.

In the description of the different facies it seems useful to make a distinction between critical and typical constituent minerals and associations. Critical constituents or associations are such whose occurrence is already a criterion of the rocks belonging to a certain facies, i. e. minerals or associations only stable under those definite conditions of that facies. As typical, on the other hand, we designate generally all the minerals stable in a facies, they may either have originated at the crystallization of that facies, in which case we call them congenetic, or they may be stable relics.

Besides the typical minerals rocks may contain minerals not stable in their association. They are either unstable relics or posterior products.[1]

Although the names of the facies here used are provisional they must still be used in a strictly defined sense. With the sanidinite facies we thus class all the pyrometamorphic rocks containing, when silicate-rocks, regular associations of some of the following silicatic type minerals:

Sanidine (homogeneous potash soda feldspars), plagioclase, sillimanite, cordierite, clinoenstatite-diopside (homogeneous mixtures of clinoenstatite, clinohypersthene and diopside), wollastonite, olivine. The sanidine and the clinoenstatite-diopside are critical.

With the hornfels facies we understand the groups of contact-metamorphic rocks which show the same mineral facies as the contact-metamorphic rocks of the inner contact-zones in the Christiania region, regardless of their composition or structure or premetamorphic development, and excluding such so-called hornfelses which show some other mineral development. Type minerals are orthoclase, plagioclase, andalusite, cordierite, biotite, hypersthene-enstatite (orthopyroxene), diopside, wollastonite, grossularite-andradite, olivine. The association of hypersthene and diopside is critical.

---

[1] The term secondary products, frequently used to designate the things here called the posterior products, seems to me less adequate, as the rocks, after their main mineral formation, have often passed through several stages of posterior changes, that might then be spoken of as secondary, tertiary, quaternary etc.

The amphibolite facies is that metamorphic facies the rocks of which follow those rules of mineral association which I found in the metamorphic rocks in the Orijärvi region. They always, when their composition allows it, contain some mineral of the amphibole group. Type minerals are microcline, plagioclase, muscovite, andalusite, cordierite, almandite, anthophyllite and cummingtonite, diopside, wollastonite, grossularite-andradite, olivine.

The greenschist facies includes metamorphic rocks containing some of the following minerals:

Albite, sericite, chlorite, talc, serpentine, epidote, calcite, dolomite. The associations of potash mica with chlorite and of epidote with albite are especially characteristic.

The definition of the eclogite facies can hardly be put in a few words. The typical silicate minerals are garnets (pyrope-almandite-grossularite), monoclinic pyroxenes (clinopyroxenes) being isomorphic mixtures of some of the silicates diopside, jadeite, pseudojadeite ($(Ca, Mg, Fe) Al_2 Si_4 O_{12}$) and a few others in small amounts, hypersthene-enstatite (orthopyroxene), olivine, disthene. The garnets with more than 30 mol. % pyrope and the jadeite and jadeite-bearing pyroxenes are critical of this facies. Rutile is a characteristic minor constituent. Many eclogites contain biotite, but in other cases no potash-mineral seems to have been stable.

Certain minerals are common to several facies. Thus quartz and calcite are stable under widely various conditions. A limestone may show the same mineral (calcite) in all the facies. Diopside exists in all but the greenschist facies.

There are, of course, gradual transitions between the facies; the transition points between definite minerals or associations are different, and combination of temperatures and pressures may have different effect upon the changes.

In rocks belonging to a definite facies, the relations between their chemical and mineralogical composition may be expressed graphically, so that the mineral associations appear directly from the diagrams based on the chemical analyses. Such a graphical solution is practicable, if the number of the components controlling the mineral composition can be reduced to an extent allowing of a diagrammatic expression. The principle of the Osann triangle has proven most useful, but it allows only three components to be represented simultaneously, while the rocks have a much larger number. By rational choice and restriction I have, however, succeeded in finding a method of expression useful for the majority of rocks of more common composition, and having excessive silica, so that always the compounds highest in silica can be formed. In one corner of the triangle is located the alumina not combined with potash or soda, in another the lime and in the third one magnesia with ferrous iron. The minor constituents are, after subtracting the amounts of $(Al, Fe)_2O_3$, CaO and $(Mg, Fe)O$ combined in them, left out of consideration.

By this method all the common silicate minerals excepting those containing soda or potash can be expressed, and some salient features of the mineral paragenesis in the five facies of rocks appear clearly, most of the characteristic and variable minerals being compounds of ferro-magnesia, lime or alumina. In all the rocks of the hornfels, the greenschist and the amphibolite facies, with the exception of the alkaline series (in which $(K_2O + Na_2O) > Al_2O_3$), soda is only present in the form of albite.[1] But in the eclogite facies soda in all kinds of rocks enters into the jadeite exclusively. Potash, in the non-alkaline rocks, may enter into the potash feldspar or into the micas. In the last-named mineral the proportions of alumina in excess of the ratio $K_2O : Al_2O_3 = 1 : 1$ will be expressed in the triangle projection under consideration, and only the potash feldspar compound is entirely lacking.

I shall quote from my previous paper (Sambandet p. 129) the rules of calculation of the A, CF values which are used in the following diagrams for non-alkaline rocks with excessive silica.

---

[1] Concerning the presence of some soda in common amphiboles and micas and pyroxenes cf. „Sambandet" p. 36 and 124. — The soda amphibole, glaucophane, and paragonite, the soda mica, occur in non-alkaline rocks belonging to some little known facies probably related to the eclogite facies (cf. p. 176).

„Previous to calculating the values intended for use in plotting the triangle projections, it is appropriate to make the correction for the ilmenite, magnetite and titanite already in the percentages of weight. The amount of ilmenite is estimated by the geometrical method, and the amount of FeO, which, with sufficient accuracy, may be taken for 50 % of the ilmenite, is subtracted from the total FeO. In the same way the Fe_2O_3 (70 %) and the FeO (30 %) in the magnetite are subtracted from the total percentages of Fe_2O_3 and FeO respectively. An analogical method is applied to the titanite whose CaO-content (30 %) is subtracted from the total CaO."

„Thereupon the molecular proportions are calculated from the percentages, and may be used without recalculation into mol.-%. SiO_2 and H_2O are left out of consideration. The further treatment is as follows: Fe_2O_3 is added to Al_2O_3. FeO, MgO and MnO are added together. The value of CO_2 is subtracted from that of CaO to eliminate the calcite present. The mol. value of P_2O_5, multiplied by 3.33, is subtracted from CaO to eliminate the apatite."

Furthermore Na_2O and K_2O are subtracted from Al_2O_3, the former being thereupon left out of consideration. The remaining molecular numbers of $Al_2O_3 - (K_2O + Na_2O) = A$, ,$CaO = C$ and $MgO + MnO + FeO = F$ are re-calculated to 100 %.

The Hornfels Facies.

Fig. 1 represents the hornfels facies. The Roman figures designate the location of the ten hornfels classes. The Arabic figures are the points of the analyses of the hornfelses of classes 1—7 taken from GOLDSCHMIDT's „Kontaktmetamorphose".

1. Andalusite-cordierite-hornfels (p. 148). $A_1 = 69$; $C = 0$; $F = 31$. ($A_1 = 68$; $C = 0$; $F = 32$).
2. Plagioclase-andalusite-cordierite-hornfels (p. 154). $A_1 = 48$; $C = 28$; $F = 24$ (mode not known).
3. Plagioclase-cordierite-hornfels (p. 156). $A_1 = 32$; $C = 11$; $F = 57$. ($A_1 = 51$; $C = 19$; $F = 30$).
4. Plagioclase-hypersthene-cordierite-hornfels (p. 162). $A_1 = 27$; $C = 11$; $F = 62$. ($A_1 = 48$; $C = 26$; $F = 26$).
5. Plagioclase-hypersthene-hornfels (p. 169). $A_1 = 28$; $C = 21$; $F = 51$. ($A_1 = 37$; $C = 31$; $F = 32$).
6. Plagioclase-diopside-hypersthene-hornfels (p. 176). $A_1 = 22$; $C = 34$; $F = 44$. ($A_1 = 24$; $C = 40$; $F = 36$).
7. Plagioclase-diopside-hornfels (p. 188). $A_1 = 6$; $C = 47$; $F = 47$. ($A_1 = 5$; $C = 50$; $F = 45$).

If only the typical minerals were present, the points representing the analyses would be located within those fields or at those lines, whose corner points correspond to the composition of the constituent minerals. The hornfelses, however, also contain biotite, a mineral present in most the hornfelses of classes 1—7. The quantity of ferrous oxide and magnesia

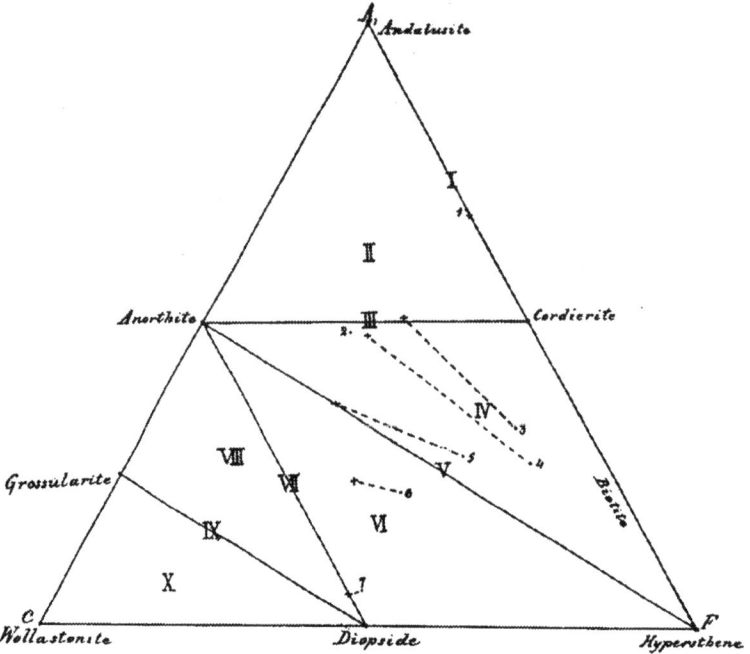

Fig. 1. The A,CF-projection of the hornfels facies.

in this mineral influences the location of the plotting points, but its existence depends upon the presence of water which does not appear in the triangle. All the points deviate from their proper positions in a direction towards the biotite field.

I have therefore also calculated the A,CF-values after having subtracted the oxides entering into the biotite, whose percentages were taken from the modes given by Goldschmidt. The values thus corrected have been quoted above within paranthesizes and plotted on the triangle (fig. 1) with crosses united to the respective points by dotted lines. It may be seen that the new values, from which the biotite has been eliminated, show perfect agreement with the theory.

The A,CF-triangle gives an instructive synopsis of the associations possible and actually existing in the rocks of the hornfels facies. The whole field has been divided into five smaller triangles. A point plotted from a rock analysis is located whithin that triangle the corners of which correspond to the composition of their minerals, and no other minerals represented in the figure are possible in that association. Thus we have a graphical illustration of the application of phaserule, that three minerals at most may occur together as rockmaking constituents in a three-component system.

The hornfels facies seems to be restricted to the inner parts of the contact-metamorphic aureoles around laccolitic igneous bodies. Towards the outer contact zones this facies gradually passes into another metamorphic facies which so far is little known but seems to be related to the amphibolite facies. The latter is represented in the contact products of most batholitic masses, and transitional forms between the hornfels and the amphibolite type occur frequently (e. g. the Eker gneiss in Harz).[1] In the contact products of hypabyssal bodies and volcanic products types transitional to the sanidinite facies may be expected.

Rocks of very various composition and premetamorphic development, metamorphosed in the hornfels facies, have been described. In the Christiania region there occur sedimentogeneous hornfelses of argillaceous and sandy materials, and contact-marbles, and besides many kinds of eruptivogeneous hornfelses.

The Sanidinite Facies.

Rocks and mineral associations of the kind that I here call the sanidinite facies have been very much studied in long times. Such petrographical standard works as A. LACROIX'

---

[1] V. H. ERDMANNSDÖRFFER, Der Ekergneiss im Harz. Jahrb. d. preuss. geol. Landesanstalt, 1909, XXX.

treatise on inclusions in the volcanic rocks[1] and R. BRAUNS' work on the sanidinites of the Laach Lake district[2] deal with this facies. BRAUNS was the first to define the characters of the pyrometamorphism and its relations to the other kinds of mineral genesis.

There seems to exist no analyses of pyrometamorphic rocks, owing to their manner of occurrence as non-homogeneous small inclusions. Thanks to experimental investigations in the laboratory, the stability relations are, however, better known than in any other facies.[3]

Fig. 2 shows quite schematically the rules of paragenesis for rocks with an excess of silica.

Aluminium silicate in pyrometamorphic rocks is always sillimanite, and experimental work has proved that this is the only stable form at all temperatures[4] below the melting point. (Whether the andalusite in the hornfels facies is truly stable or presents a false equilibrium (cf. Sambandet p. 141) has not yet been ascertained.)

Very remarkable is the difference in the pyroxene. While, in the hornfelses, the magnesium silicate forms a distinct phase, the orthopyroxene, it crystallizes here in solid solution with the diopside in all proportions. This has been shown experimentally by N. L. BOWEN.[5]

In pyrometamorphic rocks evidence is lacking, probably because no pyroxenes from rocks with adequate bulk composition have been studied.[6] In the corresponding igneous facies

---

[1] A. LACROIX, Les Enclaves des Roches Volcaniques. Macon 1893.
[2] R. BRAUNS, Die Kristallinen Shiefer des Laacherseegebietes. Bonn 1912.
[3] Summaries are given by: H. E. BOEKE, Grundlagen der physikalisch-chemischen Petrographie. Berlin 1915. — PAUL NIGGLI, Neuere Mineralsynthesen I, Fortschritte der Min., Kristallogr. und Petrogr. Band 3, 1915. Id., ibid. II, Band 6, 1919.
[4] G. A. RANKIN and F. E. WRIGHT, The Ternary system $CaO-Al_2O_3-SiO_2$. Am. Journ. Sc., XXXIX, 1915, p. 1.
[5] N. L. BOWEN, The Ternary system Diopside-Forsterite-Silica. Am. Journ. Sc. XXXVIII, 1914, p. 239.
[6] I have examined all the analyses of pyroxenes published in HINTZE's (II, p. 1104—1113) and DOELTER's (II, 1, p. 507—566) hand-books. — In picrites, basalts, andesites and diabases, the magnesium diopsides

the entstatite-augites were described in a well-known paper by W. WAHL.[1] They are very common in meteorites which are, even in other respects, very typical representatives of the same facies.

Synthetic wollastonite is able to take up a maximum of 17 per cent of diopside.[2] Analyses of natural wollastonite often

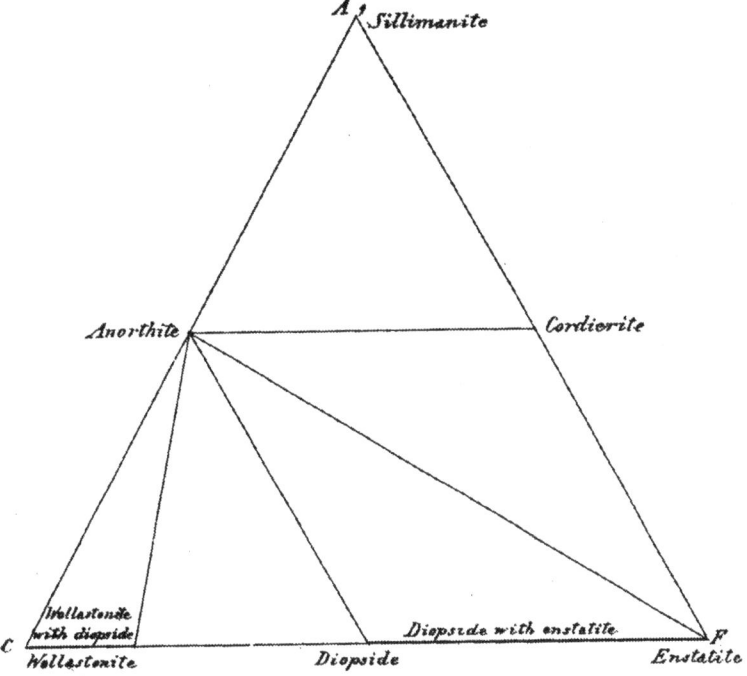

Fig. 2. The A,CF-projection of the sanidinite facies. The thick lines designate solid solutions.

show some magnesia,[3] but no general difference between those from volcanic products and deep-seated rocks can be inferred.

(enstatite-augites) are fairly frequent, but no reliable analysis of such minerals from any metamorphic rock was found, excepting the highly aluminous fassaite.

[1] W. WAHL, Die Enstatitaugite. T. M. P. M. 26, 1907.
[2] J. B. FERGUSON and H. E. MERWIN, Wollastonite and related Solid Solutions in the Ternary System Lime—Magnesia—Silica. Am. Journ. Sc. XLVIII, 1919, p. 165.
[3] C. HINTZE, Mineralogie II, p. 1015.

The solid solutions may, therefore, exist in other facies, too. No garnets are stable in the sanidinite facies. The wollastonite may therefore occur in association with anorthite, against the rules of the hornfels facies. An example of this association was recently described by H. A. BROUWER from ejected volcanic blocks from Java.[1]

The character of the alkali feldspar is also a very distinctive feature of the two facies compared. In the hornfelses orthoclase and albite occur, in the sanidinites both these compounds form solid solutions. The stability-relations of the alkali feldspars, with special regard to their geological mode of origin, have been critically studied by E. MÄKINEN.[2]

Still a radical difference exists in the modifications of silica. Its stable forms in the sanidinite facies are tridymite and cristobalite.[3] In the hornfels facies only quartz has been met with.

In the natural occurrences the sanidinite-rocks show poor approximation to true equilibria. Relics from various other facies are almost invariably present.

As appears from the previous outline, the divergencies between the two facies are great. They apparently are not so much due to a difference in the pressure than in the temperature, this having been considerably depressed, in the hornfels facies, by the pressure which has prevented the volatile components from escaping.

The Greenschist Facies.

The rocks of the greenschist facies belong to the uppermost zone of the lithosphere and chiefly originate under considerable stress. It is mainly owing to the last-named circumstance, that in them may be noted more variability in the

---

[1] H. A. BROUWER, Studien über Kontaktmetamorphose in Niederl.-Ostindien. Centralbl. f. Min. etc. 1920, 37.

[2] EERO MÄKINEN, Über die Alkalifeldspäte, Geol. För. i Stockh. Förh., 39, 1917, p. 121—184.

[3] CLARENCE N. FENNER, The Stability Relations of the Silica minerals. Am. Journ. Sc. XXXVI, 1913, p. 331—384. — See also N. L. BOWEN, ibid. XXXVIII, 1914. p. 218, and J. B. FERGUSON and H. E. MERWIN, ibid. XLV, 1918, p. 417—426.

mineralogical characters than in any other rocks, and instead of one facies there will probably be necessary to separate several.[1] Chloritoid, paragonite, disthene, staurolite, garnets, glaucophane and certain fibroid amphiboles are all regarded as typomorphic of the epi-rocks, but probably no one of them belongs to the greenschist facies.

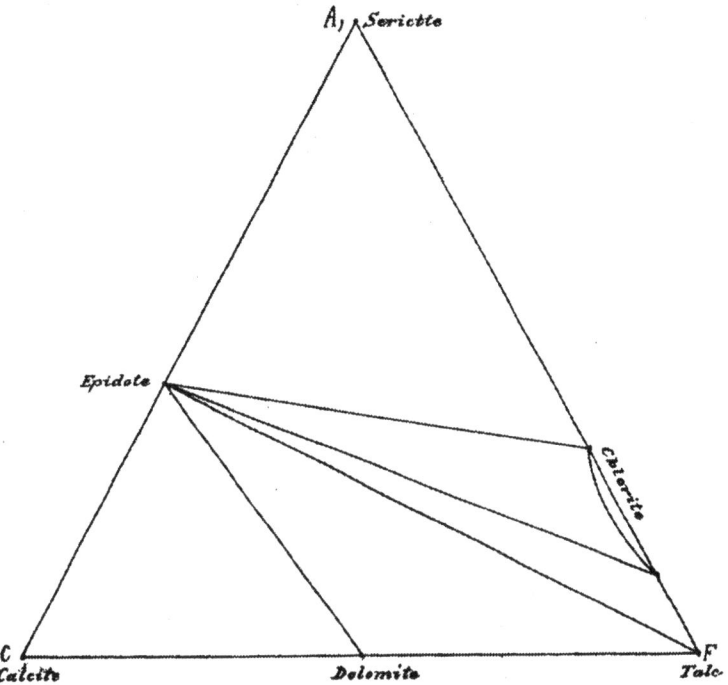

Fig. 3. The A,CF-projection of the greenschist facies.

Greenschist-rocks are fairly common, especially in the Alpine mountain-zones. They have been studied in a large number of localities which we do not need to enumerate here.

Of the typical minerals of this facies named above all but the albite are represented in fig. 3.

I am at present studying rocks of the greenschist facies from eastern parts of Fennoscandia, where they occur in manyfold

---

[1] Cf. NIGGLI, op. cit.

development, but I do not yet possess analyses of them. The greater part of the rocks of the crystalline schists of third order in GRUBENMANN's classification conform to this facies, and a large number of analyses are given in his text-book, but unfortunately the mineral composition of the rocks analyzed has not been given. Therefore no analyses have been plotted on this triangle.

The paragenetic relations are here different from those in all the other facies because of the appearance of carbonates. Now, the calcite and dolomite are apparently stable in all the different facies, in contact with any of the silicates or quartz. In metamorphosed limestones there occur very many minerals, the calcite apparently playing the rôle of an indifferent medium. Likewise the carbonates, in the rocks of the greenschist facies, are quite regularly found in association with sericite and chlorites; the boundary lines of the association fields do not seem to signify the notice "no admission for outsiders" that they do in the other facies-triangles.

This seeming controversy has a very natural explanation: In all the other facies and in the greenschist facies in the rocks the projection points of which lie within the field sericite-epidote-talc, the carbonates are indifferent only so far as the carbon dioxide cannot escape. As soon as this is possible, the carbonates will interact with the silicates and the free silica under formation of lime or ferro-magnesia silicates.

But in all the rocks of the greenschist facies containing more lime than necessary to form epidote with the alumina available, the carbonates are stable in the presence of free silica, under all circumstances. If, on the other hand, lime silicates are brought under the conditions of the greenschist facies in the presence of carbon dioxide, they will transform themselves into carbonates and epidote.

Petrographical experience shows, moreover, that there exists a range of physical conditions, were the epidote and its relatives (zoisite etc.) do not originate from carbonate-bearing sediments, and the only silicates in such rocks are fine scaly micaceous minerals. This is a further facies to which probably belong many of the lime-phyllites and certainly a large part of the products of the atmospheric weathering. We may name it the **lime-phyllite facies**.

V. M. Goldschmidt[1] has by means of Nernst's theoreme calculated the equilibrium curve $CaCO_3 + SiO_2 = CaSiO_3 + CO_2$, and pointed out that the curves for the equilibria of the formation of aluminous and ferromagnesian lime silicates run at lower temperatures than the calcite-wollastonite curve.

In Eastern Fennoscandia I have had good opportunities of studying the different stages of combination of lime and silica[2]. Starting from the eastern boundary of the pre-Cambrian area one finds at first limestones and marles of the lime-phyllite and the greenschist facies and then going towards the west, meets with tremolite-limestone, and further the diopside and wollastonite set in representing a crystallization under successively higher temperatures. All the three last-named minerals in the limestone and lime-silicate-rocks belong to the amphibolite facies. The reactions between the lime-magnesia carbonates and the silica are in fact the most sensitive indicators of variations in the pressure and temperature in the rocks that we know.

The Amphibolite Facies.

In my previous paper (Sambandet) I dealt with this facies in the rocks of the Orijärvi region in south-western Finland. In fig. 4 all the rock-analyses there published have been plotted on the A,CF-triangle and, moreover, analyses of Archaean metamorphic rocks in Finland and Sweden from the following publications:

P. J. Holmquist, The Archaean Geology of the coast-regions of Stockholm. G. F. F. 32, 1910.

H. E. Johansson, Die eisenerzführende Formation in der Gegend von Grängesberg. G. F. F. 32, 1910.

---

[1] V. M. Goldschmidt, Die Gesetze der Gesteinsmetamorphose. Vid.-Selsk. Skr., Mat.-Naturv. Kl. 1912, N:o 22. p. 12.

[2] Pentti Eskola, Victor Hackman, Aarne Laitakari ja W. W. Wilkman, Suomen kalkkikivi. With an English summary by P. E.: Limestones in Finland. With 75 figures and a map. Suomen Geologinen Toimisto, Geoteknisiä Tiedonantoja N:o 21, 1919.

H. E. JOHANSSON, The Flogberget Iron Mines. G. F. F. 32, 1910.

EERO MÄKINEN, Ytterligare om kontakten vid Naarajärvi i Lavia. G. F. F. 36, 1914.

HJ. SJÖGREN, H. E. JOHANSSON and NAIMA SAHLBOHM, Chemical and Petrographical Studies on the Ore-bearing Rocks of Central Sweden G. F. F. 36, 1914.

EERO MÄKINEN, Ein archäisches Konglomeratvorkommen bei Lavia in Finnland. G. F. F. 37, 1915. — Über Uralitporphyrit aus Pellinge in Finnland. G. F. F. 37, 1915.

PER GEIJER, Falutraktens berggrund och malmfyndigheter. Sveriges G. U., ser. C, N:o 275, 1917.

I studied, moreover, the available Swedish and Finnish petrographical literature, in order to find out to what degree the rules of association would have general application to the Archaean of Fennoscandia. Many valuable statements were found in the writings of the authors named above and still in some papers by A. E. TÖRNEBOHM, H. BÄCKSTRÖM and A. GAVELIN. In general, notes on paragenetic relations are rare in the literature.

In the amphibolite facies, in its typical development, the correspondence between the chemical and mineralogical composition is exceedingly regular, so that the mineralogical composition can be calculated from the chemical analysis with great exactness. An especial simplification of the relations is due to the fact that water always seems to have been present in quantities sufficient to allow of the formation of the hydrated minerals of this facies. Thus e. g. the quantity of the mica is controled only by the proportions of potash and ferro-magnesia and alumina. In those rocks in which the quantity of potash is sufficient to form mica whith all the $(Mg,Fe)O$ and Al_2O_3 available, the amount of micas is controled exclusively by the proportions of the latter oxides. The excessive potash then goes to form potash feldspar. In these rocks our A,CF triangle expresses quantitatively the proportions of the chief mineral constituents except potash and soda feldspar and quartz.

In fig. 4 the different mineral associations have been marked with different signs, so that the reader may convince himself that the points of the analyses really lie within those fields at whose corner points their constituent minerals are located.[1]

In rocks containing insufficient amounts of K_2O to form micas there appear some of the following minerals: Andalusite, cordierite, anthopyllite. The amounts of these in proportion to the micas are controled by the amount of potash and do not

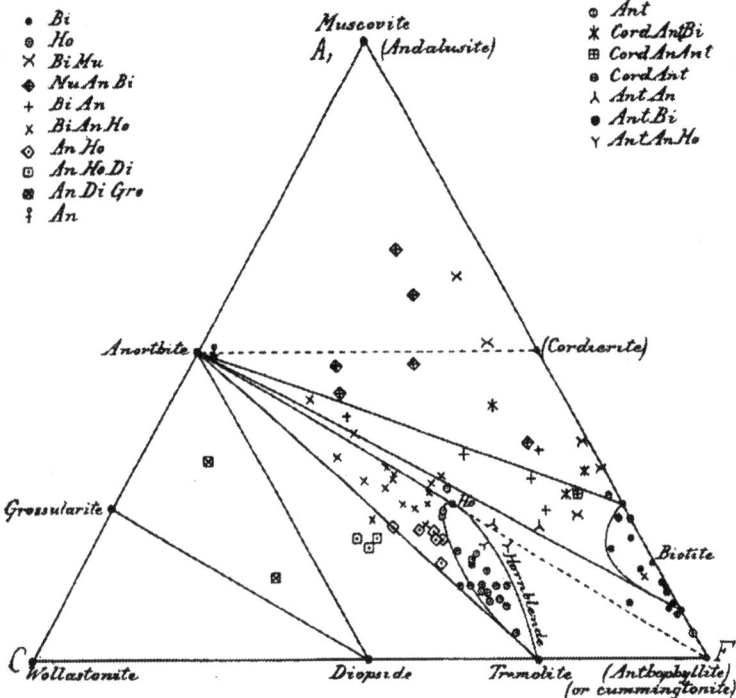

Fig. 4. The A,CF-projection of the amphibolite facies. All the associations whose designations are explained in the upper left hand side are such having excessive potash, but no excess of alumina. The location of the plotting-points, for these rocks, is an exact measure of the mineralogical composition. Those associations whose designations are explained on the upper right hand side are such having a deficiency in potash. For these the location of plotting-points is no perfect indication of the mineralogical composition.

---

[1] There are a few exceptions, e.g.: a biotite-hornblende-plagioclase-rock is located within the an-di-ho-field, and a biotite-feldspar-rock within the bi-mu-an-field. All such cases may probably be ascribed to analytical errors, the rocks being leptites, very poor in compounds under consideration, and the errors appearing magnified. Misleading results

therefore appear quantitatively in the A,CF-triangle. But I have found another triangle projection („Sambandet" 43 and 128), by which they may be expressed in a quantitative manner, and in those also the theory has proved wholly conformable to the facts.

It appears from what has been said above, that the potash feldspars should not occur in association with andalusite, cordierite or anthophyllite. In the Orijärvi region it was found that in fact they do not, and this fact had struck me long time before the above theories were built up. The same relation has been noted by P. GEIJER[1] from the Falun region and several other sulphide-ore fields in Central Sweden. In other districts in Finland and Sweden, however, microcline appears to be stable with cordierite or andalusite (or sillimanite which in such cases often takes the place of andalusite, a fact of much significance). This must be explained so that we have in such cases to do with another though closely related facies, in which the micas did not originate when the relations of the bases would have allowed it. This subdivision is especially common in the corresponding igneous facies (cordierite-granite etc.).

The Eclogite Facies.

Doubtless members of the eclogite facies are only the olivine- and enstatite-rocks, olivine-pyrope-rocks, eclogites, chloromelanitites and jadeitites.

The eclogite facies comprises but a part of the rocks which U. GRURENMANN has classed under the kata-rocks, or the rocks of the deepest zone. Neither soda-lime nor potash feldspars have been observed in congenetic association with the critical minerals of this facies, viz. eclogite-garnet and monoclinic pyroxenes of the diopside-jadeite series. Thus all kinds of gneiss

---

could, of course, also follow, if rock-minerals with variable composition (hornblende, biotite etc.) would happen to have an extraordinary composition, outside of their fields plotted on the triangles on basis of existing analyses.

[1] Op. cit. (Sveriges G. U., ser. C., n:o 275, 1917).

must be excluded. In the same way no wollastonite seems to exist here.

The eclogitic micaschists (micaschisti eclogitici) of FRANCHI[1] are certainly very closely related to the eclogite facies. Their constituents are soda pyroxene, garnet, mica (near sericite) and quartz, besides amphibole and feldspar as posterior products. Colourless mica, as well as biotite, generally occur in all kinds of eclogite, but it is difficult to say, if it is strictly congenetic.

The question, whether the quartz is typical in the eclogite facies, must also be left open.

Calcite almost certainly may occur in the eclogite facies.

In the most cases where eclogites occur, it seems that their facies is restricted to comprise only variations between definite limits of the rock composition, a feature only observed in this facies, and ascribable, as will be discussed somewhat more closely in the chapter dealing with the igneous eclogites, to the fact that most eclogites have originated at temperatures above the melting temperatures of many other rocks. Such high temperatures, however, are probably not a material condition for the formation of the eclogite rocks, while these rocks certainly cannot originate except under very high pressures, and therefore it is well-nigh possible that quartz may belong to the lower temperature portion of the eclogite facies, whilst in the higher temperature portions, solid silica could not exist.

Despite the existence of only two chief phases, garnet and pyroxene, in the eclogite rocks (to which, however, often is added disthene), their composition shows comparatively wide variation owing to solid solutions.

The A,CF triangle is not well adapted to express the rules of association in this facies, as soda and an equivalent amount of alumina, omitted in this triangle, enter here into the pyroxene to form solid solutions with lime- and ferro-magnesia compounds. For the sake of comparison with the other facies I have, how-

---

[1] S. FRANCHI, Über Feldspaturalitisierung der Natron-Thonerde-Pyroxene aus den eklogitischen Glimmerschiefern der Gebirge von Biella (Graiische Alpen). N. J. Min. etc. 1902, II, p. 112—126. Idem, Giacimenti alpini ed apenninici de roccie giadeitiche. Atti del congresso intern. di scienze storiche. Roma 1904.

ever, constructed this triangle here also (fig. 5), and most available analyses of eclogites and their minerals, including those of Norwegian examples (eclogites 18—23) made for my investigation, have been plotted on it.

The portions of the triangle which correspond to compositions actually occurring in the eclogite facies comprise but two

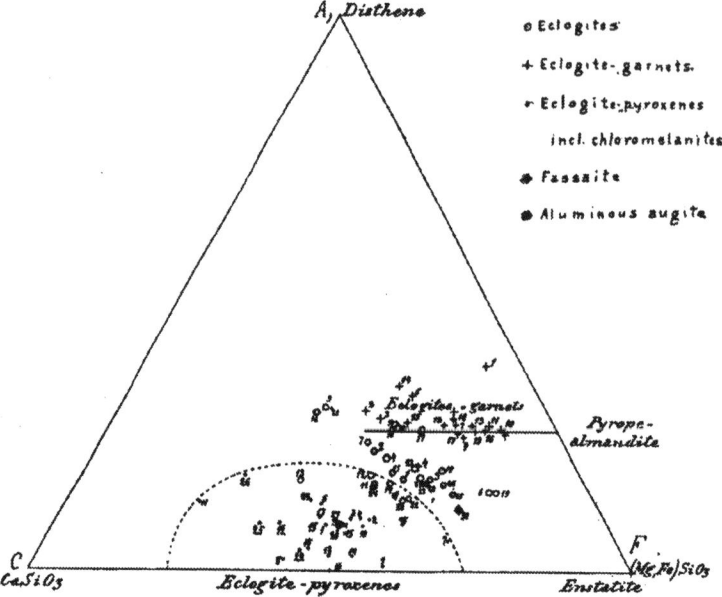

Fig. 5. The A,CF-projection of the eclogite facies.[1]

fields: garnet-clinopyroxene-disthene and garnet-orthopyroxene-clinopyroxene. It is peculiar of this facies that monomineralic rocks occur frequently, as e. g. the olivine-rock, enstatite-rock, garnet-rock, chloromelanitite, jadeitite. The metasilicate, enstatite, plays a much less important rôle than the orthosilicate, olivine. Most of the combinations possible have been observed, but the olivine alone and the combination garnet-clinopyroxene are most common.

---

[1] Figures 5, 6 and 7 have been lent from my treatise on eclogites under preparation, and fuller explanation of the figures will be given there.

On the A,CF-projection (fig. 5) I have plotted no other rock-analyses than such of eclogites. As appears from their location between the garnet and the pyroxene fields, they mostly consist almost exclusively of garnet and clinopyroxene. Some of them, however, contain appreciable amounts of disthene, as eclogites 9 and 16. Others, as 19, a Norwegian eclogite, are olivine-bearing. The olivine-rocks and enstatite-rocks, if plotted on this triangle, would be located almost exactly on the F-point. For the jadeitites, mostly consisting of very pure jadeite only, this graphical method is not at all adaptable.

The close relation between eclogites and chloromelanitites and jadeitites, and gradual transition between them, was pointed out by Damour[1] and L. S. Penfield[2] and especially by Laura Hezner[3]. In the Norwegian eclogites, as I have found, the pyroxene varies between almost pure diopside and a chloromelanite containing 8 pct. Na_2O. Pyroxenitic portions of such eclogites are true chloromelanitites.

Whilst the enstatite, olivine and disthene in the eclogite rocks are the same as in the other facies where they occur, the garnets and clinopyroxenes in the eclogite rocks exhibit quite peculiar characters in their ability of forming solid solutions or isomorphic mixtures. These relations may be made clear here.

The eclogite-garnets present an isomorphic series of almandite and pyrope in all proportions uptil 75 mol. % pyrope, and with small amounts of grossularite. As seen in fig. 6 in which all the available analyses of almandites, pyropes and grossularites have been plotted on a triangle, the garnets in granites, gneisses and mica-schists, cordierite-rocks, gabbros and amphibolites show a limited ability of taking up pyrope in solid solution, the maximum amount being 33 mol. % $Mg_3Al_2Si_3O_{12}$.[4]

---

[1] Damour, quot. in Groths Zeitschr. Kryst. 6, p. 291.
[2] L. S. Penfield, Am. J. Sc. 46, 1893.
[3] Laura Hezner, N. J. B.-B. XX, 1914.
[4] There are two exceptions: the garnet from amphibolite n:o 4 („hornblende-gneiss" from Namaputra, Zs. Kr. 36, 421) and that from granite n:o 8 („Lithoklasen einer grossen Pegmatitmasse" from Ruwenzori, Zs. Kr. 50, 512). The only descriptions of the occurrences accessible to me, viz. those in the summaries of Zs. Kr., are not sufficiently detailed to

Here the garnets from Norwegian labradorite-rocks have been regarded as eclogite-garnets, this being justifiable because they are closely related to the eclogites and, by clustering together with pyroxenes etc., show transitions into true eclogite in the Bergen and Sogn areas in Norway.

Pyropes only exist in rocks belonging to the eclogite facies. As long ago shown by C. DOELTER[1] and

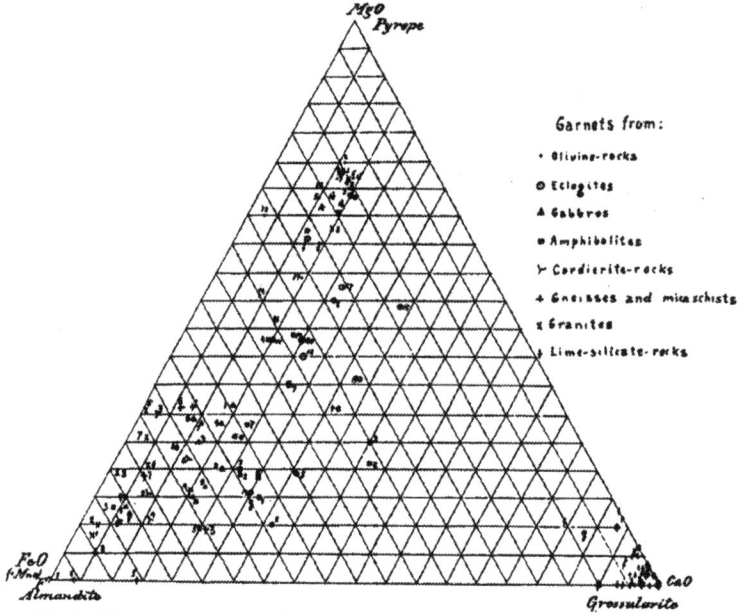

Fig. 6. The proportions of FeO, MgO and CaO in rock-forming garnets.

later verified by many others, the pyrope in serpentinized olivine-rocks is a primary magmatic mineral. Garnets rich in pyrope has also been met with in many eclogites regarded as metamorphic. Other modes of occurrence are not known. The garnets occurring in cordierite- or anthophyllite-bearing rocks or in amphibolites, rich in magnesia as these rocks may be, nevertheless are always almandites.

---

be of any real value in the present discussion. The hornblende-gneiss from Namaputra, whose garnet is a precious variety, would seem to be a changed eclogite.

[1] C. DOELTER, T. M. M. 1873.

The eclogite-garnet may as well be almandite (as the eclogite-garnet n:o 12, from Vanelvsdalen, Söndmöre, Norway) as pyrope (n:o 10 from Almklovdalen, Norway, n:o 16 from Jagersfontain, South-Africa, and n:o 21 from Böhrigen, Saxony), or also grossularite (as n:o 8, from Frankenthal, Silesia).[1] Most eclogite-garnets are mixtures of all the three compounds in proportions not met with in the garnets from any other rocks.

The ability of taking up grossularite in solid solution seems, however, to be limited in the eclogite-garnets (see fig. 6).

All the eclogite-garnets are low in titanium, a very remarkable feature.

Garnets containing more than 75 mol. % of pyrope are not known. Whether purer pyropes are unstable even in the eclogite facies or the material has never been sufficiently poor in iron, is still uncertain.

The eclogite-pyroxenes exhibit no less discrepancy as compared with the pyroxenes of the other facies, than the garnets.

Fig. 5 shows the ratios of ferro-magnesia, lime and the amount of alumina not combined with soda in the jadeite molecules. As appears, most eclogite-pyroxenes contain such an excess of alumina. They also show great variation in their CaO : $(Mg, Fe)O$ ratio and still in the amount of silica. In all the cases, however, their composition may be expressed in terms of the silicates proposed by Tschermak,[2] to which the pseudojadeite, $R''R_2'''(SiO_8)_4$, of Clarke[3] must be added. In all, there are then the following silicates: — $Ca(Mg, Fe)Si_2O_6$, $CaSiO_3$, $(Mg, Fe)SiO_3$, $(Mg, Fe)(Al, Fe)_2SiO_6$, $Ca(Al, Fe)_2SiO_6$, $NaAlSi_2O_6$, $NaFeSi_2O_6$ (in small amounts), $(Ca, Mg, Fe)(Al, Fe)_2(SiO_3)_4$.

Fig. 7 represents the same pyroxenes as fig. 5, but here also the total alumina, inclusive that combined with soda in jadeite, has been shown, and the two points representing one and the same analysis have been united by a line. The great

---

[1] H. Traube, N. J. 1889, 1, 197. Though chemically identical with a grossularite-diopside-hornfels the rock would seem from description to be genetically a true eclogite, embedded in serpentine.

[2] G. Tschermak, T. M. P. M. 32, 1914.

[3] F. W. Clarke in R. Heber Bishop, Investigations and Studies in Jade, p. 133 New York 1906.

variation in the amount of jadeite appears clearly. As an example, a jadeite (no: 28) has also been plotted; its ACF-point is located nearly at the corner A.

All the other silicates except diopside and jadeite occur in rather small amounts only. There are, however, some analyses which point to a considerable admixture of the pseudojadeite. Such are e. g. analyses of jadeitic stone implements n:os 21[1] and 26[2], both from the Alps. I have also found such a highly aluminous pyroxene in a fine-grained eclogite from Saltkjael in Selje, Norway (point 29).

It seems that earlier investigators have not been aware of the bearing which the existence of such remarkable pyroxenes, high in lime and alumina, has on the problem here dealt with. These pyroxenes characterize the eclogite facies and have the same relation to the anorthite as the jadeite has to albite.

Besides in the eclogite rocks aluminous pyroxenes occur in volcanic rocks (near the sanidinite facies). These differ from the eclogite-pyroxenes in being low in jadeite; their aluminous compound may be mostly computed as $MgAl_2SiO_6$. Fe_2O_3 is high, and Na_2O, when present, enters into the aegirite. They are often rich in TiO_2, while the eclogite-pyroxenes are low, like all the silicate minerals in the eclogites.

In fig. 5 and 7 has been plotted, for the sake of comparison, a pyroxene from a leucite-tephrite containing 9.04 % Al_2O_3, (point 31), taken at random from table XII in „Quantitative Classification".

Point 30 represents a fassaite from the Fassa Valley, the only example of aluminous pyroxene (10.10 % Al_2O_3) known in contact-metamorphic formations. It is an enstatite-augite, containing more MgO than any eclogite-pyroxene analyzed.

As a summary it may be stated that the pyroxene compounds are greatly miscible in the volcanic rocks, much less miscible in the rocks of the hornfels and the amphibolite facies, but in the eclogite rocks their ability to form solid solutions reaches its maximum, though the diopside and enstatite are not miscible in all proportions here.

---

[1] LAURA HEZNER, N. J. Min. etc. B.-B. XX, p. 141.
[2] S. FRANCHI, Boll. R. Comitato geol. d'Italia, 1900, n:o 2.

U. Grubenmann remarks[1] that the presence of disthene in the eclogites does not mean any excess of alumina in the rock. This is true, if the feldspar ratio, $(K_2, Na_2, Ca)O : Al_2O_3 = 1$, is regarded as a standard. Such a view, however, has no meaning, as feldspars do not exist here. We may well say that the appearance of disthene may be attributed to an amount of

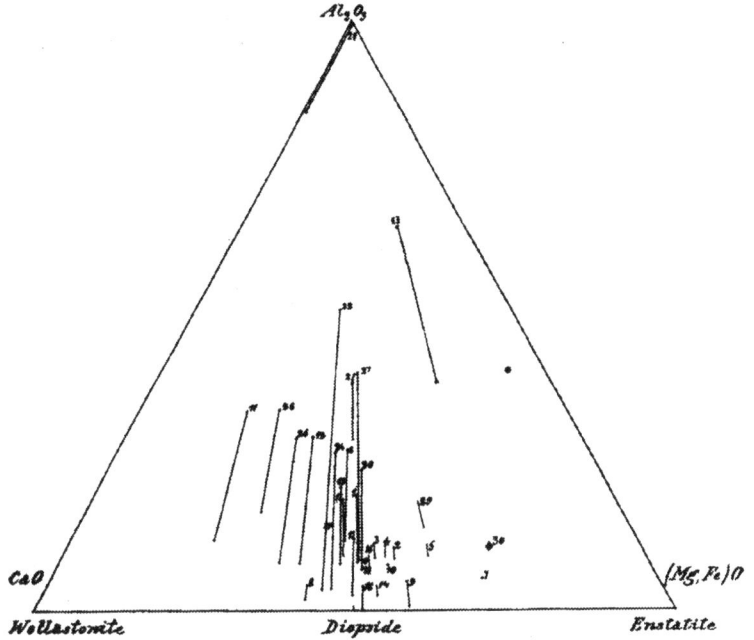

Fig. 7. The A,CF-projection of the eclogite-pyroxenes, incl. chloromelanites and jadeite (n:o 28). For each pyroxene is plotted also the ACF-projection, showing the proportion of the total Al_2O_3. — The corresponding values are combined by lines. The designations are the same as in fig. 5.

alumina excessive of that which may possibly enter into the combination of garnet and pyroxene present.

Rutile is the titanium mineral of the eclogite facies besides ilmenite, the garnets and pyroxenes being generally almost devoid of titanium. In other facies (greenschist, amphibolite, hornfels) the rutile occurs only when the concentration

---

[1] Die Kristallinen Schiefer, 1910, p. 196.

of TiO_2 in proportion to FeO and CaO is very high. In rocks of eclogitic composition the occurrence of rutile is altogether critical and its alteration into titanite or ilmenite, when an eclogite is amphibolitized, is a striking phenomenon in many eclogites from Norwegian occurrences.

Another facies closely related to that of eclogite is one characterized by eclogite-garnet, glaucophane, lawsonite and paragonite (?). It might be designed as the hydrated eclogite facies and probably belongs to a region of somewhat lower temperature. These remarkable rocks are associated with eclogites and jadeitites on the island Syra, in the Alps and in California.

In the Norwegian eclogites no glaucophane-bearing varieties are known, but primary hornblende of the common kind and green colour is fairly frequent. Still more often a similar hornblende occurs as a posterior product. This happens when the eclogite has been brought under the conditions of the amphibolite facies, and comprises the processes well-known under the terms kelyphitization (SCHRAUF) and feldspathuralitization (FRANCHI).[1] In both these the alteration begins in the pyroxenes and also makes a more rapid progress in these minerals than in the garnets. Very often we find eclogites in which all the pyroxene has been altered into hornblende in myrmekite-like intergrowth with plagioclase. In the igneous Norwegian eclogites the development often has been the following: When the crystallization of the garnet was completed and liquid magma was still present, the conditions of the amphibolite facies set in, and now hornblende and plagioclase crystallized directly from the magma around the garnet. The latter may thus be an unstable relic, if its field of stability does not continue into lower temperatures than that of the pyroxene.

---

[1] S. FRANCHI, loc. cit. N. J. 1902, II.

Editor's note: the section on Igneous Facies has been omitted.

CHAPTER 4: METAMORPHISM OF LIMESTONE AND THE PETROGENETIC GRID

Bowen, N. L. (1940) Progressive metamorphism of siliceous limestone and dolomite. *Journal of Geology*, **48**, 225–274.

Bowen's 1940 paper is included in my list of landmark papers for two reasons. For the first time it presented a lucid, systematic treatment of the phase petrology of progressive metamorphism involving devolatilization reactions, introducing students to paired triangular composition diagrams that show changes in compatibilities of phases caused by passage across a univariant reaction curve on a pressure-temperature diagram. It provided a lesson in the application of Gibbs' phase rule that is just as valid today as it was then. It was a bold foray by a master igneous petrologist into theoretical metamorphic petrology, with the metamorphism of siliceous limestone and dolomite as its subject.

Second, albeit on the very last page, it introduced the concept of the petrogenetic grid. Nowadays, the petrogenetic grid is so basic to metamorphic petrology that we tend to forget that there was a time before it came into everyday use. In comparison to contemporary grids with their network of lines of contrasting dP/dT slopes, Bowen's grid is unimpressive! There is a start to everything; he was surely on the right track.

Aside from a study by Niggli (1916) showing that the orthosilicate Ca_2SiO_4 (later known as larnite) represents higher temperatures than wollastonite + calcite, no experimental work had been done on any of the decarbonation reactions discussed by Bowen. The reaction calcite + quartz = wollastonite + CO_2 had been calculated by Goldschmidt (Chapter 1). It is remarkable how few field observations relevant to decarbonation reactions had been made prior to 1940 (pp. 226–227).

The basic assumption underlying Bowen's treatment of the progressive metamorphism of sedimentary carbonate rocks is that the order of stable reactions with increase in temperature should correspond to progressive decarbonation, i.e. to declining net contents of CO_2 in mineral assemblages saturated in CO_2 vapour. Down-temperature release of vapour is very unusual in metamorphic reactions (the blueschist-to-greenschist reaction in metabasites, for example); I am not aware of any such exception for decarbonation reactions. Thus, in the compositional triangle $CaO-SiO_2-CO_2$ used for siliceous limestones, successive steps of metamorphism are deduced from the sequence of 2-mineral joins broken in moving away from the CO_2 corner (p. 235). In the compositional tetrahedron $CaO-MgO-SiO_2-CO_2$ ($CMS+CO_2$) used for siliceous dolomites, successive stable assemblages are represented by sets of 3-mineral triangles open to the CO_2 vapour corner at the apex of the tetrahedron, that move progressively downwards away from CO_2 (p. 236).

We should note that the axis labelled P in all Bowen's pressure-temperature diagrams refers to the "external" pressure applied to the system; we might call this the pressure on the solids P_s. The pressure of CO_2 vapour is only required to be equal to the external pressure along the univariant curve for the mineral reaction involved. This treatment differs from our familiar experimental P-T diagrams where, unless otherwise stated, $P_{vapour} = P_s$ everywhere, and a vapour-undersaturated assemblage such as Wo + Cal + Qtz would not be possible at low temperature. With only three system components in siliceous limestones, the triangular diagram (Fig. A) can show the compatibility or incompatibility of minerals such as Wo with pure CO_2. However, in Figs B and C, Bowen chose not to show the minerals incompatible with pure CO_2; instead, the critical assemblage not compatible with pure CO_2 at low temperature is given for each reaction as "a)" in the text. On the other hand, assemblages in metamorphosed siliceous dolomites (Figs 1, 2, 3 etc.) are represented in triangles after projection from CO_2. Thus, *all* 3-mineral triangles shown coexist with pure CO_2, and critical assemblages incompatible at low temperature with pure CO_2 are again listed in the text as "a)".

Bowen recognized (p. 230) that CO_2 may entirely escape from the divariant high-temperature assemblage (e.g. Wo + Qtz or Wo + Cal) — but the variance remains unchanged because one phase and one component are lost, a degeneracy in modern terms. He would be delighted to know that wollastonite occurs in the H_2O-rich, CO_2-poor low-T environments of skarns and rodingites.

Bowen inferred the sequence of basic steps in the metamorphism of siliceous limestone to be: the formation of (1) wollastonite, (2) spurrite and (3) larnite. The order of steps 1 and 2 was later confirmed by high-PT laboratory experiments (Harker and Tuttle, 1956; Tuttle and Harker, 1957). But, as Bowen suspected, his list was incomplete. A phase with the composition $Ca_3Si_2O_7$ had been synthesized at high temperature, but it could not be placed in the sequence because its stability with respect to larnite + wollastonite, a solid-solid reaction, was not known. Tilley (1942) discovered this phase at the famous Scawt Hill locality in County Antrim, and called it rankinite. Shortly after that, a phase equivalent to spurrite + CO_2, namely $Ca_5Si_2O_7(CO_3)_2$, was found at Carlingford in Northern Ireland by Nockolds (1947), who named it tilleyite after his mentor. Accordingly, Bowen's list of reactions was extended to five index minerals by Tilley (1951), in the order: (1) wollastonite (2) tilleyite (3) spurrite (4) rankinite (5) larnite.

As his tentative phase diagram showed (Fig. 14), Bowen thought that the intersection of decarbonation curves on the PT-diagram would be very unusual. No information was available in those days to indicate otherwise. Murphy's law was about to change that. In the course of high-PT experimental phase equilibrium work on four reactions involving minerals in Tilley's sequence, Zharikov and Shmulovich (1969) found the dP/dT slope of the curve for the reaction: tilleyite = spurrite + CO_2 to be flatter than the other curves, creating one, possibly two, invariant points. Their experimental data were fitted by Joesten (1974) to plots of ln K vs. $1/T$ so as to derive a complete petrogenetic grid for the system at 600–1150°C and 1–1000 bars. The PT-diagram of Joesten (1974, Fig. 3) has four stable invariant points and an indifferent crossing. What this means is that there are actually at least six possible isobaric reaction sequences rather than the single ones of Bowen (1940) and Tilley (1951). Tilley did not acknowledge that his sequence might not be unique; with the advantage of hindsight, we can see from the outset that the chemographic arrangement of phases on the triangular diagram allows different choices that are still in accord with Bowen's basic tenet. Tilley's sequence is compatible with Joesten's diagram at $P < 140$ bars. Above 140 bars, rankinite and spurrite change places in the sequence, and above 500 bars larnite and spurrite change places. I confess on first reading I did not realize the implications of invariant points for Bowen's approach to progressive devolatilization reactions. Joesten (1976) used his phase diagram to infer conductive heat flow across the narrow aureole around gabbro and syenite in the Christmas Mountains, Texas, and convective overturn in the intrusion.

Much of the above concerns sanidinite-facies carbonate rocks that few readers will ever encounter; nonetheless, Bowen's paper is well worth reading as an instructive exercise in metamorphic phase equilibria.

In the case of metamorphosed siliceous dolomites, Bowen wisely decided to ignore the possible occurrence of the hydrous minerals talc, serpentine and anthophyllite, on the grounds that they are relatively unimportant and would create the same problems that he was to encounter with tremolite. Tremolite had to be included because it was already well known to be an important mineral in metamorphic siliceous dolomites. However, in order to plot the tremolite-forming reaction in the CMS triangle by projection from the CO_2 corner of the tetrahedron (Fig. 1), it was necessary to ignore its water content. For this purpose, Bowen chose for anhydrous tremolite the composition $CaMg_3Si_4O_{12}$, a "formula often given in old texts". This made "tremolite" compositionally equivalent to a pyroxene made of equal parts of diopside and enstatite – thus making it graphically co-linear with the two pyroxenes (which real tremolite is not) in the triangle CaO-MgO-SiO_2 (Fig. D). It is unclear why he did not instead choose $Ca_2Mg_5Si_8O_{23}$, which is tremolite (as in equation 1a, p. 239) minus H_2O. The basic topological relations of minerals in the projection triangle are not affected by this choice of anhydrous composition, however, and so the arrangement of reactions on the phase diagram is not affected.

Having chosen an anhydrous "tremolite", all the minerals of concern in the metamorphism of siliceous dolomites can be expressed in the four-component system CMS+CO_2. As for siliceous limestones, all decarbonation reactions among these minerals at any given pressure have a maximum temperature where $P_{CO_2} = P_{solids}$. They form an array of 13 sub-parallel curves on the P-T diagram (Fig. 14) and – anticipating Greenwood (Chapter 6) – a corresponding array of curves with positive slopes on an isobaric, two-volatile diagram of T vs. X_{CO_2}. The table of steps (Table 1, p. 257) lists the critical reactant assemblage that must be eliminated in order to indicate passage across the associated univariant reaction curve. Presence of the product mineral assemblage is not the same thing because it can occur below the univariant curve at lower pressures of CO_2, in the "fluid deficient" region, something that Bowen correctly assumed is likely in nature. The logic here is carefully stated at the top of page 260. This treatment of course contrasts with how we traditionally interpret prograde dehydration reactions, as in the Barrovian sequence (Chapter 2), where the fluid is predominantly H_2O at high pressure, and reactants and products of the isogradic reaction tend to be mutually exclusive in the field. In terms of the production of new minerals, the 13 steps became 10 (p. 260): three in siliceous limestones and seven in siliceous dolomites.

Bowen had no choice but to include tremolite, although naturally there were consequences (footnote 29 on page 260 is prophetic). First, in reality it added the component H_2O; to maintain a phase-rule variance of one, this required an additional phase. His suggestion (p. 239) regarding the existence of an aqueous solution coexisting with vapour was prescient. A T–X_{CO_2} diagram calculated by Bowers and Helgeson (1983, Fig. 10) showed that reaction 1a (p. 239) in the presence of immiscible brine and CO_2-rich vapour (a likely situation at low temperature in nature) is indeed isobarically invariant, thus univariant in P and T. Second, unlike anhydrous "tremolite" which behaves like a pyroxene, the presence of real tremolite in a reaction ensures that it will tend either to zero or to infinite temperature as the vapour approaches pure CO_2 in composition (see Chapter 6). Korzhinskii (1959) first pointed out the impossibility of predicting a unique sequence of minerals with increasing grade, unless the composition of the fluid is considered in addition to pressure and temperature. It is now well known that the initial prograde reactions in siliceous dolomites with increasing proportions of CO_2 relative to H_2O are those forming talc + calcite, tremolite + calcite, and diopside (e.g. Bucher and Frey, 1994).

Tilley (1948) found that talc and then tremolite formed outside the zone of forsterite in contact metamorphosed marble in Skye (cf. Harker 1904), meaning that Bowen's problem of an apparent early, metastable formation of forsterite from dolomite + quartz (p. 240), with or without enstatite as an intermediate step, disappeared. But forsterite remained troublesome for Bowen. What we know now is that Bowen's reaction 3 (Tr + Cal + Qtz = Di) occurs at lower temperature than reaction 2 (Tr + Dol = Fo + Cal), except perhaps at very low values of X_{CO_2} (e.g. Bucher and Frey, 1994). Thus, Bowen got reactions 2 and 3 in the wrong order. Was this a breakdown in the underlying logic of progressive decarbonation? No. If you examine Figs 2 and 3, you will find that reactions 2 and 3 affect bulk compositions (projected from CO_2) that are not mutually overlapping, and so it is not possible to say from first principles which composition is effectively more decarbonated.

The same kind of ambiguity also affects the relative placement of reactions 5 and 6 (5: Dol = Per + Cal; 6: Cal + Qtz = Wo). Bowen (p. 246) realized that clearly non-overlapping bulk compositions were involved in this case and that additional information was needed. He was led to place reaction 5 first because of the "accumulated knowledge of metamorphic rocks" (p. 246), even though he was aware of ambiguity in the field evidence (footnote 22, p. 248), and tried to hedge his bet by speculating on an indifferent crossing of the curves for reactions 5 and 6 at a pressure higher than shown in Fig. 14. Both laboratory experiments (Harker and Tuttle, 1955, 1956) and thermodynamic calculations (Weeks, 1956) later showed that reaction 6 takes place at a substantially lower temperature than 5.

We shall not concern ourselves here with the higher steps (above 6), which are encountered, much less frequently than the lower-grade steps, in silica-deficient, low magnesium rocks under the conditions of the sanidinite facies.

The major point about Bowen's paper is that it was a major advance in terms of how to deduce and display the sequence of reactions in a specific metamorphic rock-type as it underwent progressive devolatilization. The few errors he made were in those areas where theory left ambiguities in the exact order of reactions. Had he been armed with the field information we now possess, he would have been much closer to placing the reactions in the correct order. Aside from his inclusion of tremolite, Bowen's study did not explicitly consider a role for H_2O in the metamorphism of limestones. This had to wait until later when, in the steps of Korzhinskii (1959), P.J. Wyllie (1962) and H.J. Greenwood (1962) addressed the issue in detail. Greenwood (1962) presented a thermodynamic framework for metamorphism in the presence of a two-component fluid; we review his contribution in Chapter 6. We have known from this time on that the metamorphism of limestone and dolomite is influenced as much by the proportions of H_2O and CO_2 in the pore fluid as by temperature and pressure, and that the mole fraction X_{CO_2} can vary in these rocks virtually all the way from zero to one.

Bowen in this paper was the first person to show that the reactions of progressive metamorphism could be represented graphically as sub-parallel univariant lines running at a steep angle to the temperature axis so as to form a "petrogenetic grid". Assemblages other than those involved in the specific reaction curve persist across it, and no one bulk-composition is likely to show all the steps represented. His grid (Fig. 14, p. 256) contained only two, relatively straight, solid-solid reaction lines, both involving polymorphic inversion. He predicted that a large number of solid-solid reaction curves for more complex equilibria would eventually be determined, although the thought at that time was that they would be nearly vertical in the PT-diagram. It was not long, however, before more modest dP/dT slopes were proposed for solid-solid reactions; for example, the empirical PT-diagrams for the aluminosilicates of Miyashiro (1949) and Schuiling (1957), the flat-lying curves for the jadeite reactions calculated by Adams (1953), and additional reaction slopes calculated by Thompson (1955, Table 3).

In its basic form, the petrogenetic grid, unlike thermobarometry (Chapter 10), provides only "yes-no" answers to possible PT-conditions of mineral assemblages, but through the years it has been immensely valuable to metamorphic petrology in all its manifestations. The arrangement in space of univariant curves and invariant points in a given compositional system is required to conform to constraints elucidated by F.A.H. Schreinemakers in 1915–1925, a useful account of which was provided by Zen (1966). Schreinemakers' rules helped countless petrologists to explore the topologies of phase diagrams in systems incompletely constrained by experimental, thermodynamic, or field information. Bowen probably had little idea of the complexity of the "road maps" that would eventually constitute modern petrogenetic grids. In a metamorphic sequence, whether arranged in space or time, the invariant points on the PT-grid provide an instant constraint on the possible geothermal gradient involved in terrane, an idea exploited by many, including Carmichael (1978) with his "bathograds".

The use of petrogenetic grids blossomed in the second half of the 20th century as more and more reaction brackets were obtained in high-PT laboratory experiments, initially targeting minerals of end-member composition. As many metamorphic minerals occur in nature in relatively pure form, for example, jadeite, albite, aragonite, calcite, lawsonite, kaolinite, pyrophyllite, brucite, the aluminosilicates, serpentine, muscovite, diopside, tremolite, talc, grossular, wollastonite and prehnite, the univariant PT-curves limiting the stability fields of these minerals could be assembled onto primitive, complex-system grids for immediate use as desired. Constraints from these phase-equilibrium experiments found their way into thermodyamic databases, such as that of Helgeson *et al.* (1978; see Chapter 11), a step that enabled much broader application of the stability relations of minerals. Rock-types containing minerals with near-endmember compositions, such as limestones, dolomites, metaperidotites, and metabauxites, were soon adequately served by relatively simple-system phase diagrams. At a later stage, minerals with important amounts of solid solutions were depicted in PT diagrams with the aid of composition isopleths determined in laboratory experiments for specific phase assemblages, for example, X_{Fe} in cordierite, garnet and biotite (Hensen and Green, 1973; Holdaway and Lee, 1977), X_{Al} in orthopyroxene (Harley 1984), and X_{Si} in phengite (Massonne and Schreyer, 1987). Petrogenetic grids for metapelites continued to evolve through many generations (Spear and Cheney 1989). Constraints were derived from many sources: Fe-Mg and tschermaks (AlAl for MgSi) exchange data, mineral assemblages observed in the field, experiments on end-member minerals, and miscellaneous thermodynamic data, to produce progressively more accurate and useful grids. The grid for metabasites has been more of a challenge. The last decade of the 20th century saw the use of internally consistent thermodynamic datasets for minerals in the preparation of full-system PT grids, for example, Powell and Holland (1990) and Will *et al.* (1990) for pelites and metabasites. The most recent development has involved the use of the internally consistent thermodynamic database to construct phase-assemblage diagrams for specific whole-rock compositions, which can be much more useful than full-system diagrams (see Chapter 8 on pseudosections).

References

Adams, L.H. (1953) A note on the stability of jadeite. *American Journal of Science*, **251**, 299–308.

Bowen, N.L. (1940) Progressive metamorphism of siliceous limestone and dolomite. *Journal of Geology*, **48**, 225–274.

Bowers, T.S. and Helgeson, H.C. (1983) Calculation of the thermodynamic consequences of non-ideal mixing in the system H_2O-CO_2-NaCl on phase relations in geologic systems: metamorphic equilibria at high pressures and temperatures. *The American Mineralogist*, **68**, 1059–1075.

Bucher, K. and Frey, M. (1994) *Petrogenesis of Metamorphic Rocks*. Springer-Verlag, Berlin, 318 pp.

Carmichael, D.M. (1978) Metamorphic bathozones and bathograds: a measure of the depth of post-metamorphic uplift on the regional scale. *American Journal of Science*, **278**, 769–797.

Greenwood, H.J. (1962) Metamorphic reactions involving two volatile components. *Carnegie Institution of Washington, Yearbook*, **61**, 82–85.

Harker, A. (1904) The Tertiary igneous rocks of Skye. *Memoir of the Geological Survey of Scotland*, 144–151.

Harker, R.I. and Tuttle, O.F. (1955) Studies in the system CaO-MgO-CO_2, Part 1. *American Journal of Science*, **253**, 209–224.

Harker, R.I. and Tuttle, O.F. (1956) Experimental data on the P_{CO_2}-T curve for the reaction: calcite + quartz = wollastonite + carbon dioxide. *American Journal of Science*, **254**, 239–256.

Harley, S.L. (1984) An experimental study of partitioning of Fe and Mg between garnet and orthopyroxene. *Contributions to Mineralogy and Petrology*, **86**, 359–373.

Helgeson, H.J., Delaney, J.M., Nesbitt, H.W. and Bird, D.K. (1978) Summary and critique of the thermodynamic properties of rock-forming minerals. *American Journal of Science*, **278-A**, 1–229.

Hensen, B.J. and Green, D.H. (1973) Experimental study of the stability of cordierite and garnet in pelitic compositions at high pressures and temperatures, part 3. *Contributions to Mineralogy and Petrology*, **38**, 151–166.

Holdaway, M.J. and Lee, S.M. (1977) Fe-Mg cordierite stability in high-grade pelitic rocks based on experimental, theoretical, and natural observations. *Contributions to Mineralogy and Petrology*, **63**, 175–198.

Joesten, R. (1974) Local equilibrium and metasomatic growth of zoned calc-silicate nodules from a contact aureole, Christmas Mountains, Big Bend Region, Texas. *American Journal of Science*, **274**, 876–901.

Joesten, R. (1976) High-temperature contact metamorphism of carbonate rocks in a shallow crustal environment, Christmas Mountains, Big Bend region, Texas. *American Mineralogist*, **61**, 776–781.

Massonne, H. and Schreyer, W. (1987) Phengite geobarometry based on the limiting assemblage with K-feldspar, phlogopite, and quartz. *Contributions to Mineralogy and Petrology*, **96**, 212–224.

Miyashiro, A. (1949) The stability relations of kyanite, sillimanite and andalusite, and the physical conditions of the metamorphic processes (in Japanese with English abstract). *Journal of the Geological Society of Japan*, **59**, 218–223.

Niggli, P. (1916) Gleichgewichte zwischen TiO_2 und CO_2, sowie SiO_2 in Alkali, Kalk-alkali und Alkali-Aluminatschmelzen. *Zeitschrift für anorganische Chemie*, **98**, 241–326.

Nockolds, S.R. (1947) On tilleyite and its associated minerals from Carlingford, Ireland. *Mineralogical Magazine*, **28**, 151–158.

Powell, R. and Holland, T.J.B. (1990) Calculated mineral equilibria in the pelite system, KFMASH (K_2O-FeO-MgO-Al_2O_3-SiO_2-H_2O). *American Mineralogist*, **75**, 367–380.

Schuiling, R.D. (1957) A geo-experimental phase-diagram of Al_2SiO_5 (sillimanite, kyanite, andalusite). *Proceedings of the Koninklijke Nederlandse Akademie van Wetenschappen*, Series B, **60**, 220–226.

Spear, F.S. and Cheney, J.T. (1989) A petrogenetic grid for pelitic schists in the system SiO_2-Al_2O_3-FeO-MgO-K_2O-H_2O. *Contributions to Mineralogy and Petrology*, **101**, 149–164.

Thompson, J.B. (1955) The thermodynamic basis for the mineral facies concept. *American Journal of Science*, **253**, 65–103.

Tilley, C.E. (1942) Tricalcium disilicate (rankinite), a new mineral from Scawt Hill, Co. Antrim. *Mineralogical Magazine*, **26**, 190–196.

Tilley, C.E. (1948) Earlier stages in the metamorphism of siliceous dolomites. *Mineralogical Magazine*, **28**, 272–276.

Tilley, C.E. (1951) A note on the progressive metamorphism of siliceous limestones and dolomites. *Geological Magazine*, **88**, 175–178.

Tuttle, O.F. and Harker, R.I. (1957) Synthesis of spurrite and the reaction wollastonite + calcite = spurrite + carbon dioxide. *American Journal of Science*, **255**, 226–234.

Weeks, W.F. (1956) A thermochemical study of equilibrium relations during metamorphism of siliceous carbonate rocks. *Journal of Geology*, **64**, 245–270.

Will, T.M., Powell, R., Holland, T.J.B. and Guiraud, M. (1990) Calculated greenschist facies mineral equilibria in the system CaO-MgO-FeO-Al_2O_3-SiO_2-H_2O-CO_2. *Contributions to Mineralogy and Petrology*, **104**, 353–368.

Wyllie, P.J. (1962) The petrogenetic grid, an extension of Bowen's petrogenetic grid. *Geological Magazine*, 99, 558–69.

Zen, E-an (1966) Construction of pressure-temperature diagrams for multicomponent systems after the method of Schreinemakers – a geometric approach. *USGS Bulletin* **1225**.

Zharikov, V.A. and Shmulovich, K.I. (1969) High temperature mineral equilibria in the system CaO-SiO_2-CO_2. *Geochemistry International*, **6**, 853–869.

PROGRESSIVE METAMORPHISM OF SILICEOUS LIMESTONE AND DOLOMITE

NORMAN L. BOWEN
University of Chicago

ABSTRACT

The metamorphism of siliceous dolomitic limestone is considered with the aid of a composition tetrahedron which suggests that the changes may be referred to thirteen steps or grades of increasing decarbonation, taking place at successively higher temperatures at any given pressure. At the temperature appropriate to each step a certain phase assemblage becomes unstable and each step is characterized by the disappearance of its appropriate phase assemblage. The survival of that phase assemblage in a rock is rigorously indicative of the fact that the step in question has not been attained.

In most of the steps, to wit ten, the *disappearance* of its phase assemblage is accompanied by a more conspicuous phenomenon, the *appearance* of a new phase, a metamorphic mineral, but this appearance of a new phase cannot be regarded as rigorously indicative of the accomplishment of the step since the new phase can be produced otherwise. Nevertheless, if certain additional conditions are fulfilled, the new phase will appear only when the temperature of the appropriate step is attained and the metamorphic minerals then become indicators of grade of metamorphism. In the order of the rising temperature steps at which they are produced the ten minerals are tremolite, forsterite, diopside, periclase, wollastonite, monticellite, akermanite, spurrite, merwinite, and larnite. Examination of their natural occurrence suggests that they are for the most part produced under conditions which permit their use as temperature indicators.

INTRODUCTION

Under certain conditions quartz and calcite crystallize side by side and appear to constitute a stable assemblage, but, when the temperature is raised, these minerals react with each other to form wollastonite and carbon dioxide. If there is indeed an equilibrium reaction, the temperature at which the reaction proceeds must depend upon the external pressure (pressure of CO_2). Goldschmidt was the first to point out the thermodynamic relations of this reaction and its importance as an indicator of pressure-temperature conditions during rock metamorphism.[1] All subsequent studies of metamorphic rocks tend to confirm the view that the reaction represents a reversible equilibrium, although it has not proved possible to make a satisfactory laboratory study. Lime and silica form other compounds in addition to the metasilicate, and of these the orthosilicate is the most important from the present point of view, since it is the only one that has been identified in rocks. Niggli in 1916 studied experimentally the equilibrium in alkali carbonate melts containing SiO_2 and CaO, and concluded that the calcite-orthosilicate assemblage represents a higher temperature equilibrium at any given pressure than either the calcite-quartz assemblage or the calcite-wollastonite assemblage.[2] The application of these relations to rock metamorphism was not then apparent since calcium orthosilicate was not known in rocks. With its discovery by Tilley at Scawt Hill came indications of the applicability of Niggli's results to rock metamorphism, for the unusually high-temperature character of the Scawt Hill metamorphism was suggested by Tilley.[3]

There has been, then, some recognition of degrees of metamorphism of siliceous limestone, depending upon temperature; and, when we turn to the more complex case of siliceous dolomite, we find a like situation. The earliest developments of such concepts need not concern us. It will suffice to recall Eskola's establishment of four grades of metamorphism of siliceous (or silicified) magnesian limestone in western Massachusetts. These are represented, respectively, by quartz-limestone, tremolite limestone, diopside limestone. and wollastonite limestone in an ascending temperature series.[4] In addition, it is recognized by Niggli,[5] Tilley,[6] and others that monti-

[1] V. M. Goldschmidt, "Die Gesetze der Gesteinsmetamorphose," *Vid.-Selsk. Skr. I Math.-Naturv. Kl.* (1912), No. 22.

[2] P. Niggli, *Zeits. f. anorg. Chemie*, Vol. XCVIII (1916), p. 305.

[3] C. E. Tilley, *Min. Mag.*, Vol. XXII (1929), p. 86.

[4] Pentti Eskola, "On Contact Phenomena between Gneiss and Limestone in Western Massachusetts," *Jour. Geol.*, Vol. XXX (1922), p. 283.

[5] U. Grubenmann and P. Niggli, *Die Gesteinsmetamorphose*, Vol. I (Berlin: Gebrüder Borntraeger, 1924), p. 383.

[6] "Contact Metamorphic Assemblages in the System, CaO-MgO-Al_2O_3-SiO_2," *Geol. Mag.*, Vol. LXII (1925), p. 366.

cellite assemblages represent a higher grade of metamorphism of magnesian limestone than any of the assemblages discussed by Eskola in connection with his Massachusetts studies. But in spite of this general knowledge of grades of metamorphism of the rocks under consideration and very detailed treatment of some aspects of it, such as that by Goldschmidt, there appears to be no examination of the whole subject with adequate emphasis upon the progressive character of the metamorphism, in other words, upon degree or grade of metamorphism. The present paper is an attempt to supply the need of such systematic treatment.[7] Some aspects of the discussion will undoubtedly require revision in the light of better knowledge. In so far as they have been apprehended, questionable points have been noted and in part debated at the appropriate places.

PROGRESSIVE METAMORPHISM OF SILICEOUS LIMESTONE

The reaction between calcite and quartz to form wollastonite may be represented by the equation

$$CaCO_3 + SiO_2 \rightleftharpoons CaSiO_3 + CO_2, \qquad (A)$$

---

[7] This material was first brought together with the purpose of presenting to students as a logical whole, with a connecting thread of fundamental law, a subject that otherwise is bare, arid description. When this had been done, it was thought that the paper had enough novel aspects to warrant its presentation at the Minneapolis meeting of the Geological Society of America. Upon hearing the paper, Professor A. F. Buddington called my attention to a contribution by Korjinski to the International Geological Congress in Moscow and has since sent me a printed extract (D. S. Korjinski, *Mem. Soc. Russe de Min.*, Vol. XLVI [1937], p. 385). Korjinski there treats briefly the general subject of the stability range of minerals, emphasizing the control of depth (pressure). One aspect of his problem is concerned with the decomposition of silicates by CO_2 at various pressures, which is substantially the reverse of our present problem. Referring especially to lime-alumina silicates, he invokes some of the same principles and adopts a method of approach which, in some respects, resembles that adopted in the present paper. Korjinski's extract is so brief that one may do him an injustice in offering criticism based on it, but he appears to assume that there is always an unlimited supply of CO_2 in the depths and thus reaches conclusions as to the existence of certain silicates in the depths which are at variance with the facts. In any case, the problem of progressive metamorphism of carbonates cannot be regarded as the simple inverse of his problem and is more profitably attacked in the general manner that is followed in the present paper.

and inspection of this equation shows that we have to deal with a three-component system because three is the least number of substances in terms of which the composition of all four of the phases can be expressed. A system of three components existing in four phases has one degree of freedom, as the ordinary equation of the phase rule $P + F = C + 2$ shows.[8] With only one degree of freedom the fixing of any one variable fixes all others, so that if the pressure is fixed the temperature is also fixed. In other words, these four phases can coexist in equilibrium with one another at only one temperature if the pressure (of CO_2) is given. Therefore a pressure-temperature curve can be drawn showing for each pressure (of CO_2) the temperature at which all four phases are in equilibrium. Such a curve is shown in Figure A and is known as a curve of univariant equilibrium (one degree of freedom). For all values of the temperature and pressure which do not lie on this curve, not more than three phases can coexist in equilibrium, and the curve of univariant equilibrium thus separates two areas of divariant equilibrium. We may readily decide which phases can exist in each of these fields by turning to the equation for the reaction. Quartz and calcite represent the low-temperature pair at any pressure; wollastonite and CO_2, the high-temperature pair. But in association with calcite and quartz either wollastonite or CO_2 (but not both) can occur on the low-temperature side, and in association with wollastonite and CO_2 either calcite or quartz (but not both) can occur on the high-temperature side. In

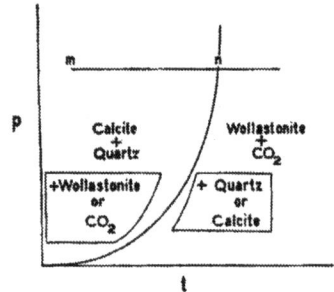

FIG. A.—Pressure-temperature diagram and paired composition diagrams of reaction equation A.

[8] In this equation $P=$ the number of phases in a system at equilibrium, $C=$ the number of components, and $F=$ the number of degrees of freedom.

other words, we have as possible three-phase assemblages all possible combinations of the four phases taken three at a time, which gives four three-phase assemblages, two of which are low-temperature assemblages and two are high-temperature assemblages, thus:

a) calcite + quartz + wollastonite
b) calcite + quartz + CO_2
} low temperature

c) wollastonite + CO_2 + quartz
d) wollastonite + CO_2 + calcite
} high temperature

These relations are shown on the diagram (Fig. A), the low-temperature assemblages being placed on the low-temperature side of the univariant curve and the high-temperature assemblages on the high-temperature side. The diagram may now be put to service in attacking the problem of metamorphism of siliceous limestone (or calcareous quartzite), as Goldschmidt did long ago. We need concern ourselves only with the case where calcite and quartz alone are present in the rock, which exists at a pressure and temperature indicated by the point m on the diagram, that is, a low temperature and a rather high pressure. If now the temperature is raised, the applied pressure will remain unchanged at the initial value and no new phases will form, the change of condition being given by the horizontal line mn, which indicates only a rise of temperature at constant pressure. But when the temperature has risen to n, the vapor pressure of the reacting system has now reached a value equal to the external pressure, and some vapor (CO_2) will form and with it some wollastonite. At the fixed external pressure the four phases can coexist at only one temperature, and the temperature will therefore remain constant at n until one phase of the low-temperature assemblage, calcite or quartz, disappears. Only when the reaction is complete can the temperature rise above n, which will be represented by movement along mn produced, the pressure remaining at the fixed value of the external applied pressure. Ordinarily all or nearly all the vapor of CO_2 will escape from the system, and the product will then consist of wollastonite with either quartz or calcite in excess, according to their relative proportion in the original rock. It is especially noteworthy that, although we are dealing with a three-component system, the product of metamorphism would consist of only two solid phases. For this special type of metamorphism, then, the number of *mineral* phases formed is one less than the number of components and not equal to the number of components as Goldschmidt has shown it must ordinarily be to satisfy divariance (random temperature and pressure). We have here one *mineral* phase less than usual because one phase on the high-temperature side is the gas CO_2 which is lost to the system.[9]

Certain aspects of the chemical relations in such a system are best pictured with the aid of a composition diagram which for this three-component system may be shown on triangular coordinates. Figure A shows two such diagrams in which CaO, CO_2, and SiO_2 are taken as components. The first triangular diagram of Figure A represents the low-temperature assemblages. In it all phases which can coexist at these temperatures are joined by straight lines. There is, therefore, a line joining SiO_2 and $CaCO_3$ and another joining $CaCO_3$ and wollastonite. The rest of the triangle toward the CaO corner has no significance in the present connection and is shown merely in dotted outline. The two triangles bounded by full lines on this first triangular diagram of Figure A bring out clearly the two low-temperature, three-phase assemblages, calcite + quartz + wollastonite and calcite + quartz + CO_2 which have already been mentioned. The diagram is thus perfectly general, in accordance with phase-rule requirements, but the really significant feature is that quartz and calcite can coexist in equilibrium under the low-temperature conditions to which this diagram applies. The quartz-calcite join is thus the line on this diagram of special significance and the only one of practical importance since the three-phase assemblages with either some wollastonite or some CO_2 are probably seldom represented in any natural rock. Nevertheless, to make the reasoning all inclusive, it is necessary to show these three-phase areas (triangles) in order that it

---

[9] When the composition is such that a quartz-wollastonite rock is the product of metamorphism, CO_2 is, of course, no longer a component of the system, and the number of components is reduced to two, which is the same as the number of phases, in correspondence with the usual case for which Goldschmidt has stated his Mineralogical Phase Rule.

may be realized that, if there were a limestone containing some detrital wollastonite and if it were recrystallized (metamorphosed) under these low-temperature conditions, quartz, calcite, and wollastonite could recrystallize side by side in perfect equilibrium with one another.

We may now suppose that a rise of temperature takes place under constant external pressure corresponding with movement along the line mn of the P-T diagram in Figure A, which we have already considered. When the temperature has reached the value n, calcite and quartz, we found, can no longer coexist; one of them must disappear. The phases which can now coexist are shown in the second large triangle of Figure A, in which again pairs of coexistent phases are joined by straight lines. The join calcite-quartz has disappeared and in its place appears the join wollastonite-CO_2 which crosses it. These are the four reacting phases of equation A, and it is a general relation for such a univariant reaction that in a composition diagram the line joining the two phases on one side of the equation must cross the line joining the two phases on the other side of the equation. The reacting phases are said to bear a "crossed relation." The small subsidiary triangle of Figure A enables us to picture more clearly what happens to mixtures of quartz and calcite (and indeed such mixtures with some wollastonite in addition) when they pass through the temperature n of the P-T diagram. Any mixture such as x which lies between a and SiO_2 on the dotted join $CaCO_3$-SiO_2 will, as a result of loss of CO_2, have its composition changed in a manner represented by the line xx', which points directly away from the CO_2 corner of the triangle. The final product of metamorphism will be x', a mixture of wollastonite and quartz. Any mixture such as y, lying between a and $CaCO_3$, will have its composition changed to y', a mixture of wollastonite and calcite. The mixture a itself which contains calcite and quartz in chemical equivalents would, of course, be transformed into wollastonite alone (a'). It is clear, then, that the constitution of the metamorphic product can be determined with the aid of the two large triangular diagrams of Figure A simply by considering movement of the composition in a direction away from the CO_2 corner of the triangle until the first join of coexistent phases for the new conditions is encountered. The two large triangular diagrams of Figure A may conveniently be called the paired diagrams of the reaction given in equation A.[10] We shall have occasion to use such paired diagrams a great deal and to refer to them as such. The paired diagrams and the considerations just brought out demonstrate that in such materials metamorphism is essentially *decarbonation*, and, as the discussion proceeds, it will become clear that progressive metamorphism is essentially progressive decarbonation.

At this point it is desirable to emphasize the central feature of this whole discussion of equilibrium in its application to actual rocks. The presence of calcite and quartz in a rock indicates that the temperature at which these become incompatible at the pressure then prevailing was not exceeded. On the other hand, the presence of wollastonite does not indicate that that temperature was exceeded, since wollastonite can, as we have seen, be present in one of the low-temperature, three-phase assemblages. Of course, if it were perfectly certain that the rock initially contained only quartz and calcite, the formation of some wollastonite would, with some qualification, indicate that the temperature in question had been exceeded. This will be discussed more fully at a later point, especially with reference to the qualification referred to.

We may now turn to a consideration of the next step in the progressive metamorphism (progressive decarbonation) of siliceous limestone. In the first step a reaction took place as a result of the fact that calcium carbonate was no longer stable in contact with pure SiO_2. In this next step (at a higher temperature for the same pressure) calcium carbonate becomes unstable in contact with calcium metasilicate (wollastonite) and reacts with it to produce spurrite and CO_2 in accordance with the equation

$$3CaCO_3 + 2CaSiO_3 \rightleftharpoons 2Ca_2SiO_4 \cdot CaCO_3 + 2CO_2 . \qquad (B)$$

---

[10] Thus equation A, the univariant reaction (P-T) curve of Fig. A and the paired triangular diagrams of Fig. A all express the same facts, each with its own emphasis. This system of referring to the equation, the figure, and the P-T curve of any reaction by the same letter or number will be followed throughout the paper.

As before we have four three-phase assemblages, thus:

$a)$ calcite + wollastonite + spurrite } low temperature
$b)$ calcite + wollastonite + CO_2

$c)$ spurrite + CO_2 + calcite } high temperature
$d)$ spurrite + CO_2 + wollastonite

Again there are paired diagrams which show these stable phase assemblages at lower temperatures and at higher temperatures, respectively. The lower-temperature diagram of this pair is the same as the higher-temperature one of the previous pair (Fig. A) which illustrated the preceding step of metamorphism. The new pair is given in Figure B. The join calcite-wollastonite of the first diagram of Figure B disappears in the second diagram of that figure where it is replaced by the join spurrite-CO_2, and these two joins have as before the crossed relation. Compositions which originally lay on the join calcite-wollastonite are moved over, as a result of loss of CO_2, to their appropriate points on the join spurrite-wollastonite or the join spurrite-calcite, the construction for which is readily obtained by drawing lines directly away from the CO_2 corner in a manner already discussed for the previous step. Here, as before, the coexistence of calcite and wollastonite in a rock indicates a low temperature (in relation to this reaction), but only with some qualification can the presence of spurrite be regarded as indicating a high temperature.

A final step in the progressive metamorphism of siliceous limestone comes if the temperature is raised to the still higher value (at the same pressure) at which wollastonite and spurrite react to give larnite and CO_2 according to the equation

$$CaSiO_3 + 2Ca_2SiO_4 \cdot CaCO_3 \rightleftharpoons 3Ca_2SiO_4 + CO_2. \qquad (C)$$

The four three-phase assemblages are:

$a)$ wollastonite + spurrite + larnite } lower temperatures
$b)$ wollastonite + spurrite + CO_2

$c)$ larnite + CO_2 + wollastonite } higher temperatures
$d)$ larnite + CO_2 + spurrite

The paired diagrams for this reaction, showing the stable-phase assemblages below and above the reaction temperature, are shown in Figure C. The join wollastonite-spurrite disappears and is replaced by the join larnite-CO_2 which has the crossed relation to it. Again the change of composition involved can be pictured by a movement directly away from the CO_2 corner of the triangle whereby compositions represented by points on the wollastonite-spurrite join will move to points either on the spurrite-larnite join or on the larnite-wollastonite join.

Further comment is necessary on this step in metamorphism as here depicted. It makes larnite and wollastonite capable of co-

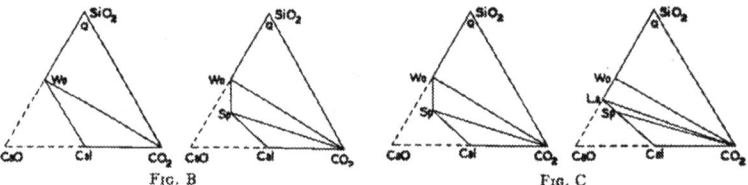

Fig. B.—Paired composition diagrams of reaction equation B
Fig. C.—Paired composition diagrams of reaction equation C

existence. Experimental studies show that Ca_2SiO_4 and $CaSiO_3$ do not occur together at high temperatures but combine to form the compound $3CaO \cdot 2SiO_2$. The 3:2 compound has, however, never been found in nature, and, in the metamorphic haloes around chalk nodules at Scawt Hill, larnite and wollastonite zones come together without any interposition. Perhaps the 3:2 compound has a minimum temperature of existence which is higher than any of the temperatures of metamorphism.

With advance of knowledge the above discussion of three steps in the progressive metamorphism of limestone may require revision, but it appears to represent present knowledge and may serve to emphasize principles. At any given pressure calcite and quartz can coexist in equilibrium only up to a certain temperature, calcite and wollastonite can coexist up to a somewhat higher temperature, spurrite and wollastonite up to a still higher temperature, and above this none of these pairs can exist. Each of these pairs is thus an indicator of the degree or grade of metamorphism or temperature of metamorphism at a given pressure. Each step in metamorphism is a successive step in decarbonation. If a triangular diagram is drawn with CaO, SiO$_2$, and CO$_2$ at its corners and calcite is joined by a line to each of the phases with which it can coexist, then the successive steps in decarbonation (i.e., the successive steps in metamorphism) can be deduced entirely from geometrical relations in the triangle by observing the joins successively encountered with motion away from the CO$_2$ corner, i.e., decarbonation. We shall now proceed to apply this concept to the more complex case of the metamorphism of siliceous magnesian limestone.

PROGRESSIVE METAMORPHISM OF SILICEOUS DOLOMITE

For the purpose of studying the metamorphism of siliceous dolomite or dolomitic limestone it is convenient to use a tetrahedron which has as the base the CaO-MgO-SiO$_2$ triangle and CO$_2$

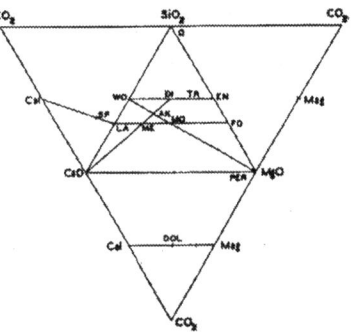

Fig. D.—Skeleton diagram of tetrahedron, illustrating composition of phases in metamorphosed, siliceous, dolomitic limestone.

as its apex, the tetrahedron that would be formed from Figure D if the three outer triangles were folded up until their three CO$_2$ apices met at a common point—the CO$_2$ apex of the tetrahedron. On the base should be plotted the points representing the compositions of all the Ca-Mg silicate compounds known to form in metamorphosed carbonate rocks of the kinds here under consideration,[11] and in their appropriate places the points for spurrite, dolomite, and calcite. By considering the relative positions in this tetrahedron of the various planes that are formed by joining calcite and dolomite (or MgO) to the various silicate phases on the base, the steps in the metamorphism (decarbonation) of siliceous carbonate rocks in their correct order can be deduced partly from geometrical relations, though

[11] Magnesite or any carbonate richer in magnesia than dolomite will not be considered.

these must be supplemented by knowledge of possible silicate assemblages.

In our consideration of the metamorphism of siliceous limestone we found that two mineral phases and CO_2 were formed at each step and that lines radiating from the CO_2 corner of the composition triangle encounter successively the lines joining the pairs of minerals formed in each successive step. In the present case we have one more component; therefore, three mineral phases and CO_2 are ordinarily formed at each step in metamorphism, and lines radiating from the CO_2 apex of the tetrahedron will, in the same manner, encounter successively the planes (triangles) formed by joining the compositions of the three mineral phases formed at each successive step of metamorphism (decarbonation). It serves no useful purpose to make a drawing of the composition tetrahedron showing these planes. Their relative position can be visualized readily only with the aid of a model.[12] We can, however, with the tetrahedral model ever before us as an aid, draw a series of diagrams which show the planes successively encountered and therefore the phases formed in various compositions at successive steps in metamorphism. These diagrams represent certain triangular planes within the tetrahedron and are best shown as projected upon the base of the tetrahedron, with the CO_2 apex as the center of projection, so that each diagram becomes an equilateral triangle, the point for $CaCO_3$ coinciding with the CaO point, the point for $MgCO_3$ coinciding with the MgO point, and the point for dolomite lying halfway between these.

The highest (closest to the CO_2 apex) of the series of planes within the tetrahedron is that representing no decarbonation and is the plane determined by the points calcite, dolomite, and SiO_2. This is shown in the first diagram of Figure 1. It indicates that at low temperatures calcite, dolomite, and quartz can coexist in equilibrium,

[12] Students who became interested in this problem constructed tetrahedra to aid in its solution. One device adopted proved particularly simple and effective. It consisted in making up a cardboard tetrahedron with one face lacking. This missing face is taken as the MgO-CO_2-SiO_2 face, with which compositions we are not here concerned. The absence of this face renders accessible the whole interior of the tetrahedron so that the compositions of the phases can be plotted on the interior side of the other faces and colored threads can be strung from point to point. These lines (threads) determine the planes encountered in the decarbonation of any composition within the tetrahedron. Among the more active students may be mentioned H. Stocking and F. E. Tippie.

and with metamorphism at the lowest temperatures they can recrystallize side by side. At a higher temperature, however, they react, and, depending upon the temperature (at a given pressure) a series of products may result. If the temperature is raised only a moderate amount, the product is a tremolite-bearing rock. It is an unfortunate feature of the metamorphism of siliceous dolomite that the very first step exhibits unusual features which make it impossible to represent it rigorously with the aid of the tetrahedron we have been considering. Tremolite contains water as an essential constituent, and its composition cannot therefore be represented in this tetrahedron. In order to plot tremolite, it has been necessary to neglect its water content, and in the plots the composition $CaMg_3(SiO_3)_4$, often given in old texts, is used as the tremolite point. When this is done, triangular planes with tremolite at a corner are the first planes encountered in the tetrahedron in moving from

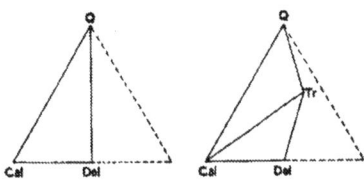

FIG. 1.—Paired composition diagrams of reaction equation 1.

points in the calcite-dolomite-silica plane in a direction directly away from the CO_2 apex (decarbonation). The tetrahedron, even with this simplification of the composition of tremolite, thus accurately represents the first product of metamorphism of siliceous dolomite (or dolomitic lime), which is well known to be tremolite. The planes encountered are two in number: the tremolite-quartz-calcite plane and the tremolite-dolomite-calcite plane. In the tetrahedron they are not co-planar, but, when projected in the manner described above, they appear as shown in the second diagram of Figure 1. Figure 1 thus shows the paired diagrams for this first reaction in metamorphism. At some temperature dolomite and quartz become incompatible with each other, reacting to form calcite and tremolite. The dolomite-quartz join of the first diagram disappears, and its place is taken by the join tremolite-calcite. These two joins have the crossed relation.[13]

[13] The crossed relation of the two joins in these projections is a reflection of the crossed relation of two planes in the space tetrahedron. These planes are the calcite-dolomite-silica plane and the calcite-tremolite-CO_2 plane.

If tremolite had the simplified anhydrous composition we have here used in plotting it, the reaction involved in this first step in metamorphism would be represented by the equation

$$3CaMg(CO_3)_2 + 4SiO_2 \rightleftharpoons CaMg_3(SiO_3)_4 + 2CaCO_3 + 4CO_2. \quad (1)$$

As written, there are five reacting phases in a system of four components, and this would represent a univariant equilibrium. Just as in the simpler systems first considered there would, at any pressure, be a definite temperature at which all five phases could coexist, and, if at any pressure the temperature of a siliceous dolomite were raised, no change would occur until a definite temperature was reached. At this temperature the reaction would proceed until either dolomite or quartz had disappeared, that one surviving which was in excess of chemical equivalence in the original rock.

Again, if this equation accurately represented the reaction, we could by inspection of the equation write down the five possible four-phase assemblages that could occur above and below the reaction temperature at a given pressure. They would be:

a) dolomite + quartz + tremolite + calcite ⎫
b) dolomite + quartz + tremolite + CO_2 ⎬ lower temperatures
c) dolomite + quartz + calcite + CO_2 ⎭

d) tremolite + calcite + CO_2 + dolomite ⎫ higher
e) tremolite + calcite + CO_2 + quartz   ⎭ temperatures

As we have found in the simpler systems, here again the number of *minerals* formed upon metamorphism is one less than the number of components. The total number of phases is, of course, equal to the number of components (four), but one of them is always CO_2 (see the two higher-temperature assemblages), so that the number of minerals is three and our triangular projections in which CO_2 is not shown can accurately present the three coexistent mineral phases produced upon metamorphism in this four-component system.

Now unfortunately, as already noted, the theoretical aspects of the case are not quite so simple, because the formula for tremolite is $Ca_2Mg_5H_2(SiO_3)_8$, and water is a component of the reacting system. The reaction is correctly written thus:

$$5CaMg(CO_3)_2 + 8SiO_2 + H_2O \rightleftharpoons Ca_2Mg_5H_2(SiO_3)_8 + 3CaCO_3 + 7CO_2. \quad (1a)$$

The appearance of water in the equation has added a component to the system, making the number of components five, but has not necessarily increased the number of phases; the H_2O and CO_2 would form only one vapor phase, and the number of phases is only five. The system is therefore not univariant but divariant, and we cannot say, on theoretical grounds, that if we started with dolomite and silica (with some water), under a certain pressure, no formation of tremolite and CO_2 could occur until a definite temperature was attained and that the reaction would go to completion at this temperature. Theoretically, the reaction could take place over a range of temperatures, and not until another metamorphic mineral formed in addition to tremolite would we have a univariant condition. Actually no phase other than tremolite and calcite does normally appear to form in dolomites in the early (lowest temperature) stages of metamorphism. The observed relations are thus simpler than the above reasoning would suggest, and this is because one possibility has been neglected. If a liquid phase were present in addition to the vapor and the four solids, the six coexistent phases would constitute a univariant system and at any fixed temperature all the variables would have fixed values. The composition of the liquid phase, the partial pressures of both CO_2 and H_2O in the vapor, and therefore the total vapor pressure would all have definite values. At any fixed external pressure, then, a wet siliceous dolomite would experience no change until a definite temperature were reached, at which temperature the reaction of equation 1a would proceed freely to the right until one of the phases—dolomite, quartz, or water (solution)—was exhausted. It is no doubt because the pressure-temperature conditions are ordinarily such as to permit the presence of a liquid solution that tremolite formation in the first step of metamorphism has all the attributes of a univariant reaction. A like remark applies to the other steps in which tremolite is one of the participating phases.

In taking the simplified anhydrous formula for tremolite, we have reduced the number of components by one, but we have also eliminated the possibility of a liquid phase[14] and thus reduced by one the number of possible phases. The conditions for univariancy are thus not changed. The use of the anhydrous formula for tremolite and of diagrams based upon it therefore involves in this respect no departure from fundamental principles and is justified by its convenience.

There are of course pressures at which wet siliceous dolomite might be heated that would not be adequate to enable retention of water (solution) up to temperatures at which formation of tremolite would proceed. Under such conditions the rock would simply be dehydrated and with further heating would suffer "dry" metamorphism, a condition perhaps corresponding with that prevailing during thermal metamorphism.[15] In thermal metamorphism of siliceous dolomite it is generally accepted that forsterite is the first phase formed. If this is true, then the reaction whereby it is formed cannot be an equilibrium reaction, for the phase assemblages involved should have represented in them only phases which, individually, can exist in equilibrium with dolomite and quartz. Obviously forsterite does not qualify since it cannot coexist with quartz. It would seem necessary to suppose that enstatite would be the first phase formed, with true equilibrium, in the thermal metamorphism of siliceous dolomite and that the first step involved would be represented by a diagram exactly like Figure 1, except for the substitution of enstatite for tremolite. Forsterite does indeed seem to be the first phase formed in the thermal metamorphism of dolomite (a failure of equilibrium).[16] The appearance of forsterite

[14] It would perhaps be preferable to say a "participating liquid phase," because in any of the reactions under consideration, including those involved in the metamorphism of siliceous limestone already discussed, a liquid phase may be present, which facilitates reaction yet is not essential to it and for the purposes of this discussion may be neglected because nonparticipating.

[15] We may interpret "dry" to mean not that no water is present but merely that no liquid water is present.

[16] Enstatite formation has been claimed in the ancient Grenville limestones of the Adirondacks (C. H. Smythe, Jr., *School Mines Quar.*, Vol. XVII [1895–96], p. 334). Modern studies do not confirm this observation.

in conditions where enstatite might be expected may be explained by assuming that, at the low temperatures involved, the first reaction step is too sluggish to be realized in the usual case, and only when the temperature is raised to a significantly higher value can any appreciable reaction occur. At that temperature forsterite formation is the necessary reaction.[17]

The complications that may be introduced by the hydrous character of tremolite in the usual first step of metamorphism may be further considered by setting down the possible phase assemblages of the true reaction represented by equation 1a. They are:

a) dolomite + quartz + solution + tremolite + calcite } lower
b) dolomite + quartz + solution + tremolite + vapor } temperatures
c) dolomite + quartz + solution + calcite + vapor }

d) tremolite + calcite + vapor + dolomite + quartz } higher
e) tremolite + calcite + vapor + dolomite + solution } temperatures
f) tremolite + calcite + vapor + quartz + solution }

When the phase assemblages are deduced with the aid of equation 1, which assumes the simplified composition of tremolite, the characteristic feature of the reaction is that dolomite and quartz cannot coexist above the temperature of the reaction (at given pressure). In the assemblages just listed the characteristic feature is that the three phases—dolomite, quartz, and solution—cannot coexist above the reaction temperature and, if the pressure is too low or the actual quantity of water available is too small (sediment too dry), that the first of these three phases to disappear at the reaction will be the solution, in which case dolomite and silica will survive the reaction, a condition exemplified by assemblage d. Thus, in the ultimate analysis, incompatibility of quartz and dolomite above a definite reaction point is a necessity only when conditions are such as to permit the existence of a pore solution both below and above the reaction temperature. If the rock "boils dry" below the reaction, no tremolite is formed; if it "boils dry" at the reaction point, some tremolite forms. In both circumstances quartz and dolomite sur-

[17] It is possible that the association enstatite-calcite is not a stable one and that dedolomitization by way of enstatite would be as definitely a failure of equilibrium as is that by way of forsterite.

vive to react only at a somewhat higher temperature with production of forsterite.[18] This is, as we have seen, not an equilibrium reaction. Production of enstatite is the expected equilibrium, but it is not usually realized. We find, then, that in the special circumstances discussed, there may be a discrepancy between the conclusions reached by assuming a "dry" formula for tremolite and those reached by using its true formula. No discrepancy of this kind appears at later steps involving tremolite for the reason that H_2O is on the left-hand side of the equation of the reaction we have just discussed, but it is on the right-hand side of the other reactions involving tremolite. The relation can be verified by setting down the phase assemblages possible at the subsequent reaction points, using the true formula for tremolite, but this will not be done here. Thus, although there are certain conditions, as discussed previously, which import an exception to the statement that quartz and dolomite become incompatible at a definite temperature (at given pressure), there are no exceptions to the statement that tremolite and dolomite become incompatible at a definite temperature in the step next to be considered nor to the statement that tremolite, forsterite, and quartz become incompatible at a definite temperature in the subsequent step. It will not be necessary, therefore, to discuss the significance of the true composition of tremolite in connection with those steps. The simplified composition which can be plotted in the tetrahedron gives the same result and will be used without further comment in discussing these steps.

The next step in metamorphism is indicated by the set of triangular planes next encountered in the tetrahedron by lines coming from the CO_2 apex (decarbonation). The set of planes is shown in the second diagram of Figure 2, which demonstrates that this step is due to the reaction of tremolite and dolomite to form forsterite and calcite (with CO_2). This is the relation indicated when tremolite is assigned the composition $CaMg_3(SiO_3)_4$ and so plotted, but, as we have seen, no difference in the deduced phase equilibria is

---

[18] It is perhaps a condition such as this that has given rise to metamorphism which is mineralogically intermediate between thermal and regional as discussed by Alfred Harker from Skye in his text, *Metamorphism* (London: Methuen & Co., 1932), pp. 85 and 86.

introduced by making this simplifying assumption. The simplified equation then is:

$$CaMg_3(SiO_3)_4 + 5CaMg(CO_3)_2 \rightleftharpoons 6CaCO_3 + 4Mg_2SiO_4 + 4CO_2. \quad (2)$$

The phase assemblages are:

$a)$ tremolite + dolomite + calcite + forsterite ⎫
$b)$ tremolite + dolomite + calcite + CO_2 ⎬ lower temperature
$c)$ tremolite + dolomite + forsterite + CO_2 ⎭

$d)$ calcite + forsterite + CO_2 + tremolite ⎫ higher
$e)$ calcite + forsterite + CO_2 + dolomite ⎬ temperature

Again only three *mineral* phases can coexist after this step in metamorphism, and our triangular diagram accurately depicts the mineral assemblages. Figure 2 thus shows the paired diagrams for this reaction. The assemblage tremolite + calcite + quartz survives. The join dolomite-tremolite disappears and is replaced by the join calcite-forsterite, which has the crossed relation to it. The fundamental feature of this step is that tremolite and dolomite can coexist below it but not above it.

The next set of triangular planes within the tetrahedron is shown in projection in the second diagram of Figure 3, which demonstrates that the next step in metamorphism involves the formation of diopside according to the equation

$$CaMg_3(SiO_3)_4 + 2CaCO_3 + 2SiO_2 \rightleftharpoons 3CaMgSi_2O_6 + 2CO_2. \quad (3)$$

The phase assemblages are:

$a)$ tremolite + calcite + quartz + diopside ⎫ lower temperature
$b)$ tremolite + calcite + quartz + CO_2 ⎭

$c)$ diopside + CO_2 + tremolite + calcite ⎫
$d)$ diopside + CO_2 + tremolite + quartz ⎬ higher temperature
$e)$ diopside + CO_2 + calcite + quartz   ⎭

The assemblages calcite-forsterite-dolomite and calcite-forsterite-tremolite survive this step. Again only three mineral phases can coexist in any assemblage after this reaction. We have here three different mineral assemblages capable of being formed, whereas in former reactions there have been only two. In other respects this step is somewhat different from the earlier ones, though fundamental factors are the same. For example, the paired diagrams for this reaction (Fig. 3) show that three minerals react to produce one mineral, and in consequence of this fact the composition points of the three reacting minerals—calcite, tremolite, and quartz—form a triangle with the point for the product of their reaction, diopside, inside that triangle. In the other reactions two minerals reacted to

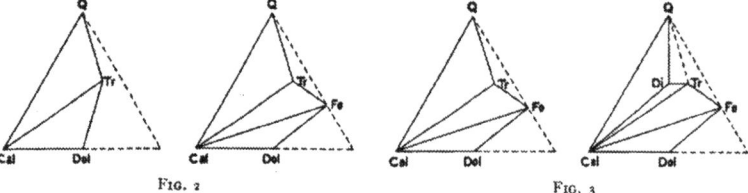

FIG. 2.—Paired composition diagrams of reaction equation 2
FIG. 3.—Paired composition diagrams of reaction equation 3

form two other minerals, and the two joins concerned had the crossed relation. The significant feature of this step of metamorphism is the coexistence of the *three* phases—calcite, tremolite, and quartz—below it but not above it.

One feature of Figure 3 as drawn may require comment. In all diagrams there are certain broken lines which lie at compositions with which we are not concerned here and are shown as broken lines for that reason. Since attention is confined to siliceous dolomite and dolomitic lime, more magnesian material being excluded from consideration, most of these lines emanate from the MgO corner of the diagram, but the line joining quartz and tremolite in Figure 3 is in the same class. With diopside present, no tremolite can appear together with quartz unless the original carbonate was more magnesian than dolomite, as a glance at the figure will show. A highly siliceous dolomite or a dolomitic quartzite will here give a diopside-quartz rock, and such a rock is in its final state with respect to metamorphic processes. It will not be affected at any of the later steps.

Indeed this is obviously true of any of the mixtures when they have been converted entirely to silicates, since decarbonation is complete and all the steps involve decarbonation.[19]

The next set of planes in the tetrahedron, encountered in our graphical decarbonation, is shown in projection in the second diagram of Figure 4. The change from the first diagram of Figure 4 consists in the disappearance of the join calcite-tremolite and the appearance of the join diopside-forsterite, indicating a reaction according to the equation

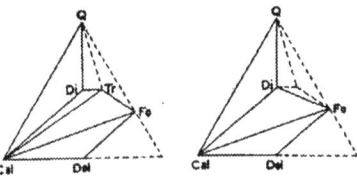

FIG. 4.—Paired composition diagrams of reaction equation 4.

$$2CaCO_3 + 3CaMg_3(SiO_3)_4 \rightleftharpoons 5CaMgSi_2O_6 + 2Mg_2SiO_4 + 2CO_2 . \quad (4)$$

The phase assemblages possible are:

a) calcite + tremolite + diopside + forsterite ⎫
b) calcite + tremolite + diopside + CO_2 ⎬ lower temperature
c) calcite + tremolite + forsterite + CO_2 ⎭

d) diopside + forsterite + CO_2 + calcite ⎫ higher
e) diopside + forsterite + CO_2 + tremolite ⎭ temperature

The fundamental feature of this step is, of course, the possibility of the coexistence of calcite and tremolite below it but not above it. We already had a step at which tremolite and calcite became incapable of coexistence if quartz were present also. The conditions involved in both of these steps are to be carefully distinguished from the conditions under which tremolite becomes unstable as such. With this we are not concerned here, for, though tremolite can still exist at higher temperatures at the given pressure, it cannot occur in materials which are no more magnesian than dolomite, and to such we have confined our discussion.[20] The assemblage *e* illustrates this point.

[19] Reactions among the silicates themselves are, of course, not impossible at higher temperatures, but no such reactions seem to exist except one to be mentioned later.

[20] The instability of tremolite and related amphiboles in higher grades of metamorphism has, however, much importance for the subject in general (see C. E. Tilley,

It is a possible assemblage on the high-temperature side of the reaction step now under consideration, but it is confined to compositions having MgO in excess of CaO (molal). This assemblage is shown in Figure 4 to make clear the connection with the preceding figure, but broken lines are used because it concerns compositions excluded from consideration, and the assemblage is omitted in subsequent figures. Indeed, with this step completed, we have no further need to consider tremolite. We gladly speed the parting guest, for all subsequent reactions are rigorously of four components, and we no longer need to indulge in the approximations which tremolite has rendered necessary.[21]

It is important to emphasize that the step whose consideration we have just completed is the final step in what is usually termed "regional metamorphism." Metamorphism is carried farther only under the higher temperature (or lower pressure) conditions of thermal metamorphism.

We now come to a step in metamorphism which is concerned with a reaction in the $CaO-MgO-CO_2$ system. It occurs at the temperature at which dolomite becomes unstable at the given pressure. We have already had a step in which dolomite became unstable in contact with tremolite; now dolomite as such becomes unstable. It should be emphasized that, although this reaction involves the dropping-down from one plane to another on lines emanating from the CO_2 apex (decarbonation) in exactly the same manner as in each preceding step, there is, nevertheless, nothing in the geometrical relations in the tetrahedron or indeed in the nature of things to determine that this will be the next step. A reaction in some other part of the tetrahedron might be the next, as far as can be determined from these general considerations, and it is solely the accumulated knowledge of metamorphic rocks which leads us to place

---

Geol. Mag., Vol. LX [1923], p. 418; E. Posnjak and N. L. Bowen, *Amer. Jour. Sci.*, Vol. XXII [1931], pp. 203–14).

[21] It may not be amiss here to state some of the considerations which led to the exclusion of the more magnesian compositions from study. In the first place, they are quantitatively much less important, but more cogent was the fact that three hydrous phases—talc, serpentine, and anthophyllite—would have appeared, and each would have introduced the same difficulties that made tremolite unwelcome.

the decomposition of dolomite next. This geological evidence will be mentioned in discussing the step immediately following.

The reaction concerned in the present step is:

$$CaMg(CO_3)_2 \rightleftharpoons CaCO_3 + MgO + CO_2, \qquad (5)$$

a four-phase reaction in a three-component system and therefore univariant. The phase assemblages are:

a) dolomite + calcite + periclase
b) dolomite + calcite + CO_2 } lower temperature
c) dolomite + periclase + CO_2

d) calcite + periclase + CO_2 } higher temperature

As a result of the disappearance of dolomite, the dolomite-forsterite join must disappear and compositions in the calcite-dolomite-forsterite plane drop down to positions in the calcite-periclase-forsterite plane as illustrated in Figure 5. Henceforth it becomes necessary to use full lines in the MgO corner of the successive triangles. The other phase assemblages, namely, calcite-forsterite-diopside and calcite-forsterite-silica, remain unchanged.

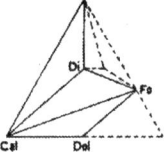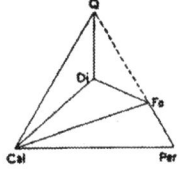

FIG. 5.—Paired composition diagrams of reaction equation 5.

The next step in metamorphism involves the formation of wollastonite in appropriate compositions and is, in one major respect, analogous to the last. It is concerned with a reaction in one of the three-component systems, in this case the system $CaO\text{-}SiO_2\text{-}CO_2$, and since, as before, these compositions are off by themselves on a side of the tetrahedron, their relations cannot be deduced from the positions of overlapping planes within the tetrahedron. Therefore, the fact that the reaction involving the formation of wollastonite is the next step is determined entirely from recorded knowledge of metamorphic rocks. More specifically, metamorphic aureoles indicate that dolomite is transformed to a periclase marble under conditions which permit beds of quartzose limestone to recrystallize as such without formation of a wollastonite marble.[22] Both these reactions precede those involving the formation of the minerals of lower silica ratio to be discussed later. Especially is it true that the step involving wollastonite formation must precede that in which there is formation of akermanite, for otherwise akermanite would find itself joined to quartz—an impossible association.

The reaction is then as follows:

$$CaCO_3 + SiO_2 \rightleftharpoons CaSiO_3 + CO_2. \qquad (6)$$

This is, of course, the same reaction already considered in connection with the metamorphism of siliceous limestone. It is important to note, therefore, that all the steps we have hitherto considered in the metamorphism of siliceous dolomitic limestone take place at lower temperatures (indicate a lower grade of metamorphism) than does the formation of a wollastonite marble from a siliceous limestone. Siliceous limestone suffers no change under conditions which have already brought about, progressively, several steps in the metamorphism of siliceous dolomitic limestone or dolomite. This contrast is, of course, in accord with the experimental demonstration that magnesian carbonates lose their CO_2 with greater ease than does calcium carbonate.

The new diagram is the second diagram of Figure 6. All compo-

[22] In point of fact the evidence of rocks upon this question is somewhat conflicting. To metamorphism at moderate depth the above statement applies. Thus, in Skye, Tertiary granites have developed periclase in dolomitic limestones under no great cover, but no wollastonite has developed even about flint nodules (Harker, *Tertiary Igneous Rocks of Skye*, p. 150). On the other hand, in the Grenville limestones of the Canadian shield wollastonite is produced at some contacts (though usually in this deep-seated metamorphism the diopside stage is not passed) without the appearance of periclase (or its alteration product, brucite) at these contacts. The deep-seated metamorphism is complicated by notable introduction of foreign material such that often too much silica would be present for the formation of periclase even though it would otherwise be possible. Thus diopside appears instead, or, with introduction of alumina, spinel. Nevertheless, it does appear that evidence of the formation of periclase would be more frequent (brucite marbles are not unknown, see M. F. Goudge, "Limestones of Ontario," *Canada Bur. Mines Bull. 781* [1938], pp. 130–32) if it really could take place before wollastonite formation under these deep-seated conditions. It is suggested, therefore, that, while periclase formation appears to precede that of wollastonite at shallow depths, the reverse order may prevail at greater depths. This possibility is further discussed in connection with the diagram of P-T curves.

sitions lying in the triangle quartz-diopside-wollastonite have reached their final state. Naturally only those assemblages which have calcite can suffer further metamorphism involving decarbonation.

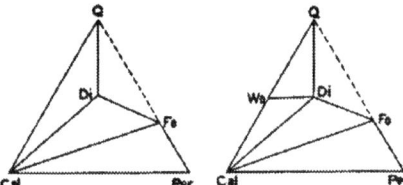

FIG. 6.—Paired composition diagrams of reaction equation 6.

Editor's note: Six pages of the original test are omitted. The story continues on page 255.

THE THIRTEEN STEPS

Reasoning based on a combination of observed mineral associations and theoretical phase-equilibrium relations thus leads to a picture of the metamorphism of carbonate rocks as capable of exhibiting thirteen steps. All these steps are univariant reactions (if we conveniently forget that there may be some departure from this condition in those early steps which involve tremolite), resulting in the evolution of CO_2, and the temperature at which they will occur therefore depends upon the applied pressure of CO_2. For each of the reactions a curve can be drawn which gives the temperature at which the reaction will go forward at any pressure—a P-T curve. In Figure 14 is shown, in purely schematic manner, the relation between the several P-T curves for these reactions. The curve for the reaction taking place at any step always lies, at any pressure, at a *lower* temperature than does the curve for the reaction of the next higher step, which is, indeed, the reason why it is here called a higher step, attention being focused upon the temperature aspect. It is clear also that the curve for any step lies at a *higher* pressure at any given temperature than the curve for the next higher step. Now, although at each step a number of different assemblages can exist below it and a number above it, some assemblages persisting through several, nevertheless each step has a characteristic feature, which indeed depends upon the reaction taking place at that step. In all cases this fundamental feature is of the same general character; a certain phase assemblage can exist below the step but not above it. Every step is thus characterized by the disappearance of a phase assemblage, each its own assemblage, and, although this fact has been pointed out in the discussion of each reaction, it is desirable to bring together here these fundamental features of each step.[26]

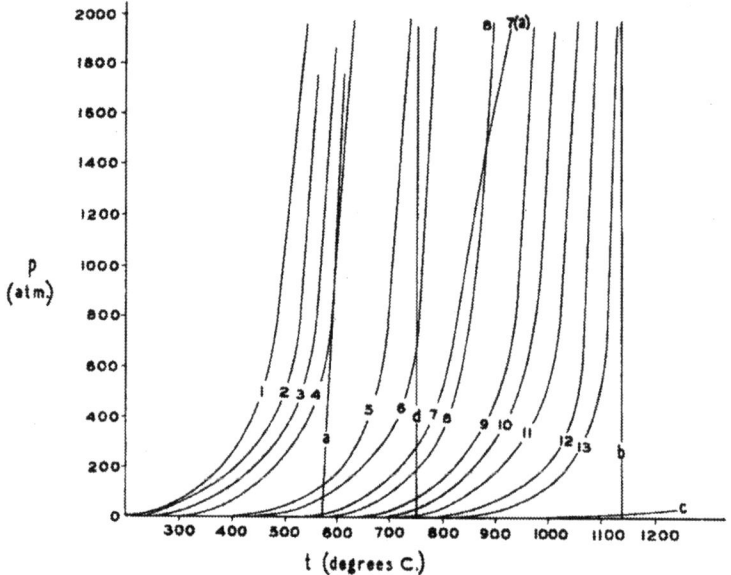

FIG. 14.—Curves *1–13* are pressure-temperature curves of the reactions of equations 1–13, respectively, and of Figs. 1–13, respectively. Curve *a* is the inversion curve α-quartz to β-quartz. Curve *b* is inversion curve wollastonite to pseudowollastonite and indicates approximate extreme temperatures of magmas. Curve *c* is dissociation pressure curve of $CaCO_3$. Line *d* is approximate upper temperature of granitic magmas.

The assemblages which can exist below each step but not above it are as shown in Table 1. At any given pressure we have, thus, thirteen points on a thermometer scale, and they are absolutely fixed, given that the compositions of all the mineral phases lie within

[26] It should perhaps be pointed out that there is no individual composition that could show all thirteen steps. The discussion applies to a carbonate formation in which different beds show a considerable range of proportions of calcite, dolomite, and silica, though certain calcite-rich beds could experience nearly all the steps.

the system under consideration. Nothing can detract from the complete generality of this aspect of the character of any step, viz., that a certain phase assemblage must disappear if that step has been attained and therefore that the survival of that assemblage indicates the failure of attainment of the conditions corresponding with that step. It matters not what other minerals are or were present, what solutions were circulating or what gases were streaming by, the persistence of, for example, wollastonite and calcite in interlocking association indicates that the conditions of step 10 were not at-

TABLE 1

TABLE OF STEPS

To left of curve *1*, i.e., below step 1	Dolomite and quartz
To left of curve *2*, i.e., below step 2	Dolomite and tremolite
To left of curve *3*, i.e., below step 3	Calcite, tremolite, and quartz
To left of curve *4*, i.e., below step 4	Calcite and tremolite
To left of curve *5*, i.e., below step 5	Dolomite
To left of curve *6*, i.e., below step 6	Calcite and quartz
To left of curve *7*, i.e., below step 7	Calcite, forsterite, and diopside
To left of curve *8*, i.e., below step 8	Calcite and diopside
To left of curve *9*, i.e., below step 9	Calcite and forsterite
To left of curve *10*, i.e., below step 10	Calcite and wollastonite
To left of curve *11*, i.e., below step 11	Calcite and akermanite
To left of curve *12*, i.e., below step 12	Spurrite and wollastonite
To left of curve *13*, i.e., below step 13	Spurrite and akermanite

tained. What is meant by "the conditions of step 10" is to be read from curve *10* of Figure 14. The implications of these inflexible relations can, indeed, be carried further. We started with siliceous limestone and dolomitic limestone and developed our discussion by considering changes in materials of fixed bulk composition (except with respect to CO_2), but in doing so it is to be hoped that no one will assume that we have joined the ranks of those who believe that there is no introduction of material in metamorphism. Although reached by that procedure, the conclusions indicated by the table of steps (Table 1) are applicable to any sort of material. The limestone might have been originally quite pure $CaCO_3$, the silica and magnesia introduced, and along with them many other materials, orthoclase and albite, say, or fluorine-bearing phases such as chondrodite, yet the facts indicated in the table of steps will still apply. Indeed, the original rock need not have been a carbonate rock at all, but to any kind of a "soaked" rock or a vein-stuff the association limitations of the table of steps apply with equal force. Again the limestone might have been initially very impure, having argillaceous material as well as silica, yet the resulting aluminous phases, such as grossularite, vesuvianite, anorthite, etc., would not affect the maximum temperature of persistence (at fixed pressures) of the phase assemblages given in the table of steps. The amount of alumina present (or added) might, of course, be so great as to prevent the formation of some of the reference phases in all stages of metamorphism. This is, however, a different matter and leads to no ambiguity, for it is the presence of a certain mineral assemblage that is used as a criterion of grade; the absence of the assemblage is of no significance in that connection.

There is, of course, one class of material for which the conclusions reached require modification, and this is material which can enter into solid solution in the mineral phases formed. If, for example, ferrous iron replaces some of the magnesia of the original dolomite or if ferrous material were introduced in the metamorphic process, then the reaction which produces monticellite in step 7 would not take place at the same temperature as it would with pure nonferrous materials nor would it take place at a fixed temperature. It would be spread through a range of temperature, at a given pressure, because the reaction is no longer univariant. This would be true of every step in which the material might participate if that step involved magnesioferrous phases. It would not be true of steps 6, 10, and 12 provided the initial material had the appropriate composition to experience these steps.[27] Apparently, if we may judge from the evidence of rocks, the amount of spreading-out of these steps from such causes is not serious, at least with moderate content of iron or other similar material,[28] and probably does not cause overlapping and con-

[27] Although wollastonite takes $FeSiO_3$ and indeed $MgSiO_3$ into solid solution at high temperatures, the amount of such solution at most metamorphic temperatures is very small, and wollastonite may for all practical purposes of the present discussion be regarded as of fixed composition.

[28] Alumina has already been mentioned as not affecting equilibrium in so far as it goes to form separate phases, such as grossularite, etc., but it will, of course, affect equilibrium when it enters into solid solution in any of the phases of the reference

fusion of steps. Of course, if the amount of foreign material present, either originally or newly added, is such that, say, the pyroxene of step 7 could not appropriately be referred to as diopside but is perhaps more appropriately to be called augite or hedenbergite, the scheme here propounded would be entirely inapplicable. The reason for stressing the fixed character of these steps in the "pure" materials is that here we have some definite points to which we may tie geologic temperatures—points whose relative position only is suggested by this study but whose absolute position may be susceptible of laboratory determination. And when we speak of points it will be realized, of course, that these steps could be assigned definite temperature values only if the pressure were known, so that the aim of the laboratory study would obviously be the determination of the P-T curve for each of these reactions, the curves that are shown schematically in Figure 14. When this had been done for the "pure" mixtures of the system upon which attention is here focused, the direction and magnitude of the changes brought about when other substances enter into solid solution in the "pure" phases could then be sought. The pressure-temperature relations ascertained would furnish valuable information not only on temperatures but also on pressures prevailing during various geologic processes, igneous, metamorphic, and hydrothermal. In the meantime we have, from the theoretical development alone, some indications as to relative temperatures and pressures.

THE METAMORPHIC MINERALS

The reactions at which certain phase assemblages must disappear furnish, as we have seen, evidence as to temperatures, but it is in

---

assemblages. The most important effect of this kind is its formation of gehlenite, which goes into solid solution in akermanite; indeed the mineral most commonly encountered in metamorphic rocks is closer to gehlenite than to akermanite. The step involving formation of melilite is thus considerably more complicated than has been indicated in the theoretical development where only the akermanite end-member could be considered possible. There is a further complication in the indications that akermanite may be incapable of existence at low temperatures (N. L. Bowen, J. F. Schairer, and E. Posnjak, *Amer. Jour. Sci.*, Vol. XXVI [1933], p. 276), but this is not unconnected with the situation just mentioned because apparently gehlenite-akermanite solid solutions can exist at lower temperatures than can akermanite itself. Even with these complications it is apparent that the melilite phase behaves in the metamorphic process in approximately the manner deduced for the end-member of this phase, akermanite.

some measure of a negative character. The existence of a certain phase assemblage shows that a certain temperature *has not been attained*, and the conclusion is absolutely rigorous, given "purity" of the phases. It would be gratifying if we could find in these reactions some temperature evidences of a positive character, in short, if we could say that a certain temperature *has been attained*. Analysis of the situation shows that we cannot make statements of this character that can be rigorously supported by theoretical considerations, and yet the reactions are not without some value in this capacity.

Each of the reaction steps is characterized by the disappearance of a phase assemblage and some of them by the first appearance of a phase. The question is, "Can the *appearance* of a phase be used as a criterion of temperature in the same manner that the *disappearance* of a phase assemblage serves in this capacity?" and the answer to this question has been anticipated in the concluding sentence of the preceding paragraph.

The mineral phases produced in the metamorphism of siliceous (or silicified) dolomitic limestones are listed below in the order of their production at the thirteen steps. There are only ten of them because some of the steps are concerned with reactions between phases earlier produced. They are:

1. Tremolite
2. Forsterite
3. Diopside
4. Periclase
5. Wollastonite
6. Monticellite
7. Akermanite
8. Spurrite
9. Merwinite
10. Larnite[29]

The arrangement is in the order of their production with rising temperature, but it must be strongly emphasized that there is nothing inherently of a high-temperature character in the minerals with the higher numbers. All indications are that any of these minerals is stable at ordinary temperatures.[30] It is only with respect to their

[29] As an aid in remembering the order of these ten minerals the following ten-word jingle is suggested:
> Tremble, for dire peril walks,
> Monstrous acrimony's spurning mercy's laws

[30] This statement does not, of course, apply to Ca_2SiO_4 in the form larnite, for its inversion to a stable low-temperature form can be induced merely by the blow of a hammer. The significant point is that the compound Ca_2SiO_4 is stable at low temperatures, as are all other compounds listed.

production in metamorphism of carbonates, assumed as purely thermal or substantially so, that these minerals can be thus arranged in an ascending temperature series. In the discussion of the several steps in metamorphism the mineral assemblages have been given, and in each case the assemblage *a* comprises all the *mineral* phases involved in the reaction. The phase which we have spoken of as produced by the reaction is perfectly stable in contact with all the other mineral phases concerned in the reaction at temperatures *below* the reaction.

We could affirm that the formation of a certain mineral indicated the attainment of a certain temperature only in the special instances where we could be perfectly sure that the initial material was a siliceous dolomitic limestone and was metamorphosed by heat alone or with no more than an insignificant amount of immobile pore solution to facilitate reaction. And indeed our conclusion would in such circumstances be as rigorous as one based on the *failure of disappearance* of certain phase assemblages. However, in any individual case we may not make these assumptions as to the simplicity of conditions. There may have been circulating solutions, and here is where the case based on failure of disappearance of a phase assemblage manifests its superiority over the case based on the *appearance* of a phase. Evidence of the presence of circulating solutions of whatever character can, as we have found, affect in no manner the significance to be attached to the survival of a phase assemblage, indeed evidence of that kind would strengthen the case, since the solutions would indubitably facilitate such reactions between phases as could have occurred under the prevailing pressure-temperature conditions. But in connection with the significance to be attached to the appearance of a new phase the mere possibility of the presence of circulating solutions is of importance because such solutions could introduce a phase or even induce its formation at temperatures far below that at which it would be produced in that simple type of metamorphism assumed in the foregoing discussion, the simplicity being there assumed, of course, for the purpose of discovering laws or generalizations.

It has already been shown that the metamorphic minerals formed in the higher steps are not inherently of a high-temperature character, so the fact that solutions of the right kind might introduce and deposit one or more of these minerals at relatively low temperatures need not be enlarged upon, but the fact that solutions could induce the formation of these minerals at relatively low temperatures from materials present in the rock may require some comment. In this case, indeed, the solution need not be of any highly specialized character. There is only one requirement, viz., that the solution can dissolve CO_2, and it therefore must not be saturated with CO_2 for the prevailing conditions. We might assume in an extreme and wholly imaginary case a solution of caustic alkali which could freely absorb CO_2 and reduce its pressure to a negligible value. In the presence of such a solution even the "highest" of the reactions could evidently take place at relatively low temperature. Or we might assume as another extreme the circulation of pure water. It has significant solvent power for CO_2 and, if water were circulating with sufficient rapidity (in sufficient quantity), the effective pressure of CO_2 would be reduced sufficiently to produce significant lowering of the temperature below that at which each of the steps of metamorphism would "normally" proceed.[31] In any of these cases it is necessary to assume circulation of solutions; in some cases, very vigorous circulation. The mere presence of a minute amount of immobile pore solution, however potent it might be in absorbing CO_2, would not affect the simple picture originally presented, for a minute amount of solution would have its powers in that direction exhausted when only a negligible amount of reaction of the rock constituents had occurred, and any significant reaction would depend upon the factors presented in the initial discussion.

Now, when we examine the actual occurrences of the ten minerals of the metamorphic series, we find that the limitations upon their use as temperature indicators are perhaps not so important as this lengthy discussion might suggest. On the whole, it would appear that conditions ordinarily approach more closely the simple picture involving only insignificant amounts of solution rather than a con-

[31] A streaming-through of gases accomplishing a washing-out of the CO_2 would plainly have the same effect.

dition involving free circulation of great quantities of solution, or else that such solutions as circulate have relatively little power of carrying away CO_2, perhaps because they are already saturated with it or nearly so. If, for example, we examine the occurrences of these metamorphic minerals, we find that at the relatively low temperatures (and high pressures) of regional metamorphism the diopside stage is never passed; at the somewhat higher temperatures of the contact zones about granites, which come from the coolest of magmas, the wollastonite stage may be attained but not passed. It is only at the hotter contacts of syenitic or granodioritic masses that we begin to get members higher in the series, and the highest members are associated only with basic, for the most part basaltic, rocks. Temperature is, of course, not the only controlling factor. As the P-T curves show, a low pressure acts in the same direction as a high temperature; therefore, we may expect to find the higher members of the series at the contacts of hypabyssal intrusives, and these expectations are borne out by the actual occurrences. It is not necessary to adhere to the view that mere depth and therefore weight of overburden fixes the pressure in order to see significance in the depth factor as exerting a control over the mineral character. Porosity of the rocks, existence of open fissures, and like factors will play a role in fixing the effective pressure of CO_2, but they are in large measure functions of depth. We find, therefore, that broad general correspondence of the mineral characters with depth-temperature relations which has just been pointed out, but in the detailed relations about any individual intrusive mass simplicity may be lacking. There may be no simple arrangement, with an inner zone showing the "highest" minerals and successive outer zones showing progressively "lower" minerals. Local channels, proximity to which provided relatively easy escape for CO_2 and perhaps relatively easy ingress of hot solutions, might lead to the formation of localized deposits of high-grade minerals even in outer portions of a metamorphic aureole.

Editor's note: Three pages of the original test are omitted at this point.

THE PRESSURE-TEMPERATURE CURVES

It is perhaps desirable to discuss some of the general relations of the P-T curves of Figure 14. They are curves of univariant equilibrium and, except in some limiting cases, represent equilibrium between five phases in a four-component system. The question naturally arises as to whether any of these curves meet at univariant points. The first four curves are purely schematic and are concerned with equilibria that involve alternately quartz and forsterite. Two adjacent curves could intersect, therefore, only at a point where forsterite and quartz are in equilibrium. There is no temperature, at ordinary pressure, where quartz and forsterite can coexist. They combine to form enstatite, and the volume relations are such as to discourage belief that high pressure would favor formation of forsterite and quartz. It may be concluded, therefore, that none of the first four curves intersects each other at any pressure. The curve a will intersect each of the curves 1 and 3 at some point, and each point will be an invariant point (sextuple), but these points will not be discussed in detail. The point of intersection of curve a with curve 6 displays the same general relations and is controlled by the same principles and will be chosen for discussion instead. Since curve 6 is concerned with a limiting case of ternary equilibrium in which quartz is one of the phases, the intersection with curve a will be an invariant (quintuple) point. At this point five phases are in equilibrium—calcite, wollastonite, α-quartz, β-quartz, and CO_2. From it should emanate five curves of univariant equilibrium between four phases as follows:

 d) calcite + wollastonite + α-quartz + β-quartz
 e) calcite + CO_2 + α-quartz + β-quartz
 f) wollastonite + CO_2 + α-quartz + β-quartz
 g) calcite + wollastonite + α-quartz + CO_2
 h) calcite + wollastonite + β-quartz + CO_2

Curves d and e will coincide with each other and with that portion of curve a of Figure 14 which lies at lower temperatures than the point of intersection. Curve f coincides with that portion of curve a which lies at higher temperatures than the point of intersection. Curves g and h are the two portions of curve 6 of Figure 14, the part which lies below and the part which lies above the temperature of intersection, respectively. There should indeed be a slight change of slope of curve 6 of Figure 14 at this point of intersection. The five curves about the quintuple point, by reason of the fact that two of the phases, α- and β-quartz, have the same composition,[32] are thus reduced to three, the curve of equilibrium of α- and β-quartz which passes without discontinuity through the point and the two curves that are shown in Figure 14 as the one curve 6. The point of intersection is nonetheless a quintuple point.

Curve 5 is the dolomite-dissociation curve based on rough determinations by Eitel,[33] and curve 6 is the well-known calculated curve of Goldschmidt.[34] Neither can be regarded as established with sufficient accuracy to warrant insistence on the plotted relations. It has, indeed, already been suggested that they cross; that at low pressures the dolomite dissociation comes at lower temperatures than wollastonite formation, whereas at high pressures the wollastonite equilibrium runs at lower temperatures. The point at which they cross, if there is such a point, may lie at higher pressures than any indicated on Figure 14. In any case, it is to be noted that such a point is of an entirely different nature from those points just discussed and is in no sense a unique equilibrium point. It is merely the point at which the P-T curves of two wholly independent chemical systems happen to cross. They are the projection on the P-T plane of the P-T-X (pressure, temperature, composition) equilibria of two systems having wholly independent compositions. The point at which they might cross has thus no chemical significance any more than has the point where the curve a crosses a curve of equilibrium in which quartz has no part.

All the other curves of Figure 14 are purely schematic, except b and c. The latter is the determined curve of dissociation of calcium carbonate up to the temperature where melting phenomena intervene. In view of the position of this curve it is not surprising that

[32] G. W. Morey and E. D. Williamson, *Jour. Amer. Chem. Soc.*, Vol. XL (1918), pp. 59–84.

[33] W. Eitel, *Neues. Jahrb. f. Min.*, Beilage Band LI (1925), S. 477–93.

[34] It is firmly embedded in the literature that the Goldschmidt calculation gives a pressure of approximately 1 atmosphere at 500° C. As a matter of fact, log p is approximately 1 at 500° C. and the pressure is somewhat more than 12 atmospheres.

calcium carbonate in nature never suffers simple dissociation into lime and CO_2. The curve does suggest, of course, that a block of limestone in a basic lava flow at, say, 1150° C., would be dissociated unless it was immersed to a depth of more than some 200 feet below the surface of the liquid lava. The block would inevitably have suffered additive metamorphism as it was heated up in the magma with the development of various silicates, but, if any calcite survived, it would be dissociated, provided the conditions were those suggested above. Any lime that may have been so formed would be so reactive that it would enter into combination during the cooling and crystallization of the magma, and the evidence that it had ever formed would undoubtedly be lost.

Curve b is the inversion curve of wollastonite to pseudowollastonite and is shown as a vertical line because the densities of the two forms are sensibly the same, at least at the ordinary temperature, and pressure would therefore have a negligible effect upon the temperature of transformation. The curve b is also intended to indicate the approximate upper limit of temperature of magmas in the accessible part of the earth—a temperature which far exceeds that prevailing during most episodes of metamorphism (or mineral formation of any kind) except perhaps in some examples of pyrometamorphism induced by basic magmas. Here the temperature may have equaled this value or perhaps surpassed it in moderate degree, for lavas may possibly attain somewhat higher temperatures under surface conditions than they ever attain at other levels of the accessible crust.[35] The reasons for taking this temperature as a probable upper limit of temperatures available in the accessible crust are the nonoccurrence of pseudowollastonite in natural rocks and the failure of evidence that wollastonite is ever a pseudomorph after pseudowollastonite.

Curves 7 and 8 are concerned, respectively, with monticellite and with akermanite formation. Akermanite is itself probably unstable at all temperatures of these curves so that curve 8 can be regarded only as a schematic presentation of the relative conditions of formation of melilite, a solid solution containing akermanite. Curve 7

[35] Bowen, *The Evolution of the Igneous Rocks* (Princeton: Princeton University Press, 1928), p. 184.

represents the coexistence of the five phases—calcite, diopside, forsterite, monticellite, and CO_2—but, since experimental evidence shows that monticellite and diopside do not coexist at high temperatures, the curves 7 and 8 are shown as intersecting at an invariant (sextuple) point where curve 7 ends and curve 7a, an independent curve with a different slope, begins and runs to higher temperatures, the phases in equilibrium being calcite, forsterite, monticellite, akermanite, and CO_2. The curve 8 would, however, be continuous through the point because both parts of it are nominally concerned with the equilibrium diopside + calcite ⇌ akermanite + CO_2, and the other phases present (monticellite below the invariant point and forsterite above it) have no part in the reaction. This is indeed much ado about a curve and an intersection which probably do not exist as such, akermanite being presumably unstable at these temperatures; but, since the curve is here used to represent the more complex melilite equilibrium, it has been considered desirable to show and discuss this intersection, because it expresses diagrammatically the fact that equilibria involving coexistence of monticellite and diopside occur below a certain temperature, above which they are replaced by equilibria involving coexistence of forsterite and akermanite (melilite).

Curve 9 is concerned with a subsidiary reaction which requires no special discussion. Curve 10 refers to a reaction in the three-component system resulting in the formation of spurrite. Concerning it and the higher temperature curves, it can be said only that they represent the apparent relative positions of these reactions. Spurrite occurs in such a manner as to suggest that it is formed in the ascending cycle in the same general manner as the other minerals of the metamorphic series under discussion and not during retrograde metamorphism as are most of the minerals (probably including the closely related tilleyite) associated with the species formed in the reactions of these higher curves. In 1916 Niggli studied equilibrium in carbonate melts and interpreted his results as indicating that there were certain conditions under which Ca_2SiO_4 and calcite were in equilibrium with each other and CO_2 (about 850° C. and 1 at. CO_2).[36] But spurrite is evidently stable under some conditions, and,

[36] Niggli, *Zeits. f. anorg. Chemie*, Vol. XCVIII (1916), p. 305.

if we accept Niggli's interpretation, it is necessary to conclude that there is a point at which calcite, spurrite, and Ca_2SiO_4 are in equilibrium. This relation would give rise in the ternary system to a quintuple point at which the five phases—wollastonite, calcite, larnite, spurrite, and CO_2,—would be in equilibrium. From the quintuple point would emanate the five curves of univariant equilibrium concerned with the five assemblages of four phases each, as follows:

1. Wollastonite, calcite, larnite, spurrite
2. Wollastonite, calcite, larnite, CO_2
3. Wollastonite, calcite, spurrite, CO_2
4. Wollastonite, larnite, spurrite, CO_2
5. Calcite, larnite, spurrite, CO_2

Two arrangements of these curves which permit the coexistence of larnite (the mineral name is here used loosely to indicate the compound Ca_2SiO_4 in any form) and calcite at atmospheric pressure are given in Figure 15, a and b, each curve being numbered to indicate its equilibrium assemblage in accordance with the numbered assemblages above. Curves 1 and 5 actually form one continuous curve which coincides with the curve of binary equilibrium, calcite + larnite $\rightleftarrows$ spurrite, and is nearly vertical.

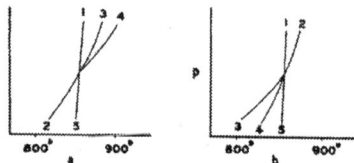

FIG. 15.—Pressure-temperature curves about a hypothetical quintuple point—calcite, spurrite, wollastonite, larnite, CO_2.

But if Figure 15, a, represented the facts, then it would not have been possible to prepare spurrite (or a crystalline substance having the same composition) from its components at 350° C., which was done by Shepherd,[37] since the curve 1, 5 would represent the minimum temperatures (nearly constant at all pressures) of existence of spurrite. And if Figure 15, b, were valid, then it would not have been possible to make spurrite (or an isomer) at 1380°, as was done by Eitel,[38] for the curve 1, 5 would indicate the maximum temperature of existence of spurrite. Experimental evidence is thus apparently contradictory. At

[37] In F. E. Wright, *Amer. Jour. Sci.*, Vol. XXVI (1908), p. 551.
[38] *Neues Jahrb. f. Min.*, Beilage Band XLVIII (1923), S. 63.

present it seems best to accept the evidence that spurrite, or an isomer, is stable throughout the whole temperature range of our diagram and that under no conditions is larnite or some form of Ca_2SiO_4 in equilibrium with calcite. This is accomplished in the diagram by drawing an independent curve of the spurrite-wollastonite-calcite equilibrium which never intersects any curve involving larnite. Revision may be necessary with increased knowledge.

On the other curves of the diagram little comment is needed. With advancing knowledge the relations suggested may require notable revision. This would be especially true if investigation should prove that merwinite and larnite are not different phases but different concentrations of the same phase, a possibility that is suggested by the remarkable similarity of their properties. The similarity is, however, no greater than that displayed, say, by monticellite and forsterite, and these are undoubtedly distinct phases, so that the suggested drastic modification of this part of the equilibrium diagram may not be necessary.

OF FACIES

The curves of Figure 14 indicate clearly that, aside from certain possible complications already fully discussed, the formation of any given metamorphic mineral of the series will require a higher temperature if the pressure be low than if the pressure be high. The formation of a certain mineral, such as monticellite, for example, might be taken as indicative of the attainment of a certain grade of metamorphism, or the rock containing it might be described as belonging to a monticellite facies, if we adopt the language of the facies classification. When reactions such as those here considered are involved, it is plain that the development of a certain facies cannot be taken as indicating a definite set of conditions. At a low pressure a monticellite facies would be developed at a comparatively low temperature; at a higher pressure it would be developed only at a higher temperature, and no definite statement about either one of these conditions could be made without rather definite knowledge of the other. The concept of facies as thus used may therefore be a rather indefinite one as far as assignment of conditions of formation is concerned. It is often said, too, that two kinds of rocks are isofacial with each other, and in so far as they are found side by side and evidently formed under identical conditions they are, indeed, for that occurrence isofacial. The term is, as so used, simply a truism, and any attempt to extend the concept of isofaciality must be carried out with caution. We cannot, for example, make the assumption that, because the materials of these two rocks formed certain assemblages of minerals under these conditions, therefore when the one assemblage is found elsewhere the other assemblage is to be expected there also, if any rock material of the same bulk composition as this second assemblage exists there. On the contrary, the bulk composition which had given the second assemblage might now be found as a totally different assemblage. This is a necessary consequence of the possible difference in the forms of the P-T curves of the reactions the two different materials are capable of experiencing, a fact that is illustrated in highly simplified form in the relation between curve b and curve a of Figure 14. These two curves intersect at about 100 atmospheres, and this fact indicates that below 100 atmospheres a quartzite made up entirely of α-quartz could form at the same temperature as a wollastonite marble. The two rocks could thus form side by side; in other words, they would be isofacial. But above 100 atmospheres a quartzite composed of α-quartz could form at the same temperature as a quartz marble, and these two would be isofacial. It is plain, therefore, that we cannot make any such statement as "wollastonite marble is isofacial with an α-quartz quartzite," and in general we must be very careful about stating that any rock is isofacial with another rock.

This limitation upon the use of such terms arises from the fact that the P-T curves of different equilibria have different slopes, a relation which is not, however, a disadvantage but is rather a distinct advantage in attacking questions of petrogenesis. In addition to the curves a and b of Figure 14, which refer to equilibria between solids only, we may eventually be able to determine a large number of such curves, perhaps for more complex equilibria, but still for equilibria between solids only. These curves will be nearly vertical and thus will cut across the curves involving a gas equilibrium. Of these latter it may eventually be possible to add many more, depicting the dissociation of hydrates, sulphides, oxides, and other compounds. The intersecting curves of the two classes will thus cut the general P-T diagram up into a grid which we may call a petrogenetic grid. With the necessary data determined by experiment we might be able to locate very closely on the grid both the temperature and the pressure of formation of those rocks and mineral deposits of any terrane that were formed at a definite stage of its history, provided always that a sufficient variety of composition of materials occurred in the terrane to permit adequate cross reference. The determinations necessary for the production of such a grid constitute a task of colossal magnitude, but the data will be gradually acquired, and we shall thus slowly proceed toward an adequate knowledge of the conditions of formation of rocks and mineral deposits. It was probably with considerations of this kind in mind that Eskola proposed his concept of the facies. Undue emphasis upon either pressure or temperature is avoided, as is especially any assumption regarding concomitant variation of temperature with pressure (depth). Rather is it necessary to be able to view mineral formation as affected by pressures and temperatures that may vary for the most part quite independently of each other, though in the depths, where pressures are high, there can, of course, be none but comparatively high temperatures.

CHAPTER 5: COMPOSITION PROJECTIONS

Thompson, J. B. (1957) The graphical analysis of mineral assemblages in pelitic schists.
The American Mineralogist, **42**, 842–858.

The usefulness of composition, or compatibility, diagrams in the analysis of metamorphic mineral parageneses was recognized long ago by Eskola (1915). He reduced the dimensionality of composition space (to use the modern terminology) by projecting minerals and rocks from H_2O and SiO_2 onto triangular ACF and A'KF diagrams. The A'KF diagram (where A' is the alumina in excess of that coupled with K_2O, Na_2O and CaO in K-feldspar, albite, and anorthite) was designed as better suited to non-calcareous rocks such as metapelites, metagranites and metagreywackes. Both ACF and A'KF diagrams were used extensively in textbooks up to and through the 1960s (e.g. Turner, 1948; Ramberg, 1952; Barth, 1952; Fyfe *et al.*, 1958; Turner and Verhoogen, 1960; Winkler, 1965, 1967; Turner, 1968). The shortcomings of A'KF diagrams were obvious to all: an important stable join from aluminosilicate to biotite was precluded by links from muscovite to garnet, staurolite, chloritoid and cordierite, and the lumping together of FeO and MgO under "F" rendered impossible the coexistence of familiar ferromagnesian mineral pairs in the various facies series of metamorphism.

The sensitivity of metamorphosed argillaceous sediments (metapelites) to the conditions imposed on them makes them, among all lithotypes, the best indicators of regional gradients in metamorphic intensity or grade. By the middle of the twentieth century, metamorphic terranes worldwide had been mapped with pelitic isograds representing the successive appearance (or disappearance) of index minerals across them (see Chapter 2). However, the compositional complexity of pelitic rocks and most of their constituent minerals was clearly frustrating attempts to devise a suitable two-dimensional visual aid that could provide an understanding of the reactions involved in their metamorphism. To judge from the volume of papers on pelitic schists that appeared in the years following the publication of Jim Thompson's landmark paper in 1957, it is clear that an important barrier to progress had been removed. Now it was possible to characterize the reactions in a satisfactory manner, and thereby account for the influence of temperature and pressure on the phase assemblages observed.

For pelites it was patently inappropriate to combine the components FeO and MgO − pelite minerals vary widely in their Fe/Mg ratios − but how then were we to avoid having to use a multidimensional diagram, such as a tetrahedron (or worse), to show essential compositional variations? Thompson took advantage of the fact that, when modelling pelites free of SiO_2 and H_2O in the tetrahedron AKFM (where F is FeO and M is MgO), the vast majority occupy compatibility volumes (irregular tetrahedra) and triangles that all share muscovite as a corner phase. This meant that projection of compositions away from that of muscovite onto a convenient surface like the AFM plane (the image plane) preserves the phase relations of the tetrahedron. At that time, projection from a complex mineral composition rather than from a simple component like H_2O or SiO_2 was novel.

The triangular (Fig. 5) rather than the parallel-edge form (Fig. 7) of what soon became known as the Thompson AFM-projection has been the most widely used by petrologists. Like ordinary 3-component systems that include solid solutions, the diagram contains sequences of 3-phase triangles (divariant, taking H_2O as a phase), 2-phase, trivariant fields of infinite quasi-parallel tie-lines, and one-phase, quadrivariant fields. Any one Thompson projection is theoretically uniquely determined by the attendant intensive parameters of temperature, pressure and the activity or chemical potential of H_2O. The term "humidity" (p. 844) did not catch on because geochemists have come to use a sliding standard-state for minerals and fluids the pure phase at the T and P of interest so that activity and humidity are the same. Changes in intensive properties shift the locations of 3-phase triangles and 2-phase fields in directions that are in large measure predictable. The sense of movement of phase-fields (effectively left or right because of formula constraints on most of the minerals) is determined by the reaction stoichiometry (including H_2O unless it is conserved) and the relative Fe/Mg ratios of the minerals involved. Most of the minerals in pelitic schists show consistent differences in Fe/Mg ratio (siderophile tendency), although in 1957 the sense of partition of Fe and Mg was not well known among garnet, staurolite and chloritoid, or among cordierite, chlorite and biotite. The effect of pressure on the diagram can be predicted if the Clapeyron slope dP/dT of the reactions involved is known. These shifts correspond to "continuous" reactions which involve three projected phases ± muscovite (or orthoclase in an alternative diagram for high-temperature pelites), quartz and H_2O. The projection can only function correctly if the chemical potential of H_2O is uniform across the diagram. Thompson argued (p. 845) that we should consider the H_2O humidity (activity) to be externally controlled.

When, as a result of changes in T or P, a 2-phase field thins down to a line and two adjacent 3-phase fields make contact, there follows a reaction involving 4 AFM projection phases. The reaction corresponds to a univariant line on the petrogenetic PT-grid and produces a change in the topology of the diagram, hence the term "discontinuous" reaction. It involves a tie-line flip and a compatibility change: A + B = C + D. A reaction that leads to the appearance or disappearance of a mineral in the interior of the diagram (A = B + C + D) is also discontinuous. In both cases, the reaction will be witnessed by a significant range of bulk compositions, making it, as recognized by Thompson, an ideal candidate for an isogradic reaction. No other diagram in general use in petrology illustrates these kinds of phase relations for mineral solutions with the same clarity.

While the visual advantage of the Thompson AFM-projection was obvious, its quantitative use was hindered for some years by the need to prepare mineral separates for chemical analysis (e.g. Phinney, 1963; Albee, 1965; Hounslow and Moore, 1967). Carmichael (1970) used these authors' chemical data and the optical properties of chlorite to construct a sequence of schematic Thompson projections clarifying the isogradic reactions across a terrane of Barrovian metamorphism in Ontario. He showed that the staurolite and kyanite isograds in pelites of appropriate bulk composition correspond to tie-line flips in the AFM-projection caused by progress of discontinuous reactions consuming muscovite and quartz, and evolving H_2O: respectively, Chl + Grt + Ms + Qtz → Bt + St + H_2O, and St + Chl + Ms + Qtz → Ky + Bt + H_2O. Neither reaction, of course, concerns the stability limits *per se* of Chl, St, Grt or Ky. The sillimanite isograd involves no change in topology, being in theory the polymorphic transition from Ky to Sil.

It was not until the electron microprobe became widely available in the late 1960s that phase fields in the AFM-projection, based on the analysis of samples of diverse bulk composition, were located routinely and with precision. This development resolved some of the ambiguities in the partition of Fe and Mg among the AFM minerals, for example, garnet, chloritoid and staurolite (Rumble, 1971; Albee, 1972; Grambling, 1983). Guidotti et al. (1975) showed that the Crd-Bt join passes to the right (Mg-rich side) of Chl, so that in low-pressure metamorphism Chl disappears upgrade inside the AFM diagram via a reaction terminal to Chl (see Spear 1993, Fig. 10-10). Albee (1972) illustrated the prograde sequence of reactions in Barrovian pelites from the garnet to the Grt + Crd zone, showing in Thompson projections and isobaric pseudobinary T-X_{Fe-Mg} sections the dependence of phase assemblages on bulk composition and temperature. His diagrams showed clearly how high-Al and low-Al pelites might have different sequences of index minerals. Among other things, this clarified the low-grade occurrence of kyanite in high-Al rocks, something observed long ago: for example, kyanite with chloritoid and muscovite in Unst (Read 1934), and kyanite in the biotite zone in New Hampshire (Chapman, 1939).

The biotite isograd is unusual in that in many (but not all) pelites it corresponds not to a discontinuous but to a continuous reaction: Chl + Kfs = Ms + Bt + Qtz + H_2O (Mather, 1970). It is an effective isograd reaction because the 3-phase triangle Chl-Kfs-Bt apparently moves swiftly from the Fe to the Mg side of the Thompson diagram with increase in temperature.

Greenwood (1975) performed an extremely valuable service by presenting a generalized method using simple matrix multiplication that enabled projection of any desired combination of phases and components in any system, not just pelitic systems. As a result, composition projections became much more widely used (see below). In an original variation on the AFM-projection for pelites, he projected from Qtz, Ky and Ms onto the plane H_2O-FeO-MgO (Greenwood, 1975, Fig. 2), a step that changed the status of H_2O from an intensive to extensive variable, thus permitting simultaneous consideration of H_2O-saturated and H_2O-deficient assemblages at the same P and T. Greenwood emphasized the requirements of "legal" projections: (1) the equilibrium state of the system is uniquely determined by specifying P, T and the proportions of the components, and (2) the phases or components from which the projection is made must have their chemical potentials fixed at a constant value over the entire diagram. In many systems in nature, a projection phase (even muscovite) is found to vary in composition in different assemblages at constant T and P. Potential misalignments arising from this can be minimized if an image plane for the projection is chosen to lie between the projection phase (e.g. a solid solution such as biotite or chlorite) and all the solid solution minerals to be projected (Thompson, 1979; Powell and Sandiford, 1988; Will, 1998, p. 174). Linear algebraic methods for projections were further discussed by Spear et al. (1982) and Spear (1993).

A.B. Thompson (1976a) took advantage of the predictability in movement of 3-phase triangles in the Thompson projection to construct a comprehensive set of projections and isobaric T-$X_{(Fe-Mg)}$ and isothermal P-$X_{(Fe-Mg)}$ sections for middle grades of a Barrovian facies series. For a particular bulk composition, he showed that an individual mineral, e.g. biotite, may reverse its trend of Fe-Mg enrichment upgrade, depending on the associated AFM phases and the continuous reactions relating them. He then showed (Thompson, 1976b) that the available experimental and thermochemical data were sufficient to make these diagrams quantitative. One glance at the T-$X_{(Fe-Mg)}$ diagrams of Albee (1972) or Thompson (1976a,b) makes it obvious why the phase petrology of pelites could not be properly understood before J.B. Thompson's paper in 1957.

Compositional phase diagrams have proven invaluable for the characterization of isogradic reactions, identifying assemblages that probably represent equilibrium states, and determining whether local differences among assemblages reflect changes in grade or differences in bulk composition. Properly drawn projections, from the relatively simple Thompson variety to those with multiple projection phases, provide a test of the assumptions underlying their construction. Ideally, the projection should show regular element partitioning between phases and an absence of crossing tie-lines. If we can account for the crossing of tie-lines between phases, then we will have learned something about metamorphic controls and processes. More often than not, writers have reported samples of pelitic schist to contain more KFMASH minerals (>3 on the projection) than would ordinarily be expected in an equilibrium assemblage. Possible explanations for this and crossing tie-lines in general may be obvious or quite subtle. Obvious cases have been attributed to the presence of non-AFM components in the minerals projected, such as Ca and Mn in garnet, Zn in staurolite, and Ti or Fe^{3+} in biotite. Increasing Ca and Mn in garnet moves the 3-phase triangle Grt-Bt-Ky to the right across the Thompson projection (Hodges and Spear, 1982). In fact, low-grade garnet largely owes its presence to the stabilizing effect of Mn and Ca (Spear and Cheney, 1989; Symmes and Ferry, 1992). In other cases the extra phase can sometimes be shown to be relict or alternatively a late-stage reaction product, or the sample was sitting on an isograd! Especially interesting are those instances in which the crossing tie-lines cannot be explained by any of these possibilities. When they are found as a result of comparing samples on a layer-to-layer or outcrop-to-outcrop scale, local variation in the chemical potential of H_2O is indicated, and the relations may be clarified by projecting onto a plane containing H_2O (Rumble, 1978; Grambling, 1981; Dickenson, 1988). This violates Thompson's original proposal for the external control of H_2O humidity, of course, although in terms of kJ/mol differences in chemical potential of H_2O it might not be much. The stable coexistence in one sample of 4 or more pure KFMASH minerals on the image plane is indicative of a reaction relationship among the minerals present, and internal buffering of the H_2O potential (e.g. Lang and Rice, 1985; Giaramita and Day, 1991). These observations might have relevance to the issue of fluid infiltration during metamorphism (Chapter 12). Sample-to-sample variation in redox state theoretically moves the effective whole-rock AFM composition across the diagram rather than shifting the phase fields (Thompson, 1972), except when the redox variation is large enough to influence the activity of H_2O (Rumble, 1978).

A comprehensive account of the mineralogical changes in pelitic rocks across the three principle facies series was provided by Spear (1993). His use of the Thompson projection coupled with AKF and AKM end-member compatibility diagrams keyed to divariant fields in the calculated petrogenetic PT grid (Spear and Cheney, 1989) makes this a very helpful reference, notwithstanding later refinements (e.g. Pattison et al., 2002).

The generality of projection methods using matrix multiplication (Greenwood, 1975) proved to be especially helpful for viewing phase compatibilities in chemically complex systems such as metabasites and metagreywackes, where the compositional dimensions could be reduced by appropriate selection of a subset of commonly occurring minerals as projection points. In addition to H_2O, CO_2 and Qtz, authors have chosen to project from various customized combinations of chlorite, epidote, prehnite, lawsonite, hematite, titanite, hornblende, albite and plagioclase (e.g. Harte and Graham, 1975; Brown, 1977; Holland, 1979; Laird, 1980; Nakajima, 1982; Spear, 1982; Brown and Ghent, 1983; Liou et al., 1985; Springer et al., 1992; Broecker and Day, 1995, and many others).

Even in a carefully drawn, 2-dimensional graphical representation of a system with many components, it may be difficult to distinguish valid tie-line intersections from artifacts of the projection. Greenwood (1967) pointed out that linear algebraic dependencies among minerals in multicomponent systems are entirely equivalent to graphical intersections in simple phase diagrams, and that the search for possible mass balances may be found more rigorously by means of linear programming and regression analysis (Greenwood, 1968), thereby freeing us from the limitations of two-dimensional plots. These techniques were used by Lang and Rice (1985) and Giaramita and Day (1991), among others. Fisher (1989) showed that the algebraic technique known as singular value decomposition (SVD) can be applied in a straightforward fashion for the same purpose. Complete chemical analyses of all minerals present are required. For example, Lang (1991) used SVD to demonstrate the absence of a mass-balance relationship between St and St + Ky assemblages in metapelites near Baltimore, Maryland, despite contradictory indications in the AFM projection. Todd and Evans (1994) employed SVD to specify the isochemical reaction of amphibolite to two-pyroxene granulite caused by outcrop-scale infiltration of CO_2 above a marble layer.

Further major developments in the art of composition projections hardly seem likely. Clearly, projections will continue to have value in the display and analysis of mineral parageneses in metamorphic rocks. Nevertheless, there perhaps remain possibilities for creative research in this area, given the right kind of problems, especially for the mathematically inclined.

References

Albee, A.L. (1965) Phase equilibria in three assemblages of kyanite-zone pelitic schists, Lincoln Mountain quadrangle, central Vermont. *Journal of Petrology*, **6**, 246–301.

Albee, A.L. (1972) Metamorphism of pelitic schists: reaction relations of chloritoid and staurolite. *Geological Society of America Bulletin*, **83**, 3249–3268.

Barth, T.F.W. (1952) *Theoretical Petrology*. John Wiley and Sons, Inc., New York, 387 pp.

Broecker, M. and Day, H.W. (1995) Low-grade blueschist facies metamorphism of metagreywackes. *Journal of Metamorphic Geology*, **13**, 61–78.

Brown, E.H. (1977) Phase equilibria among pumpellyite, lawsonite, epidote, and associated minerals in low-grade metamorphic rocks. *Contributions to Mineralogy and Petrology*, **64**, 123–136.

Brown, E.H. and Ghent, E.D. (1983) Mineralogy and phase relations in the blueschist facies of the Black Butte and Ball Rock areas, northern California Coast Ranges. *American Mineralogist*, **68**, 365–372.

Carmichael, D.M. (1970) Intersecting isograds in the Whetstone Lake area, Ontario. *Journal of Petrology*, **11**, 147–181.

Chapman, C.A. (1939) Geology of the Mascoma quadrangle, New Hampshire. *Geological Society of America Bulletin*, **50**, 127–180.

Dickenson, M.P. (1988) Local and regional differences in the chemical potential of water in amphibolite facies pelites. *Journal of Metamorphic Geology*, **6**, 365–381.

Eskola, P. (1915) On the relation between the chemical and mineralogical composition in the metamorphic rocks of the Orijärvi region. *Bulletin de la Commission Géologique de Finlande*, **44**, (English summary pp. 109–145).

Fisher, G.W. (1989) Matrix analysis of metamorphic mineral assemblages and reactions. *Contributions to Mineralogy and Petrology*, **102**, 69–77.

Fyfe, W.S., Turner, F.J. and Verhoogen, J. (1958) *Metamorphic Reactions and Metamorphic Facies*. Geological Society of America Memoir, **73**, 259 pp.

Giaramita, M.J. and Day, H.W. (1991) Buffering in the assemblage staurolite-aluminum silicate-biotite-garnet-chlorite. *Journal of Metamorphic Geology*, **9**, 363–378.

Grambling, J.A. (1981) Kyanite, andalusite, sillimanite, and related mineral assemblages in the Truchas Peaks region, New Mexico. *American Mineralogist*, **66**, 702–722.

Grambling, J.A. (1983) Reversals in Fe-Mg partitioning between chloritoid and staurolite. *American Mineralogist*, **68**, 373–388.

Greenwood, H.J. (1967) The N-dimensional tie-line problem. *Geochimica et Cosmochimica Acta*, **31**, 465–490.

Greenwood, H.J. (1968) Matrix methods and the phase rule in petrology. *23rd International Geological Congress, Prague*, **6**, 267–279.

Greenwood, H.J. (1975) Thermodynamically valid projection of extensive phase relationships. *American Mineralogist*, **60**, 1–8.

Guidotti, C.V., Cheney, J.T. and Comatore, P. (1975) Coexisting cordierite + biotite + chlorite from the Rumford Quadrangle, Maine. *Geology*, **3**, 147–148.

Harte, B. and Graham, C.M. (1975) The graphical analysis of greenschist to amphibolite facies mineral assemblages in metabasites. *Journal of Petrology*, **16**, 347–370.

Hodges, K.V. and Spear, F.S. (1982) Geothermometry, geobarometry and the Al_2SiO_5 triple point at Mt. Moosilauke, New Hampshire. *American Mineralogist*, **67**, 1118–1134.

Holland, T.J.B. (1979) High water activities in the generation of high pressure kyanite eclogites in the Tauern Window, Austria. *Journal of Geology*, **87**, 1–27.

Hounslow, A.W. and Moore, J.M. Jr. (1967) Chemical petrology of Grenville schists near Fernleigh, Ontario. *Journal of Petrology*, **8**, 1–28.

Laird, J. (1980) Phase equilibria in mafic schist from Vermont. *Journal of Petrology*, **21**, 1–37.

Lang, H.M. (1991) Quantitative interpretation of within-outcrop variation in metamorphic assemblage in staurolite-kyanite-grade metapelites, Baltimore, Maryland. *The Canadian Mineralogist*, **29**, 655–671.

Lang, H.M. and Rice, J.M. (1985) Regression modeling of metamorphic reactions in metapelites, Snow Peak, northern Idaho. *Journal of Petrology*, **26**, 889–924.

Liou, J.G., Maruyama, S. and Cho, M. (1985) Phase equilibria and mineral parageneses of metabasites in low-grade metamorphism. *Mineralogical Magazine*, **49**, 321–333.

Mather, J.D. (1970) The biotite isograd and the lower greenschist facies in the Dalradian rocks of Scotland. *Journal of Petrology*, **11**, 253–275.

Nakajima, T. (1982) Phase relations of pumpellyite-actinolite facies metabasites in the Sanbagawa metamorphic belt in central Shikoku, Japan. *Lithos*, **15**, 267–280.

Pattison, D.M., Spear, F.S., Debuhr, C.L., Cheney, J.T. and Guidotti, C.V. (2002) Thermodynamic modeling of the reaction muscovite + cordierite → Al_2SiO_5 + biotite + quartz + H_2O: constraints from natural assemblages and implications for the metapelitic petrogenetic grid. *Journal of Metamorphic Geology*, **20**, 99–118.

Phinney, W.C. (1963) Phase equilibria in the metamorphic rocks of St. Paul Island and Cape North, Nova Scotia. *Journal of Petrology*, **4**, 90–130.

Powell, R. and Sandiford, M. (1988) Sapphirine and spinel relationships in the system $FeO-Al_2O_3-SiO_2-TiO_2-O_2$ in the presence of quartz and hypersthene. *Contributions to Mineralogy and Petrology*, **98**, 64–71.

Ramberg, H. (1952) *The Origin of the Metamorphic and Metasomatic Rocks*. University of Chicago Press, 317 pp.

Read, H.H. (1934) The metamorphic geology of Unst in the Shetland Islands. *Geological Society of London Quarterly Journal*, **90**, 637–688.

Rumble, D. III (1978) Mineralogy, petrology and oxygen isotopic geochemistry of the Clough Formation, Black Mountain, western New Hampshire, U.S.A. *Journal of Petrology*, **19**, 317–340.

Spear, F.S. (1982) Phase equilibria of amphibolites from the Post Pond Volcanics, Mt. Cube quadrangle, Vermont. *Journal of Petrology*, **23**, 383–426.

Spear, F.S. (1993) *Metamorphic Phase Equilibria and Pressure-Temperature-Time Paths*. Mineralogical Society of America Monograph, 799 pp.

Spear, F.S. and Cheney, J.T. (1989) A petrogenetic grid for pelitic schists in the system $SiO_2-Al_2O_3-FeO-MgO-K_2O-H_2O$. *Contributions to Mineralogy and Petrology*, **101**, 149–164.

Spear, F.S., Rumble, D.III and Ferry, J.M. (1982) Linear algebraic manipulation of n-dimensional compositional space. Pp. 53–104 in: *Characterization of Metamorphism through Mineral Equilibria* (J.M. Ferry, editor). Reviews in Mineralogy, **10**, Mineralogical Society of America, Washington, D.C.

Springer, R.K., Day, H.W. and Beiersdorfer, R.E. (1992) Prehnite-pumpellyite to greenschist facies transition, Smartville Complex, near Auburn, California. *Journal of Metamorphic Geology*, **10**, 147–170.

Symmes, G.H. and Ferry, J.M. (1992) The effect of whole-rock MnO content on the stability of garnet in pelitic schists during metamorphism. *Journal of Metamorphic Geology*, **10**, 221–237.

Thompson, A.B. (1976a) Mineral reactions in pelitic rocks: I. Prediction of P-T-X(Fe-Mg) phase relations. *American Journal of Science*, **276**, 401–424.

Thompson, A.B. (1976b) Mineral reactions in pelitic rocks: II. Calculation of some P-T-X (Fe-Mg) phase relations. *American Journal of Science*, **276**, 425–454.

Thompson, J.B. (1972) Oxides and sulfides in regional metamorphism of pelitic schists. *24th International Geological Congress, Montreal. Proceedings Section 10*, 27–35.

Thompson, J.B. (1979) The Tschermak substitution and reactions in pelitic schists (in Russian), In: *Problems of Physiochemical Petrology* Vol. 1 (V.A Zharikov, W.I. Fonarev and S.P. Korikovskii, editors). Moscow Academy of Science, Moscow, Russia.

Todd, C.S. and Evans, B.W. (1994) Properties of CO_2-induced dehydration of amphibolite. *Journal of Petrology*, **35**, 1213–1239.

Turner, F.J. (1948) *Mineralogical and Structural Evolution of the Metamorphic Rocks*. Geological Society of America Memoir **30**, 342 pp.

Turner, F.J. (1968) *Metamorphic Petrology*, 1st edition. McGraw-Hill, New York, 403 pp.

Turner, F.J. and Verhoogen, J. (1951, 1960) *Igneous and Metamorphic Petrology*. McGraw-Hill, New York, 694 pp.

Will, T.M. (1998) *Phase Equilibria in Metamorphic Rocks*. Lecture Notes in Earth Sciences, **71**. Springer-Verlag, New York, 315 pp.

Winkler, H.G.F. (1967) *Petrogenesis of Metamorphic Rocks*, Springer-Verlag, New York, 236 pp.

THE GRAPHICAL ANALYSIS OF MINERAL ASSEMBLAGES IN PELITIC SCHISTS*

James B. Thompson, Jr., *Harvard University, Cambridge, Massachusetts.*

Abstract

The principal minerals used as metamorphic indicators in pelitic schists are, to a first approximation, phases in the system: SiO_2-Al_2O_3-MgO-FeO-K_2O-H_2O. At a given pressure, temperature and humidity (activity of H_2O), quartz-bearing assemblages in this system are dependent on the relative amounts of Al_2O_3, MgO, FeO, and K_2O. These assemblages may be shown three-dimensionally in the tetrahedron: Al_2O_3-MgO-FeO-K_2O. Assemblages which include muscovite as well as quartz may be represented successfully in two dimensions by projecting the compositions of minerals onto some suitable plane with the composition KAl_3O_5 as point of projection. This projection is applicable to rocks over a wide range in bulk composition and through wide range of metamorphic conditions. It also provides a unique characterization of the mineral facies observed in pelitic schists. The principal variation in composition of most of the minerals involved is given by the ratio $MgO/MgO+FeO$. This ratio may commonly be determined with sufficient accuracy by simple measurements of optical or other physical properties.

Introduction

In the classic work of Barrow (1893, 1912) in the Scottish Highlands, and in numerous field and petrographic studies since that time, regional metamorphic zoning has been described and mapped in terms of the successive appearance of certain minerals or mineral assemblages in aluminous schists. Such schists are generally formed by the metamorphism of pelitic sediments. The principal indicator minerals in the order of their appearance in the area studied by Barrow, are: chlorite, biotite, almandite, staurolite, kyanite and sillimanite. Other minerals of pelitic schists have been used as indicators by various authors. A chloritoid isograd has been drawn in the Taconic area by Balk (1953; see also Tilley, 1925). Heald (1950) has drawn an isograd on the first appearance of the pair: orthoclase+sillimanite.

A pelitic schist is not a simple chemical system and the mineralogic variations that have been observed do not lend themselves readily to graphical analysis. Certain components such as ZrO_2, CO_2, S or P_2O_5 are not present in the above mentioned "indicators" in measurable amounts and may be dismissed if we simply omit from consideration those phases in which such components appear in measurable amounts—regarding these latter phases, in effect, as the crucible in which the others react. Even with eliminations of this sort, however, we are still faced with a polyphase, multicomponent system and rather arbitrary simplifi-

---

* Contribution from the Department of Mineralogy and Petrography, Harvard University No. 382.

cation is necessary. Possible consequences of this simplification will be discussed at the end of this paper.

THE CHOICE OF COMPONENTS

To a first approximation the indicator minerals listed above may be regarded as phases in the six-component system: SiO_2-Al_2O_3-MgO-FeO-K_2O-H_2O, a system which also includes minerals such as muscovite, quartz, andalusite, cordierite and others commonly associated with the above indicators. It is apparent that this is but an approximation when we consider the known content of Fe_2O_3 or TiO_2 in a typical biotite, CaO or MnO in garnet, or Na_2O in muscovite or potassic feldspar, even though in many instances it is possible to give the compositions of these minerals in terms of the six components listed above and leave very little of their substance unaccounted for. The success of such an approximation may be measured by its applicability which will be discussed more fully below. However, it must be remembered that in any specific natural occurrence an apparently anomalous mineral association may be related to an unusual abundance, in one or more of the minerals, of some neglected component.

It may seem of little value that the system is now narrowed down to six components, since even at a specified temperature and pressure a three-dimensional model is needed to represent completely the phase assemblages in a system of as few as four components. Though this is true enough, we can profit from the fact that we are not interested in *all possible* assemblages in the system, but in only those that occur in rocks under present consideration. Specifically, the schists and gneisses in which mineral zones have been mapped almost invariably contain quartz, and we are not seriously restricting our field of observation if we omit from the present discussion all assemblages where quartz is not present. Now, since quartz is virtually pure SiO_2, it is clear that an increase or decrease in the amount of SiO_2 in a quartzose rock will be reflected simply in an increase or decrease of quartz, other things being equal, and that the minerals present other than quartz are dependent on the relative amounts of components other than SiO_2.

THE GRAPHICAL STATUS OF H_2O

The restriction that quartz be present thus has the effect of reducing the compositional variables by one so that we are left, for graphical purposes, with a five-component system: Al_2O_3-FeO-MgO-K_2O-H_2O. Were a phase corresponding to any of these other components commonly present we could eliminate it as we did SiO_2, but this does not appear to be the case. It is possible that a fluid phase made of nearly pure H_2O was present in such rocks at one time or another, but it is not a phase that we observe in typical thin sections of a schist or gneiss, nor are there "holes" where such a phase might have been. There is to be sure, even in a schist or gneiss, evidence of H_2O in excess of that combined in the crystalline phases, and evidence that this extra H_2O is concentrated along the grain boundaries, but it would be unwarranted to assume from this that such an intergranular film may be regarded, even approximately, as either pure H_2O chemically or like a bulk aqueous fluid in its physical properties. We shall, however, eliminate H_2O from our graphical analysis, but for rather different reasons.

It is apparent, from comparison of chemical analyses of pelitic sediments (Clarke, 1924, pp. 516–518 and p. 552) with those of typical schists and gneisses derived therefrom (Clarke, p. 553 and pp. 625–626) that such rocks lose H_2O when metamorphosed, and, from consideration of the minerals involved, that the higher the grade of the metamorphism, the greater the loss.

Whether a rock will tend to gain or lose H_2O at a given pressure and temperature is dependent on the activity, a_{H_2O}, or chemical potential, μ_{H_2O}, of H_2O in the rock (or its intergranular film) relative to its immediate surroundings.

If a_{H_2O} or μ_{H_2O}, or the humidity η,[1] in the rock is higher than in its surroundings, the rocks will tend to lose H_2O, and vice versa. In an open system of this sort the stable phase-assemblage will be dependent on the

---

[1] $a_{H_2O} = e^{\dfrac{\mu_{H_2O} - \mu_{H_2O}^{*}}{RT}}$ where $\mu_{H_2O}^{*}$ is the chemical potential of H_2O in some standard

or reference state. The *relative humidity of* meteorologists may be regarded as an activity of H_2O expressed in per cent (rather than on a fractional basis as is usual in chemical literature) and where the standard state is water-saturated air (or distilled water) at the P and T in question. A more useful standard state, for geologic purposes, would be the stable form of pure H_2O at the P and T in question. This would be virtually identical with the meteorological standard state at low pressures and temperatures and have the advantage of being independent of other components present at high temperatures and pressures. The activity of H_2O with standard state so defined will be referred to henceforth as simply the "humidity." More formally:

$$\eta \equiv e^{\dfrac{\mu_{H_2O} - \mu_{H_2O}^{0}}{RT}}$$

where $\mu_{H_2O}^{0}$ is μ_{H_2O} for the stable form of pure H_2O at P, T. There are no other "humidities" in standard geological parlance with which it may be confused, and it is virtually identical with the meteorological relative humidity in a surface or near-surface environment. Furthermore, the intuitive connotations commonly associated with the word "humidity" are qualitatively correct with respect to η but not with respect to μ_{H_2O}. Specifically: a rise in temperature at constant P and η will, in general, tend to dehydrate the system and a rise in temperature at constant P and μ_{H_2O} will tend to hydrate the system.

temperature and pressure, the *activities* or *chemical potentials* of those components to which the system is open, and upon the *relative amounts* of those components to which the system is closed, or, in our specific case, on P, T, η, and the relative amounts of Al_2O_3, MgO, FeO and K_2O. This includes as a limiting case, the possible occurrence of pure H_2O as a separate phase. Such an occurrence would correspond to unit (or 100 per cent) humidity at any given pressure and temperature. This would be the practical maximum value for the humidity at the given pressure and temperature owing to the choice of the stable form of the pure substance (at P, T) as standard state. A humidity less than unity (or 100 per cent) might be visualized, graphically, as the situation which would prevail if a hypothetical form of H_2O, more stable than the actual stable one at the P and T in question, were present in the system. In general the graphical analysis of mineral assemblages neglecting the amount of H_2O relative to other components seems to work. This is consistent either with the presence of pure H_2O as a phase at the time the assemblage formed, or with the assemblage having reached its final state in accord with some externally controlled humidity.

The Tetrahedron: Al_2O_3-MgO-FeO-K_2O

Compositions with respect to the remaining four components, Al_2O_3, MgO, FeO, and K_2O, may be plotted in the tetrahedron of Fig. 1. The portion Al_2O_3-$KAlO_2$-FeO-MgO or "A-K-F-M" has been employed by Barth (1936, also Osberg, 1952) for the illustration of mineral assem-

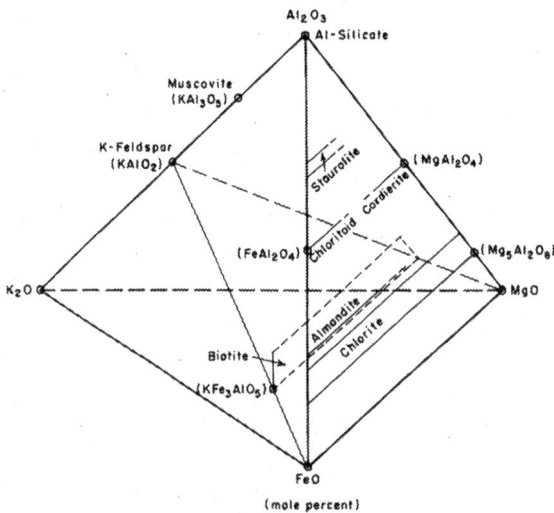

Fig. 1. The system SiO_2-Al_2O_3-MgO-FeO-K_2O-H_2O showing phases stable with quartz at P, T, η, as plotted in tetrahedron Al_2O_3-MgO-FeO-K_2O.

blages in schists. The "A-K-F" diagram of Eskola (1915) is similar but treats (MgO+FeO) as one component. Barth (p. 819) gives practical convenience as the reason for disregarding H$_2$O, but not without misgivings (p. 822). Eskola and Osberg assume H$_2$O to have been in excess. We have justified its elimination on a different basis which includes the latter as a limiting case.

A tetrahedral diagram is cumbersome for written presentation. Barth employed a perspective view developed by H. von Philipsborn (1928), and Osberg showed the Al$_2$O$_3$-FeO-MgO face with biotite projected onto the line FeO-MgO, the point of projection apparently being indefinitely removed from FeO-MgO in the plane KAlO$_2$-FeO-MgO. Barth's diagram is difficult to visualize particularly if several assemblages are to be shown, and Osberg's diagram, which resembles one that we shall describe below, does not clarify the relationships of potash feldspar when it appears in association with aluminous minerals such as almandite or cordierite.

The usefulness of the "A-K-F" diagram of Eskola, as extensively employed by Turner (1948) and other authors, is impaired by the lumping of MgO and FeO as one component. Though Fe^{++} and Mg substitute for each other readily (and perhaps nearly ideally) in solid solutions, they definitely do not behave as one component. Eskola (1915, p. 123) was, to some extent, aware of this, and the point has been clarified and emphasized by Bowen (1925). The ratios MgO/(MgO+FeO), in pairs of coexisting mineral phases typically differ by easily measurable amounts. Furthermore the nature of any system of the type: X-MgO, at any given P and T is commonly quite different from the analogous system: X-FeO. One may, for example, compare the systems: SiO$_2$-MgO (Bowen and Andersen, 1914) and SiO$_2$-FeO (Bowen and Schairer, 1932) at any P and T where both are not entirely liquid.

The approach outlined below will make use of the fact that muscovite, like quartz, is nearly ubiquitous in metamorphosed pelites and rocks of comparable composition. The principal exceptions occur in the inner parts of certain contact aureoles and in the highest grades of regional metamorphism. Consequently we have relatively little interest in assemblages which do not include muscovite. Muscovite, however, is unlike quartz in that its composition is not that of a simple limiting component such as SiO$_2$. The solution of the problem is thus more complicated and may be regarded, graphically, as a projection onto some appropriate surface of the compositions of phases in equilibrium with muscovite.

The Representation of Phases Occurring with Muscovite

The compositions of most natural muscovites, with respect to the components considered, lie at or very near the point labeled "KAl$_3$O$_5$"

in Fig. 1. At any given P, T and η the range of composition of muscovite will thus be a small volume at and near "KAl_3O_5," the surface of which will be terminus to all tie lines, compatibility triangles and compatibility tetrahedra for those assemblages in this system which include muscovite in addition to quartz. The surface of the volume for muscovite compositions should resemble, in general features, that of a wind-faceted pebble, and be characterized by smooth surfaces (termini of tie lines), edges (termini of triangles) and corners (termini of tetrahedra). The opposite ends of the tie lines, opposite edges of the triangles, and opposite faces of the tetrahedra will also constitute a surface giving the compositions of phases compatible with muscovite. The latter surface, consisting of three-phase triangles (three phases, that is, in addition to quartz and muscovite), bundles of two-phase tie lines, and one-phase areas, will be topologically like an ordinary three-component compatibility diagram except that points upon it will not, as a rule, be co-planar. We may, however, construct a map of such a surface by some suitable projection of it. An important consideration in so doing is to avoid, as far as possible, points of projection that would lead to overlapping of different portions of the projected surface. Thus in a simplified, two-dimensional analogue (Fig. 2), F would be a suitable point of projection for phases in equilibrium with phase α, but Q would not. In some instances it may not be possible to avoid overlap in projection. As shown in Fig. 2 this is most likely where the ubiquitous phase has a wide range of compositional variability. Fortunately, muscovite, in terms of the components considered shows but minor variation in composition from $KAl_3Si_3O_{10}(OH)_2$, or "KAl_3O_5" in the tetrahedron of Fig. 1. This means that tie lines from other phases to muscovite will be very nearly radial to "KAl_3O_5," and that the choice of this as point of projection will make any overlap highly unlikely.

A convenient surface to receive the projection is the plane: Al_2O_3-MgO-FeO. Many of the indicator minerals have compositions already lying in or very near it. The essential features of the projection are shown in Fig. 3. Compositions in the sub-tetrahedron A-P-F-M will project onto the area A-F-M, those in the volume N-O-P-F-M onto the extension of the plane A-F-M beyond the line F-M, and those in the sub-tetrahedron B-N-O-P onto the extension of the plane AFM beyond A. Compositions in the plane N-O-P will project to points indefinitely removed from A. We might, for convenience, define compositions in the volume A-F-M-N-O-P as projecting in the *positive* sense onto A-F-M and those in the volume B-N-O-P as projecting in the *negative* sense onto A-F-M. Though Fig. 3 aids in visualization the actual operation is more readily accomplished by formula as indicated in Fig. 4. Mineral assemblages typical of the lower sillimanite zone in west-central New Hampshire are shown in Fig. 5. Nearby rocks in the kyanite zone have similar assemblages with kyanite taking the place of sillimanite. The "negative" part of the projection has been shown in Fig. 5, but is trivial, for many purposes, since each possible assemblage is represented in the positive part even though all bulk compositions are not.

Another convenient plane upon which to project is the plane that includes the edge MgO-FeO and is parallel to the edge Al_2O_3-K_2O (Fig. 6).

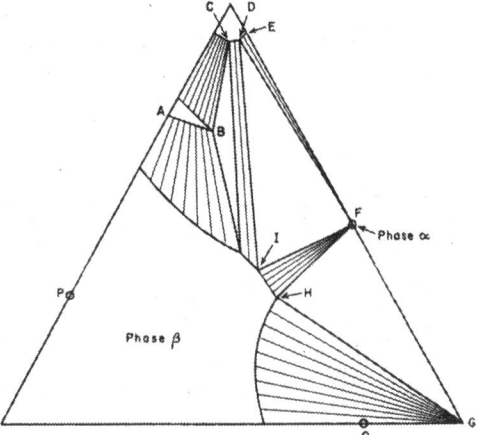

FIG. 2. The phases in equilibrium with phase α, a fixed compound, are shown by the line running G-H-I-D-E. If the compositions of these phases are projected through point F into some straight line, no portions of G-H-I-D-E will be superimposed in the projection. Phases in equilibrium with β, a solution, are given by the line running A-B-C-D-F-G. Projection through a point such as P would result in the portions A-B and B-C being partly superimposed. If a phase has an extreme range of solution it may not be possible in some instances to choose a point such that there will be no overlap. If the phase is a fixed compound, lying on the periphery of the triangle (or surface of the tetrahedron as in the example in the text), then no difficulty can arise if the point of projection be chosen as the composition of the compound, as all tie lines must then be radial to that point.

In this case all compositions in the tetrahedron project in the positive sense, but the lack of resemblance to a standard triangular diagram is perhaps confusing. The method of plotting is shown in Fig. 7 with the projected fields of several minerals indicated, as in Fig. 4. In Fig. 8 the sets of assemblages shown in Fig. 5 are plotted again in the alternative projection. The two projections are entirely equivalent, the first more easily adapted to triangular coordinates and the second to orthogonal coordinates. The first provides relatively more "working room" in the

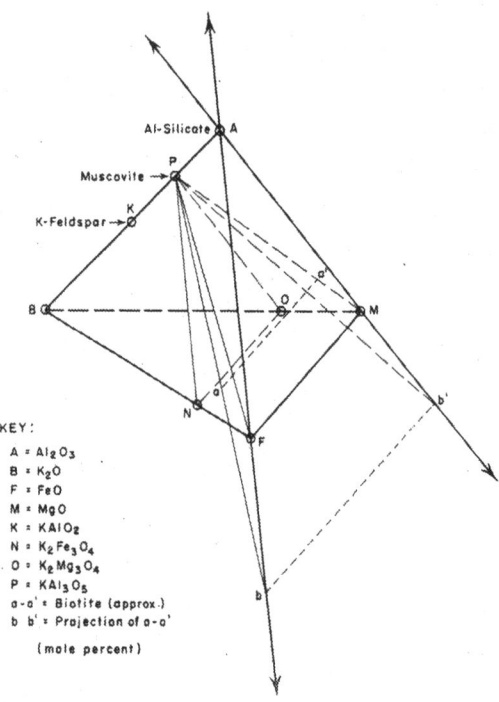

FIG. 3. The tetrahedron Al_2O_3-K_2O-FeO-MgO showing projection through idealized muscovite composition onto plane determined by Al_2O_3, FeO and MgO.

less aluminous range of composition, and the second does the same for the more aluminous range of composition.[1]

Although our concern has been mainly with plotting the compositions of coexistent minerals, a rock composition may also be plotted provided the content of all mineral phases not appearing in the projection has first been subtracted from the total bulk composition. This may be

[1] In a way the above procedures are analogous to the conventional method of presenting high-temperature experimental data in a system such as "FeO"-SiO_2 (Bowen and Schairer, 1932). Actually, as the authors emphasize, this study was of ternary equilibria in the system Fe-Si-O, and the "binary" diagram represents phases in equilibrium with an iron crucible. Both wustite and liquid, in this system, contain more oxygen than in a purely ferrous system, and their compositions are projected, by calculation, onto the line FeO-SiO_2. The calculation of Bowen and Schairer is equivalent to projection along lines parallel to the edge Si-O. Allen and Snow (1955), use a different calculation in their work in the system Ca-Fe-Si-O, and their "ternary" diagram is a projection, through oxygen, onto the plane CaO-FeO-SiO_2, of phases in equilibrium with iron. The relative amount of "Fe_2O_3" in the phases in the above systems is commonly small, hence the method of projection makes little practical difference. A closer analogy to our procedure would be to project through Fe as that is the composition of the ubiquitous phase.

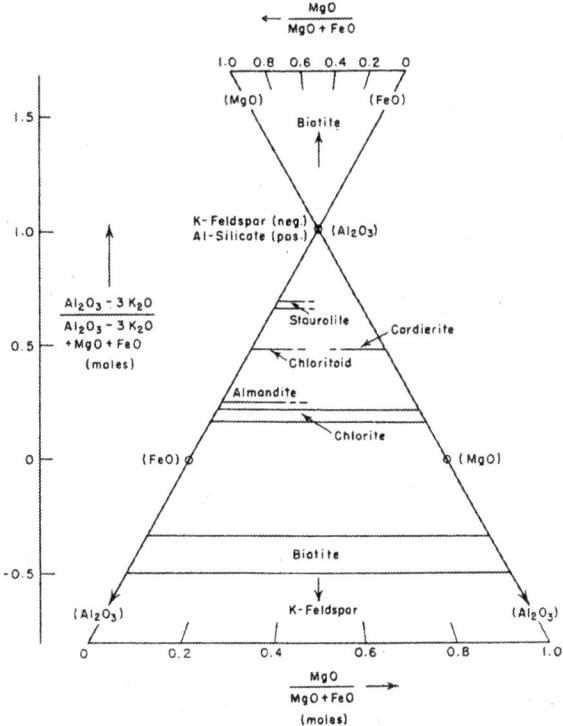

FIG. 4. System SiO$_2$-Al$_2$O$_3$-MgO-FeO-K$_2$O-H$_2$O showing possible phases in equilibrium with quartz and muscovite at P, T, η. Projection is through point KAl$_3$O$_5$ of Fig. 1 onto plane Al$_2$O$_3$-MgO-FeO.

accomplished by direct calculation from the mode, or by making a norm-like calculation based on an assumed distribution of the pertinent components. Though previous authors have emphasized calculations of this sort, the value of making them is doubtful except perhaps for purposes of "predicting" the mineral composition of a rock under metamorphic conditions other than those observed. We can see, in thin section, what phases and assemblages are actually present and, if necessary, can calculate the bulk composition from the mode. For purposes of finding out about the physical environment during metamorphism, however, the important consideration is with regard to the relative compositions of coexisting phases, not their relative amounts. The bulk composition of a given assemblage or the relative modal abundance of the phases, may thus, for present purposes at least, be regarded as simply a pre-metamorphic accident (except with regard to H$_2$O), though admittedly of prime importance in determining the pre-metamorphic nature of the rock.—A plea might also be inserted at this point for the reporting, in petrographic

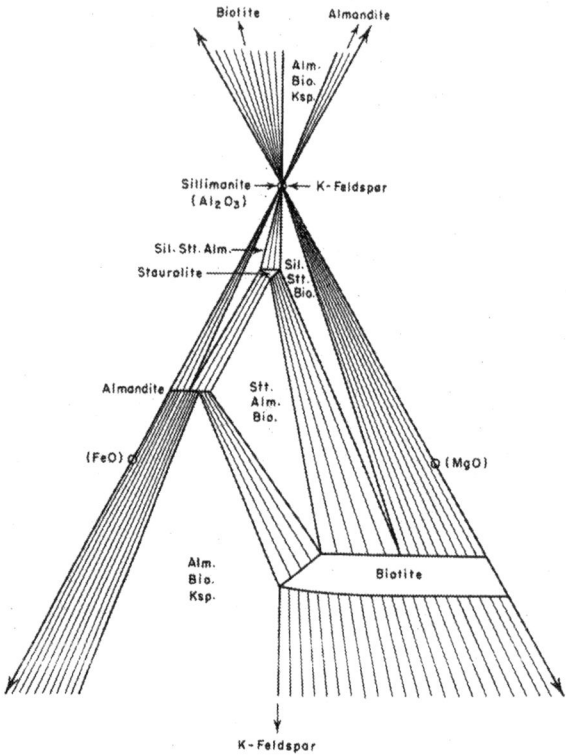

FIG. 5. Phases with muscovite and quartz in the system SiO_2-Al_2O_3-MgO-FeO-K_2O-H_2O as observed (schematically) in the lower sillimanite zone of west-central New Hampshire. Projection as in Figure 4.

descriptions of metamorphic rocks, of specific assemblages rather than just averaged modes. Slides from several specimens may contain mutually incompatible assemblages and much information of physico-chemical value may be lost in the averaging.

THE EFFECTS OF OTHER COMPONENTS

There are two possibilities with regard to a component not considered in the above analysis. One possibility is that the extra component is not present in measurable proportions in any of the phases with which we have been concerned. In this case the mutual equilibrium among these phases is not affected by the presence of the extra component. This component must then be present in some other phase or phases than those appearing in our projection. The total number of coexisting phases containing SiO_2-Al_2O_3-MgO-FeO-K_2O-H_2O and any one other component is limited by the phase rule. Hence, if more than one of these phases contains the extra component, the possible number of phases pertinent to our projection must be correspondingly limited. As long as any of the phases in the projection remain, however, their mutual equilibria are independent of the presence of the extra component and the phases that contain it.

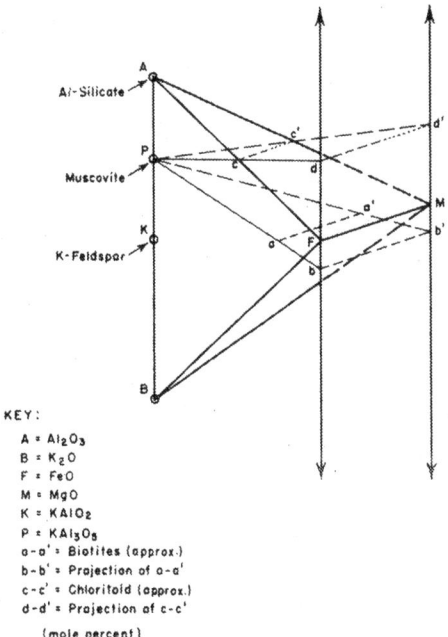

Fig. 6. The tetrahedron Al_2O_3-K_2O-FeO-MgO showing projection through idealized muscovite composition onto plane passing through FeO and MgO and parallel to the line Al_2O_3-K_2O.

The other possibility is that the neglected component *does* enter one of the phases of concern. In this case the mutual equilibria of these phases are affected, and those phases that incorporate the extra component may be stabilized so as to appear in assemblages where they would not otherwise be present. The most likely offenders will be discussed individually.

Na_2O may occur in significant proportions in either potassic feldspar or muscovite and to a lesser and perhaps negligible extent in biotite. The point at which the feldspar would project would not, however, be altered by this substitution, and the other projected phases contain little or no Na_2O. The problem of anomalous stabilization of muscovite or potassic feldspar is also cancelled by the fact that Na_2O is commonly sufficiently abundant that a fundamentally sodic phase is present. This is typically albite in high-grade assemblages or low-grade assemblages poor in alumina, and typically paragonite in the more highly aluminous low-grade assemblages. Where the albite or paragonite contains no extraneous component other than soda the variance of the assemblage is the same as it would be if no soda were present. Quantitatively, however, equilibria involving muscovite or potassic feldspar would be somewhat displaced relative to the corresponding equilibria in a soda-free system.

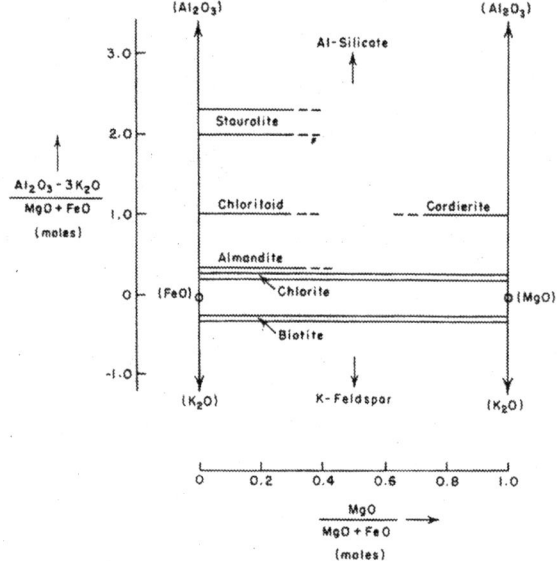

FIG. 7. System SiO_2-Al_2O_3-MgO-FeO-K_2O-H_2O showing possible phases in equilibrium with quartz and muscovite at P, T, η. Projection is through point KAl_3O_5 of Fig. 1 onto plane parallel to edges Al_2O_3-K_2O and MgO-FeO.

The principal difficulty arises with the occurrence of a calcic plagioclase rather than albite.

CaO is also troublesome through the occurrence of a grossularite component in the garnet, and MnO may be significant in either garnet or chloritoid. Occurrences of four of the projected phases in one assemblage are known, though not common, and several examples are known to the writer. In some of these the garnets have been shown to have a significant content of CaO, or MnO, or both. It is probably best, for purposes of the projections discussed herein, to either disregard, or at least treat with caution, any garnet in which the sum of components other than pyrope and almandite is more than a few per cent.

In plotting compositions of minerals containing extraneous components the writer does not simply subtract the CaO or MnO, but also Al_2O_3, K_2O, etc. such that the material subtracted represents an inde-

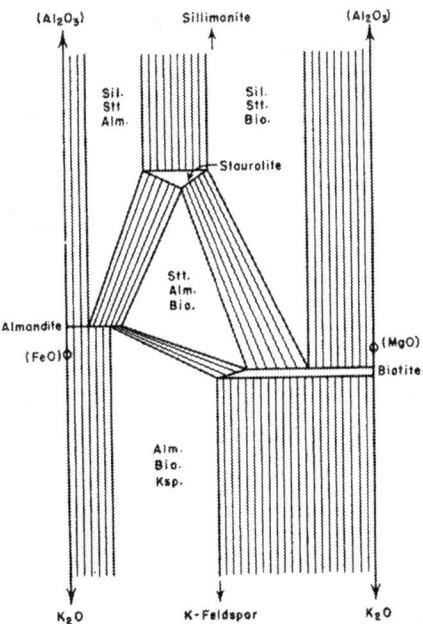

FIG. 8. Phases with muscovite and quartz in the system SiO_2-Al_2O_3-MgO-FeO-K_2O-H_2O as observed (schematically) in the lower sillimanite zone of west-central New Hampshire. Projection as in Fig. 7 and equivalent to Fig. 5.

pendently variable component of the mineral itself. Thus in a garnet analysis[1] yielding:

(a)	$Ca_3Fe_2^{+++}(SiO_4)_3$	2.4	mole per cent
(b)	$Ca_3Al_2(SiO_4)_3$	5.5	mole per cent
(c)	$Mn_3Al_2(SiO_4)_3$	2.1	mole per cent
(d)	$Fe_3^{++}Al_2(SiO_4)_3$	71.7	mole per cent
(e)	$Mg_3Al_2(SiO_4)_3$	18.3	mole per cent

the plotted composition would be on the basis of (d) and (e) recalculated to 100 per cent.

It might be pointed out that if the systems were closed to H_2O and that component not present in excess, then coexistence of four of the projected phases should be much more common than it is. As a rule it is not possible to vary the H_2O-content of most mineral assemblages without the appearance of a new phase. This is not consistent with an inherited bulk-content of H_2O, but is consistent with an externally controlled humidity.

[1] The garnet from the Gassetts schist, Chester, Vermont.

Iron Oxides

Fe_2O_3, as a component, has been omitted from consideration though typically present in small amounts in most ferromagnesian minerals. Ferric or partly ferric phases such as magnetite, hematite, or stilpnomelane are also fairly common in metamorphosed pelites.

The writer has considered the possibility that the activity of oxygen might, like the humidity, be externally controlled. In such a case Fe_2O_3 would behave graphically like FeO, and both magnetite and hematite could be treated as phases in the projection, plotting at the point marked "FeO" in either Fig. 4 or Fig. 7. Stilpnomelane would then project somewhere between the line "FeO-MgO" and the range of chlorite. In general, however, the analysis of pelitic assemblages has been more successful when the system has been treated as closed to oxygen, and Fe_2O_3 kept as a separate component graphically. This is emphatically the case in the common occurrence reported by James and Howland (1955) of *both* magnetite and hematite in metamorphosed iron formation. In any occurrence of this sort the activity of oxygen at any given pressure and temperature is fixed internally, not externally, owing to the presence of two phases in the two-component system Fe-O.

Mineral Facies in Pelitic Schists

In order to construct a complete projection for the indicator minerals at any locality it is necessary that the rocks at that locality show enough variation in bulk composition that all three-phase assemblages are represented (five-phase assemblages counting quartz and muscovite). Owing to the normal variation, from bed to bed, in many clastic sediments, it is quite commonly possible to construct a major portion of such a projection on the basis of one outcrop or group of adjacent outcrops, and it is indeed wise in attempting to do so to find an area where rocks of considerable heterogeneity are intimately interstratified. Where a complete or nearly complete construction can be made it can be said to characterize the *mineral facies* for the pelitic schists at that locality. It is thus not any one assemblage but the set or "ensemble" of assemblages that defines the facies. Only then can we specify that any two rocks of identical composition (except for H_2O) from separated isofacial localities, will have the same mineral composition as required by the definition of Eskola (1915, p. 114). Clearly, many assemblages may occur in more than one facies, and one assemblage is in general insufficient to determine the facies other than within gross limits.

The compositions at the corners of the three-phase fields will, in general, show *continuous variation* from place to place in response to variation in P, T and η at the time of formation of the rocks. In addition

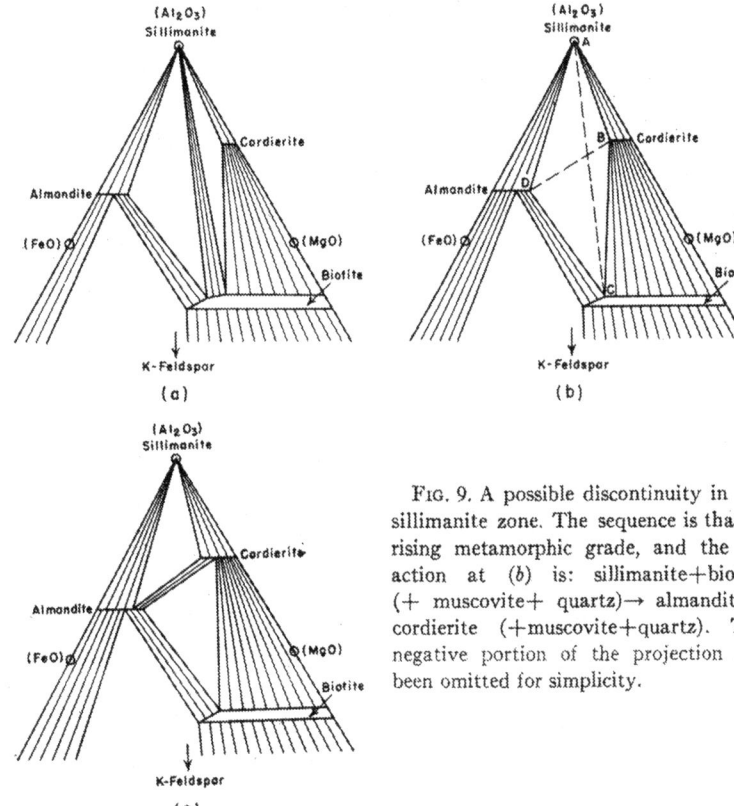

FIG. 9. A possible discontinuity in the sillimanite zone. The sequence is that of rising metamorphic grade, and the reaction at (b) is: sillimanite+biotite (+ muscovite+ quartz)→ almandite+ cordierite (+muscovite+quartz). The negative portion of the projection has been omitted for simplicity.

to such continuous variation in facies there will be discontinuities in facies involving changes in the basic topology of the projection. These may involve the appearance of a new phase, the disappearance of an old one, or a change in the compatible associations as in Fig. 9.

The discontinuities in facies are probably the most suitable changes upon which to draw isograds since they are independent, if observed at all, of local variation in the ratios of the critical components. In Fig. 9b all compositions in the quadrilateral A-B-C-D would show the discontinuity. An isograd based on the simple shift of the boundary of a three-phase field across some specific bulk composition, on the other hand, is clearly dependent on the composition selected as well as on the externally controlled variables: P, T and η.

A discontinuity in facies in this system represents conditions of either univariant or bivariant equilibrium. Where H_2O is gained or lost, as is commonly the case, the equilibrium is bivariant. The dependence of bivariant equilibria upon the externally controlled variables has been discussed by the author in a previous paper (Thompson, 1955). Where H_2O is not actively involved the equilibrium is univariant. Examples of univariant equilibria would be a change from kyanite to sillimanite or a reaction such as: sillimanite+biotite (+quartz+muscovite)→almandite+cordierite (+quartz+muscovite). In the latter example almandite or cordierite but not both (barring extraneous components) may be present on the left hand side (Fig. 9a), and biotite or sillimanite, but not both, on the right hand side (Fig. 9c). This equilibrium is univariant because muscovite and biotite are the only hydrous phases involved as well as the only potassic phases. In both, to a first approximation, H_2O is equal to $2K_2O$, hence, no water is gained or lost in the above reaction.

A complete sequence of projections representing all changes from the chlorite zone to the sillimanite zone in a metamorphic area must clearly, on geometric grounds, involve many independent discontinuities in facies. Furthermore, the sequence determined in one area should not be taken as one that should necessarily be applicable to any other metamorphic terrane. A different path with respect to P, T and η is quite likely to pass through a different sequence of facies.

The number of possible facies is certainly large, even if we regard as topologically "the same" those which may be transformed into one another by continuous variation, and it is not possible to indicate them all without more knowledge than we now have of the compositions of coexisting minerals. Knowledge of the relative values of the ratio $MgO/(MgO+FeO)$ in coexisting pairs limits the topological possibilities and a partial listing in order of increasing tendency to concentrate FeO is:

(1) Cordierite
(2) Biotite
(3) Staurolite
(4) Almandite

Chlorite is probably more "siderophile" than cordierite and less so than biotite. Chloritoid is definitely more siderophile than biotite, chlorite, or cordierite, but is of uncertain status relative to staurolite and almandite. Even if it were known that chloritoid came between staurolite and almandite in the above series, however, the relative topology of these three phases would still be ambiguous, as the chloritoid composition could still lie on either side of the tie line joining the staurolite and garnet occurring with it. Inversions in siderophile tendency are unlikely owing to the near ideality of the substitution of Fe^{++} for Mg. Once the topologic relationships have been determined it will be possible, at least in principle, to establish the possible facies and clarify their sequential relations to one another. The author will present chemical data bearing on this matter in a subsequent paper.

Acknowledgments

The ideas contained herein were presented in part at a symposium on the role of volatiles in geological processes, at Luray, Virginia, in October, 1955, and in part at a symposium on metamorphic facies at Washington, D. C., in March, 1956. Both symposia were sponsored by the Geophysical Laboratory. The writer is grateful to the participants for their stimulating comment and discussion. Recent students of the writer have uncovered much pertinent petrographic information, and have detected many flaws with their questions. The writer is also indebted to John L. Rosenfeld and Marland P. Billings for critical review of the manuscript.

References

ALLEN, W. C., AND SNOW, R. B. (1955), The orthosilicate-iron oxide portion of the system CaO-"FeO"-SiO_2: *Jour. Amer. Ceram. Soc.*, **38**, 264–280.

BALK, R. (1953), Structure of graywacke areas and Taconic Range, east of Troy, New York: *Bull. Geol. Soc. Amer.*, **64**, 811–864.

BARROW, G. (1893), On an intrusion of muscovite-biotite gneiss in the south-east Highlands of Scotland: *Geol. Soc. London, Quart. Jour.*, **49**, 330–358.

BARROW, G. (1912), On the geology of lower Dee-side and the southern Highland border: *Geol. Assoc. Proc.*, **23**, 268–284.

BARTH, T. F. W. (1936), Structural and petrologic studies in Dutchess County, New York, Pt. II: *Geol. Soc. Amer. Bull.*, **47**, 775–850.

BOWEN, N. L. (1925), The mineralogical phase rule: *Jour. Wash. Acad. Sci.*, **15**, 280–284.

BOWEN, N. L., and ANDERSEN (1914), The binary system MgO-SiO_2: *Amer. Jour. Sci.*, 4th ser., **37**, 487–500.

BOWEN, N. L., AND SCHAIRER, J. F. (1932), The system FeO-SiO_2: *Amer. Jour. Sci.*, 5th ser., **24**, 177–213.

CLARKE, F. W. (1924), The data of geochemistry, fifth edition: *U. S. Geol. Survey Bull.* **770**, 841 pp.

ESKOLA, P. (1915), On the relation between chemical and mineralogical composition in the metamorphic rocks of the Orijärvi region: *Comm. Géol. Finlande*, Bull. **44**, 145 pp.

HARKER, A. (1939), Metamorphism, second edition, revised: Methuen, London, 362 pp.

HEALD, M. T. (1950), Structure and petrology of the Lovewell Mountain quadrangle, New Hampshire: *Geol. Soc. Amer. Bull.*, **61**, 43–90.

JAMES, H. L., AND HOWLAND, A. L. (1955), Mineral facies in iron- and silica-rich rocks: *Geol. Soc. Amer. Bull.*, **66**, 1580–1588.

JUURINEN, A. (1956), Composition and properties of staurolite: *Ann. Acad. Sci. Fennicae:* Ser. A-III, **47**, 53 pp.

OSBERG, P. H. (1952), The Green Mountain anticlinorium in the vicinity of Rochester and East Middlebury, Vt.: *Vermont Geol. Surv.*, Bull. **5**, 127 pp.

PHILIPSBORN, H. VON (1928), Zur graphischen Behandlung quartenär systeme: *N. Jahrb. f. Min. Geol. Paläont.* Beil-Bd. **57**, Abt. A. 973–1012.

THOMPSON, J. B. (1955), The thermodynamic basis for the mineral facies concept: *Am. Jour. Sci.*, **253**, 65–103.

TILLEY, C. E. (1925), Petrographical notes on some chloritoid rocks: *Geol. Mag.*, **62**, 309–318.

TURNER, F. J. (1948), Mineralogical and structural evolution of the metamorphic rocks: *Geol. Soc. Amer.* Memoir **30**, 342 pp.

CHAPTER 6: MIXED VOLATILES

Greenwood, H.J. (1962) Metamorphic reactions involving two volatile components. *Carnegie Institution of Washington, Yearbook,* **61**, 82–85.

Among the gas species that constitute metamorphic pore fluids, H_2O and CO_2 are the most important, not simply because they commonly dominate all others, but also because they are reactive: they combine with minerals to form hydrates and carbonates. Metamorphic rocks such as pure and impure metacarbonates, calcareous pelites, metaironstones, and metaperidotites may contain both hydrates (e.g. talc, amphiboles, micas, serpentines, brucite, humites, etc.) and carbonates (e.g. calcite, aragonite, dolomite, ankerite, magnesite, siderite). Equilibrium among these and other minerals is a function of temperature, pressure, and the chemical potentials of H_2O and CO_2.

Bowen (1940) understood the role of H_2O in diluting the concentration of CO_2 in the pore-fluid of meta-carbonate rocks, thereby lowering the temperatures of decarbonation reactions. But, as we saw in Chapter 4, Bowen encountered problems when he tried to accommodate the hydrate tremolite in the siliceous dolomite system. Many years elapsed before laboratory experiments on magnesite (Harker, 1958) and calcite + quartz (Walter *et al.*, 1962) calibrated the temperature effects of non-reactive gases such as H_2O and Ar on carbonate breakdown reactions. CO_2 was shown to have an analogous effect on the dehydration of brucite (Walter *et al.*, 1962). Mindful of the enormous volumetric differences between rock and pore fluid, Wyllie (1962) constructed phase diagrams comparing the dissociation behaviour of carbonate minerals in the presence of an inert gas component, in closed systems, and in systems open to the transfer of fluid components. It was becoming clear that the past assumption of many petrologists equating the pressure of a single volatile species to total pressure could no longer be considered generally valid.

In the landmark article presented here, a report of work in progress, Hugh Greenwood showed that the isobaric $\partial T/\partial X$(fluid) slopes of mineral reactions involving reactive and non-reactive gases in binary fluids are determined, for ideal gas mixtures, simply by their reaction stoichiometries and reaction entropies.

Prior to this article, the graphical treatment of reactive, two-volatile metamorphic systems took the form of reaction lines drawn in isobaric, isothermal diagrams with the chemical potentials of the volatile species as abscissa and ordinate (e.g. Korzhinskii 1959; Zen, 1961). Clearly, much more useful in the context of metamorphic field gradients is a phase diagram depicting equilibrium reactions as a function of at least one of the determining variables temperature and pressure, as well as the proportions of the two principal gas species. Greenwood presented an expression for the slope $\partial T/\partial X_2$ at constant pressure for reaction curves in the presence of an ideal two-component, one-phase fluid (Fig. 22), where X_1 and X_2 are the mole fractions of gas components 1 and 2, and $X_1 + X_2 = 1$. The gas components are conventionally taken to be the molecular species H_2O and CO_2, not H-O-C. He simplified the logic statements in his paper by making the stoichiometric coefficients of the gas components numerically analogous to their mole fractions; this is not a necessary step, but of course in any expression of thermodynamic equilibrium all the extensive properties must match the balanced reaction. What Greenwood showed was that various types of crystal-vapour reaction give rise to a network of curves with contrasting $\partial T/\partial X_2$ slopes that constitute first-order constraints on the topology of the phase diagram for any two-volatile system, especially when taken in conjunction with the rules of Schreinemakers. A corresponding expression for the slope $\partial P/\partial X_2$ at constant temperature, and a derivation of these equations, was given in a subsequent contribution (Greenwood 1967). The generalized T-X_{CO_2} diagram in that paper (Fig. 1), showing the five distinctive reaction types, is the one that has most frequently been reproduced.

Greenwood's basic T-X_{CO_2} diagram must count as one of the most elegant and profound of all phase diagrams in petrology. Its enormous relevance to metamorphic reactions was obvious, although in the short 1962 report he largely confined himself to basically one observation: the possibility that isograds based on mixed-volatile reactions would map out in a crossed fashion resembling the T-X diagram. Not many years later, Carmichael (1970) found this to be the case in the Whetstone Lake area, Ontario.

The new T-X_{CO_2} diagram showed that it was not so easy to predict the sequence of potential isogradic reactions taking place with increasing temperature in the progressive metamorphism of something like a siliceous dolomite, as Bowen had tried to do. The sequence will differ according to the kind of fluid behaviour that applies in each terrane. Two end-member models can be envisaged in the context of the T-X_{CO_2} diagram: (1) internal buffering of the pore fluid composition in the presence of an isobaric (or isothermal) univariant mineral assemblage, for example, calcite + wollastonite + quartz; and (2) external control of fluid composition as a result of pervasive and overwhelming fluid infiltration, with prevailing isobaric (or isothermal) divariant assemblages. Greenwood (1967) thought that, more often than not, internal buffering of the fluid composition was likely to be the case in nature. Figure 10 in that paper was the first published T-X_{CO_2} diagram showing possible buffering paths during progressive metamorphism, and how these might depend on fluid/rock proportions and the consumption of reactant minerals. Phase diagrams in the same paper also showed the limited stability of hydrate minerals like serpentine in CO_2-bearing hydrous fluid, and the extraordinary lowering of the temperature for others, such as anthophyllite, in the presence of increasing proportions of CO_2.

Greenwood's 1962 article initiated a flurry of activity in two related areas of research: laboratory investigation of the T, P and X_{CO_2} conditions of mineral-fluid equilibria in simple systems approximating siliceous dolomites and ultramafic rocks, and field investigations of the petrology of metamorphic rocks involved in progressive two-volatile reactions. In laboratories worldwide including Greenwood's own, experimental studies determined X_{CO_2} at known T and P for specific reactions. From these investigations we were able to build full petrogenetic T-X_{CO_2} grids for metamorphic systems evolving two volatile components (e.g. Skippen, 1974), and, equally important, acquire valuable constraints for subsequent use in thermodynamic databases of minerals (Chapter 11). Kerrick (1974) provided a comprehensive review of the topic. The experimental results reconciled the apparent disagreements in earlier studies regarding the first isogradic reaction in siliceous dolomites: forming tremolite + calcite (Bowen, 1940), talc + calcite (Tilley, 1948) or diopside (Engel and Engel, 1960).

The progress of isobaric univariant reactions in exclusively fluid-buffered systems is predictably small, to the point where it may be hard to detect at all, as shown quantitatively for metamorphosed siliceous dolomites by Greenwood (1975). Abrupt mineral appearances or disappearances (potential isograds) in fluid-buffered metamorphism can be expected instead at isobaric points in the T-X_{CO_2} diagram, i.e. across the corresponding univariant curves in the PT diagram. Major reaction progress of isobaric univariant assemblages, to the point where one mineral is eliminated, is to be expected in the case of external control of fluid composition by means of massive fluid infiltration. Significant progress of reactions that have very steep or vertical slopes in the T-X_{CO_2} diagram necessarily implies infiltration of reactive fluid, because they cannot be driven by the addition (or loss) of heat.

Field studies in many locations (e.g. Trommsdorff, 1972; Kerrick et al., 1973; Moore and Kerrick, 1976; Ferry, 1976; Rice, 1977a,b; Suzuki, 1977; Jansen et al., 1978; and many others, see Rice and Ferry, 1982, Table 1) agreed with Greenwood's (1967, 1975) prediction: that the frequency of occurrence of isobaric univariant assemblages and, in some cases, the field distribution of isobaric invariant assemblages, were consistent with control of the pore fluid by buffering as shown on the T-X_{CO_2} diagram. However, when obvious progress of the isobaric univariant reaction can be demonstrated, we would expect the calculated equilibrium fluid composition in the buffering model to match the composition of the fluid evolved. Rice and Ferry (1982, Table 2) found that this was seldom the case. Fluid infiltration had to play a role as well. The magnitude of possible infiltration by an external fluid while the buffer assemblage still held requires a measurement of reaction progress. John Ferry accepted this research challenge and published an extended series of contributions beginning in 1978 and continuing to the present day (see Chapter 12). Ferry was a direct beneficiary of the process implications in Greenwood's T-X_{CO_2} diagram.

Considering that most metamorphic reactions involve the gain or loss of a fluid component, it is not hard to see that the thermodynamic basis for phase equilibria in mixed volatile systems (Eugster and Skippen, 1967) and its graphical expression (Greenwood 1962, 1967) have great relevance to our understanding of the controls of metamorphism. Whereas the qualitative implications of T-X(fluid) diagrams were important, their quantitative calculation, as described for example by Skippen and Carmichael (1977) and Ferry and Burt (1982), soon became essential for most studies. This required thermodynamic data for minerals and fluid species (see Chapter 11), including the mixing properties of the fluid. Skippen (1975) showed that isothermal-isobaric plots of mixed-volatile reactions in terms of the log fugacity of two fluid species, analogous to Korzhinskii's μ-μ diagrams, allowed the display of mixed-volatile reaction boundaries where gas pressures totalled less than the system pressure.

Much as the Thompson AFM-projection became a necessary adjunct to any petrological study of metapelites, so the T-X_{CO_2} diagram in later years became the visual aid of choice in most papers on metacarbonate rocks.

Because the dP/dT slopes of devolatilization reactions are mostly positive, the isothermal P-X_{CO_2} diagram resembles an upside-down copy of the T-X_{CO_2} diagram. Happily, the recognition of quasi-isothermal decompression in the history of collisional belts (Chapter 9) rescued the P-X_{CO_2} diagram from total obscurity. It was used by Franz and Spear (1983) for carbonate rocks in the Tauern window, Austria, and it may be that others have used it.

Greenwood's graphical treatment of mixed volatile systems of course applies equally well to reactions of minerals with other binary gas mixtures. Under reducing conditions, hydrogen becomes a significant fluid species in the system OH, in which case reaction temperatures can usefully be plotted against the proportions of H_2 and H_2O. Greenwood (1975) depicted the redox phase relations of the ferrous mica annite on such a diagram. As he indicated, from the point of view of the forces driving reactions, this made more sense than the conventional T-$\log f_{O_2}$ diagram.

To accommodate redox as well as dehydration and decarbonation reactions in metamorphic iron-formation, Frost (1979) plotted isobaric diagrams using the variables T, T-$\log f_{O_2}$, and X_C, where $X_C = X_{CO_2} + X_{CO} + 1/3 X_{CH_4}$. Connolly (1995) expanded on this approach by showing that characterization of C-O-H fluid in terms of the single compositional variable X_O^f, where $X_O^f = n_O/(n_O + n_H)$ and n_O and n_H are atomic proportions of oxygen and hydrogen in the fluid in projection from C, enabled construction of P-T-X_O^f diagrams that are analogous to the familiar P-T-X_{CO_2} diagrams of Greenwood. This approach avoided problems arising from having to express the C-O-H fluid in terms of the molecular species present, of which there are several, depending on the fO_2. His isobaric T-X_O^f fluid diagrams simultaneously portrayed redox, dehydration, and decarbonation in a model system relevant to graphitic pelites and metamorphosed iron-formation (Connolly, 1995). Connolly also used his diagram to depict reactions relevant to serpentinized peridotite simultaneously under relatively high and low redox conditions in the system C-O-H-Fe-MgO-SiO_2 (Connolly, 1994).

The isobaric T-X_{CO_2} diagram was successfully applied to model mineral zones and isogradic reactions (Skippen,

1974; Greenwood, 1975) in numerous examples of the contact metamorphism of metacarbonate, ultramafic and meta-ironstone rocks. For regional terranes, however, multiple T-X_{CO_2} diagrams were needed to cover the range in pressure. Experiments (Eggert and Kerrick, 1981) and calculations (Trommsdorff and Evans, 1977, Fig. 8; Franz and Spear, 1983, Fig. 6; Evans and Guggenheim, 1988, Fig. 20) began to show that the P, T and X_{CO_2} coordinates of some mixed-volatile reactions underwent dramatic, correlated changes within the PT range of metamorphism. It was clearly time to learn more about mixed-volatile reactions in the P-T plane.

Acting as if on cue, the authors of three papers published in the same year (Baker *et al.*, 1991; Connolly and Trommsdorff, 1991; Carmichael, 1991) discussed the interesting chemographic/topological issues that arise when locating mixed-volatile reactions on a PT phase diagram. Reaction curves on the mixed-volatile PT-diagram are projections from the full system P-T-X_f diagram; the more familiar invariant points on the isobaric T-X_f diagram represent the intersection of the full-system univariant PT-lines with the isobaric plane in that diagram. The trouble with the univariant reactions is that, unlike those in one-volatile systems, their stoichiometries (mass balances) vary continuously with P and T, in concert with changes in X_{CO_2} in reactions in adjacent divariant fields. The stoichiometry determined from a linear combination of any two isobaric univariant curves intersecting at a point on the T-X_f diagram (Greenwood, 1975) is only valid for the fluid composition at that point on that diagram. This is broadly analogous to univariant reactions in systems involving multiple mineral solid solutions, such as the system FMAS as described by Hensen (1971, see Chapter 8). As a result, reaction coefficients can change sign (at singular points) along the PT-curve, altering the curve labels, i.e. which minerals are on the 'high' and 'low' side of the curve. These effects impart strong curvature to some of the univariant lines, giving enhanced value to the PT projection as a petrogenetic grid. Fluid compositions change along the curves from high X_{CO_2} at low pressure to very low values at high pressure, a consequence of the generally contrasting dP/dT slopes of dehydration and decarbonation reactions (Carmichael, 1991). Good examples are provided by the univariant assemblage Tlc-Mag-En-Fo in the MSH system and Tlc-Tr-Qtz-Cal-Do in the CMAS system (Baker *et al.*, 1991, Figs 2 and 10). Labour-intensive calculations are no longer necessary (to my relief!) to calculate the univariant PT curves in mixed-volatile systems. Carmichael (1991) outlined a simple algorithm for the iterative calculation of the X(fluid) of a univariant curve as a function of T and P. For a complete PT phase diagram, computer programs such as THERMOCALC (Powell and Holland, 1988) and VERTEX (Connolly and Trommsdorff, 1991) are obviously much faster. Baker *et al.*, (1991) showed that the mixed-volatile PT-projection can be used as a template for pseudosection or phase-assemblage diagrams (Chapter 8).

It would seem safe to predict that mixed-volatile phase diagrams, supported by the measurement of mineral modes and compositions, will continue to play a central role in the quantitative study of fluid buffering and transport in metamorphism.

References

Baker, J., Holland, T.J.B. and Powell, R. (1991) Isograds in internally buffered systems without solid solutions: principles and examples. *Contributions to Mineralogy and Petrology*, **106**, 170–182.

Bowen, N.L. (1940) Progressive metamorphism of siliceous limestone and dolomite. *Journal of Geology*, **48**, 225–274.

Carmichael, D.M. (1970) Intersecting isograds in the Whetstone Lake area, Ontario. *Journal of Petrology*, **11**, 147–181.

Carmichael, D.M. (1991) Univariant mixed-volatile reactions: Pressure-temperature phase diagrams and reaction isograds. *The Canadian Mineralogist*, **29**, 741–754.

Connolly, J.A.D. (1994) Phase diagrams for graphitic rocks. Pp. 128–156 in: *Phase Diagram Applications in Earth Materials Science* (R. Funiciello, C.A. Ricci and V. Trommsdorff, editors). VII Summer School Proceedings, University of Siena, Siena,

Connolly, J.A.D. (1995) Phase diagram methods for graphitic rocks and application to the system C-O-H-FeO-TiO$_2$-SiO$_2$. *Contributions to Mineralogy and Petrology*, **119**, 94–116.

Connolly, J.A.D. and Trommsdorff, V. (1991) Petrogenetic grids for metacarbonate rocks: pressure-temperature phase-diagram projection for mixed volatile systems. *Contributions to Mineralogy and Petrology*, **108**, 93–105.

Eggert, R.G. and Kerrick, D.M (1981) Metamorphic equilibria in the siliceous dolomite system: 6 Kb experimental data and geologic applications. *Geochimica et Cosmochimica Acta*, **45**, 1039–1049.

Engel, A.E. and Engel, C.G. (1960) Progressive metamorphism and granitization of the major paragneiss, northwest Adirondack Mountains, New York. *Geological Society of America Bulletin*, **71**, 1–58.

Eugster, H.P and Skippen, G.B (1967) Igneous and metamorphic reactions involving gas equilibria. Pp. 492–520 in: *Researches in Geochemistry* **2** (P.H. Abelson, editor). Wiley, New York.

Evans, B.W. and Guggenheim, S. (1988) Talc, pyrophyllite, and related minerals. Pp. 225–294 in: *Hydrous Phyllosilicates* (S.W. Bailey, editor). Reviews in Mineralogy, **19**, Mineralogical Society of America, Washington, D.C.

Ferry, J.M. (1976) Metamorphism of calcareous sediments in the Waterville-Vassalboro area, south-central Maine: mineral reactions and graphical analysis. *American Journal of Science*, **276**, 841–882.

Ferry, J.M. and Burt, D.M. (1982) Characterization of metamorphic fluid composition through mineral equilibria. Pp. 207–262 in: *Characterization of Metamorphism through Mineral Equilibria* (J.M. Ferry, editor). Reviews in Mineralogy, **10**, Mineralogical Society of America, Washington, D.C.

Franz, G. and Spear, F.S. (1983) High pressure metamorphism of siliceous dolomites from the central Tauern window, Austria. *American Journal of Science*, **283-A** (Orville Volume), 396–413.

Frost, B.R. (1979) Mineral equilibria involving mixed volatiles in a C-O-H fluid phase: the stabilities of graphite and siderite. *American Journal of Science*, **279**, 1033–1059.

Greenwood, H.J. (1967a) Mineral equilibria in the system MgO-SiO$_2$-H$_2$O-CO$_2$. Pp. 542–567 in *Researches in Geochemistry* **2** (P.H. Abelson, editor). Wiley, New York.

Greenwood, H.J. (1975) Buffering of pore fluids by metamorphic reactions. *American Journal of Science*, **275**, 573–593.

Harker, R.I. (1958) The system MgO-CO$_2$-argon, and the effect of inert pressure on certain types of hydrothermal reaction. *American Journal of Science*, **256**, 128–138.

Hensen, B.J. (1971) Theoretical phase relations involving cordierite and garnet in the system MgO-FeO-Al$_2$O$_3$-SiO$_2$. *Contributions to Mineralogy and Petrology*, **33**, 191–214.

Jansen, J.H.B., Kraats, A.H., Rijst, H. and Schuiling, R.D. (1978) Metamorphism of siliceous dolomites at Naxos, Greece. *Contributions to Mineralogy and Petrology*, **67**, 279–288.

Kerrick, D.M. (1974) Review of metamorphic mixed volatile (H$_2$O-CO$_2$) equilibria. *American Mineralogist*, **59**, 729–762.

Kerrick, D.M, Crawford, K.E. and Randazzo, A.F. (1973) Metamorphism of calcareous rocks in three roof pendants in the Sierra Nevada, California. *Journal of Petrology*, **14**, 303–325.

Korzhinskii, D.S. (1959) *Physicochemical Basis of the Analysis of the Paragenesis of Minerals*. Consultants Bureau, Inc., New York, 142 pp.

Moore, J.M. and Kerrick, D.M. (1976) Equilibria in the siliceous dolomites of the Alta aureole, Utah. *American Journal of Science*, **276**, 502–524.

Powell, R. and Holland, T.J.B. (1988) An internally consistent thermodynamic dataset with uncertainties and correlations: 3. Applications to geobarometry, worked examples and a computer program. *Journal of Metamorphic Geology*, **6**, 173–204.

Rice, J.M. (1977a) Progressive metamorphism of impure dolomitic limestone in the Marysville aureole. *American Journal of Science*, **277**, 1–24.

Rice, J.M. (1977b) Contact metamorphism of impure dolomitic limestone in the Boulder aureole, Montana. *Contributions to Mineralogy and Petrology*, **59**, 237–259.

Rice, J.M. and Ferry, J.M. (1982) Buffering, infiltration, and the control of intensive variables during metamorphism. Pp. 263–326 in: *Characterization of Metamorphism through Mineral Equilibria* (J.M. Ferry, editor). Reviews in Mineralogy, **10**, Mineralogical Society of America, Washington, D.C.

Skippen, G. (1974) An experimental model for low pressure metamorphism of siliceous dolomitic marble. *American Journal of Science*, **274**, 487–509.

Skippen, G.B. (1975) Thermodynamics of experimental sub-solidus systems including mixed volatiles. *Fortschrift in Mineralogie*, **52**, 75–99.

Skippen, G.B. and Carmichael, D.M. (1977) Mixed volatile equilibria. Pp. 109–125 in: *Application of Thermodynamics to Petrology and Ore Deposits* (H.J. Greenwood, editor). Mineralogical Association of Canada, Short Course Handbook, **2**, 230 pp.

Suzuki, K. (1977) Local equilibrium during the contact metamorphism of siliceous dolomites in Kasuga-mura, Japan. *Contributions to Mineralogy and Petrology*, **61**, 79–89.

Tilley, C.E. (1948) Earlier stages in the metamorphism of siliceous dolomites. *Mineralogical Magazine*, **28**, 272–276.

Trommsdorff, V. (1972) Change in T-X during metamorphism of siliceous dolomite rocks of the central Alps. *Schweizerische Mineralogische und Petrographische Mitteilungen*, **52**, 567–571.

Trommsdorff, V. and Evans, B.W. (1977) Antigorite-ophicarbonates: Phase relations in a portion of the system $CaO-MgO-SiO_2-H_2O-CO_2$. *Contributions to Mineralogy and Petrology*, **60**, 39–56.

Walter, L.S., Wyllie, P.J. and Tuttle, O.F. (1962) The system $MgO-CO_2-H_2O$ at high pressures and temperatures. *Journal of Petrology*, **3**, 49–62.

Wyllie, P.J. (1962) The effect of 'impure' pore fluids on metamorphic dissociation reactions. *Mineralogical Magazine*, **33**, 9–25.

Zen, E-an (1961) The zeolite facies: An interpretation. *American Journal of Science*, **259**, 401–409.

Metamorphic Reactions Involving Two Volatile Components

H. J. Greenwood

Many metamorphic reactions involve more than one volatile or mobile component and are therefore influenced by pressure, temperature, and the composition of the coexisting fluid phase. The equilibrium relationships may be portrayed in a variety of ways, for example, by plotting the chemical potentials of the mobile components against one another at constant temperature and pressure (Korzhinskii, 1959; Zen, 1961). Alternatively, the situation may be represented on an isobaric T-x diagram, on which are plotted the temperature and the composition of the coexisting fluid phase, the other components being regarded as nonvolatile or immobile. This kind of diagram has some advantages over the chemical potential, or μ_i versus μ_j, diagram, not the least of which is its direct use of the measurable variables temperature, pressure, and composition.

Equations have been derived for the equilibrium boundaries between reacting phase assemblages in such systems. These have the same form as the usual expressions for crystal-liquid equilibria, but they do not carry the restriction that the relative proportions of the two volatile components are limited by the proportions of the other components. The effect of removing this restriction is to make stable many reactions that would normally be regarded as metastable.

The slope of an equilibrium boundary for a reaction taking place at constant pressure in the presence of a one-phase binary fluid having zero enthalpy of mixing is

$$\left(\frac{\partial T}{\partial x_2}\right)_P = \frac{RT}{\Delta S}\left(\frac{\nu_2}{x_2} - \frac{\nu_1}{x_1}\right)$$

where x_2 is the mole fraction of component 2 in the fluid and ν_2 is the stoichiometric coefficient of component 2 in the reaction. All reactions that take place in such systems can be expressed by the general relation

$$aA \rightarrow bB + \nu_1 + \nu_2$$

in which a moles of solid phases A react to give b moles of solid phases B and ν_1 and ν_2 moles of the volatile components 1 and 2, respectively. The equation of the reaction should be written so that

$$|\nu_1| + |\nu_2| = 1 \quad \text{and} \quad \nu_1 + \nu_2 \gtrless 0$$

to make the stoichiometric coefficients equivalent to the mole-fraction composition of the gas given off in the reaction.

Inspection of these equations reveals several points of interest to metamorphic petrology. If $\nu_2 = 0$,

$$\left(\frac{\partial T}{\partial x_2}\right)_P = \frac{RT}{\Delta S}\left(-\frac{1}{x_1}\right) \quad (1)$$
$$< 0$$

If $\nu_1 = 0$,

$$\left(\frac{\partial T}{\partial x_2}\right)_P = \frac{RT}{\Delta S}\left(\frac{1}{x_2}\right) \quad (2)$$
$$> 0$$

If $\nu_1 > 0$ and $\nu_2 > 0$,

$$(\partial T/\partial x_2)_P = 0 \quad (T_{\max})$$

where $\nu_1 = x_1, \nu_2 = x_2$. (3)

If $\nu_1 = -\nu_2$, equal amounts of components 1 and 2 appear on opposite sides of the reaction, and their entropies tend to cancel, making ΔS for the reaction small and $(\partial T/\partial x_2)_P$ correspondingly large. Accordingly, as

$$\Delta S \rightarrow 0, \quad (\partial T/\partial x_2)_P \rightarrow \infty \quad (4)$$

If $-1 < \nu_1 < 0, 1 > \nu_2 > 0$
$$(|\nu_1| < |\nu_2|),$$

$$+\infty > \left(\frac{\partial T}{\partial x_2}\right)_P > \frac{RT}{\Delta S}\left(\frac{1}{x_2}\right) \quad (5)$$

If $1 > \nu_1 > 0, -1 < \nu_2 < 0$
$$(|\nu_1| > |\nu_2|),$$

$$-\infty < \left(\frac{\partial T}{\partial x_2}\right)_P < \frac{RT}{\Delta S}\left(-\frac{1}{x_1}\right) \quad (6)$$

The importance of these rather terse statements to metamorphic petrology can best be appreciated by examining some geologically interesting reactions that lend themselves to this treatment. Equation 1 describes a reaction in which only component 1 is given off (H_2O in fig. 22). As an example of such a reaction we might take

$$\underset{\text{Tremolite}}{Ca_2Mg_5Si_8O_{22}(OH)_2} \rightarrow \underset{\text{Diopside}}{2CaMgSi_2O_6} +$$
$$\underset{\text{Enstatite}}{3MgSiO_3} + \underset{\text{Quartz}}{SiO_2} + H_2O$$

Equation 2 describes a reaction in which only component 2 is given off (CO_2 in fig. 22). Example (see fig. 22):

$$\underset{\text{Magnesite}}{MgCO_3} \rightarrow \underset{\text{Periclase}}{MgO} + CO_2$$

Equation 3 describes a reaction in which both volatile components are given off, such as

$$\tfrac{1}{4}\underset{\text{Tremolite}}{Ca_2Mg_5Si_8O_{22}(OH)_2} + \tfrac{3}{4}\underset{\text{Calcite}}{CaCO_3} +$$

$\tfrac{1}{2}SiO_2 \rightarrow \tfrac{1}{4}CaMgSi_2O_6 + \tfrac{3}{4}CO_2 + \tfrac{1}{4}H_2O$
Quartz — Diopside

(See fig. 22, T_{max} at $x_{CO_2} = 0.75$.)

Equation 4 describes a reaction such as

$Mg(OH)_2 + CO_2 \rightarrow MgCO_3 + H_2O$
Brucite — Magnesite

(See fig. 22, vertical boundary.)

Equation 5 describes a reaction such as

$\tfrac{5}{8}CaMg(CO_3)_2 + \tfrac{7}{8}SiO_2 + \tfrac{1}{8}H_2O \rightarrow$
Dolomite — Quartz

$\tfrac{1}{8}Ca_2Mg_5Si_8O_{22}(OH)_2 + \tfrac{3}{8}CaCO_3 + \tfrac{7}{8}CO_2$
Tremolite — Calcite

Equation 6 describes a reaction such as

$4H_4Mg_3Si_2O_9 + 9CaCO_3 + 5CO_2 \rightarrow$
Serpentine — Calcite

$Ca_2Mg_5Si_8O_{22}(OH)_2 +$
Tremolite

$7CaMg(CO_3)_2 + 7H_2O$
Dolomite

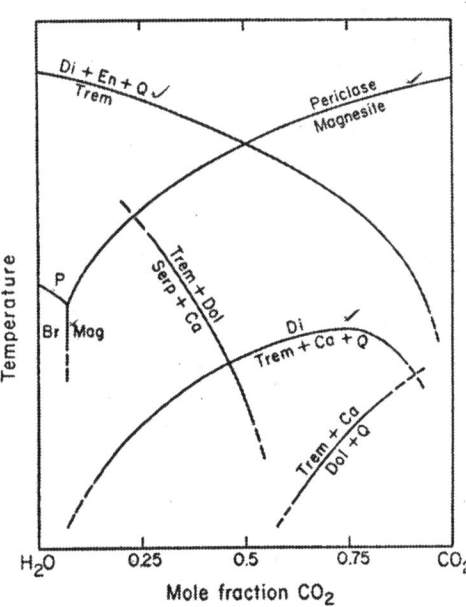

Fig. 22. Diagrammatic sketch illustrating the six types of crystal-vapor equilibrium reactions in binary gas mixtures. H_2O is component 1, and CO_2 is component 2, of the equations.

We may, for the sake of discussion, regard these reactions as models of metamorphic isograds. The most obvious feature is that an isograd defined on the basis of a reaction that evolves one volatile component may cross an isograd defined on the basis of a reaction that evolves the other volatile component. In addition, a plot like figure 22 may be regarded as a map of an area that has a gradient in the proportions of CO_2 and H_2O across it at a large angle to the thermal gradient. If such an area could be found in the field, containing rocks of suitable compositions, it should be possible to demonstrate the crossing of isograds. In reactions like the formation of diopside from tremolite, calcite, and quartz, it is clearly of great importance to know something of the composition of the fluid phase in equilibrium with the minerals before coming to any conclusion about the temperature of metamorphism, even assuming some knowledge of the total pressure.

Reactions like those described by equations 4, 5, and 6 are perhaps the most interesting of all when they are regarded as isograds. Their steep slopes in T-x plots like figure 22 show that the progress of many such reactions is affected more by the composition of the coexisting fluid phase than by either temperature or pressure. This observation leads directly to the concept of an isograd that is essentially neither isotherm nor isobar but that provides a firm limit on the composition of the fluid with which the minerals of the rock could have been in equilibrium. It cannot be too strongly urged, therefore, that when an isograd is under discussion the chemical reaction be precisely defined.

Experiments are now under way that will fix the positions of reactions of the sort just discussed in the system MgO-CaO-SiO_2-H_2O-CO_2. The apparatus is essentially the same as was used in an earlier investigation of the system $NaAlSi_2O_6$-H_2O-argon (Greenwood, 1961) in which the solid phases are held in open capsules in a bomb containing a mixture of CO_2 and H_2O. Pressure and temperature are measured, and the composition of the gas is analyzed at the end of each run. The stability of wollastonite has been studied rather fully, and preliminary data are now available on a number of other equilibria. Figure 23 shows the stability relations of wollastonite in mixtures of CO_2 and H_2O at 1000 and 2000 bars. All the reactions shown represent reversals of the equilibrium. The data are in good agreement with those of Harker and Tuttle (1956), assuming that the CO_2 and H_2O mix ideally. This apparent close approach to ideal mixing is probably illusory, because it seems likely that the gas mixture contains three

give diopside occurs at a lower temperature than the wollastonite reaction.

$$\tfrac{1}{4}Mg_3Si_4O_{10}(OH)_2 + \tfrac{3}{4}CaCO_3 + \tfrac{1}{2}SiO_2 \rightarrow$$
$$\text{Talc} \quad \text{Calcite} \quad \text{Quartz}$$
$$3CaMgSi_2O_6 + \tfrac{1}{4}H_2O + \tfrac{3}{4}CO_2$$

According to equation 3 this reaction curve must pass through a maximum in temperature where $x_{CO_2} = 0.75$. At a total pressure of 1000 bars the temperature of this maximum has been determined to be $600° \pm 25°C$, at least 25° lower than the wollastonite curve at this composition, confirming the field observation that diopside can be formed at lower temperatures than wollastonite.

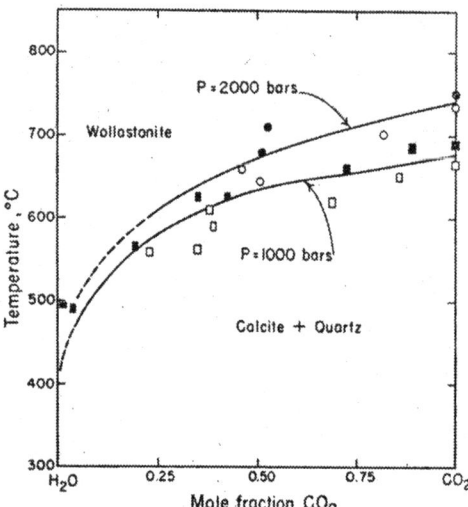

Fig. 23. Stability relations of calcite, quartz, and wollastonite in mixtures of H_2O and CO_2. Circles, 2000 bars; rectangles, 1000 bars.

molecular species rather than two. Reaction between CO_2 and H_2O to produce H_2CO_3 could easily produce the same effect as ideal mixing of CO_2 and H_2O on a solid-gas equilibrium. The accumulation of more data on mineral equilibria in the mixtures will allow direct estimation of the extent of reaction between H_2O and CO_2. In addition to the wollastonite reaction, preliminary runs indicate that the reaction of talc, calcite, and quartz to

CHAPTER 7: BLUESCHISTS

Ernst, W.G. (1971) Do mineral parageneses reflect unusually high-pressure conditions of Franciscan metamorphism? *American Journal of Science*, **270**, 81–108.

It may seem incredible to us now, but in 1971 (Ernst, 1971a) it was necessary for someone to argue forcibly the proposition that glaucophane-schist facies metamorphism in the Franciscan of California (and elsewhere) reflected nothing other than unusually high lithostatic pressures, i.e. high P/T, or unusually low geothermal gradients. Let us follow what some of the giants of metamorphic petrology had said up to that time about blueschists (as we now call them), especially with respect to pressure.

Because of their density, Eskola (1929, 1939) regarded glaucophane schists as higher-pressure equivalents of greenschists. Harker (1939), not exactly a follower of Eskola, denied the existence of a glaucophane-schist facies, arguing that such rocks were merely greenschists rich in Na (either original or introduced). Turner (1948) supported Harker and viewed the glaucophane-schist facies as substantially equivalent to the greenschist and albite-epidote amphibolite facies; consistent with earlier work in California (e.g. Taliaferro 1943), he thought that glaucophane-bearing rocks were produced by the metasomatic action of soda, iron and magnesia near ultrabasic intrusions. In his textbook, Barth (1952) placed "glaucophane schists (?)" at higher P than greenschists on his facies diagram, but thought that the underlying cause was probably high shearing stress, rather than high hydrostatic pressure; he also speculated that some glaucophane schists formed metastably at lower pressure. Ramberg (1952) thought that glaucophane schists were very rare, needing special physical conditions and perhaps also unusual bulk chemical compositions. De Roever (1955) explained textures indicating jadeite + quartz formation from albite in low-grade metasediments to be a reflection of high confining pressure, not shearing stress, citing thermodynamic calculations made by Adams (1953). In Fyfe *et al.* (1958), Turner accepted the glaucophane-schist facies as representing higher P than the greenschist facies, but still advocated a role for a saline pore fluid. Miyashiro and Banno (1958) concurred with Eskola's idea of the glaucophane-schist facies. Miyashiro (1961) suggested that the great depths needed for the jadeite-glaucophane type of metamorphism might be reached in locally and temporarily thickened crust during orogeny; he also suggested that, in some cases, orogenic compression possibly played a role in producing the high pressure. His recognition of paired coeval high-pressure and low-pressure metamorphic belts predated plate tectonics. Essene *et al.* (1965) accepted high P/T (5–10 kbar at 150 to 300°C) as the explanation for regional belts of glaucophane-schist rocks as described, for example, by McKee (1962) and Ernst (1965). However, they still argued for the role elsewhere of hot solutions released by serpentinization reactions in adjacent ultramafic bodies.

By this time, several experimental studies had confirmed the high P/T nature of aragonite as well as jadeite + quartz.

Those who accepted the experimental evidence had to propose downwarping of the geosynclinal prism to depths of as much as 30 to 40 km (e.g. Miyashiro, 1961; Ernst, 1965). The absence of an acceptable geodynamic model capable of taking rocks to great depths of burial at low temperatures (and returning them to the surface) prevented universal acceptance of high lithostatic pressure as fundamental to the glaucophane-schist facies. It did not help that Circumpacific metamorphic belts, such as the Franciscan of California, contained rocks that in many critical places were in a confused state of structural juxtaposition – what was later termed mélange – and apparently lacked consistent and meaningful field relationships. So, in spite of experiments showing the high-P/T stability of minerals like jadeite, aragonite and lawsonite, alternative explanations for glaucophane-schist minerals were refined rather than rejected right on through the 1960s. These included metasomatism (Essene *et al.*, 1965); tectonic overpressure (Blake *et al.*, 1969); high fluid pressure (Brothers, 1970); and metastable crystallization from earlier high free-energy phases such as strained calcite and high albite (Gresens, 1971; Newton and Fyfe, 1976).

In the paper presented here, Gary Ernst does a masterly job of marshalling the diverse field, microscopic, phase equilibrium, geomechanical, and geochemical evidence that rendered the alternative hypotheses unlikely, and high lithostatic pressure as the preferred causal mechanism for producing blueschist-facies mineral assemblages. This was the first of many papers by Ernst (for example, 1971b, 1973, 1975, 1977) that related the structures, ages, rock associations and metamorphic zonation of blueschists to their formation in active subduction zones. He noted especially the overlapping ages of protoliths and metamorphism and the polarity in their distributions in the field. By the start of the 1970s the association of blueschist terranes with the known or inferred locus of downturn of subducted lithospheric slabs, and the P-T consequences of this, had become clear and generally accepted. Isacks *et al.* (1968) had pointed out that lithospheric underthrusting at a velocity of 5 to 15 cm/y implied a downgoing slab that would be much cooler than its surroundings. Calculated isotherms beneath an island arc system (Oxburgh and Turcotte, 1970, 1971) depicted a cold dense slab descending to P-T conditions ideal for blueschist metamorphism, and also accounted for the paired metamorphic belts described by Miyashiro (1961). Dewey and Bird (1970) connected plate tectonics to the growth of mountain belts, recognizing Cordilleran-type mountain belts that result from the consumption of oceanic lithosphere, and collision-type belts resulting from the impact of a continent margin with another continent or an island arc.

What have we learned about blueschists since then? Contrasting retrograde P-T-$time$ paths have shown that a useful distinction can be made between accretionary "Franciscan" and collisional "Alpine" types of blueschist belts. The former are related to decoupling from the descending oceanic lithosphere and refrigeration during gradual return to the surface, whereas the latter involve perhaps a halt to convergence, isostatic recovery, and partial overprinting of high P/T assemblages by epidote-amphibolite and greenschist-facies minerals (Ernst, 1988). The contrasting tectonic environments involved were found to correlate with distinctive geological associations and properties for these belts, such as the protolith package, protolith vs. metamorphic age-gap, structural styles, and mélange vs. coherent stratigraphy.

Patrick and Day (1995) reviewed space and time aspects of blueschist rocks in the western Cordillera of America, where much of the classical work on blueschists had been done. Maruyama et al. (1996) presented a comprehensive review of the salient features of all the world's 250 blueschist belts. They emphasized contrasts in the protolith packages of the Alpine (A) and Franciscan (B or Benioff) tectonic types (Bally 1981): platform carbonates, peraluminous shales, bimodal volcanics, basement granites, garnet peridotite (A) vs. reef limestones, greywackes, cherts, MORB volcanics, absence of crystalline basement rocks and spinel peridotite (B). In addition to the different P-T-t paths and corresponding presence or absence of significant back-reaction, they noted that maximum pressures are high (45 kbar) for A type and low (12 kbar) for B type belts. In my view this last figure is likely to exclude some of the eclogitic rocks characteristic of the B type, such as in the Franciscan and the Sanbagawa terrane (Aoya, 2001). It also depends on the assignment of blueschist belts to A or B. Some blueschist terranes contain protoliths consistent with both A and B types (e.g. Margarita Island, Venezuela, Stöckhert et al., 1995). The Bantimala eclogites, South Sulawesi, Indonesia, occur in sheared serpentinite associated with turbidite and mélange containing sandstone, shale, chert and basalt (all B type characteristics), formed at an estimated 18 to 24 kbar, and suffered a refrigerated return path (Miyazaki et al., 1996). Ernst (2003) resolves this problem by recognizing tectonic regimes transitional between A and B types, in which subduction of both oceanic and continental lithosphere took place. It is understandable, given the density of eclogitized ocean crust, that deeply B-subducted crust does not readily return to the surface, in contrast to A-subducted crust, with its large content of meta-granitoid material, which remains less dense than upper-mantle peridotite. According to Maruyama et al. (1996), 80% of blueschist belts are B type. Circum-Pacific blueschist belts do not belong exclusively to the B category: A-type blueschists with all the appropriate characteristics occur in New Caledonia and in the Seward Peninsula-Brooks Range-Yukon belt. Conversely, some B-type belts are now present in continental interiors. The Tethyan belt accounts for a large proportion of the A-type blueschists.

Whereas pure versions of jadeite, lawsonite and aragonite can be readily synthesized, this has proven until very recently not to be the case for the typomorphic mineral glaucophane (Jenkins, 2003). After as many as nine previous efforts, reliably reversed brackets on the stability of glaucophane (Gln + Qtz = Tlc + Ab) appear to have been obtained (for example, 13 kbar at 600°C, Corona and Jenkins, 2005). This is in good agreement with a half-bracket by Holland (1988), which probably means that little or no adjustment will be needed to the thermodynamic data currently in use for glaucophane (e.g. Guiraud et al., 1990; Will et al., 1998; Carson et al., 1999). Thermodynamic data support a distinction between higher P/T lawsonite blueschist and lower P/T epidote blueschist (Evans, 1990). A common observation made in many A-type blueschist-eclogite belts (Forbes et al., 1984; Droop, 1985; Schliestedt, 1986) is the presence of box-shaped pseudomorphs composed of epidote, quartz and paragonite (or margarite) enclosed in garnet. These have been interpreted as formerly lawsonite, a clear indication of a P-T path oblique to the classical trajectory associated with B-type subduction models.

In the many instances of outcrop-scale co-occurrence of blueschist with greenschist or eclogite lithologies in both A and B types, we have learned how to distinguish between those caused by incomplete and selective overprint in the course of evolving P-T conditions (Holland and Ray, 1985; Schliestedt and Matthews, 1987; Bröcker 1990; Breeding et al., 2003), and those caused by whole-rock compositional variation, usually in Fe/Mg or Fe^{3+}/Al ratios (e.g. Schliestedt, 1986: Owen, 1989). Blue glaucophane-riebeckite solid solutions in the 30–70% range called crossite are typical of relatively oxidized metamorphosed pillow-basalts, in some cases occurring amidst rocks of the greenschist facies. The easily identified crossite, with optic plane ⊥(010), has been mentioned repeatedly in the petrological literature, but sadly it lost its place in the nomenclature of amphiboles (Leake et al., 1997).

The Al-rich compositional maturity of fine-grained shelf sediments that become entrained in A-type belts produces blueschist mineral assemblages that were unknown prior to 1970. They include minerals such as Mg- and Fe-carpholite, sudoite, Mg-chloritoid, and Mg-staurolite, as well as pyrophyllite and kyanite (e.g. Chopin and Schreyer, 1983: Theye et al., 1992; Goffé and Oberhänsli, 1992; Vidal et al., 1992; El-Shazly, 1995). The presence of these minerals has greatly facilitated estimates of P and T in A-type terranes. They seldom occur in B-type belts. The K-white-mica in these rocks is phengite, but this changes to uniaxial, high-Si phengite-3T in associated silicic rocks, where it occurs in company with biotite (ordinarily absent in blueschists) and highly ordered albite (or jadeite) and microcline (e.g. Evans and Patrick, 1987).

The discovery of ultra-high pressure metamorphism (UHPM) in (Chopin, 1984, Chapter 14) shifted the emphasis of many petrologists away from blueschists in favour of these more extreme and rewarding targets. UHPM belts are collisional in character, hence Alpine in nature, and blueschist-facies metamorphism was part of their prograde path (Chopin, 1984; Coleman and Wang, 1995; Compagnoni et al., 1995; Hacker and Peacock, 1995), if not their retrograde history. Eclogitic rocks associated with B-type blueschists do not contain kyanite or any other indications of conditions approaching UHPM.

The link between blueschists and zones of convergence gives rocks of the blueschist facies a notable feature not shared with other metamorphic facies, namely a frequency of occurrence with geological time that increases from the Proterozoic to the Cenozoic (De Roever, 1956). Youngest of them (2.7–3.0 Ma) is A-type blueschist on D'Entrecasteaux Island, Papua-New Guinea (Hill and Baldwin, 1993). It is generally accepted that the absence of Archaean blueschists (Liou *et al.*, 1990) may be attributed to the time it took for Earth to cool sufficiently to allow the lithosphere to sink into the underlying asthenosphere and start the modern episode of plate tectonics and subduction (Ernst, 1972). Maxima and minima in the amount of blueschist produced with time are believed to correlate with rates of ocean-floor spreading (Maruyama *et al.*, 1996). It is true that the survival of blueschists over great lengths of geological time is problematic (England and Richardson, 1977, Chapter 9): they need to be rapidly exhumed, but not so efficiently that they all get eroded (Draper and Bone, 1981). Barrientos and Selverstone (1993) pointed out that epidote blueschists, unlike other blueschist-facies assemblages such as lawsonite blueschist, depend on aqueous fluid infiltration and fluid-consuming reactions for their conversion to chloritic greenschists, so that blueschist preservation does not necessarily require rapid uplift. Based on Neoproterozoic ages for first the appearance of ophiolites, blueschist-facies metamorphic rocks, and UHPM terranes, Stern (2005) argued that it was not until then that subduction began. There is truth on both sides of this question.

References

Adams, L.H. (1953) A note on the stability of jadeite. *American Journal of Science*, **251**, 299–308.

Aoya, M. (2001) *P-T-D* path of eclogite from the Sanbagawa belt deduced from combination of petrological and microstructural analyses. *Journal of Petrology*, **42**, 1225–1248.

Bally, A.W. (1981) Thoughts on the tectonics of folded belts. Pp. 13–32 in: *Thrust and Nappe Tectonics* (N.J. Price and K. McKay, editors). Special Publication 9, Geological Society of London.

Barrientos, X. and Selverstone, J. (1993) Infiltration vs. thermal overprinting of epidote blueschists, Ile de Groix, France. *Geology*, **21**, 69–72.

Barth, T.F.W. (1952) *Theoretical Petrology*. John Wiley and Sons, Inc., New York, 387 pp.

Blake, M.C., Irwin, W.P. and Coleman, R.G. (1969) Blueschist-facies metamorphism related to regional thrust faulting. *Tectonophysics*, **8**, 237–246.

Breeding, C.M., Ague, J.J., Bröcker, M. and Bolton, E.W. (2003) Blueschist preservation in a retrograded, high-pressure, low-temperature metamorphic terrane, Tinos, Greece: Implications for fluid flow paths in subduction zones. *Geochemistry Geophysics Geosystems*, **4**, 1–11 (9002, doi: 10.1029/2002GC000380).

Bröcker, M. (1990) blueschist-to-greenschist transition in metabasites from Tinos Island (Cyclades, Greece): Compositional control or fluid infiltration? *Lithos*, **25**, 25–39.

Brothers, R.N. (1970) Lawsonite-albite schists from northernmost New Caledonia. *Contributions to Mineralogy and Petrology*, **25**, 185–202.

Carson, C.J., Powell, R. and Clarke, G.L. (1999) Calculated mineral equilibria for eclogites in $CaO-Na_2O-FeO-MgO-Al_2O_3-SiO_2-H_2O$: application to the Pouébo Terrane, Pam Peninsula, New Caledonia. *Journal of Metamorphic Geology*, **17**, 9–24.

Chopin, C. (1984) Coesite and pure pyrope in high-grade blueschists of the Western Alps: a first record and some consequences. *Contributions to Mineralogy and Petrology*, **86**, 107–118.

Chopin, C. and Schreyer, W. (1983) Magnesiocarpholite and magnesiochloritoid: two index minerals of pelitic blueschists and their preliminary phase relations in the model system $MgO-Al_2O_3-SiO_2-H_2O$. *American Journal of Science*, **283-A**, 72–96.

Coleman, R.G. and Wang, X. (1995) Overview of the geology and tectonics of UHPM. Pp. 1–32 in: *Ultrahigh Pressure Metamorphism* (R.G. Coleman and X. Wang, editors). Cambridge University Press, Cambridge, UK.

Compagnoni, R., Hirajima, T. and Chopin, C. (1995) Ultra-high-pressure metamorphic rocks in the Western Alps. Pp. 206–243 in: *Ultrahigh Pressure Metamorphism* (R.G. Coleman and X. Wang, editors). Cambridge University Press, Cambridge, UK.

Corona, J.C. and Jenkins, D. (2005) Experimental investigation of the reaction: glaucophane + 2 quartz = 2 albite + talc. *Geological Society of America, Abstracts with Programs*, **37**, No. 7, p. 90.

De Roever, W.P. (1955) Genesis of jadeite by low-grade metamorphism. *American Journal of Science*, **253**, 283–298.

De Roever, W.P. (1956) Some differences between post-Paleozoic and other regional metamorphism. *Geologie en Mijnbouw*, **18**, 123–127.

Dewey, J.F. and Bird, J.M. (1970) Mountain belts and the new global tectonics. *Journal of Geophysical Research*, **75**, 2625–2647.

Draper, G. and Bone, R. (1981) Denudation rates, thermal evolution and preservation of blueschist terranes. *Journal of Geology*, **89**, 601–613.

Droop, G.T.R. (1985) Alpine metamorphism in the south-east Tauern Window, Austria: 1. *P-T* variations in space and time. *Journal of Metamorphic Geology*, **3**, 371–402.

El-Shazly, A.E. (1995) Petrology of Fe-Mg carpholite-bearing metasediments from NE Oman. *Journal of Metamorphic Geology*, **13**, 379–396.

England, P.C. and Richardson, S.W. (1977) The influence of erosion upon the mineral facies of rocks from different metamorphic environments. *Journal of the Geological Society of London*, **134**, 201–213.

Ernst, W.G. (1965) Mineral parageneses in Franciscan metamorphic rocks, Panoche Pass, California. *Geological Society of America Bulletin*, **76**, 8798–8914.

Ernst, W.G. (1971a) Do mineral parageneses reflect unusually high-pressure conditions of Franciscan metamorphism. *American Journal of Science*, **270**, 81–108.

Ernst, W.G. (1971b) Metamorphic zonations on presumably subducted lithospheric plates from Japan, California and the Alps. *Contributions to Mineralogy and Petrology*, **34**, 43–59.

Ernst, W.G. (1972) Occurrence and mineralogic evolution of blueschist belts with time. *American Journal of Science*, **272**, 657–668.

Ernst, W.G. (1973) Blueschist metamorphism and P-T regimes in active subduction zones. *Tectonophysics*, **17**, 255–272.

Ernst, W.G. (1975) Systematics of large-scale tectonics and age progressions in Alpine and circum-Pacific blueschist belts. *Tectonophysics*, **26**, 229–246.

Ernst, W.G. (1977) Mineral parageneses and plate tectonic settings of relatively high-pressure metamorphic belts. *Fortschritte der Mineralogie*, **54**, 192–222.

Ernst, W.G. (1988) Tectonic history of subduction zones inferred from retrograde blueschist *P-T* paths. *Geology*, **16**, 1081–1084.

Ernst, W.G. (2003) High-pressure and ultrahigh-pressure metamorphic belts – subduction, recrystallization, exhumation, and significance for ophiolite study. *Geological Society of America, Special Paper*, **373**, 365–384.

Eskola, P. (1929) Om mineralfacies. *Geologiska Föreningens i Stockholm Förhandlingar*, **51**, 157–172.

Eskola, P. (1939) Die metamorphen Gesteine. Pp. 263–407 in: *Die Entstehung der Gesteine* (T.F.W. Barth, C.W. Correns and P. Eskola, editors). Julius Springer, Berlin (reprinted in 1960 and 1970).

Essene, E.J., Fyfe, W.S. and Turner, F.J. (1965) Petrogenesis of Franciscan glaucophane schists and associated metamorphic rocks, California. *Contributions to Mineralogy and Petrology*, **11**, 695–704.

Evans, B.W. (1990) Phase relations of epidote blueschists. *Lithos*, **25**, 3–23.

Evans, B.W. and Patrick, B.E. (1987) Phengite-3T in high-pressure metamorphosed granitic orthogneisses, Seward Peninsula, Alaska. *The Canadian Mineralogist*, **25**, 141–158.

Forbes, R.B., Evans, B.W. and Thurston, S.P. (1984) Regional high-pressure metamorphism, Seward Peninsula, Alaska. *Journal of Metamorphic Geology*, **2**, 43–54.

Fyfe, W.S., Turner, F.J. and Verhoogen, J. (1958) *Metamorphic Reactions and Metamorphic Facies*. Geological Society of America, Memoir **73**, 259 pp.

Goffé, B. and Oberhänsli, R. (1992) Ferrocarpholite and magnesiocarpholite in the Bündnerschiefer of the eastern Central Alps (Grisons and Engadine Window). *European Journal of Mineralogy*, **4**, 835–838.

Gresens, R.L. (1971) Do mineral parageneses reflect unusually high-pressure conditions of Franciscan metamorphism? *American Journal of Science*, **271**, 311–318.

Guiraud, M., Holland, T.J.B. and Powell, R. (1990) Calculated mineral equilibria in the greenschist-blueschist-eclogite facies in Na_2O-FeO-MgO-Al_2O_3-SiO_2-H_2O. *Contributions to Mineralogy and Petrology*, **104**, 85–98.

Hacker, B.R. and Peacock, S.M. (1995) Creation, preservation, and exhumation of UHPM rocks: An example from the Western Alps of Italy. Pp. 159–181 in: *Ultrahigh Pressure Metamorphism*, (R.G. Coleman and X. Wang, editors). Cambridge University Press, Cambridge, UK.

Harker, A. (1939) *Metamorphism. A Study of the Transformation of Rock-Masses*. Methuen, London, 362 pp.

Hill, E.J. and Baldwin, S.L. (1993) Exhumation of high-pressure metamorphic rocks during crustal extension in the D'Entrecasteaux region, Papua, New Guinea. *Journal of Metamorphic Geology*, **11**, 261–277.

Holland, T.J.B (1988) Preliminary phase relations involving glaucophane and applications to high-pressure petrology: new heat capacity and thermodynamic data. *Contributions to Mineralogy and Petrology*, **99**, 134–142.

Holland, T.J.B. and Ray, N.J. (1985) Glaucophane and pyroxene breakdown reactions in the Pennine units of the eastern Alps. *Journal of Metamorphic Geology*, **3**, 417–438.

Isacks, B., Oliver, J. and Sykes, L.R. (1968) Seismology and the new global tectonics. *Journal of Geophysical Research*, **73**, 5855–5899.

Jenkins, D.M. (2003) Glaucophane synthesis: the role of water. *Geological Society of America, Abstracts with Programs*, **35**, abstract 34-8.

Leake, B.E., Woolley, A.R., Arps, C.E.S., Birch, W.D., Gilbert, M.C., Grice, J.D., Hawthorne, F.C., Kato, A., Kisch, H.J. Krivovichev, V.G., Linthout, K., Laird, J., Mandarino, J., Maresch, W.V., Nickel, E.H., Rock, N. M. S., Schumacher, J.C., Smith, D.C. Stephenson, N.C.N., Ungaretti, L., Whittaker, E.J. W. and Youzhi, G. (1997) Nomenclature of amphiboles: Report of the Subcommittee on Amphiboles of the International Mineralogical Association Commission on New Minerals and Mineral Names. *Mineralogical Magazine*, **61**, 295–321.

Liou, J.C., Maruyama, S., Wang, X. and Graham, S. (1990) Precambrian blueschist terranes of the world. *Tectonophysics*, **181**, 97–111.

Maruyama, S., Liou, J.G. and Terabayashi, M. (1996) Blueschists and eclogites of the world and their exhumation history. *International Geology Review*, **38**, 485–594.

McKee, E.B. (1962) Widespread occurrence of jadeite, lawsonite, and glaucophane in central California. *American Journal of Science*, **260**, 596–610.

Miyashiro, A. (1961) Evolution of metamorphic belts. *Journal of Petrology*, **2**, 277–311.

Miyashiro, A. and Banno, S. (1958) Nature of glaucophanitic metamorphism. *American Journal of Science*, **256**, 97–110.

Miyazaki, K., Zulkarnain, I., Sopaheluwakan, J. and Wakita, K. (1996) Pressure-temperature conditions and retrograde paths of eclogites, garnet-glaucophane rocks and schists from South Sulawesi, Indonesia. *Journal of Metamorphic Geology*, **14**, 549–563.

Newton, R.C. and Fyfe, W.S. (1976) High-pressure metamorphism. Pp. 101–186 in: *The Evolution of the Crystalline Rocks* (D.K. Bailey and R. Macdonald, editors). Academic Press, New York.

Owen, C. (1989) Magmatic differentiation and alteration in isofacial greenschists and blueschists, Shuksan Suite, Washington: statistical analysis of major element variation. *Journal of Petrology*, **30**, 739–762.

Oxburgh, E.R. and Turcotte, D.L. (1970) Thermal structure of island arcs. *Geological Society of America Bulletin*, **81**, 1665–1688.

Oxburgh, E.R. and Turcotte, D.L. (1971) Origin of paired metamorphic belts and crustal dilation in island arc regions. *Journal of Geophysical Research*, **76**, 1315–1327.

Patrick, B.E. and Day, H.W. (1995) Cordilleran high-pressure metamorphic terranes: progress and problems. *Journal of Metamorphic Geology*, **13**, 1–8.

Ramberg, H. (1952) *The Origin of Metamorphic and Metasomatic Rocks*. The University of Chicago Press, Chicago, USA, 317 pp.

Schliestedt, M. (1986) Eclogite-blueschist relationships as evidenced by mineral equilibria in the high-pressure metabasic rocks of Sifnos (Cycladic Islands), Greece. *Journal of Petrology*, **27**, 1437–1459.

Schliestedt, M. and Matthews, A. (1987) Transformation of blueschist to greenschist facies rocks as a consequence of fluid infiltration, Sifnos (Cyclades), Greece. *Contributions to Mineralogy and Petrology*, **97**, 237–250.

Stern, R.J. (2005) Evidence from ophiolites, blueschists, and ultrahigh-pressure metamorphic terranes that the modern episode of subduction tectonics began in Neoproterozoic time. *Geology*, **33**, 557–560.

Stöckhert, B., Maresch, W.V., Brix, M., Kaiser, C., Toetz, A., Kluge, R. and Krückhans-Lueder, G. (1995) Crustal history of Margarita Island (Venezuela) in detail: Constraint of the Caribbean plate-tectonic scenario. *Geology*, **23**, 787–790.

Taliaferro, N.L. (1943) Franciscan-Knoxville problem. *American Association of Petroleum Geologists Bulletin*, **27**, 109–219.

Theye, T., Seidel, E. and Vidal, O. (1992) Carpholite, sudoite, and chloritoid in low-grade high-pressure metapelites from Crete and the Peloponnese, Greece. *European Journal of Mineralogy*, **4**, 487–507.

Turner, F.J. (1948) *Mineralogical and structural evolution of the metamorphic rocks*. The Geological Society of America, Memoir **30**, 342 pp.

Vidal, O. and Goffé, B. and Theye, T. (1992) Experimental study of the stability of sudoite and magnesiocarpholite and calculation of a new petrogenetic grid for the system FeO-MgO-Al_2O_3-SiO_2-H_2O. *Journal of Metamorphic Geology*, **10**, 603–614.

Will, T.M. Okrusch, M., Schmädicke, E. and Chen, G. (1998) Phase relations in greenschist-blueschist-amphibolite-eclogite facies: Calculated mineral equilibria in the system Na_2O-CaO-FeO-MgO-Al_2O_3-SiO_2-H_2O (NCFMASH), with applications to the PT evolution of metamorphic rocks from Samos, Greece. *Contributions to Mineralogy and Petrology*, **132**, 85–102.

DO MINERAL PARAGENESES REFLECT UNUSUALLY HIGH-PRESSURE CONDITIONS OF FRANCISCAN METAMORPHISM?

W. G. ERNST

Department of Geology and Institute of
Geophysics and Planetary Physics,
University of California, Los Angeles, California 90024

ABSTRACT. Metamorphosed graywacke, shale, chert, and mafic volcanic rocks of the Franciscan group of California display low-grade mineral parageneses absent from metamorphic terranes of more continental affinities. Characteristic phases include glaucophane-crossite, jadeitic pyroxene, lawsonite, pumpellyite, and aragonite. Explanations of this distinctive metamorphism fall into two groups: either metasomatism or metastable recrystallization taking place under normal low-grade conditions; or low-temperature production at relatively high pressures resulting from either tectonic overpressures or very deep burial.

Although metasomatic effects are not unknown, existence of metamorphic rock compositions virtually indistinguishable from those of protoliths proves the genetic insignificance of chemical change. Metastable recrystallization is of general importance in metamorphism; however, growth of jadeitic pyroxene + quartz from albite of low structural state argues against a metastable low-pressure origin for Diablo Range metagraywackes. Experimentally determined rock strengths preclude tectonic overpressures exceeding about 1 kb for the "strong" metagraywacke during recrystallization in the presence of an aqueous fluid phase; interbedded jadeitic pyroxene-bearing metashale is even weaker. Phase equilibrium relations obtained in the laboratory are compatible with observed parageneses and suggest lithostatic pressure of 5 to more than 8 kb for the inferred 150° to 300°C temperature range. The required 20 to 30 + km depth of metamorphism is in accord with the hypothesis of Late Mesozoic accumulation and tectonic thickening of the Franciscan in one or a series of oceanic trenches contemporaneously being overridden by the North American lithospheric plate along a Benioff (subduction) zone and with geothermal gradients computed for the descending plate. Imbricate thrusting of more deeply buried, higher pressure metamorphosed rocks on the east over less intensely recrystallized, more westerly sections seems to account for some of the observed structural, temporal, and paragenetic features within the Franciscan terrane.

The subduction zone setting thus accounts for the development of relatively high-pressure, low-temperature blueschist facies mineral assemblages. Postulated alternative origins are not as obviously related to continental margins, and, were one of them to be accepted, it would be necessary to regard the striking world-wide restriction of glaucophane schists and associated rocks to plate junctions as coincidental.

STATEMENT OF THE PROBLEM

Feebly to thoroughly metamorphosed, chiefly clastic sedimentary rocks of the Franciscan group crop out extensively in the California Coast Ranges and more sparsely in Baja California and its off-shore islands as well as in southwestern Oregon. Although some parts of the terrane contain relatively coherent sections of strata, other associated units are lithologic mélanges (Hsü, 1968), due both to sedimentary slumping and tectonic disruption. Because of the paucity of fossils and the lack of widespread marker horizons, the structural relations and both stratigraphic and tectonic thicknesses of the Franciscan are incompletely known. This metamorphosed—and in part chaotically deformed—eugeosynclinal terrane has been thrust beneath the roughly contemporaneous, less deformed, only feebly recrystallized Great Valley sequence (for example, see Irwin, 1960, 1964, 1966; Dickinson, 1966; Blake and others, 1967, 1969; Bailey and Blake, 1969; Page, 1970). The tectonic juxtaposition evidently took place adjacent to and within a Late Mesozoic subduction zone marking the margins of North American and Pacific lithospheric plates (Hamilton, 1969; Ernst, 1970; Page, 1970). For descriptions of the lithologies and the regional geologic relationships, the reader is referred to the detailed study of the California Coast Ranges by Bailey, Irwin, and Jones (1964). The areal extent of these units is shown in figure 1.

In situ metamorphosed Franciscan rocks exhibit a systematic set of characteristic mineral assemblages which, to some extent, are areally distinct and hence are inferred to reflect a progressive metamorphic sequence. Quartz-, phengite-, and chlorite-bearing metaclastic rocks, principally graywackes, micrograywackes, and black shales, display the critical phase compatibilities:

A. laumontite + albite ± calcite;

B. pumpellyite and/or lawsonite + albite ± calcite;

C. lawsonite ± pumpellyite + albite ± aragonite;

D. lawsonite + jadeitic pyroxene ± aragonite.

The less abundant, interlayered, pod-like masses and somewhat larger piles of metamorphosed pillow lavas concomitantly have developed characteristic zeolitized and pumpellyitized greenstone and lawsonite + blue amphibole and/or omphacite-bearing assemblages. The metamorphic petrology of a number of specific areas has been described by many workers (for example, Bloxam, 1956, 1959, 1960; McKee, 1962a, 1962b; Coleman, 1961, 1965, 1967a; Coleman and Lee, 1962, 1963; Ghent, 1965; Ernst, 1965, in press; Ernst and Seki, 1967; Ernst and others, 1970, chaps. 3 and 4; Blake, Irwin, and Coleman, 1967, 1969; Blake and Cotton, 1969; Bailey and Blake, 1969; Raymond, 1970).

The origin of these phase assemblages of low metamorphic grade has puzzled petrologists since before the turn of the century (a general summary has been presented by Turner, 1968, p. 289-295). Basically, explanations regarding the petrogenesis of mafic blueschists, jadeitic pyroxene-bearing metaclastics, and their mineralogically less exotic associated lithologies can be divided into two groups. (1) The phase assemblages may reflect low-temperature metasomatism accompanying the emplacement and serpentinization of alpine-type peridotites common in most glaucophane schist terranes, or they may represent metastable recrystallization. (2) Alternatively, the dense minerals—such as sodic amphibole, sodic pyroxene, lawsonite, and aragonite, which have been demonstrated in the laboratory to be favored by high pressures—are postulated to have formed during low-temperature recrystallization under high pressures resulting from deep burial or due to the generation of tectonic overpressures substantially exceeding the lithostatic value. Four different schools of thought have thus evolved, two of which favor relatively low-pressure recrystallization, the other two relatively high-pressure recrystallization.

METASOMATIC PRODUCTION OF OBSERVED ASSEMBLAGES

Taliaferro (1943, p. 159-182) ascribed the formation of glaucophane schists and related rocks in the California Coast Ranges to metasomatism accompanying the intrusion of ultramafic plutons. Other workers have also been impressed by the nearly world-wide association of blueschists with alpine-type peridotites and by the peculiar bulk compositions of some glaucophane schists. For these reasons, many investigators of Franciscan mineral parageneses (for example, Brothers, 1954; Chesterman, 1960; Bloxam, 1966; Fyfe and Zardini, 1967; Gresens, 1969) have invoked metasomatism or at least catalytic interaction of an active pore solution with condensed assemblages as the causative agent in the production of glaucophane schists and kindred metamorphics (see also Essene, Fyfe, and Turner, 1965). Most, but not all, have linked the fluid emanations to nearby mafic or ultramafic bodies.

There is no question that the effects of metasomatism are present in metamorphosed Franciscan rocks and that in some cases exchange of material has taken place between serpentinized peridotites and wall rocks or inclusions; examples have been well documented, for instance, by Ransome (1894, p. 222-226), Davis (1918, p. 275-278), Bloxam (1960-1966), Coleman (1961, 1967b, p. 28-31), and Barnes and O'Neil (1969). The fundamental question, however, is not whether local or regional metasomatism has taken place, but whether chemical exchanges between preexisting rocks and a chemically active pore fluid are required to produce the blueschist mineralogic suite. One may further inquire whether the presence of alpine-type serpentinized peridotites is necessary for the formation of such lithologies.

The answer to the second question is most probably negative. On a global basis, it is true that glaucophane schists and ultramafics commonly are spatially associated, as in the Alps, the Caribbean, and New Caledonia, as well as the California Coast Ranges. However, the near absence of coeval serpentinized peridotites in the Shuksan belt of blue amphibole-bearing schists in Washington State (Misch, 1959, 1966, 1969), in the glaucophanitic metabasalt + phyllite complex of Calabria, southern Italy (Hoffman, 1970), and the very limited occurrence of ultramafics in the Sanbagawa glaucophane schist belt of southwestern Japan (Saito and others, 1960, geologic map; see also Miyashiro, 1966, fig. 3) all argue against a genetic link between ultramafic masses and glaucophane schists. Moreover, the extensive emplacement of ultramafic plutons in some orogenic areas, such as the Appalachian belt, evidently has failed to produce blueschist-type phase assemblages. As is discussed below, the general association of blueschists and serpentinized ultramafic rocks may be the

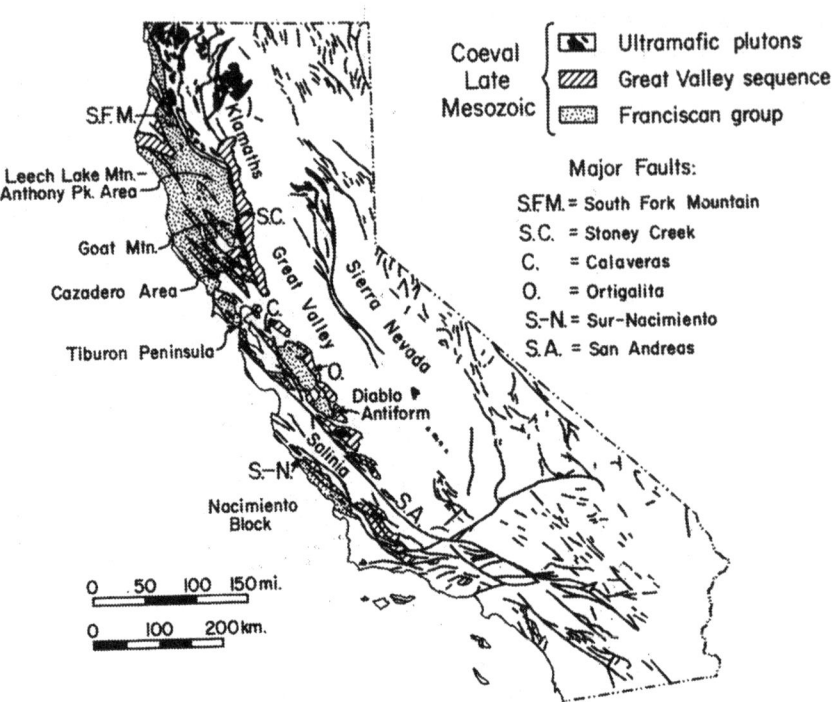

Fig. 1. Regional distribution of Franciscan group and Great Valley sequence rocks in the California Coast Ranges. Note that although the depositional ages of the largely sedimentary and metasedimentary series are reasonably well known, the "age" of the cold ultramafic plutons is considered to be the time of emplacement (California Coast Ranges only). The South Fork Mountain, Stoney Creek, Ortigalita, and Sur-Nacimiento faults are viewed as segments of the Coast Range thrust (Bailey, Blake, and Jones, 1970). The juxtaposition of Salinia, Sierran-type basement, and the Nacimiento Block of Franciscan-type basement with the Diablo Range is thought to be the result of large scale Tertiary strike-slip motion along the San Andreas fault (Hill and Dibblee, 1953).

result, rather, of tectonic interaction between crust and mantle at the suture zone marking convergence of lithospheric plates.

To carry the argument specifically to the Franciscan parageneses, the most remarkably recrystallized portion of this terrane with regard to mineralogy is exposed in the Diablo antiform (Bloxam, 1956; McKee, 1962a, 1962b; Ernst, 1965, in press; Ernst and Seki, 1967; Ernst and others, 1970; Bailey and Blake, 1969); except for the Red Mountain sill[1] (Maddock, 1964, Saad, 1969) this portion of the eugeosynclinal mélange is characterized by a near absence of alpine-type peridotites (for example, see fig. 4; Bailey, Irwin, and Jones, 1964, pl. 1). Those serpentinites that do occur are extremely small bodies and clearly are not commensurate with the areal extent of thoroughly recrystallized metamorphic

---

[1] Diablo Range localities mentioned in the text are indicated in figure 4, those exclusive of the Diablo Range are located in figure 1.

rocks.[2] Bailey and Blake (1969) have made a similar observation for the northern Coast Ranges.

Furthermore, viewing the Franciscan terrane as a whole, ultramafic rocks are exposed chiefly along the west side of the Great Valley, particularly marking the Stoney Creek fault zone (see fig. 1), and although the underlying eugeosynclinal assemblage has been converted in many places to blueschist-type compatibilities, the tectonically overlying Great Valley strata are either weakly laumontized or are practically unmetamorphosed. According to Bailey, Irwin, and Jones (1970), the serpentinized peridotites are the lowest member of the overthrust plate, hence could not have interacted metsomatically with the Franciscan rocks without similarly affecting the Great Valley section. Moreover, as will be described in the section dealing with tectonic overpressures, the Great Valley (+ basal ultramafic) and Franciscan terranes in general have been juxtaposed too late (60-90 m.y. B.P.) to account for an early period (120-150 m.y. ago) of relatively high-pressure Franciscan recrystallization.

Of the authors who have invoked metasomatism associated with serpentinized peridotites, only Taliaferro (1943, figs. 4-6) provided diagrammatic sketches of the inferred field relations; the metamorphic rocks described as bordering a serpentinite sill on Tiburon Peninsula, San Francisco Bay, are now known to be tectonic blocks and not parts of a distinct aureole (Dudley, ms). No maps purporting to show metamorphic aureoles surrounding similar plutons have even been published.

We may conclude that a *close spatial correlation between ultramafic plutons and Franciscan metamorphic rocks does not exist.*

The more general question of regional metasomatism still remains. At the turn of the century, Washington (1901) and Smith (1907, p. 224-240) clearly demonstrated that among blueschists from various terranes, and the Franciscan in particular, rock bulk compositions need not have changed significantly during the recrystallization. Based on the equivalence of bulk chemistry and on the distinctive phase petrology, Eskola (1939) recognized blueschists and kindred lithologic assemblages as representatives of a separate mineral facies, the glaucophane schist (or blueschist) facies.

With regard to the Franciscan, it is clear that some metavolcanics and metasediments have developed the characteristic glaucophane schist phase assemblages without substantial introduction or removal of material (see chemical analyses presented by Coleman and Lee, 1963, tables 2-4; Coleman, 1965, table 4; Bailey, Irwin, and Jones, 1964, tables 1, 4, 14; Ernst, 1965, tables 10, 12, in press, table 4; Ernst and others, 1970, tables 1, 2, 5-9). The isochemical nature of the metamorphism of certain rocks of metabasaltic composition has been illustrated previously (Ernst, 1959, fig. 26, 1963a, fig. 1; Coleman and Lee, 1963, fig. 19). However, metasedimentary rocks are far more abundant than the mafic meta-

[2] In the Panoche Pass area of the Diablo Range (Ernst, 1965, pl. 1), one of three mapped patches of jadeitic metagraywacke does seem to be associated fortuitously with serpentinite (no metasomatic effects were observed in the country rock adjacent to this body), but the other two jadeitic patches definitely are not.

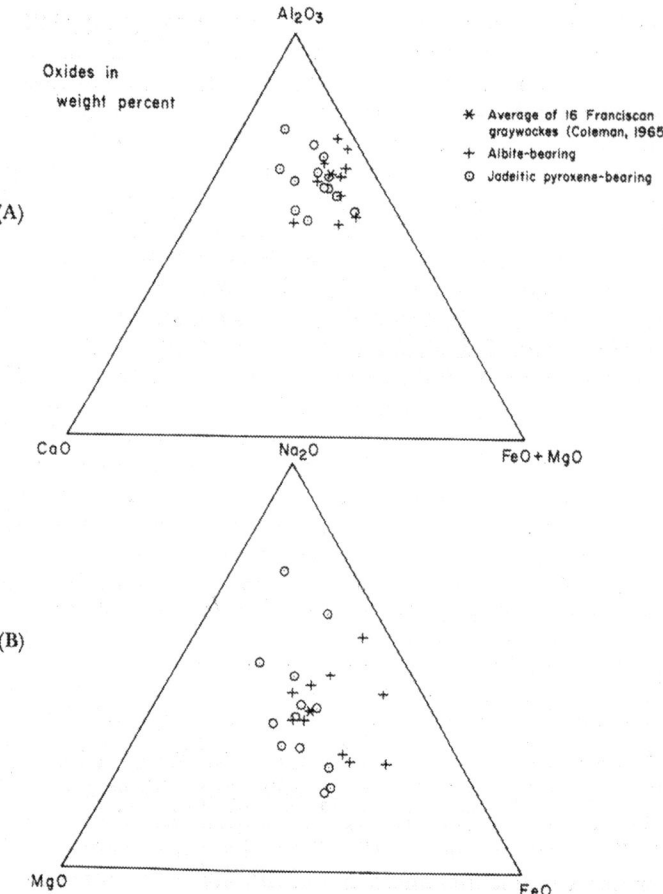

Fig. 2. Chemical range of Franciscan metagraywackes with regard to the oxides: (A) Al_2O_3, CaO, FeO + MgO; and (B), Na_2O, MgO, FeO. Albite-bearing metagraywacke gravimetric analyses are from Ernst (1965, table 12) and Ernst and others (1970, tables 2, 6, and 7); jadeitic pyroxene-bearing metagraywacke gravimetric analyses are from Bloxam (1956, table 2; 1960, table 1), Coleman (1965, table 4), Ernst (1965, table 12), and Ernst and others (1970, tables 6 and 7).

volcanics in the California Coast Ranges, and although the megascopic evidence of recrystallization is not as obvious, a striking metamorphic mineralogy has been produced; thus, because of their volumetric importance, it is critical for the discussion to determine whether or not these rocks exhibit the effects of chemical exchange.

A range of metaclastic bulk compositions as a function of phase petrology is presented in this report as figures 2 and 3; although not complete, a broad overlap in chemical ranges is obvious. For such rocks, physical rather than chemical differences must have produced the con-

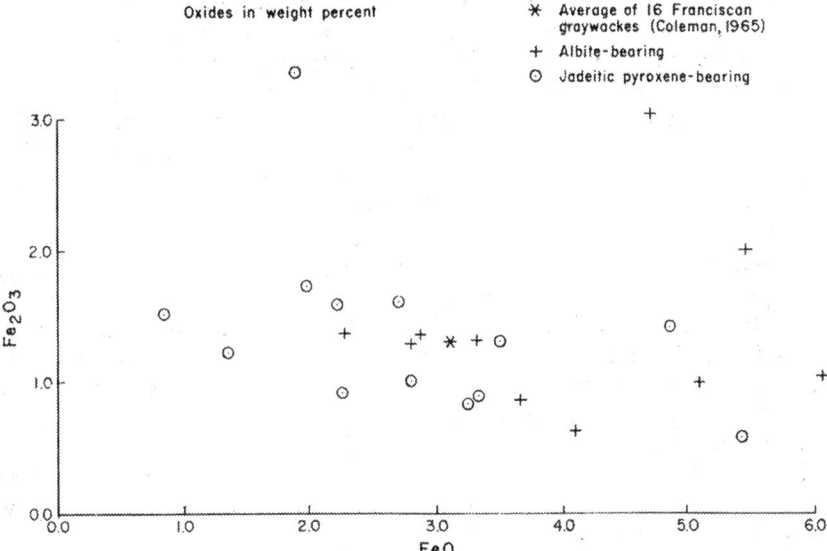

Fig. 3. Proportions of FeO and Fe_2O_3 for conventionally analyzed Franciscan metagraywackes; samples are the same as those illustrated in figure 2.

trasting assemblages. Even in outcrops exhibiting monomineralic veins and other evidence of local metasomatism, no new phases absent from the host rock have been formed. Thus, although metasomatic effects are apparent in some exposures, *chemical changes cannot be held responsible for generation of the distinctive mineralogy of blueschists and related rocks* but merely account for the observed differences in the phase proportions, which necessarily reflect bulk chemistry.

METASTABLE RECRYSTALLIZATION OF OBSERVED ASSEMBLAGES

The problem of metastability is encountered in the investigation of virtually all metamorphic processes. Undoubtedly this phenomenon is of major significance in considering low-temperature parageneses, for reaction rates under such conditions are of critical importance in determining the exact nature of the assemblage produced from a relatively unstable protolith. The principal constraint operative in such a recrystallization process is that under the metamorphic conditions the observed phase associations must have possessed a lower total Gibbs free energy than the initial reacting material. Several different varieties of metastable recrystallization can be envisioned:

1. low-pressure production of metastable blueschist-type phase compatibilities from initial mineral assemblages of even higher Gibbs free energy;
2. low-pressure stability of what would otherwise be a metastable blueschist assemblage as a result of greatly increased surface

energy (hence molar G) of the low-pressure, inherently stable phase association due to a particularly active pore fluid;

3. high-pressure persistence of metastable low-pressure phase assemblages except where reaction rates were sufficient to allow their conversion to more stable blueschist-type compatibilities.

The first two explanations obviate the necessity of postulating the attendance of high pressures during Franciscan metamorphism, whereas the third ascribes the somewhat irregular development of glaucophane schists and allied rocks to local catalysis under more uniform blueschist facies conditions. Let us consider each of these three hypotheses in turn.

1. Fyfe and Newton (in press) have advocated a proposal, advanced originally by Hlabse and Kleppa (1968), that accounts for the apparently high-pressure production of jadeitic pyroxene in metagraywackes by the breakdown of metastable high albite at moderate pressures. Calorimetric work by Hlabse and Kleppa showed that albite of high structural state (that is, Al and Si disordered among the tetrahedral sites) possesses a molar entropy exceeding that of ordered, low albite by 3.5 entropy units. At modest temperatures, the stable equilibrium jadeite + quartz = low albite lies at pressures 2 to 3 kb higher than the metastable equilibrium jadeite + quartz = high albite (compare Newton and Smith, 1967, fig. 6, with Hlabse and Kleppa, 1968, fig. 1).

In what manner could metastable disordered albite possibly have been present in the Franciscan rocks? Either it might have accumulated as clastic grains of volcanic albite, or it might have represented an intermediate stage in the recrystallization of more calcic detrital plagioclase of various structural states. An implication of the first alternative concerns the possible provenance of high albite: although the Klamath and Sierran terranes easily could have been the source of great amounts of largely plutonic, ordered, Ca-bearing plagioclase as well as volcanic, somewhat disordered calcic plagioclase derived from the superjacent extrusives, occurrences of albite-bearing rhyolites required in vast quantities by the first alternative are practically non-existent in the geologic record of California. X-ray work presented by Seki, Ernst, and Onuki (1969, tables 19 and 20) for 83 samples of Franciscan metagraywackes and metavolcanic rocks indicates that sodic plagioclases from the Diablo Range and its northern extension are uniformly of low structural state. Such feldspars exhibit textural evidence of an early stage of albitization of preexisting detrital grains of intermediate plagioclase, hence the second alternative must be considered.

In general, if intermediate plagioclase recrystallizes to disordered $NaAlSi_3O_8$ as a transitional stage preceding the formation of low albite, it is strange that high albite has not been described from a variety of graywackes and arkoses (for example, see Hawkins, 1967, table 5), nor has jadeitic pyroxene been widely reported as a constituent of metamorphosed feldspathic sandstones. A far commoner situation involves the observed production of ordered albite from more calcic plagioclase, as is true of the more deeply buried portions of the Great Valley sequence, reported by Dickinson, Ojakangas, and Stewart (1969). Inasmuch as the Franciscan and Great Valley strata seem to have had a common source, it is also unlikely that occurrences of jadeitic pyroxene derived from metastable high albite would be confined to the former sequence.

Photomicrographs illustrating the partial or complete replacement of preexisting sodic plagioclase in Franciscan clastic rocks are presented in plate 1. A complex, fine-grained intergrowth of jadeitic pyroxene + minor lawsonite + quartz which faithfully pseudomorphs plutonic—hence ordered—sodic plagioclase subhedra in a granitic pebble from a Panoche Pass metaconglomerate is shown in plate 1-A (Ernst, 1965). A similar metaconglomerate has been studied west of Pacheco Pass (see also Fyfe and Zardini, 1967); here the degree of replacement of plutonic albite in the granitic pebbles is incomplete (see samples X-39, X-30A, X-39B, X-40, X-92, X-93B, X-93C, X-93D, X-207A of Seki, Ernst, and Onuki, 1969, tables 6 and 20; Ernst and others, 1970, fig. 9). In plate 1-B clastic grains of albitized plagioclase exhibit rare acicular sprays of jadeitic pyroxene in a metagraywacke that crops out near the Calaveras Reservoir (Ernst, in press). Like other Diablo Range metasedimentary rocks, these Pacheco Pass and Calaveras Reservoir sodic plagioclases have refractive indices less than that of Canada balsam; whole rock X-ray examination of the illustrated specimens yields $Cu_{K\alpha}$ two θ separations between 131 and 1$\bar{3}$1 peaks of 1.11 and 1.15 respectively, indicating the presence of virtually pure, ordered albite.

Plate 1 also illustrates the general lack of shearing observed in some jadeitic pyroxene-bearing metaclastics, particularly in rocks of the Diablo Range; in many of these samples, recrystallization has taken place with preservation of the original detrital textures. Therefore, accumulated strain cannot be called upon to have promoted mechanically-induced twinning in albite (which thereby would have attained an elevated molar Gibbs free energy relative to strain-free albite).

In summary, *there is no evidence to suggest that disordered albite, or highly strained albite, was an important constituent of Franciscan metaclastic rocks at the time of intense metamorphism. All available data indicate the presence of ordered, low albite as in the Great Valley section; hence for jadeitic pyroxene-bearing metagraywackes of the Franciscan, the appropriate equilibrium to be considered is the one investigated experimentally by Newton and Smith (1967), rather than the lower pressure metastable reaction.*

The occurrence of apparently high-pressure aragonite in Franciscan metamorphic rocks has also been discussed by Ernst and others (1970, chaps. 3, 4, and 13) and by Fyfe and Newton (in press). These authors have pointed out that several processes may lead to metastable formation of the orthorhombic polymorph. These include: (A) precipitation of aragonite from a solution that has become supersaturated with respect to $CaCO_3$ due to the presence of cations such as Mg which inhibit growth of the stable polymorph (see Wray and Daniels, 1957; Bischoff and Fyfe,

PLATE 1

Photomicrographs of albite-jadeitic pyroxene textural relations in Diablo Range

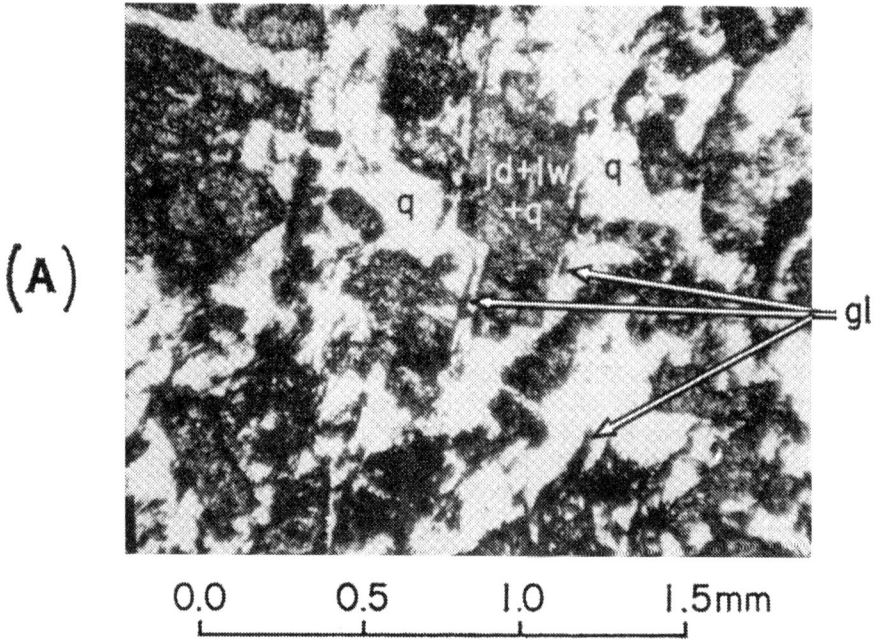

A. Granitic pebble in metaconglomerate from Panoche Pass (sample no. 185). Although the original microscopic examination failed to reveal jadeitic pyroxene, this rock lies within the so-called "jadeitic pyroxene + lawsonite isograd" (Ernst, 1965, table 8 and pl. 1), and the absence of both sodic plagioclase and sodic pyroxene was puzzling (glaucophane is abundant in this pebble, however). Reexamination of the sample during the course of the present study revealed that pseudomorphs after subhedral plutonic plagioclase consist of fine-grained intergrowths of jadeitic pyroxene, lawsonite, and quartz. Plane light.

1968); (B) production of aragonite within the calcite I stability field by metastable inversion of calcite II (Boettcher and Wyllie, 1967, 1968a); and (C) growth of aragonite from strained—hence elevated molar G—calcite within the (strain-free) calcite stability field (Newton, Goldsmith, and Smith, 1969). There is no doubt that each of these mechanisms may be operative in a specific geologic environment. However, Franciscan phase assemblages exhibit the systematic progressive changes albite + quartz $\pm$ calcite → albite + quartz $\pm$ aragonite → jadeitic pyroxene + quartz $\pm$ aragonite, precisely the sequence predictable from experimental phase equilibrium studies. Therefore, it seems unnecessary to call on metastable low-pressure crystallization of aragonite in Franciscan metamorphic rocks, especially where associated sodic pyroxene can be shown to have been produced stably at even higher pressures than required for the formation of the aragonite.

PLATE 1

metaclastics. Abbreviations are as follows: ab = albite; gl = glaucophane; jd = jadeitic pyroxene; lw = lawsonite; ms = white mica; q = quartz.

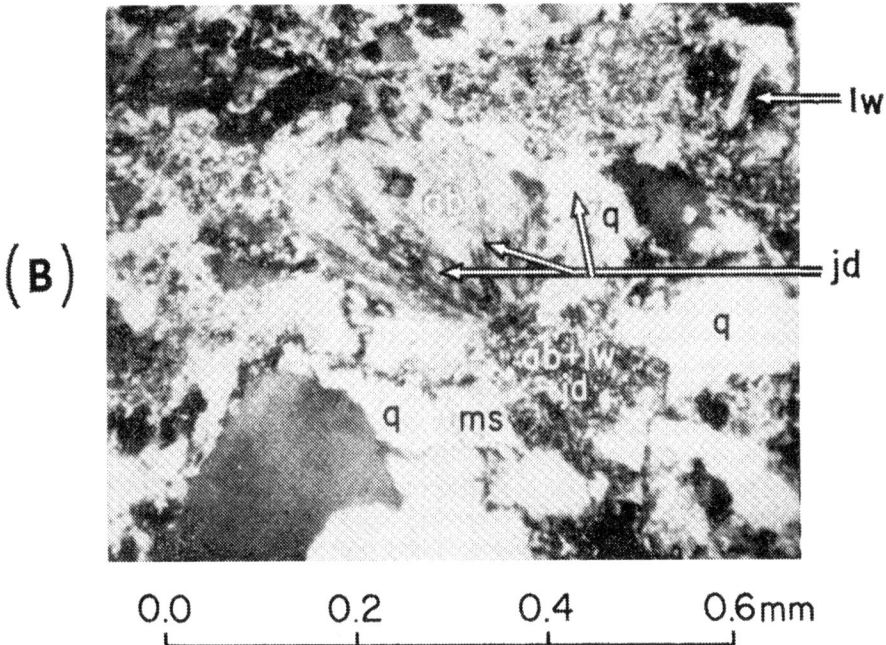

B. Detrital plagioclase grain in metagraywacke (sample no. R-325) from near Calaveras Reservoir (Ernst, in press); the low albite is partly replaced by an acicular spray of jadeitic pyroxene. Crossed nicols.

2. Gresens (1969) has directed attention to a supposed association of *in situ* Franciscan metamorphic rocks with serpentinites and erected a hypothetical solution-recrystallization model based on effects attending hydration of peridotite masses. The process of serpentinization probably results in residual concentration of the saline constituents of an initial country rock pore fluid, because an altering peridotite acts as a desiccating agent in removing H_2O from the surroundings. Oxidation of initially ferrous iron in the preexisting anhydrous ferromagnesian silicates of the ultramafic pluton is thought to have caused a relative lowering of the oxygen fugacity in the aqueous phase. Based on H_2O contents of the involved rock types, crude mass balance calculations indicate that ultramafics and host rocks necessarily would have had to have been present in roughly subequal proportions for the hypothesized effects of serpentinization to have been of more than local importance in altering the composition of the residual interstitial solutions.[3]

[3] Metagraywackes now contain about 3 to 4 wt percent H_2O^+, hydrated ultramafics up to 13 to 14 wt percent H_2O^+. If the original country rocks had lost half their H_2O resulting in only a 2:1 concentration of the residual brines during complete serpentinization of peridotite emplaced in an originally anhydrous condition, then the ultramafic material would have to have been present as an unacceptable 20 to 30 percent of the terrane (see figs. 1 and 4).

Gresens (1969, p. 99-105, and personal commun., 1970) suggests that the resultant saline, reducing hydrothermal fluid might interact with grain surfaces of the originally stable phases such as plagioclase and calcite to produce sodic pyroxene, sodic amphibole, pumpellyite, lawsonite, and aragonite. This process conceivably could take place at low pressures, but only if the surface free energy contribution to the total G of the (formerly) stable assemblage—much of it sand- and silt-sized particles—raised its total Gibbs free energy to a value in excess of that of the metastable blueschist-type compatibility. The hypothesis is as yet untested experimentally. There is no surety that a concentrated, re-

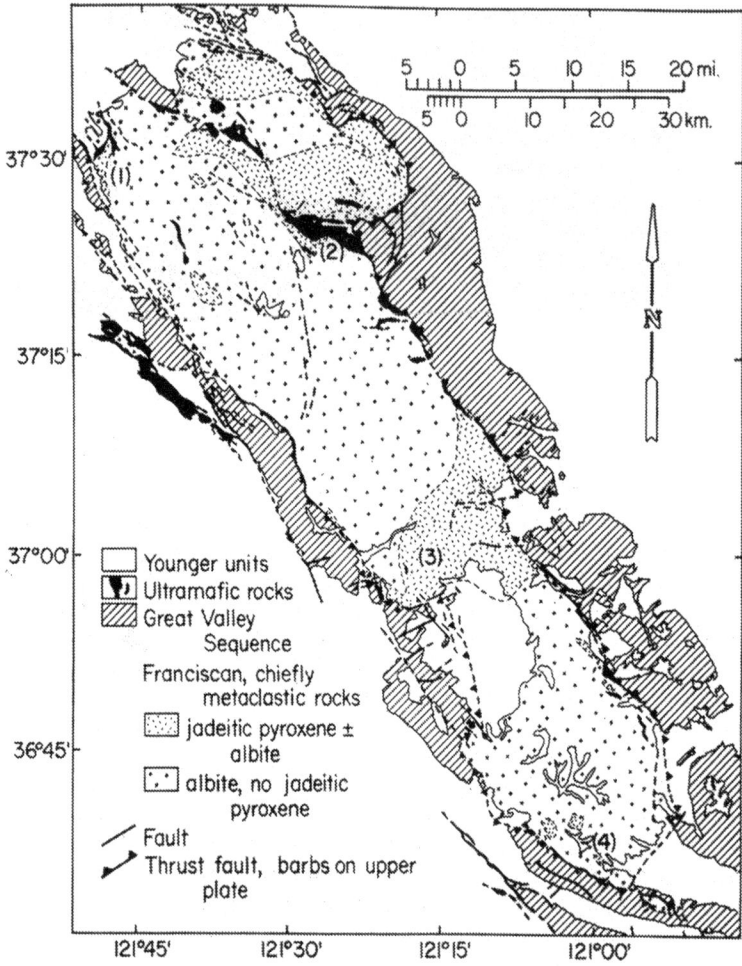

Fig. 4. Distribution of jadeitic pyroxene-bearing and jadeitic pyroxene-free metagraywackes from the Diablo Range, California Coast Ranges (simplified from Ernst, in press, fig. 7). Areas mentioned in the text and located on this map include: (1) Calaveras Reservoir; (2) Red Mountain; (3) Pacheco Pass; and (4) Panoche Pass.

ducing brine would in fact significantly raise the total G of albite-bearing phase assemblages, let alone cause it to increase to the extent that such assemblages would attain higher values of total Gibbs free energy than metastable glaucophane schist-type compatibilities for the same bulk composition.

The petrogenetic model is invoked to account for an alleged spatial and genetic association between serpentinized peridotites and the rather unique Franciscan metamorphic rocks, but such a relationship is not evident in any area known to the writer (see discussion in the previous section). As a case in point, the areal distributions of ultramafics and of jadeitic pyroxene-bearing and Na-pyroxene-lacking *in situ* metaclastic rocks of the Diablo Range are illustrated in figure 4. Serpentinized peridotites are volumetrically unimportant in this portion of the Franciscan terrane. A clear correlation between jadeitic metagraywackes and the serpentinites is not obvious.

It may also be noted that, in figures 2B and 3, proportions of the oxides Na_2O, FeO, and MgO, and the Fe_2O_3/FeO ratios of albite-bearing Franciscan metaclastic rocks exhibit *virtually* the same range of values as do those of jadeitic metagraywackes; if the latter assemblage had been produced by reaction of protoliths with a reducing saline pore fluid, one might expect sodic pyroxene-bearing metaclastics to display systematically higher Na_2O contents and lower ferric/ferrous ratios.

In conclusion, although perhaps not impossible, *available field relations, chemical and thermodynamic data do not seem to substantiate the metastable crystallization hypothesis proposed by Gresens.*

3. In the Pacheco Pass area, albite is associated with jadeitic pyroxene in metagraywackes over an extensive region (McKee, 1962a; Ernst and Seki, 1967). Because of the presence of minor Ca, Fe, and Mg in the pyroxene solid solution, this phase must coexist stably with quartz and albite over a P–T zone (Robertson, Birch, and MacDonald, 1957). Calculations by Essene and Fyfe (1967) and experimental studies by Newton and Smith (1967) demonstrate that at low temperatures and in the presence of quartz, $NaAlSi_2O_6$-rich solid solutions complete decomposition to albite-bearing assemblages at a pressure about 1 kb lower than does pure jadeite. Therefore, although widespread association of Na-pyroxene with albite in the vicinity of Pacheco Pass might reflect local domain equilibrium compatibilities produced under physical conditions within this relatively narrow P–T zone, such an explanation requires a somewhat restricted metamorphic geothermal gradient.

Textural relations described by McKee (1962a), Ernst (in press), and Ernst and others (1970) indicate that preexisting albite has been partially or completely replaced by jadeitic pyroxene + quartz in Diablo Range metagraywackes (refer also to pl. 1). Accordingly, the observed associations of these phases in some cases may reflect the metastable persistence of albite. Under this hypothesis, large tracts of the Franciscan terrane now exposed in the Diablo Range may have been subjected more uniformly to the high pressures attending blueschist facies metamorphism, but sluggish reaction rates allowed the local, incomplete preservation of metastable initial phase associations. Recrystallization to more stable mineral compatibilities would be accelerated by the catalytic effect of fluids and by the increased reaction surface produced by granulation accompanying shearing (Ernst, 1963a, p. 17). A similar suggestion to account for an inverted metamorphic sequence described in Franciscan rocks of the South Fork Mountain area, northern Coast Ranges, has been advanced by Bailey and Blake (1969), but it is not the explanation these authors favor (see next section).

To summarize, *petrographic relations indicate that blueschist-type phase associations possessed lower total Gibbs free energies than did the preexisting ones during Franciscan metamorphism. Conceivably other configurations of even lower total G could have existed, but inasmuch as traces of them have not been recognized, the glaucophane schist-type associations are believed to have represented the stable, equilibrium configuration.*

GENERATION OF OBSERVED ASSEMBLAGES BY TECTONIC OVERPRESSURES

Building on the earlier works of Irwin (1960, 1964), Kilmer (1962), and by Blake (ms), Blake, Irwin, and Coleman (1967) have described a sequence of inverted or "upside down" metamorphism in the vicinity of South Fork Mountain, northern California. Here imbricate thrust sheets of the Klamath province overlie Franciscan rocks along a low-angle, east-dipping fault. Within the Franciscan terrane, hydrous calcium-aluminum silicates, which have grown in quartzose, albite-bearing metagraywackes, are thought by these authors to exhibit a progressive west-to-east paragenesis of pumpellyite → lawsonite.

An even lower grade, laumontite-bearing association crops out to the west of the pumpellyitized metagraywacke terrane. Calcite occurs sporadically on the west, whereas aragonite is a critical phase nearer the thrust surface. As described by these investigators, the degree of textural reconstitution also gradually increases, proceeding eastward toward this low-angle fault. Blake, Irwin, and Coleman (1967) have suggested that such relations represent an inverted metamorphic sequence. They invoked the buildup of considerable stress—tectonic overpressures—and concomitantly high aqueous fluid pressures to account for the observed assemblages in the Franciscan rocks lying along the sole of the thrust[4]; the confining pressure ($P_{lithostatic}$) was thought to have been moderate, on the order of 4 kb, at the time of recrystallization. More recently, Blake, Irwin, and Coleman (1969), Blake and Cotton (1969), and Bailey and Blake (1969) have extended this tectonic-metamorphic concept to other portions of the Franciscan terrane to account for the observation that in many areas, more thoroughly recrystallized (for example, jadeitic

[4] Other authors (for example, Coleman and Lee, 1962, p. 594; Coleman, 1967a) have also called on large tectonic overpressures to explain apparently high-pressure blueschist facies mineral assemblages in the Franciscan but without specifically relating the strain buildup to structural features.

pyroxene-bearing) metaclastic sections seemingly overlie lower-pressure lithologies.

Blake, Irwin, and Coleman (1969, p. 244) have drawn attention to presumably analogous inverted metamorphic sequences in the outer metamorphic belt of Japan, the Urals, Kamchatka Peninsula, Venezuela, New Caledonia, and New Zealand and have speculated that production of blueschist facies rocks in general may require the attendance of substantial tectonic overpressures. But at least where the present author has some familiarity, namely that part of the outer metamorphic belt of Japan exposed in Shikoku, alternative interpretations are possible. Based on metamorphic and structural studies (Kawachi, 1965, 1968; Ernst and others, 1970, chaps. 7, 8, 15), it has been demonstrated that the metamorphic sequence observed in Shikoku reflects the existence of previously unrecognized recumbent folding. Tectonic emplacement of a higher grade nappe, or series of nappes, over a virtually unmetamorphosed section evidently caused rapid lithostatic pressure increment and an initial blueschist facies recrystallization in the underlying mass, followed by gradual temperature buildup within the latter during the subsequent period of erosive unloading. Structural complexities described in some of the other terranes also hint at the possibility of penecontemporaneous and postmetamorphic tectonic juxtaposition—rather than a simple case of "upside-down" metamorphism.

The generation and persistence during recrystallization of substantial tectonic overpressures require that the rocks so stressed possess considerable strength. Previous experimental results indicate that, as laboratory strain rates decrease, so do dry rock strengths (Heard, 1963). Moreover, in the presence of an aqueous fluid, the strengths of silicates are greatly lessened above moderate temperatures (Griggs and Blacic, 1965); on the base of oxygen isotope studies, Taylor and Coleman (1968, p. 1735-1737) have shown that glaucophane-schist mineral assemblages from the Cazadero area of northern California equilibrated with an ubiquitous hydrous fluid phase. Recent experimental investigation of both strong Franciscan metagraywacke and weak Franciscan metashale by Brace and others (1970) has also shown that, under the appropriate metamorphic conditions (that is, temperatures of about 150°-300°C, moderately high fluid pressures, and geologically reasonable strain rates of approximately 10^{-13} or 10^{-14}), the possibilities of generating tectonic overpressures exceeding about 1 kb on a regional scale are remote for a homogeneous, strong metagraywacke terrane; where such lithologies are intercalated with a subequal proportion of incompetent metashales (as is the case for jadeitic pyroxene-bearing metaclastics exposed in the vicinity of Pacheco Pass), the maintenance of substantial tectonic overpressures seems clearly impossible. The role of solution and recrystallization was not considered by Brace and others, but this process would be expected to cause further weakening of the rocks. Hence, although deformed rocks assuredly have been subjected to at least minor differential stress, laboratory experiments indicate that the magnitude of the effect in Franciscan metaclastics must have been quite small.

Furthermore, although the concept of tectonic overpressures has been invoked to account for an inverted, presumably contemporaneous progressive metamorphic sequence along the sole of a major thrust fault (Blake, Irwin, and Coleman, 1967, p. 7, 1969, p. 237-239, 241), the described petrologic and field relationships as yet have not been sufficiently documented either at South Fork Mountain or farther south. The complete absence of mineralogic evidence for stress buildup in the immediately overlying tectonically emplaced Great Valley section has not been explained either. Conceivably, several thrust plates of contrasting phase assemblages (hence P–T histories) could have been juxtaposed in the South Fork Mountain area as is evidently the case in Franciscan rocks of the nearby Leech Lake Mountain–Anthony Peak region (Suppe, 1969a).

In the Diablo Range where the author has been working, the spatial distribution of jadeitic pyroxene-bearing metagraywackes appears to be unrelated to proximity to the bounding Ortigalita thrust fault, as illustrated in figure 4 (see also Ernst, in press). In this area, it is unclear whether the contact between albitic and jadeitic pyroxene ($\pm$ albite)-bearing metaclastic portions of the terrane represents an isograd or an as yet unrecognized tectonic contact; it is possible that in some places this contact may represent an isograd, whereas in other localities it may be a fault.

Several radiometric studies have dealt with the time of metamorphism of thoroughly recrystallized Franciscan rocks located adjacent to the thrust surface. In the Leech Mountain–Anthony Peak region of northern California, Suppe (1969b) presented evidence for a 150 to 151 m.y. metamorphic age for a plate of jadeitic pyroxene-bearing metagraywacke, whereas overlying and underlying blocks of fossiliferous albitic metagraywacke yielded recrystallization dates ranging from 104 to 109 m.y. Reported apparent ages of metamorphism of the nearby South Fork Mountain schist range from 123 to 136 m.y. Also in northern California, coarse-grained mafic blueschists disposed along a thrust fault at Goat Mountain were metamorphosed at least as long ago as 137 to 148 m.y. (Ernst and others, 1970, chap. 14; Suppe, personal commun., 1970). Near Pacheco Pass, recrystallization occurred—or at least argon loss ceased—about 115 to 122 m.y. B.P. (Suppe, personal commun., *in* Ernst, in press). Although not clearly related to a nearby thrust fault, in place metashales associated with blueschists of the Cazadero area (Coleman and Lee, 1963) provide radiometric ages of metamorphism of 130 to 135 m.y., similar to those listed above (Lee and others, 1963). Although a spectrum of apparent ages has been presented, it would appear that metamorphism of the Franciscan rocks took place largely during (or in some areas possibly prior to) the Early Cretaceous. This is significant because Upper Cretaceous Great Valley sequence rocks constitute a large portion of the overlying plate, both in northern California and especially surrounding the Diablo Range, hence the thrust fault relations now recognized probably were produced 30 to 60 m.y. after culmination of the metamorphic event. Conceivably earlier thrusting events occurred, but the structural complexities so generated in the Franciscan itself have not been adequately deciphered to the present time.

In summary, *neither experimental strength-of-materials studies, detailed petrologic mapping, nor metamorphic-stratigraphic temporal relations lend support to the hypothesized production of tectonic overpressures on a regional scale in Franciscan metamorphic rocks.*

PRODUCTION OF OBSERVED ASSEMBLAGES BY DEEP BURIAL

Arguments favoring deep burial, on the order of 20 to 30 km or more, are derived primarily from a comparison of the observed mineral parageneses with experimentally determined phase equilibria. Although the stability relationships of pumpellyite are known only through reconnaissance work (see Hinrichsen and Schürmann, 1969), those involving laumontite and lawsonite have been studied by several investigators (Newton and Kennedy, 1963; Crawford and Fyfe, 1965; Liou, 1968, 1969). Physical conditions of the calcite-aragonite transition have been located by numerous experimentalists employing a variety of techniques (Jamieson, 1953; Clark, 1957; Crawford and Fyfe, 1964; Boettcher and Wyllie, 1967, 1968a; Newton, Goldsmith, and Smith, 1969). The equilibrium between jadeite + quartz and albite and that between jadeitic pyroxene + quartz and albite have been investigated by Birch and LeComte (1960), Newton and Smith (1967), and Boettcher and Wyllie (1968b). The P–T order-disorder relations in sodic amphiboles were studied by Ernst (1963b) (see also Papike and Clark, 1968). Equilibrium diagrams are summarized in figure 5.

As previously stated, numerous petrologic investigations of specific portions of the Franciscan terrane have documented systematic mineral parageneses. The progressive metamorphic phase assemblages developed in quartzose clastic rocks are compatible with the experimentally investigated phase equilibria. Thus the change in mineral assemblage, (A) laumonite + albite ± calcite → (B) pumpellyite and/or lawsonite + albite ± calcite → (C) lawsonite + albite ± aragonite → (D) lawsonite + jadeitic pyroxene ± aragonite, represents a sequence of pressure increment at nearly constant temperature as seen from the schematic metamorphic geothermal gradient indicated in figure 5D. Sodic amphibole—the more ordered, glaucophane II polymorph—occurs as a minor constituent in some of the originally more chloritic metaclastics, where it is associated with jadeitic pyroxene and/or aragonite. The volumetrically much less abundant mafic igneous rocks metamorphosed *in situ* display a simpler equilibrium paragenesis ranging from calcitic, albitic greenstones to glaucophane II and/or omphacite + lawsonite + sphene + aragonite (that is, blueschist-type) compatibilities, suggesting an increasing pressure sequence. *The close correspondence of the observed changes in mineralogy with the experimentally determined phase equilibria lends*

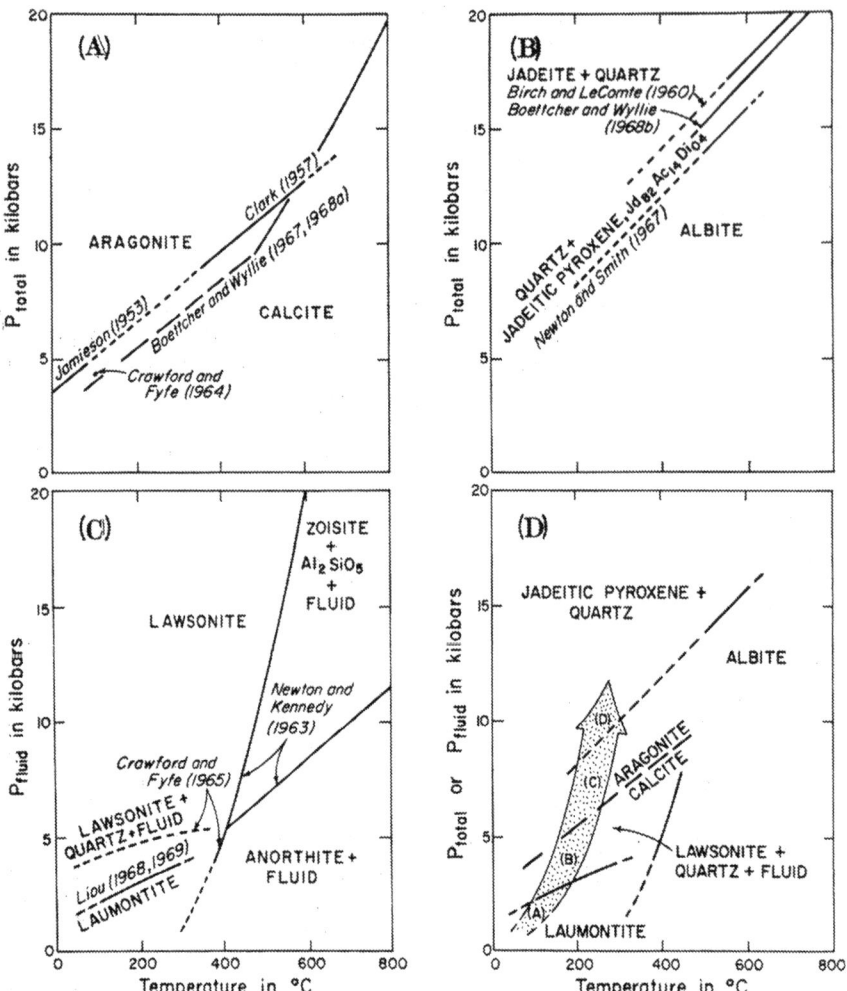

Fig. 5. Some experimentally determined or calculated phase equilibria which pertain to the observed Franciscan mineral parageneses. In (D), the inferred range of physical conditions attending the progressive metamorphism is indicated very diagrammatically.

credence to the hypothesis that the natural occurrences represent a stable sequence.

If so, judging from combination of the oxygen isotopic geothermometry presented by Taylor and Coleman (1968) and the laboratory phase equilibrium data, *the most intensely recrystallized in place Franciscan metagraywackes and associated rocks would seem to have been subjected to physical conditions of approximately 150 to 300°C at from 5 to more than 8 kb total pressure.* Provided the production of such mineral assemblages is not due fundamentally to processes involving metasomatism or metastable recrystallization as argued earlier, high pressures and low temperatures of blueschist metamorphism appear to be indicated for the Franciscan. Moreover, if the criticisms of tectonic overpressures advanced in the last section are valid, it follows that *the 5 to 8+ kb pressure necessitated by the paragenesis must involve deep burial, on the order of 20 to 30 km or more.*

Objections to such extreme depths of burial reflect the fact that nowhere has a Franciscan stratigraphic section of the required magnitude actually been measured in the California Coast Ranges, nor has a downward increase in metamorphic "grade" ever been demonstrated. Inasmuch as the depositional base of the Franciscan has not been recognized, however, and the entire section is considerably disturbed, the tectonic thickness during metamorphism (and at present as well) is unknown. Therefore, as of now, it seems premature if not incorrect to regard this deposit as of insufficient tectonic thickness to provide the required burial depths for sections of rock now exposed. Because the gross fabric of the Franciscan terrane has still to be deciphered, it also may be premature to draw the conclusion that the paragenetic sequence is "right side up" or "upside down"; in any case, the problem seems to be chiefly structural and thus far is unresolved. Nevertheless, the geologic plausibility of such great depths of burial may be appropriately questioned.

DISCUSSION

Franciscan metamorphic mineral assemblages appear to provide compelling evidence for a process that combines high lithostatic pressures with low temperatures. This author has proposed the accumulation of Franciscan, principally clastic, first-cycle debris chiefly within the confines of one or a series of oceanic trenches to account for the seemingly rapid downsinking of large volumes of material to great depths (Ernst, 1965, p. 905-910). It was thought that this mechanism would allow sufficient depression of the isotherms to preserve temperatures on the order of 150 to 300°C at 20 to 30 km below the surface. Nevertheless, lacking additional independent evidence of great depth of burial, various other models for producing the Franciscan metamorphic rocks have been persuasively advocated, namely recrystallization due to metasomatism, metastability, or tectonic overpressure. While admitting the validity and importance of those processes, the arguments presented in previous sections seem to the present author to require the rejection of these as determining factors in production of the Franciscan metamorphic mineral assemblages. *All these other hypotheses, for instance, fail to explain the contrasts in phase petrology between coeval Franciscan and Great Valley rocks of similar provenance; only profound burial of the Franciscan followed by substantial postmetamorphic uplift can account for the remarkable mineralogic differences between these tectonically juxtaposed series.*

Now, with the advent of the concept of new global tectonics (for example, Isacks, Oliver, and Sykes, 1968), the general problem of blueschist facies metamorphism needs to be reexamined. Such low-grade metamorphic belts are confined to the vicinity of present or ancient colliding plate margins—or sutures (Dewey and Bird, 1970; Dewey and Horsfield, 1970). In these zones of convergence, masses of crust and mantle lithosphere appear to have been overridden along the Benioff zone. The relatively rapid spreading rates currently observed, coupled with the inferred superposition of plates at their junctions, provides an overall mechanism for pressure increment at low temperatures: although the increase in lithostatic pressure in the overridden, downgoing slab is instantaneous, the temperature rise depends on the relatively slow process of conductive heat transfer. A dynamic, quasi-steady state is thereby produced in which the isotherms within the descending lithospheric plate exhibit a rather extreme downbowing as illustrated by the computations presented by Oxburgh and Turcotte (1968, 1970), MacKenzie (1969), and Minear and Toksöz (1970). Here, then, is a general process to explain the generation of high lithostatic pressures and low temperatures along suture zones and within the descending slab. Inasmuch as this is the locale of blueschists and related metamorphic rocks, it seems only plausible to relate their formation to an environment characterized by unusually high pressures and low temperatures.

Trenches, too, are confined to suture zones, where they evidently result from the dynamics attending a downgoing slab. But was the Franciscan deposited in a trench—or subduction zone—environment? The local geometry of plate convergence can result in a variety of types of superposition of lithospheric blocks, not all of which necessarily produce trenches. For instance, in eastern New Guinea a slab of oceanic crust + uppermost mantle evidently has been thrust over continental crust (Davies and Milson, 1969), generating relatively high-pressure mineral assemblages in the lower plate. Perhaps a similar tectonic regime gave rise to the metamorphic + ultramafic complex in New Caledonia. (This view is not shared by either Coleman, 1967a, or Brothers, 1970; both of these investigators have mapped in New Caledonia, and both regard the high-pressure mineral assemblages as having been produced by tectonic overpressures.) In Shikoku, as discussed previously, the emplacement of recumbent folds, possibly reflecting convergence between the Japanese arc and the western Pacific plate, is hypothesized to have loaded a section of thin continental crust tectonically, thereby generating relatively high-pressure mineral assemblages in the underlying rocks (Ernst and others, 1970, chap. 15). Such a process conceivably might also account for the observed early stage of Alpine blueschist facies metamorphism (for example, see van der Plas, 1959; Ellenberger, 1960; de Roever and Nijhuis, 1963; Bearth, 1966). The common characteristic of virtually all these glaucophane schist belts is that they appear to be associated with the lower plates (or at least undertucking of material adjacent to the lower plates) in the vicinity of suture zones, where lithostatic pressures would be high and temperatures low.

There are reasons for suggesting that the Franciscan was deposited in an oceanic, trench-type environment among which, however, must be included the following facts. (1) Taking into account the offset produced by Tertiary strike-slip movement on the San Andreas fault (Hill and Dibblee, 1953), the present areal distribution of the Franciscan marks a linear belt of very thick accumulation roughly parallel to but seaward of the more-or-less coeval Klamath, Sierran, and Salinian volcanic-plutonic arcs (for example, see Bailey, Irwin, and Jones, 1964). Present day trenches typically lie approximately 50 to 250 km to the oceanic side of such volcanic arcs (see Menard, 1964, chap. 5; Dickinson and Hatherton, 1967), just as the Franciscan is disposed relative to the Sierran-type terrane. (2) The depositional base of the Franciscan nowhere has been recognized, and the only visible fragments of underlying units consist of basaltic and ultramafic compositions—units typical of oceanic crust and mantle (Ernst, 1965). (3) There is clear evidence for underthrusting of the Franciscan relative to the South Fork Mountain schists and the Great Valley sequence (Irwin, 1960, 1964; Brown, 1964; Page, 1970). (4) The in-part chaotic (mélange) nature of the Franciscan (Hsü, 1968) is compatible with a process involving deposition and both penecontemporaneous and later shearing and tectonic dislocation within a subduction zone. And (5), great quantities of sediment known to have been derived from western North America and shed westward are missing from the geologic record, hence presumably were overridden by the continent and dragged down along a trench-Benioff zone complex (Gilluly, 1969.)

If Pacific sea floor + mantle underflow occurred at the western margin of North America during Late Mesozoic time, the deposition of vast amounts of clastic debris eroded from the continentalward arc would necessarily have resulted in the subduction of Franciscan-type sediments. *Models presented by Hamilton (1969, fig. 5), Page (1970, fig. 9), and Ernst (1970, figs. 3 and 4) all illustrate depths of underthrusting that satisfy the 20 to 30+ km seemingly required by the observed Franciscan metamorphic mineral assemblages. Explanations of these parageneses that call upon metasomatism, metastable crystallization, or tectonic overpressures at best seem to be only incidentally related to such events.*

DYNAMIC MODEL

Let us provisionally accept a trench-subduction zone environment for Franciscan accumulation, tectonic thickening, deformation, and recrystallization. The generalized dynamic model is presented in figure 6A, modified from Ernst (1970, fig. 3); it shows the proposed relationships among downwarped isotherms in the vicinity of the trench deposits, partial melting of the downgoing oceanic crust at much greater depths to provide the Sierran-type island arc volcanics and plutonics, and lithospheric plate convergence. The exposure of what apparently is oceanic crust along the western margin of the Great Valley (Bailey, Blake, and Jones, 1970) is illustrated in figure 6B. According to the model illustrated

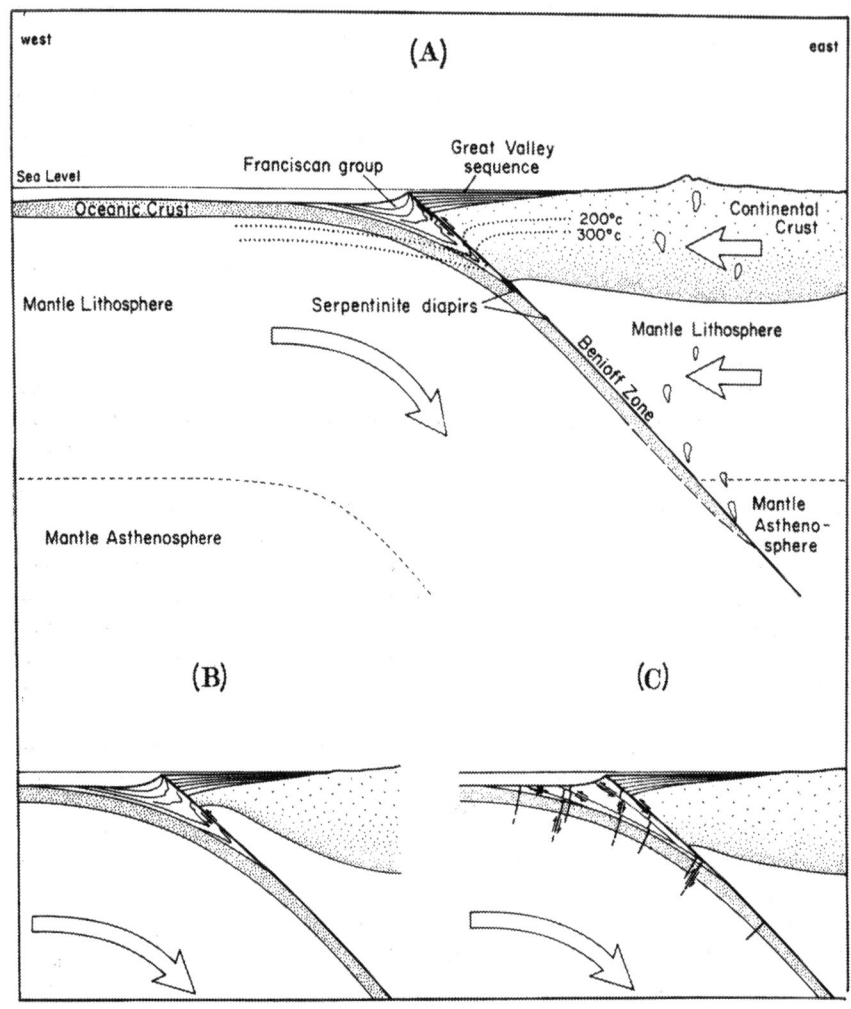

Fig. 6. Tectonic model relating Franciscan group features to a period of active plate consumption, slightly modified from Ernst (1970, fig. 3); no vertical exaggeration implied. (A) presents the overall geometry illustrating downbowing of the isotherms of the trench complex due to sea-floor spreading; basaltic oceanic crust is presumed to have been transformed to eclogite and subsequently to have undergone partial melting at profound depths to yield Sierran, island arc-type igneous rocks (see Ringwood and Green, 1966, p. 420 and fig. 9F). Whereas (A) exhibits a regional setting appropriate to the South Fork Mountain and Nacimiento-Salinia areas, (B) shows relations including a basal contact of overlying Great Valley strata with oceanic crust of the North American plate exposed directly east of the Coast Range thrust in central California as documented by Bailey, Blake, and Jones (1970). (C) presents hypothesized fault sets presumably controlled by the Benioff master shear zone (see Malahoff, 1970, fig. 11) and tensional features in the vicinity of the downturn of the oceanic lithospheric plate (Isacks, Oliver, and Sykes, 1968, fig. 7). Some of the shears subparallel to the Benioff zone may transect oceanic crust and underlying mantle.

in figure 6, Franciscan and Great Valley rocks owe their contrasting lithologies and style of deformation—but essentially common provenance—to the fact that they were laid down on the eastern Pacific and North American plates respectively; their juxtaposition along the South Fork Mountain–Stoney Creek–Ortigalita–Sur-Nacimiento thrust (that is, the Coast Range thrust of Bailey, Blake, and Jones, 1970) is thought to represent the crustal expression of a Late Mesozoic Benioff shear zone (Ernst, 1970).

It still remains to be answered how such a tectonic setting could account for the present structural relations and areal distribution of the Franciscan metamorphic mineral assemblages. Aspects of the model depicted in figure 6C may provide a partial explanation: here, hypothesized differential movements on intersecting fault sets within the subducted prism of tectonically thickened and deformed trench mélange would require the thrusting of deeper, largely older, more thoroughly recrystallized eastern rocks relatively over predominantly shallower, younger, feebly metamorphosed western sections as is observed in some portions of the Diablo Range and its northern extension. Modern day oceanic deeps are typified by what appears to be extensional rifting (normal faulting) as shown for instance by Ludwig and others (1966), Isacks, Oliver, and Sykes (1968, fig. 7), and Malahoff (1970). The last author has interpreted this system of faults in some localities as being subparallel to the Benioff zone, with the sense of movement similar to that along the juncture between the overriding and downgoing slabs (Malahoff, 1970, fig. 11). With regard to the Franciscan, we may conjecture that several sets of faults could have developed at any one time, as shown in figure 6C, leading to successive undertucking of the more oceanward, higher level blocks beneath the more continentalward, more deeply buried sections. Slight changes in the geometry of plate impingence with time would have been reflected in variations in the attitudes of the fault systems. Thus a series of imbricate thrusts—anastomosing with time—would be expected to have shuffled systematically more easterly higher-pressure Franciscan rocks over more feebly recrystallized strata on the west.

The net displacement of the Franciscan suite would have been successively deeper and deeper beneath the more normal, low-pressure Great Valley sequence during active subduction as illustrated in figure 6. However, the relationships observed today are quite different: blueschists and related high-pressure metamorphics which appear to have recrystallized at profound depths evidently have been brought up subsequently to their present level of exposure, principally along eastern portions of the Franciscan terrane. Large-scale structures, mineral parageneses, and the ages of rocks and fossils in different tectonic blocks all indicate that *the Franciscan terrane now spatially associated with the Great Valley sequence was metamorphosed in an environment far removed from the present one.* Because the predominantly metaclastic trench filling possessed a lower aggregate density than the mantle material which it displaced during subduction, diminution in the spreading rate, or cessation of underflow, of the eastern Pacific lithospheric plate would have resulted in the buoyant upward surge of the Franciscan terrane. From the present gross distribution of the lithologic assemblages now observed, it would appear that the transport direction of the ensimatic mélange involved both large western and vertical components. Such movements would be expected further to enhance the imbrication of blueschist facies blocks chiefly on the east over less deeply buried oceanward sections and would result in the asymmetric upturn of the now steeply dipping western limb of the Great Valley synclinorium (see Lachenbruch, 1962; Safonov, 1962).

In the author's opinion, such a model involving deep burial, seemingly required by the observed phase relations, in a Late Mesozoic subduction zone at the western margin of North America, more adequately explains the currently understood character of the Franciscan and its juxtaposition against other rock units than alternative hypotheses thus far advanced.

ACKNOWLEDGMENTS

The somewhat parochial point of view presented in this paper is the result of a continuing association with the Franciscan itself and with scientific colleagues also engaged in its investigation. Diverse portions of the terrane possess contrasting features; hence different opinions are quite justifiably held. The present paper has benefited from the criticisms of E. H. Bailey, M. C. Blake, Jr., and R. G. Coleman, all of the U.S. Geological Survey; R. L. Gresens, University of Washington; R. C. Newton, University of Chicago; and especially John Suppe, University of California, Los Angeles, for which the author is very much obliged. It should be noted, of course, that none of these authorities agrees very completely with the ideas expressed here.

REFERENCES

Bailey, E. H., and Blake, M. C., Jr., 1969, Late Mesozoic sedimentation and deformation in western California: Geotektonika, v. 3, p. 17-34; v. 4, p. 24-34 (in Russian).

Bailey, E. H., Blake, M. C., Jr., and Jones, D. L., 1970, On-land Mesozoic oceanic crust in California Coast Ranges: U.S. Geol. Survey Prof. Paper 700-C, p. 70-81.

Bailey, E. H., Irwin, W. F., and Jones, D. L., 1964, Franciscan and related rocks and their significance in the geology of western California: California Div. Mines and Geology Bull., v. 183, 177 p.

Barnes, I., and O'Neil, J. R., 1969, The relationship between fluids in some fresh alpine-type ultramafics and possible modern serpentinization, Western United States: Geol. Soc. America Bull., v. 80, p. 1947-1960.

Bearth, Peter, 1966, Zur mineralfaziellen Stellung der Glaukophangesteine der Westalpen: Schweizer Mineralog. Petrog. Mitt., v. 46, p. 13-23.

Birch, Francis, and LeComte, Paul, 1960, Temperature-pressure plane for albite composition: Am. Jour. Sci., v. 258, p. 209-217.

Bischoff, J. L., and Fyfe, W. S., 1968, Catalysis, inhibition, and the calcite-aragonite problem. I. The aragonite-calcite transformation: Am. Jour. Sci., v. 266, p. 65-79.

Blake, M. C., Jr., ms, 1965, Structure and petrology of low-grade metamorphic rocks, Blueschist facies. Yolla Bolly area, northern California: Ph.D. thesis, Stanford Univ., 91 p.

Blake, M. C., Jr., and Cotton, W. R., 1969, Inverted metamorphic mineral zones in Franciscan metagraywacke of the Diablo Range, Northern California [abs.]: Geol. Soc. America Abs. with Programs, v. 1, no. 2, p. 6-7.

Blake, M. C., Jr., Irwin, W. P., and Coleman, R. G., 1967, Upside-down metamorphic zonation, blueschist facies, along a regional thrust in California and Oregon: U.S. Geol. Survey Prof. Paper 515-C, p. 1-9.

——— 1969, Blueschist-facies metamorphism related to regional thrust-faulting: Tectonophysics, v. 8, p. 237-246.

Bloxam, T. W., 1956, Jadeite-bearing metagraywackes in California: Am. Mineralogist, v. 41, p. 488-496.

——— 1959, Glaucophane-schists and associated rocks near Valley Ford, California: Am. Jour. Sci., v. 257, p. 95-112.

——— 1960, Jadeite-rocks and glaucophane schists from Angel Island, San Francisco Bay, California: Am. Jour. Sci., v. 258, p. 555-573.

——— 1966, Jadeite-rocks and blueschists in California: Geol. Soc. America Bull., v. 77, p. 781-786.

Boettcher, A. L., and Wyllie, P. J., 1967, Revision of the calcite-aragonite transition, with the location of a triple point between calcite I, calcite II, and aragonite: Nature, v. 213, p. 792-793.

——— 1968a, The calcite-aragonite transition measured in the system $CaO-CO_2-H_2O$: Jour. Geology, v. 76, p. 314-330.

——— 1968b, Jadeite stability measured in the presence of silicate liquids in the system $NaAlSiO_4-SiO_2-H_2O$: Geochim. et Cosmochim. Acta, v. 32, p. 999-1012.

Brace, W. F., Ernst, W. G., and Kallberg, R. W., 1970, An experimental study of tectonic overpressure in Franciscan rocks: Geol. Soc. America Bull., v. 81, p. 1325-1338.

Brothers, R. N., 1954, Glaucophane schists from the North Berkeley Hills, California: Am. Jour. Sci., v. 252, p. 614-626.

——— 1970, Lawsonite-albite schists from northernmost New Caledonia: Contr. Mineralogy and Petrology, v. 25, p. 185-202.

Brown, R. D., 1964, Thrust-fault relations in the northern Coast Ranges, California: U. S. Geol. Survey Prof. Paper 475-D, p. 7-13.

Chesterman, C. W., 1960, Intrusive ultrabasic rocks and their metamorphic relationships at Leech Lake Mountain, Mendocino County, California: Internat. Geol Cong., 21st, Copenhagen 1960, Rept., Pt. 13, p. 208-215.

Clark, S. P., Jr., 1957, A note on calcite-aragonite equilibrium: Am. Mineralogist, v. 42, p. 564-566.

Coleman, R. G., 1961, Jadeite deposits of the Clear Creek area New Idria district, San Benito, County, California: Jour. Petrology, v. 2, p. 209-247.

——— 1965, Composition of jadeitic pyroxene from the California graywackes: U.S. Geol. Survey Prof. Paper 525-C, p. 25-34.

——— 1967a, Glaucophane schists from California and New Caledonia: Tectonophysics, v. 5, p. 479-498.

——— 1967b, Low-temperature reaction zones and alpine ultramafic rocks of California, Oregon, and Washington: U.S. Geol. Survey Bull. 1247, 49 p.

Coleman, R. G., and Lee, D. E., 1962, Metamorphic aragonite in the glaucophane schists of Cazadero, California: Am. Jour. Sci., v. 260, p. 577-595.

——— 1963, Glaucophane-bearing metamorphic rock types of the Cazadero area, California: Jour. Petrology, v. 4, p. 260-301.

Crawford, W. A., and Fyfe, W. S., 1964, Calcite-aragonite equilibrium at $100°C$: Science, v. 144, p. 1569-1570.

——— 1965, Lawsonite equilibria: Am. Jour. Sci., v. 263, p. 262-270.

Davies, H. L., and Milsom, J. S., 1969, Eastern Papua geology and gravity [abs.]: Geophys. Union Trans., v. 50, p. 333.

Davis, E. F., 1918, The radiolarian cherts of the Franciscan group: California Univ. Pub. Geol. Sci., v. 11, p. 235-432.

Dewey, J. F., and Bird, J. M., 1970, Mountain belts and the new global tectonics: Jour. Geophys. Research, v. 75, p. 2625-2647.

Dewey, J. F., and Horsfield, B., 1970, Plate tectonics, orogeny and continental growth: Nature, v. 255, p. 521-525.

Dickinson, W. R., 1966, Table Mountain serpentinite extrusion in California Coast Ranges: Geol. Soc. America Bull., v. 77, p. 451-472.

Dickinson, W. R., and Hatherton, T., 1967, Andesitic volcanism and seismicity around the Pacific: Science, v. 157, p. 801-803.

Dickinson, W. R., Ojakangas, R. W., and Stewart, R. J., 1969, Burial metamorphism of the Late Mesozoic Great Valley sequence, Cache Creek, California: Geol. Soc. America Bull., v. 80, p. 519-526.

Dudley, P. P., ms, 1967, Glaucophane schists and associated rocks of the Tiburon Peninsula, Marin County, California: Ph.D. thesis, Univ. of California, Berkeley, 116 p.

Ellenberger, F., 1960, Sur une paragénèse éphémère á lawsonite et glaucophane dans le métamorphisme alpin en Haute-Maurienne (Savoie): Soc. géol. France Bull., v. 7, p. 190-194.

Ernst, W. G., 1959, Alkali amphiboles: Carnegie Inst. Washington Year Book 58, p. 121-126.

────── 1963a, Petrogenesis of glaucophane schists: Jour. Petrology, v. 4, p. 1-30.

────── 1963b, Polymorphism in alkali amphiboles: Am. Mineralogist, v. 48, p. 241-260.

────── 1965, Mineral parageneses in Franciscan metamorphic rocks, Panoche Pass, California: Geol. Soc. America Bull, v. 76, p. 879-914.

────── 1970, Tectonic contact between the Franciscan mélange and the Great Valley sequence, crustal expression of a Late Mesozoic Benioff zone: Jour. Geophys. Research, v. 75, p. 886-901.

────── in press, Petrologic reconnaissance of Franciscan metagraywackes from the Diablo Range, Central California Coast Ranges: Jour. Petrology, in press.

Ernst, W. G., and Seki, Y., 1967, Petrologic comparison of the Franciscan and Sanbagawa metamorphic terraness: Tectonophysics, v. 4, p. 463-478.

Ernst, W. G., Seki, Y., Onuki, H., and Gilbert, M. C., 1970, Comparative study of low-grade metamorphism in the California Coast Ranges and the Outer Metamorphic Belt of Japan: Geol. Soc. America Mem. 124, 276 p.

Eskola, Pente, 1939, Die metamorphen Gesteine, *in* Barth, T. F. W., Correns, C. W., and Eskola, P. J., Die Enstehung der Gesteine: Springer, Berlin, p. 263-407.

Essene, E. J., and Fyfe, W. S., 1967, Omphacite in Californian rocks: Contr. Mineralogy and Petrology, v. 15, p. 1-23.

Essene, E. J., Fyfe, W. S., and Turner, F. J., 1965, Petrogenesis of Franciscan glaucophane schists and associated metamorphic rocks, California: Contr. Mineralogy and Petrology, v. 11, p. 695-704.

Fyfe, W. S., and Newton, R. C., in press, High pressure metamorphism, *in* Bailey, D. K., ed., Experimental Petrology: New York, Academic Press.

Fyfe, W. S., and Zardini, R., 1967, Metaconglomerate in the Franciscan formation near Pacheco Pass, California: Am. Jour. Sci., v. 265, p. 819-830.

Ghent, E. D., 1965, Glaucophane-schist facies metamorphism in the Black Butte area, Northern Coast Ranges, California: Am. Jour. Sci., v. 263, p. 385-400.

Gilluly, James, 1969, Oceanic sediment volumes and continental drift: Science, v. 166, p. 992-994.

Gresens, R. L., 1969, Blueschist alteration during serpentinization: Contr. Mineralogy and Petrology, v. 24, p. 93-113.

Griggs, D. T., and Blacic, J. D., 1965, Quartz, anomalous weakness of synthetic crystals: Science, v. 147, p. 292-295.

Hamilton, Warren, 1969, Mesozoic California and the underflow of Pacific mantle: Geol. Soc. America Bull., v. 80, p. 2409-2430.

Hawkins, J. W., 1967, Prehnite-pumpellyite facies metamorphism of a graywacke-shale series, Mount Olympus, Washington: Am. Jour. Sci., v. 265, p. 798-818.

Heard, H. C., 1963, Effect of large changes in strain rate in the experimental deformation of Yule marble: Jour. Geology, v. 71, p. 162-195.

Hill, M. L., and Dibblee, T. W., Jr., 1953, San Andreas, Garlock and Big Pine faults, California: Geol. Soc. America Bull., v. 64, p. 443-458.

Hinrichsen, von T., and Schürmann, K., 1969, Untersuchungen zur Stabilität von Pumpellyit: Neues Jahrb. Mineralogie Montasch., v. 10, p. 441-445.

Hlabse, T., and Kleppa, O. J., 1968, The thermochemistry of jadeite: Am. Mineralogist, v. 53, p. 1281-1292.

Hoffman, C., 1970, Die Glaukophangesteine, ihre stofflichen Äquivalente und Umwandlungsprodukte in Nordcalabrien (Süditalien): Beitr. Mineralogie Petrographie, v. 27, p. 283-320.

Hsü, K. J., 1968, Principles of mélanges and their bearing on the Franciscan-Knoxville Paradox: Geol. Soc. America Bull., v. 79, p. 1063-1074.

Irwin, W. P., 1960, Geologic reconnaissance of the Northern Coast Ranges and Klamath Mountains, California, with a summary of the mineral resources: California Div. Mines and Geology Bull. 179, 80 p.

―――― 1964, Late Mesozoic orogenies in the ultramafic belts of northwestern California and southwestern Oregon: U.S. Geol. Survey Prof. Paper 501-C, p. 1-9.

―――― 1966, Geology of the Klamath Mountains province, *in* Bailey, E. H., ed., Geology of Northern California: California Div. Mines and Geology Bull. 190, p. 19-38.

Isacks, B., Oliver, J., and Sykes, L. R., 1968, Seismology and the new global tectonics: Jour. Geophys. Research, v. 73, p. 5855-5899.

Jamieson, J. C., 1953, Phase equilibria in the system calcite-aragonite: Jour. Chem. Physics, v. 21, p. 1385-1390.

Kawachi, Y., 1965, Finding of overturned graded bedding in spotted crystalline schists of the Sanbagawa metamorphic zone in central Shikoku, Japan: Geol. Soc. Japan Jour., v. 72, p. 311-313.

―――― 1968, Large-scale overturned structure in the Sanbagawa metamorphic zone in central Shikoku, Japan: Geol. Soc. Japan Jour., v. 74, p. 607-616.

Kilmer, F. H., 1962, Anomalous relationship between the Franciscan formation and metamorphic rocks, northern Coast Ranges, California [abs.]: Geol. Soc. America Spec. Paper 68, p. 210.

Lachenbruch, M. C., 1962, Geology of the west side of the Sacramento Valley, California, *in* Bowen, O. E., Jr., ed., Geologic guide to the gas and oil fields of Northern California: California Div. Mines and Geology Bull. 181, p. 53-66.

Lee, D. E., Thomas, H. H., Marvin, R. F., and Coleman, R. G., 1963, Isotope ages of glaucophane schists from Cazadero, California: U.S. Geol. Survey Prof. Paper 475-D, p. 105-107.

Liou, J. G., 1968, Zeolite equilibria in the system $CaO \cdot Al_2O_3 \cdot 2SiO_2 - SiO_2 - H_2O - CO_2$, the stabilities of wairakite and laumontite [abs.]: Geol. Soc. America Program Ann. Mtg. 1968, Mexico City, p. 175.

―――― 1969, P–T stabilities of laumonite, wairakite and lawsonite [abs]: Am. Geophys. Union Trans., v. 50, p. 352.

Ludwig, W. J., Ewing, J. I., Ewing, M., Murauchi, S., Den, N., Asano, S., Hoffa, H., Hayakawa, M., Asanuma, T., Ichikawa, K., and Noguchi, I., 1966, Sediments and structure of the Japan Trench: Jour. Geophys. Research, v. 71, p. 2121-2137.

MacKenzie, D. P., 1969, Speculations on the consequences and causes of plate motions: Royal Astron. Soc. Geophys. Jour., v. 18, p. 1-32.

Maddock, M. E., 1964, Geology of the Mt. Boardman Quadrangle, Santa Clara and Stanislaus Counties, California: California Div. Mines and Geology Map Sheet 3.

Malahoff, A., 1970, Some possible mechanisms for gravity and thrust faults under oceanic trenches: Jour. Geophys. Research, v. 75, p. 1992-2001.

McKee, Bates, 1962a, Widespread occurrence of jadeite, lawsonite, and glaucophane in central California: Am. Jour. Sci., v. 260, p. 596-610.

―――― 1962b, Aragonite in the Franciscan rocks of the Pacheco Pass area, California: Am. Mineralogist, v. 47, p. 379-387.

Menard, H. W., 1964, Marine geology of the Pacific: New York, McGraw-Hill, 271 p.

Minear, J. W., and Toksöz, M. N., 1970, Thermal regime of a downgoing slab and global tectonics: Jour. Geophys. Research, v. 75, p. 1397-1419.

Misch, Peter, 1959, Sodic amphiboles and metamorphic facies in Mount Shuksan belt, Northern Cascades, Washington [abs.]: Geol. Soc. America Bull., v. 70, p. 1736-1737.

―――― 1966, Tectonic evolution of the Northern Cascades of Washington State, *in* Gunning, H. C., ed., Tectonic history and mineral deposits of the Western Cordillera: Canadian Inst. Mining Metallurgy Spec., v. 8, p. 108-148.

―――― 1969, Paracrystalline microboudinage of zoned grains and other criteria for synkinematic growth of metamorphic minerals: Am. Jour. Sci., v. 267, p. 43-63.

Miyashiro, Akiho, 1966, Some aspects of peridotite and serpentinite in orogenic belts: Japanese Jour. Geology Geography, v. 37, p. 45-61.

Newton, R. C., Goldsmith, R. J., and Smith, J. V., 1969, Aragonite crystallization from strained calcite at reduced pressures and its bearing on aragonite in low-grade metamorphism: Contr. Mineralogy Petrology, v. 22, p. 335-348.

Newton, R. C., and Kennedy, G. C., Some equilibrium reactions in the join $CaAl_2Si_2O_8 - H_2O$: Jour. Geophys. Research, v. 68, p. 2967-2983.

Newton, R. C., and Smith, J. V., 1967, Investigations concerning the breakdown of albite at depth in the earth: Jour. Geology, v. 75, p. 268-286.

Oxburgh, E. R., and Turcotte, D. L., 1968, Problem of high heat flow and volcanism associated with zones of descending mantle convective flow [abs.]: Am. Geophys. Union Trans., v. 49, p. 318.

―――― 1970, Thermal structure of island arcs: Geol. Soc. America Bull., v. 81, p. 1665-1668.

Page, B. M., 1970, Sur-Nacimiento fault zone of California: continental margin tectonics: Geol. Soc. America Bull., v. 81, p. 667-690.

Papike, J. J., and Clark, J. R., 1968, The crystal structure and cation distribution of glaucophane: Am. Mineralogist, v. 53, p. 1156-1173.

Peterman, Z. E., Hedge, C. E., Coleman, R. G., and Snavely, P. D., 1967: $^{87}Sr/^{86}Sr$ ratios in some eugeosynclinal sedimentary rocks and their bearing on the origin of granitic magma in orogenic belts: Earth and Planetary Sci. Letters, v. 2, p. 433-439.

Plas, L. van der, 1959: Petrology of the northern Adula region, Switzerland (with particular reference to glaucophane-bearing rocks): Leidse Geol. Meded, v. 24, p. 415-602.

Ransome, F. L., 1894, The geology of Angel Island: California Univ. Pub., Geol. Sci., v. 1, p. 193-240.

Raymond, L. A., 1970, Relationships between blueschists facies metamorphism, folding, and faulting in Franciscan rocks, Seegers Ranch area, northeastern Diablo range, California [abs.]: Geol. Soc. America Abs. with Programs, v. 2, no. 2, p. 133-134.

Ringwood, A. E., and Green, D. H., 1966, An experimental investigation of the gabbro-eclogite transformation and some geophysical implications: Tectonophysics, v. 3, p. 383-427.

Robertson, E. C., Birch, Francis, and MacDonald, G. J. F., 1957, Experimental determination of jadeite stability relations to 25,000 bars: Am. Jour. Sci., v. 255, p. 115-137.

Roever, W. P. de, and Nijhuis, H. J., 1963, Plurifacial alpine metamorphism in the eastern Betic Cordilleras (SE Spain), with special reference to the genesis of the glaucophane: Geol. Rundschau, v. 53, p. 324-336.

Saad, A. H., 1969, Paleomagnetism of Franciscan ultramafic rocks from Red Mountain, California: Jour. Geophys. Research, v. 74, p. 6567-6578.

Safonov, A., 1962, The challenge of the Sacramento Valley, California, in O. E. Brown, Jr., ed., Geologic guide to the gas and oil fields of Northern California: California Div. Mines and Geology Bull. 181, p. 77-98.

Saito, M., Hasimoto, K., Sawata, H., and Shimazaki, Y., 1960, Geology and mineral resources of Japan, 2d ed.: Tokyo, Japan, Geol. Survey, 304 p.

Seki, Y., Ernst, W. G., and Onuki, H., 1969, Phase proportions and physical properties of minerals and rocks from the Franciscan and Sanbagawa metamorphic terranes, a supplement to Geol. Soc. America Mem. 124: Tokyo, Japan, Japan Soc. Promotion Sci., 85 p.

Smith, J. P., 1907, The paragenesis of the minerals in the glaucophane-bearing rocks of California: Am. Philos. Soc. Proc., v. 45, p. 183-242.

Suppe, John, 1969a, Franciscan geology of the Leech Lake Mountain-Anthony Peak region, northern Coast Ranges, California [abs.]: Geol. Soc. America Abs. with Programs, 1969 (v. 1), pt. 2, p. 65-66.

―――― 1969b, Times of metamorphism in the Franciscan terrain of the northern Coast Ranges, California: Geol. Soc. America Bull., v. 80, p. 135-142.

Taliaferro, N. L., 1943, Franciscan-Knoxville problem: Am. Assoc. Petroleum Geologists Bull., v. 27, p. 109-219.

Taylor, H. P., and Coleman, R. G., 1968, O^{18}/O^{16} ratios of coexisting minerals in glaucophane-bearing metamorphic rocks: Geol. Soc. America Bull., v. 79, p. 1727-1756.

Turner, F. J., 1968, Metamorphic petrology: New York, McGraw-Hill, 403 p.

Washington, H. S., 1901, A chemical study of the glaucophane schists: Am. Jour. Sci., 4th ser., v. 11, p. 35-59.

Wray, J. L., and Daniels, F., 1957, Precipitation of calcite and aragonite: Am. Chem. Soc. Bull., v. 79, p. 2031-2034.

CHAPTER 8: PSEUDOSECTIONS

Hensen, B.J. (1971) Theoretical phase relations involving cordierite and garnet in the system MgO-FeO-Al$_2$O$_3$-SiO$_2$. *Contributions to Mineralogy and Petrology*, 33, 191–214.

The name 'pseudosection' has been applied to a specific kind of phase diagram, one designed for multicomponent systems that contain mineral solid solutions. It differs from conventional phase diagrams in that it is calculated for one specific bulk composition (*P-T* pseudosection), or for one specific compositional variable (*T-X* and *P-X* pseudosections). Because of the compositional restriction, the diagram excludes portions of projected univariant curves in the 'total' phase diagram that are irrelevant to the rock. The diagram shows instead the divariant, trivariant and higher-variance phase fields that are specific for that bulk composition. Thus, the pseudosection contains less information than the total multidimensional phase diagram and projections within it. However, pseudosections are easier to use for a metamorphic rock or group of rocks of limited compositional variability, because they show only the equilibrium assemblages that are possible in that system. Their use eliminates the problem of figuring out what elements of the total phase diagram do and do not apply to the rocks in question. The name pseudosection first appears to have been used in lectures by Roger Powell in 1986 to denote a two-dimensional section through the total phase diagram in which the phase compositions are not all in the plane of the section.

Any path across a *P-T* pseudosection or a specific *X*-path in a *T-X* or *P-X* pseudosection will show all predicted appearances and disappearances of minerals, and should, therefore, ideally replicate what might be inferred petrographically from thin sections of natural samples that followed the same path. Thus, the pseudosection is a tool we can expect to be of great value in deciphering the metamorphic evolution of a terrane and ultimately the tectonic history of an orogen. Readers might want to consult the text by Will (1998), where pseudosections are discussed in detail.

In his 1971 paper on cordierite-garnet relations in the system MgO-FeO-Al$_2$O$_3$-SiO$_2$, Hensen was the first to construct phase diagrams with these properties, although he did not use the term pseudosection. Such diagrams were slow to catch on because their quantitative calculation requires thermodynamic data for all the mineral end-members involved, and suitable comprehensive databases were not available until recently.

Hensen's paper provides the relative newcomer to metamorphic phase equilibria with an analysis using Schreinemakers' methods of the phase relations among seven minerals in the 4-component system FMAS. This constitutes a multisystem with a potential maximum of seven invariant points (according to experimental work, four are stable and three are metastable), each of which is the source of six univariant reactions. In a system such as this with mineral solid solutions in exchange equilibrium, each ferromagnesian mineral changes composition (X_{Mg}) continuously along the univariant reaction curve as a function of *T* and *P*. As a result, reaction coefficients also change continuously, so that thermodynamic reaction properties such as ΔS and ΔV, and thus Clapeyron slopes, may show significant variation along the curve. Hensen did not consider the case where a coefficient changes sign, a singular point on the curve. The phase rule requires the locus of each univariant curve (with four AFM minerals + quartz) to be independent of bulk composition. Many petrogenetic *P-T* grids for metapelites in the literature are composed of invariant points and intersecting univariant curves with these properties (e.g. Thompson, 1976; Harte and Hudson, 1979; Bickle and Archibald, 1984; Grant, 1985; Koons and Thompson, 1985; Pattison and Harte, 1985; Spear and Cheney, 1989; Powell and Holland, 1990; Spear, 1993). These diagrams do not show the *P-T* limits of specific divariant phase assemblages because this requires selection of a bulk composition. Hensen's schematic *P-T* diagram, a pseudosection (Fig. 8), required specification of a bulk X_{Mg} and limits on its relative Al-content. The univariant reaction assemblage (five minerals) is limited in length, and bordered by divariant reaction fields (four minerals) and, at intersection points, trivariant fields (three minerals).

Hensen's paper gives clear, graphical illustrations of how the locations and widths of divariant assemblages vary in *P-T* and in *P-X* sections as the triangle representing that assemblage shifts across the triangular AFM diagram (Figs 3, 4 and 5). Thus, the pseudosection diagram can be assembled graphically from a reasonably accurate set of AFM projections for different temperatures and pressures. (Note that I have made corrections to the text: Fig. 5 legend, p. L188 bottom line of text, and p. L189 line 1. Note also that whereas Hensen correctly plots in his AFM diagrams the elevations [A/(A+F+M)] of garnet and cordierite at 0.25 and 0.5 respectively, the variable used in the text is A/(F+M)$_b$ which is 1/3 for garnet and 1 for cordierite).

Albee (1972, Figs 13, 14) constructed schematic "pseudobinary *T-X*$_{Mg}$ sections" for metapelites to help clarify our understanding of the sequence and exact nature of successive metamorphic reactions relative to the bulk composition of individual rocks. Seemingly inconsistent appearances, disappearances, absences, and "persistents" of minerals had troubled previous workers on metapelites. In his textbook, Powell (1978, Figs 9-12 and 9-13) constructed schematic "pseudobinary *T-X*$_{Mg}$ sections" for Barrovian metamorphism at two A-values across the Thompson AFM projection, one above and one below the composition of chlorite. The X_{Mg} values in his *T-X*$_{Mg}$ phase-loops are not those of the minerals (Chl, Grt, St, Bt) that occur inside the phase-loops, but those of the bulk compositions corresponding to the changes in mineral assemblages. His Fig. 9-14, on the other hand, is a more complex "T-X projection", with crossing phase-loops that show the actual X_{Mg} values of minerals in the three-phase assemblages of the Thompson projection. The appearance and disappearance of minerals in a prograde metapelite sequence depend (among other things) on the bulk X_{Mg} and the relative Al contents. It is easier to visualize these changes for typical pelite compositions in the pseudobinary T-X section than in the T-X projection, which covers all bulk compositions in the AFM projection from FeMg-biotite to aluminosilicate. Of course, it is the gradual changes in mineral compositions with P and T along the univariant lines that give rise to what we see in the pseudosections.

Hudson (1980) drew schematic P-T diagrams for three whole-rock X_{Mg} values and an elevation in the Thompson AFM projection between biotite and chlorite showing uni-, di-, and tri-variant fields for Buchan-style metamorphism of pelites. Labotka (1981) took a section across the AFM projection at the elevation of the garnet-chlorite join to construct "schematic pseudobinary T-X_{Mg} sections" in order to compare low-pressure, andalusite-type metamorphism of pelites in the Panamint Mountains, California, with Barrovian-type metamorphism in the nearby Funeral Mountains.

Calculated petrogenetic grids covering extensive portions of the PT plane and allowing for mineral solid solutions became feasible only in the late 1980s when internally consistent thermodynamic databases (Chapter 11) began to contain a sufficiently complete listing of end-member mineral data. Unlike grids constructed from petrological observations and limited experimental calibration, calculated full-system grids cover all possible bulk compositions in the chosen system and depict all invariant and univariant assemblages. Their conversion to pseudosections greatly reduces their complexity and enhances their utility, provided a bulk composition is chosen that is appropriate for the application. Quantitative P-T-X pseudosections showing the occurrence of higher-variance assemblages for selected bulk compositions soon followed, and they have since been gaining in popularity.

The first of several papers showing quantitative P-T and T-X pseudosections (Guiraud et al., 1990; Will et al., 1990a,b; Powell and Holland, 1990) explored the P-T-X consequences of the Holland and Powell (1990) database. They considered the compositions of typical metapelites, metabasites, siliceous dolomites, and metaperidotites. They showed the critical role in phase-assemblage stability of the $Al_2Si_{-1}(Mg,Fe)_{-1}$ or tschermaks substitution in many common metamorphic minerals, in addition to the better-known $MgFe_{-1}$ variation. Predicted relationships were found to agree well with assemblages and mineral compositions in P-T-X paths known from natural assemblages; understandably, differences were found when the model system chosen was too "simple". In addition, it was possible for the first time to calculate realistic phase assemblages and compositions for low-grade, greenschist-facies conditions where direct experimental constraints are few.

Later contributions focused more on practical applications of predicted phase assemblages in pseudosections to thermobarometry and on possible P-T-X paths (Dymoke and Sandiford, 1992; Xu et al., 1994; Carrington and Harley, 1995; Schmädicke and Evans, 1997; Will et al., 1998; Reche et al., 1998; Vance and Mahar, 1998; Carson et al., 1999; White et al., 2000; Tinkham et al., 2001; Tinkham and Ghent, 2005; Pattison and Vogel, 2005; Brouwer and Engi, 2005). In many cases, specific high-variance fields cover substantial parts of the diagram, and are thus not very useful without further elaboration. Stüwe and Powell (1995) showed that the changing modal proportions of minerals stable in such high-variance fields in the pseudosection are able to supply additional constraints on the PT path. In addition, these fields can be contoured with isopleths for the mole-fractions of mineral end-members participating in the equilibria (e.g. Meyre et al., 1997; Powell et al., 1998). The preparation of all these diagrams has been made possible by sophisticated computer programs that were many years in the making, such as THERMOCALC (Powell and Holland, 1988), PERPLEX (Connolly 1990; Connolly and Petrini, 2002) and DOMINO (de Capitani, 1994; Biino and de Capitani, 1995). Also, the MELTS software package (Asimow and Ghiorso, 1998) has been extended to calculate subsolidus phase relations. Equilibrium states in THERMOCALC are obtained by the simultaneous solution of linearly independent equilibria; the others employ free-energy minimization codes.

At the present time, there seems to be a preference against use of the term pseudosection, as the following examples from the 2005 D.M. Carmichael issue of *The Canadian Mineralogist* indicate. Tinkham and Ghent (2005) prefer to use the term "isochemical section", short for "isochemical P-T phase-diagram section". They argue that these are true sections through the phase diagram at constant chemical composition, and are not "false" sections. Pattison and Vogl (2005) prefer the names "mineral-assemblage stability diagram" for P-T and T-X pseudosections, where X refers to the whole rock, and "T-X mineral-composition diagram" where X refers to the minerals in a conventional projection. Bucher et al. (2005) use the term "assemblage stability diagram", and Brouwer and Engi (2005) the term "equilibrium phase diagram" for a bulk composition X. For petrological phase diagrams in general, the terms "equilibrium" and "stability" are of course redundant.

Whatever we finally agree to call them, pseudosections offer a sensitivity that rivals the best of our thermobarometers, and they will be used more and more as we acquire greater confidence in the quality of the thermodynamic database and activity-composition models for the minerals. It is generally found that the uptake of minor elements such as Mn, Ti and Fe^{3+} in some of the commonest minerals necessitates their inclusion in the system composition (e.g. White et al., 2000). Because the di-, tri-, and higher-variance phase boundaries turn out to be so sensitive to bulk composition, the "effective" bulk composition must be carefully chosen to appropriately accommodate "partial" equilibrium (Loomis 1983) resulting from the presence of large, unreactive porphyroblasts, and the effects of domain equilibrium, including the separation of partial melt (Stüwe, 1997: Vance and Mahar, 1998; White and Powell, 2002; Marmo *et al.*, 2002; Tinkham and Ghent, 2005; Brouwer and Engi, 2005).

In high-grade rocks such as granulites, assemblage information tends to be more reliable than thermobarometry based on element exchanges, which generally reset on cooling (Hensen, 1987; Frost and Chako, 1989; Pattison *et al.*, 2003). Of course, if the same thermodynamic dataset and activity-composition models are used, the basic calibration from thermobarometry and pseudosection phase diagrams should be the same. In pseudosections, minor elements unevenly partitioned among the minerals shift the boundaries of net-transfer reactions. In exchange thermometry, the effects of minor constituents cancel out except when they involve non-ideal solutions (see Chapter 10 on thermometry); in which case the P-T shift may be magnified because of the small reaction entropy of the exchange reaction.

Clearly, there are tremendous advantages in being able to compute phase diagrams tailored to one's own rocks. Selecting an appropriate bulk composition and correct values for the activity of H_2O for dehydration reactions can be problematic (Xu *et al.*, 1994), but they can be overcome by trial and error. The assumption of a pure H_2O fluid is not reasonable, but then, what is? In some instances, it may be possible to constrain possible values of a_{H_2O} by using it as the X variable of the diagram and matching the sequence of assemblages with those in the field. This diagram can also be used to follow the H_2O-undersaturated path of a rock undergoing decompression (Carson *et al.*, 1999). In the range of dehydration melting, it is convenient to make H_2O an extensive variable and include its amount in the system composition (Powell *et al.*, 2005).

Phase diagrams with isopleths representing the composition of solid solution minerals in fields between end-member univariant lines have been in use for some time, but it was the seemingly modest paper by Hensen that first showed the geometry of the various multivariant assemblages that can occur in these fields, enabling first qualitative and later quantitative pseudosections to be constructed. Hensen (1987) noted that in natural rocks reaction textures resulting from divariant (continuous) reactions are much more frequently encountered than those related to univariant reactions. As refinements continue to be made in the quality of the thermodynamic database and solution models for some of the more complex metamorphic minerals in these diagrams (e.g. amphiboles), the usefulness of phase-assemblage diagrams/pseudosections will only increase.

References

Albee, A.L. (1972) Metamorphism of pelitic schists: reaction relations of chloritoid and staurolite. *Geological Society of America Bulletin*, **83**, 3249–3268.

Asimow, P.D. and Ghiorso, M.S. (1998) Algorithmic modifications extending MELTS to calculate subsolidus phase relations. *American Mineralogist*, **83**, 1127–1132.

Bickle, M.J. and Archibald, N.J. (1984) Chloritoid and staurolite stability: implications for metamorphism in the Archean Yilgarn Block, western Australia. *Journal of Metamorphic Geology*, **2**, 179–203.

Biino, G.G. and de Capitani, C. (1995) Equilibrium assemblage calculations: a new approach to metamorphic petrology. Pp. 11–53 in: Studies on metamorphic rocks and minerals of the western Alps. A volume in memory of Ugo Pognante. *Bolletin Museo Regionale Scienze Naturali, Torino*, **13**.

Brouwer, F.M. and Engi, M. (2005) Staurolite and other aluminous phases in Alpine eclogite from the Central Swiss Alps: analysis of domain evolution. *The Canadian Mineralogist*, **43**, 105–128.

Bucher, K., de Capitani, C. and Grapes, R. (2005) The development of a margarite-corundum blackwall by metasomatic alteration of a slice of mica schist in ultramafic rock, Kvesjöen, Norwegian Caledonides. *The Canadian Mineralogist*, **43**, 129–156.

Carrington, D.P. and Harley, S.L. (1995) Partial melting and phase relations in high-grade metapelites: an experimental petrogenetic grid in the KFMASH system. *Contributions to Mineralogy and Petrology*, **120**, 270–291.

Carson, C.J., Powell, R. and Clarke, G.L. (1999) Calculated mineral equilibria for eclogites in $CaO-Na_2O-FeO-MgO-Al_2O_3-SiO_2-H_2O$: application to the Pouébo Terrane, Pam Peninsula, New Caledonia. *Journal of Metamorphic Geology*, **17**, 9–14.

Connolly, J.A.D. (1990) Multivariable phase diagrams: an algorithm based on generalized thermodynamics. *American Journal of Science*, **290**, 666–718.

Connolly, J.A.D. and Petrini, K. (2002) An automated strategy for calculation of phase diagram sections and retrieval of rock properties as a function of physical conditions. *Journal of Metamorphic Geology*, **20**, 697–708.

de Capitani, C. (1994) Gleichgewichts-Phasendiagramme: Theorie und Software. Beiheft of the *European Journal of Mineralogy*, **6**, 48.

Dymoke, P. and Sandiford, M. (1992) Phase relations in Buchan facies series assemblages: calculations with applications to andalusite–staurolite parageneses in the Mount Lofty Ranges, South Australia. *Contributions to Mineralogy and Petrology*, **110**, 121–132.

Frost, B.R. and Chacko, T. (1989) The granulite uncertainty principle: limitations on thermobarometry in granulites. *Journal of Geology*, **97**, 435–450.

Grant, J.A. (1985) Phase equilibria in low-pressure partial melting of pelitic rocks. *American Journal of Science*, **285**, 409–435.

Guiraud, M., Holland, T.J.B. and Powell, R. (1990) Calculated mineral equilibria in the greenschist-blueschist-eclogite facies in $Na_2O-FeO-MgO-Al_2O_3-SiO_2-H_2O$. *Contributions to Mineralogy and Petrology*, **104**, 85–98.

Harte, B. and Hudson, N.F.C. (1979) Pelitic facies series and the temperatures and pressures of Dalradian metamorphism in E. Scotland. Pp. 323–337 in: *The Caledonides of the British Isles* (A.L. Harris, C.H. Holland and B.E. Leake, editors) Geological Society of London.

Hensen, B.J. (1971) Theoretical phase relations involving cordierite and garnet in the system $MgO-FeO-Al_2O_3-SiO_2$. *Contributions to Mineralogy and Petrology*, **33**, 191–214.

Hensen, B.J. (1987) P-T grids for silica-undersaturated granulites in the systems MAS (n + 4) and FMAS (n + 3) – tools for the derivation of P-T paths of metamorphism. *Journal of Metamorphic Geology*, **5**, 255–271.

Holland, T.J.B and Powell, R. (1990) An enlarged and updated internally consistent thermodynamic dataset with uncertainties and correlations: the system $K_2O-Na_2O-CaO-MgO-MnO-FeO-Fe_2O_3-Al_2O_3-TiO_2-SiO_2-C-H_2O-O_2$. *Journal of Metamorphic Geology*, **8**, 89–124.

Hudson, N.F.C. (1980) Regional metamorphism of some Dalradian pelites in the Buchan area, N.E. Scotland. *Contributions to Mineralogy and Petrology*, **73**, 39–51.

Koons, P.O. and Thompson, A.B. (1985) Non-mafic rocks in the greenschist, blueschist and eclogite facies. *Chemical Geology*, **50**, 3–30.

Labotka, T.C. (1981) Petrology of an andalusite-type regional metamorphic terrane, Panamint Mountains, California. *Journal of Petrology*, **22**, 261–296.

Loomis, T.P. (1983) Compositional zoning of crystals: a record of growth and reaction history. Pp. 1–60 in: *Kinetics and Equilibrium in Mineral Reactions* (S.K. Saxena, editor). Advances in Physical Geochemistry, **3**, Springer-Verlag, New York.

Marmo, B.A., Clarke, G.L. and Powell, R. (2002) Fractionation of bulk rock composition due to porphyroblast growth: effects on eclogite facies mineral equilibria, Pam Peninsula, New Caledonia. *Journal of Metamorphic Petrology*, **20**, 151–165.

Meyre, C., de Capitani, C. and Partsch, J.H. (1997) A ternary solution model for omphacite and its application to geothermometry for the Middle Adula nappe (Central Alps, Switzerland). *Journal of Metamorphic Geology*, **15**, 687–700.

Pattison, D.R.M. and Harte, B. (1985) A petrogenetic grid for pelites in the Ballachulish and other Scottish aureoles. *Journal of the Geological Society of London*, **142**, 7–28.

Pattison, D.R.M. and Vogl, J.J. (2005) Contrasting sequences of metapelitic mineral assemblages in the aureole of the tilted Nelson Batholith, British Columbia: Implication for phase equilibria and pressure determination in andalusite-sillimanite-type settings. *The Canadian Mineralogist*, **43**, 51–88.

Pattison, D.R.M., Chacko, T., Farquhar, J. and McFarlane, C.R.P. (2003) Temperatures of granulite-facies metamorphism: Constraints from experimental phase equilibria and thermometry corrected for retrograde exchange. *Journal of Petrology*, **44**, 867–900.

Powell, R. (1978) *Equilibrium Thermodynamics in Petrology*. Harper and Row, London, 284 pp.

Powell, R. and Holland, T.J.B. (1988) An internally consistent thermodynamic dataset with uncertainties and correlations: 3. Applications to geobarometry, worked examples and a computer program. *Journal of Metamorphic Geology*, **6**, 173–204.

Powell, R. and Holland, T.J.B. (1990) Calculated mineral equilibria in the pelite system, KFMASH (K_2O-FeO-MgO-Al_2O_3-SiO_2-H_2O). *American Mineralogist*, **75**, 367–380.

Powell, R., Holland, T.J.B. and Worley, B. (1998) Calculating phase diagrams involving solid solutions in non-linear equations, with examples from THERMOCALC. *Journal of Metamorphic Geology*, **16**, 575–586.

Powell, R., Guiraud, M. and White, R.W. (2005) Truth and beauty in metamorphic phase equilibria: conjugate variables and phase diagrams. *The Canadian Mineralogist*, **43**, 21–34.

Reche, J., Martinez, F.J. and Arboleya, M.L. (1998) Low- to medium-pressure Variscan metamorphism in Galicia (NW Spain): evolution of a kyanite-bearing synform and associated bounding antiformal domains. Pp. 61–79 in: *What Drives Metamorphism and Metamorphic Reactions?* (P.J. Treloar and P.J. O'Brien, editors). Special Publications, **138**, Geological Society, London.

Schmädicke, E. and Evans, B.W. (1997) Garnet-bearing ultramafic rocks from the Erzgebirge, and their relation to other settings in the Bohemian Massif. *Contributions to Mineralogy and Petrology*, **127**, 54–74.

Spear, F.S. (1993) *Metamorphic Phase Equilibria and Pressure-Temperature-Time Paths*. Mineralogical Society of America, Monograph, 799 pp.

Spear, F.S. and Cheney, J.T. (1989) A petrogenetic grid for pelitic schists in the system SiO_2-Al_2O_3-FeO-MgO-K_2O-H_2O. *Contributions to Mineralogy and Petrology*, **101**, 149–164.

Stüwe, K. (1997) Effective bulk composition changes due to cooling: a model predicting complexities in retrograde reaction textures. *Contributions to Mineralogy and Petrology*, **129**, 43–52.

Stüwe, K. and Powell, R. (1995) PT paths from modal proportions: application to the Koralm Complex. *Contributions to Mineralogy and Petrology*, **119**, 83–93.

Thompson, A.B. (1976) Mineral reactions in pelitic rocks: II. Calculation of some P-T-X(Fe-Mg) phase relations. *American Journal of Science*, **276**, 425–454.

Tinkham, D.K. and Ghent, E.D. (2005) Estimating P-T conditions of garnet growth with isochemical phase-diagram sections and the problem of effective bulk composition. *The Canadian Mineralogist*, **43**, 35–50.

Tinkham, D.K., Zuluaga, C.A. and Stowell, H.H. (2001) Metapelite phase equilibria modeling in MnNCKFMASH: the effect of variable Al_2O_3 and MgO/(MgO+FeO) on mineral stability. *Geological Materials Research*, **3**, 1–42.

Vance, D. and Mahar, E. (1998) Pressure-temperature paths from P–T pseudosections and zoned garnets: potential limitations and examples from the Zanskar Himalaya, NW India. *Contributions to Mineralogy and Petrology*, **132**, 225–245.

White, R.W. and Powell, R. (2002) Melt loss and the preservation of granulite facies mineral assemblages. *Journal of Metamorphic Geology*, **20**, 621–623.

White, R.W., Powell, R., Holland, T.J.B. and Worley, B.A. (2000) The effect of TiO_2 and Fe_2O_3 on metapelitic assemblages at greenschist and amphibolite facies conditions: mineral equilibria calculations in the system K_2O-FeO-MgO-Al_2O_3-SiO_2-H_2O-TiO_2-Fe_2O_3. *Journal of Metamorphic Geology*, **18**, 497–511.

Will, T.M. (1998) *Phase Equilibria in Metamorphic Rocks*. Lecture Notes in Earth Sciences, **71**. Springer-Verlag, New York, 315 pp.

Will, T.M., Powell, R., Holland, T.J.B. and Guiraud, M. (1990a) Calculated greenschist facies mineral equilibria in the system CaO-MgO-FeO-Al_2O_3-SiO_2-H_2O-CO_2. *Contributions to Mineralogy and Petrology*, **104**, 353–368.

Will, T.M., Powell, R. and Holland, T.J.B. (1990b) A calculated petrogenetic grid for ultramafic rocks at low pressures in the system CaO-FeO-MgO-Al_2O_3-SiO_2-CO_2-H_2O. *Contributions to Mineralogy and Petrology*, **105**, 347–358.

Will, T.M. Okrusch, M., Schmädicke, E. and Chen, G. (1998) Phase relations in greenschist-blueschist-amphibolite-eclogite facies: Calculated mineral equilibria in the system Na_2O-CaO-FeO-MgO-Al_2O_3-SiO_2-H_2O (NCFMASH), with applications to the PT evolution of metamorphic rocks from Samos, Greece. *Contributions to Mineralogy and Petrology*, **132**, 85–102.

Xu, G., Will, T.M. and Powell, R. (1994) A calculated petrogenetic grid for the system K_2O-FeO-MgO-Al_2O_3-SiO_2-H_2O, with particular reference to contact-metamorphosed pelites. *Journal of Metamorphic Geology*, **12**, 99–119.

Theoretical Phase Relations Involving Cordierite and Garnet in the System MgO—FeO—Al$_2$O$_3$—SiO$_2$

B. J. Hensen*

Department of Geophysics and Geochemistry, Australian National University

Received August 21, 1971

Abstract. Theoretical stability relations have been derived between the phases cordierite (Cd), garnet (Ga), hypersthene (Hy), olivine (Ol), sapphirine (Sa), spinel (Sp), sillimanite (Si) and quartz (Qz) in the system MgO—FeO—Al$_2$O$_3$—SiO$_2$. Natural rock data and experimental evidence suggest that the Mg/Mg + Fe^{2+} ratio (X) of coexisting ferromagnesian phases decreases as follows: $X_{Cd} > X_{Sa} > X_{Hy} > X_{Ol} > X_{Sp} > X_{Ga}$. By use of this information four stable invariant points are proposed involving the phases: Cd, Hy, Sa, Ga, Si, Qz; Cd, Sa, Ga, Sp, Si, Qz; Cd, Hy, Sa, Ga, Sp, Qz; Cd, Ga, Hy, Ol, Sp, Qz. All univariant curves in the system are nonterminal, representing the breakdown of a join rather than the stability limit of an individual phase. A detailed treatment of divariant equilibria involving two and three ferromagnesian solid solutions illustrates the potential of these equilibria as Pressure-Temperature indicators. Interactions between solid-solid reactions and dehydration reactions involving biotite in the system MgO—FeO—Al$_2$O$_3$—SiO$_2$—K$_2$O—H$_2$O have been graphically analysed. The addition of biotite to anhydrous divariant assemblages does not affect the composition of coexisting phases at constant P and T but can affect their relative proportions.

Introduction

The stability relations among the phases cordierite (Cd), garnet (Ga), hypersthene (Hy), olivine (Ol), sapphirine (Sa), spinel (Sp), sillimanite (Si) and quartz (Qz) in the system MgO—FeO—Al$_2$O$_3$—SiO$_2$ have been analysed by the Phase Rule and Schreinemakers methods[1]. The choice of the above mentioned minerals for the analysis is justified in view of their common association in natural high grade metamorphic rocks.

In most natural rocks containing one or several anhydrous ferromagnesian phases, biotite is a common constituent. Therefore the relations between solid-solid reactions, which are the main concern of the study, and the dehydration reactions involving biotite in the system MgO—FeO—Al$_2$O$_3$—SiO$_2$—K$_2$O—H$_2$O have also been considered.

The theoretical analysis has been developed chiefly as an aid in deducing experimentally observed phase relations in complex model pelitic compositions (Hensen and Green, in prep.). The proposed model phase diagram in which the relative stabilities of the ferromagnesian phases occurring in the system are represented as a function of pressure and temperature, could not have been constructed on the basis of theoretical considerations only. The information obtained from the experiments, quoted above, has been essential in selecting the significant

* Present address: Geophysical Laboratory, 2801 Upton Street, Northwest Washington, D. C. 20008, U.S.A.

[1] For a discussion of the methods of Schreinemakers the reader is referred to Schreinemakers (1965) and Zen (1966).

phase equilibria from a larger number of mathematically possible ones. The treatment given here differs from that of previous authors (e. g. Hess, 1969; Grant, 1968). It deals mainly with anhydrous equilibria and more importantly, it puts much more emphasis on the analysis of divariant equilibria, which are believed to be of particular relevance to mineral assemblages occurring in natural rocks. The importance of divariant equilibria in natural processes was first emphasized by Goldschmidt (1911), who argued that the occurrence of invariant and univariant mineral assemblages in natural rocks has a very low probability (Goldschmidt's mineralogical Phase Rule). Ramberg (1964) has drawn attention to the potential of divariant reactions as geothermo-barometers.

A theoretical analysis of part of the system $MgO-FeO-Al_2O_3-SiO_2$ has been given by Khlestov (1964) and Marakushev and Kudryavtsev (1965).

Phase Analysis

In order to analyse the phase equilibria in a multicomponent system, the chemographic relationships of the phases occurring in the system have to be known. In the present analysis the seven phases cordierite, sapphirine, hypersthene, spinel, garnet, sillimanite and quartz will be considered[2]. The relative compositions of the ferromagnesian phases (in terms of their $Mg/Mg+Fe^{2+}$ ratios (X) can be derived from experimental results and from natural rock data (Hensen and Green, in prep.). The observed relative values for the $Mg/Mg+Fe^{2+}$ ratios are: $X_{Cd} > X_{Sa} > X_{Hy} > X_{Sp} > X_{Ga}$[3]. (Note that spinel is $MgAl_2O_4-FeAl_2O_4$ solid solution only.)

Invariant Points and Univariant Curves

In a 4-component system with 7 phases there are 7 invariant points each involving 6 phases (one phase being absent in each case). The invariant points are identified by the missing phase e. g. (Z) is the invariant point from which the phase Z is absent. A maximum of 6 univariant reactions pass through every invariant point. A list of all possible invariant points and univariant equilibria is given in Table 1. At an invariant point, pressure, temperature and the compositions of all phases are uniquely fixed. In the case of univariant equilibrium one variable can be chosen (P, T or the composition of one phase, e. g. X_{Cd}), whereby all other variables are determined uniquely.

A nondegenerate univariant reaction involves five phases. It can be characterized by the phases absent from it, e. g. (X, Z) is the univariant equilibrium not involving phases X and Z and connecting invariant points (X) and (Z) on a $P-T$ phase diagram. There are two types of univariant reactions possible. For a system having seven phases, A, B, C, D, E, F, G one can have:

1. The breakdown of a phase: $A \rightleftharpoons B+C+D+E$ (F, G) terminal reaction, Hess (1969). Chemographically A lies inside the tetrahedron BCDE.

2. The breakdown of a join: $A+B \rightleftharpoons C+D+E$ (F, G) (non-terminal reaction). A lies outside the tetrahedron BCDE and a line through A and B intersects the side CDE.

---

[2] For equilibria involving Fe-rich olivine see Appendix 2.
[3] The alternative with $X_{Sp} < X_{Ga}$ is considered in Appendix 3.

Table 1. *Invariant points and univariant reactions in the system* $MgO-FeO-Al_2O_3-SiO_2$ *involving the phases cordierite, garnet, hypersthene, sapphirine, spinel, sillimanite and quartz.*

Invariant point	(Sp)		(Hy)		(Si)		(Sa)	
Phases at invariant point	Cd Sa Hy Ga Si Qz		Cd Sa Sp Ga Si Qz		Cd Sa Hy Sp Ga Qz		Cd Hy Sp Ga Si Qz	
Univariant reaction join	Hy Sa Cd Si Qz Ga Sa Qz Cd Ga Cd Hy Hy Si Si Cd Ga Sa Sa Hy Ga Ga Si Qz Qz	S Qz Cd Ga Si	Sa Qz Cd Sp	Sp Qz Cd Ga Si	Sp Sa Sa Cd Ga Ga Si Si	Sa Qz Cd Sp	Cd Sp Sa Ga Si	Cd Ga Hy Sp Qz Si	Sa Ga Hy Sp Qz	Cd Ga Hy Sp Si	Cd Hy Si Cd Sp Qz	Sp Qz Cd Ga Si	Hy Si Cd Sp Ga
plane								
Missing phase(s)[a]	(Cd) (Qz) (Hy) (Ga) (Sa) (Si)		(Sp) (Ga, Si) (Sa) (Cd) (Qz)		(Sp) (Sa) (Cd) (Ga, Hy)		(Si) (Hy) (Sp) (Qz) (Ga) (Cd)	

Invariant point	(Cd)		(Ga)		(Qz)			
Phases at invariant point	Sa Hy Sp Ga Si Qz		Cd Sa Hy Sp Si Qz		Cd Sa Hy Sp Ga Si			
Univariant reaction join	Sa Sp Cd Hy Sa Ga Hy Qz Sa Sa Ga Hy Ga Si Sp Ga Si Sp	Hy Si Sa Sp Qz	Hy Si Sa Sp Qz	Sa Qz Cd Sp	Cd Sp Hy Sa Si	Hy Si Cd Sp Qz	Sa Ga Cd Hy Si	Sa Qz Cd Hy Si	Hy Si Cd Cd Sp Ga	Cd Sp Sa Hy Si
plane								

In order to predict the type of reaction (1 or 2) for each of the univariant assemblages listed in Table 1, the chemographic relations (or relative positions of the phases in compositional space) have to be known. This knowledge is also required for the detection of degenerate equilibria.

From the relative values for the $Mg/Mg + Fe^{2+}$ ratio of the ferromagnesian phases as given above, it has been deduced that all the univariant reactions in Table 1 are of the second type and that only one degenerate reaction occurs. Therefore, all univariant reactions mark the breakdown of a join, prohibiting the coexistence of two phases, in favour of a triangular plane, allowing the coexistence of the three phases forming the apices of the triangle. The only degenerate reaction (Hy, Ga, Si) is represented by the intersection of two co-planar joins (i. e. Cd—Sp and Sa—Qz)[4].

The fact that no terminal univariant reactions (reactions marking the breakdown of a single phase) occur in the system, means that the stability of the individual phases is not limited by any of the univariant boundaries[5]. As will be shown later, the stability of several phases (e. g. Cd, Ga, Sa and Sp) is limited by divariant reactions. Since the possible invariant points and univariant reactions are known, it is feasible to construct a pressure-temperature phase diagram. The relative positions of the invariant points and univariant lines in the $P-T$ plane have to comply with Schreinemakers principles. These restrictions alone do not provide a unique solution for a $P-T$ diagram. However, by using evidence obtained experimentally, a model phase diagram can be constructed that is consistent with both experimental and theoretical constraints.

The model $P-T$ diagram is shown in Fig. 1. It is apparent from this diagram that some of the invariant points listed in Table 1 have been deduced to be metastable. The reasons for choosing the stable and metastable invariant points, the relative positions of the stable invariant points and the slopes of univariant curves emanating from them, are discussed below.

The invariant point (Sp) has been inferred to be stable from experimental evidence (Hensen and Green, 1970; in prep.). The negative slope of the reaction (Sa, Sp) and the positive slope of (Cd, Sp), both passing through (Sp), have also been deduced from the experimental data (Fig. 2).

As can be seen from the diagram, assemblages containing cordierite occur at the low pressure (high volume) side of all but one of the univariant reactions[6]. This is due to the relatively large molar volume (low density) of cordierite compared to all other ferromagnesian phases. Similarly, garnet occurs at the high pressure side of most phase boundaries. However, there are some exceptions to these general relationships. In order to explain these anomalies, e. g. reactions

---

[4] For simplicity sapphirine composition is assumed to be $Mg_2Al_4SiO_{10}$, thus disregarding more aluminous compositions.

[5] Excepting the kyanite $\rightleftharpoons$ sillimanite, phase boundary which has been disregarded here. The experimental evidence indicates that this reaction does not affect the proposed phase relations. Its intersection with the univariant curve (Sa, Sp) should give rise to the univariant boundary $Cd + Ga \rightleftharpoons Hy + Ky + Qz$.

[6] Experimental evidence indicates that the reaction (Si, Sa) shown with a negative slope in Figs. 1 and 2 actually has a positive slope with Cd and Ga occurring at the high pressure side.

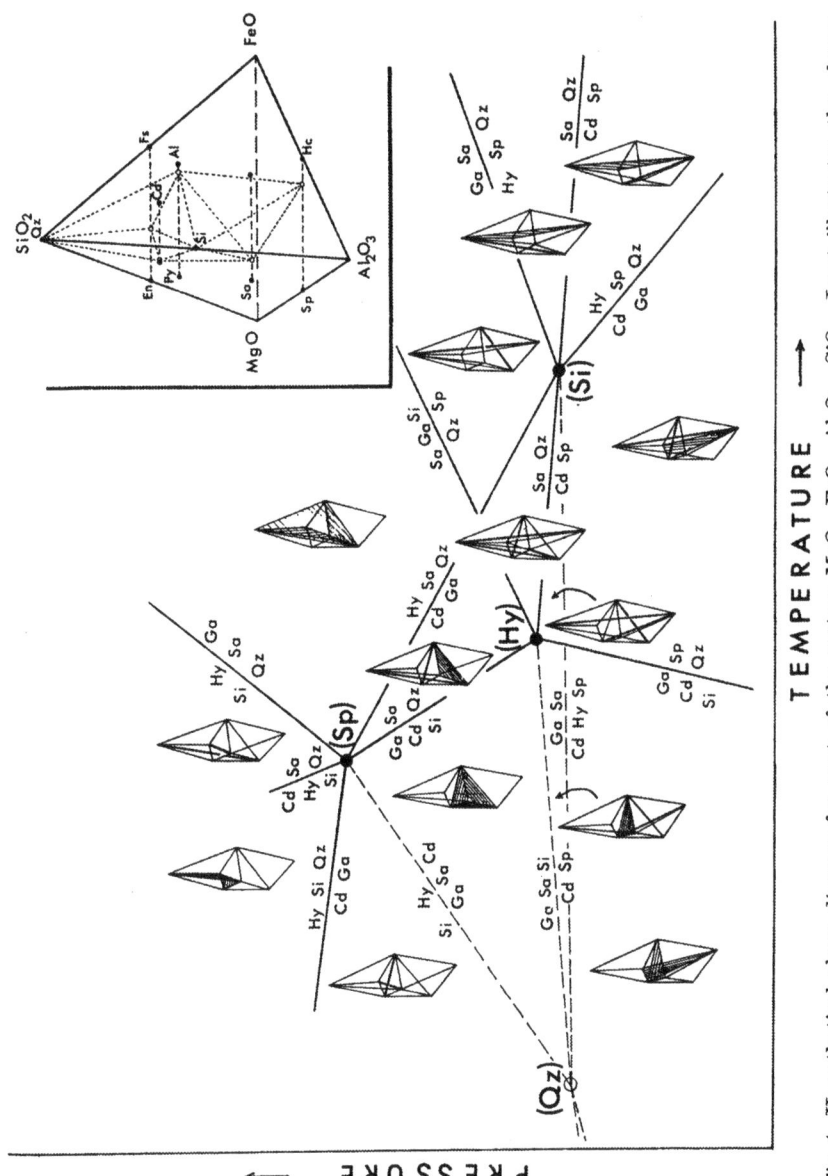

Fig. 1. Hypothetical phase diagram for part of the system $MgO-FeO-Al_2O_3-SiO_2$. Inset illustrates the chemographic relations of the phases considered in the analysis

(Hy, Sa) and (Sp, Cd), it is necessary to consider the volume change (ΔV) of a univariant reaction.

The coefficients of a univariant reaction and thus its volume change, can be determined only if the compositions of all ferromagnesian phases participating in it are known. Therefore it is not enough to determine the order of $Mg/Mg + Fe^{2+}$ ratios of the phases involved, but it is necessary to know the values of these ratios. Thus the reaction

$$Sp + Qz \rightleftharpoons Ga + Cd + Si \quad (Hy, Sa)$$

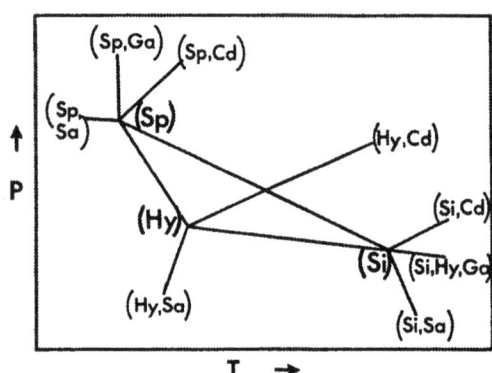

Fig. 2. Schematic $P-T$ diagram showing univariant phase boundaries. Shorthand notations as used in text. Relative slopes of boundaries in this figure have no significance (compare Fig. 1)

can have a negative ΔV only if the amount of cordierite involved in this reaction is extremely small. This will depend on the relative compositions of the ferromagnesian phases. Since the difference between X_{Sp} and X_{Ga} is probably very small and the difference between both of these and X_{Cd} is relatively large (or in other words $K_{D(Sp-Ga)} \ll K_{D(Cd-Ga)}$[7]) it is indeed possible for this reaction to have a negative volume change[8]. A similar argument can be used for reaction (Sp, Cd).

As the composition of each phase changes continuously along a univariant phase boundary and since as a result the coefficients for the reaction will vary, the volume and entropy changes of the reaction will also vary. It is therefore possible that some of the univariant boundaries are curved. In extreme cases a change in sign of the ΔV could occur by which the low volume side of a reaction becomes the high volume side. The negative slope of the reaction (Ga, Sp) and the existence and relative position of the invariant points (Hy) and (Si) are inferred by combining experimental evidence with restrictions imposed by Schreinemakers rules (Hensen and Green, in prep.).

Assuming the above deductions to be correct, the invariant points (Ga) and (Sa) and (Cd) have to be metastable for the following reasons. The invariant point (Ga) must be co-linear with the points (Hy) and (Si) because it shares with them the degenerate reaction (Ga, Hy, Si)[9]. It has to occur at the intersection of (Hy, Si) with the metastable extension of (Sp, Ga). As can be seen from Fig. 1, this intersection can only occur outside the stability field of Hy—Si. Therefore (Ga) which involves Hy and Si must be metastable.

The invariant point (Sa) has to occur at the intersection of the phase boundaries (Sp, Sa), (Si, Sa) and (Hy, Sa). This intersection lies outside the stability field of Hy—Si (Fig. 2), and as a result (Sa) which involves Hy and Si is necessarily metastable.

(Cd) could be a stable invariant point if the curves (Sp, Cd), (Hy, Cd) and (Si, Cd) intersect at high pressure. It is believed to be more likely that these curves diverge rather than converge towards higher pressure and thus a metastable point

---

[7] This information can be deduced from natural rock data.

[8] For an alternative solution see Appendix 3.

[9] (Hy), (Ga) and (Si) are not necessarily co-linear in the sense that they lie on a straight line, because the univariant phase boundaries may be curved.

(Cd) at low pressure is required (Fig. 1). Another possibility is that the three curves are almost parallel, whereby (Cd) would loose any significance.

By the present criteria, the invariant point (Qz) could be stable. It may be rendered metastable by reactions involving another phase (e. g. corundum), not considered in the present analysis. In Fig. 2 only those curves which pass through one of the stable, silica-saturated points (Hy), (Si) and (Sp) have been represented.

The univariant reactions in the phase diagram apply for all bulk compositions within the system $MgO-FeO-Al_2O_3-SiO_2$. The compositions of the ferromagnesian phases in univariant equilibrium are uniquely fixed by pressure (or temperature), and change along the phase boundaries so as to become more magnesian with increasing pressure. For a particular bulk composition, the phase assemblages will be bounded partly by portions of the univariant reactions and partly by divariant reactions (to be discussed in following sections). As pointed out before, no terminal reactions occur in the system as considered. Therefore, none of the phase boundaries in Fig. 1 delimit the stability of a single phase. However, appropriate parts of the diagram must be entirely within the stability limits of the end member phases. Some of the univariant curves of Fig. 1 terminate in invariant points of either the pure Fe or Mg systems. This only occurs if the invariant points are themselves stable in the end member systems. For example the boundary of reactions (Ga, Sp) in Fig. 1 has a relatively short length because the upper stability limit of magnesian cordierite is probably reached at a pressure only slightly in excess of the invariant point (Sp). This end point represents an invariant point in the system $MgO-Al_2O_3-SiO_2$ (compare Hensen and Essene, 1971; point (Py) in Fig. 3).

Divariant Reactions

Five divariant reactions (involving 4 phases) can be derived from each nondegenerate, univariant reaction (involving 5 phases). In Table 2 all possible divariant reactions related to the three invariant points of interest (Sp), (Hy) and (Si) have been listed.

The four phases in divariant equilibrium form a tetrahedron in compositional space. The compositions of the coexisting phases are a function of both pressure and temperature. A nondegenerate, divariant reaction can be specified by three non-participating phases, e. g. (X, Y, Z). On a $P-T$ diagram this divariant reaction will intersect the univariant phase boundaries (X, Y), (X, Z) and (Y, Z). These boundaries limit the stability field of the divariant assemblage of reaction (X, Y, Z) for any bulk composition. Take for example the reaction

$$Cd \rightleftharpoons Ga + Si + Qz \quad (Sp, Sa, Hy). \tag{1}$$

The stability field of the assemblage Cd—Ga—Si—Qz in Fig. 1 is limited by the univariant curves (Sa, Sp), marking the breakdown of the join (Cd—Ga) and by the curves (Sp, Hy) and (Hy, Sa), marking the disappearance of the assemblage Ga—Cd—Si.

Before going into the relationships between univariant and divariant assemblages two examples of non-degenerate, divariant reactions will be discussed.

Table 2. *Divariant reactions related to invariant equilibria*

Reaction	Missing phases	ΔV
I. *Invariant point* (Sp)		
Hy + Si ⇌ Ga + Qz	(Cd, Sa, Sp)	−ve
Sa + Qz ⇌ Ga + Si	(Cd, Hy, Sp)	−ve
Sa + Hy + Qz ⇌ Ga	(Cd, Si, Sp)	−ve
Sa + Hy + Si ⇌ Ga	(Cd, Qz, Sp)	−ve
Sa + Qz ⇌ Hy + Si	(Cd, Ga, Sp)	−ve
Cd ⇌ Ga + Si + Qz	(Sa, Hy, Sp)	−ve
Cd + Hy ⇌ Ga + Qz	(Sa, Si, Sp)	−ve
Cd + Ga ⇌ Hy + Si	(Sa, Qz, Sp)	−ve
Cd ⇌ Sa + Qz	(Hy, Ga, Si, Sp)	−ve
Cd + Sa ⇌ Ga + Si	(Hy, Qz, Sp)	−ve
Cd + Hy + Sa ⇌ Ga	(Si, Qz, Sp)	−ve
Cd ⇌ Hy + Si + Qz	(Ga, Sa, Sp)	−ve
Cd + Sa ⇌ Hy + Si	(Ga, Qz, Sp)	−ve
II. *Invariant point* (Hy)		
Sp + Qz ⇌ Ga + Si	(Cd, Sa, Hy)	−ve
Sa + Qz ⇌ Ga + Si	(Cd, Sp, Hy)	−ve
Sp + Qz ⇌ Sa	(Cd, Ga, Si, Hy)	−ve
Sa ⇌ Ga + Sp + Si	(Cd, Qz, Hy)	−ve
Cd ⇌ Ga + Si + Qz	(Sa, Sp, Hy)	−ve
Cd ⇌ Sp + Qz	(Sa, Ga, Si, Hy)	−ve
Cd + Sp ⇌ Sa	(Ga, Si, Qz, Hy)	−ve
Cd ⇌ Sa + Qz	(Sp, Ga, Si, Hy)	−ve
Cd + Sa ⇌ Ga + Si	(Sp, Qz, Hy)	−ve
III. *Invariant point* (Si)		
Hy + Sp + Qz ⇌ Ga	(Cd, Sa, Si)	−ve
Sp + Qz ⇌ Sa	(Cd, Hy, Ga, Si)	−ve
Hy + Sa + Qz ⇌ Ga	(Cd, Sp, Si)	−ve
Hy + Sa ⇌ Ga + Sp	(Cd, Qz, Si)	−ve
Cd ⇌ Sp + Qz	(Sa, Hy, Ga, Si)	−ve
Cd + Hy ⇌ Ga + Qz	(Sa, Sp, Si)	−ve
Cd + Hy + Sp ⇌ Ga	(Sa, Qz, Si)	−ve
Cd + Sp ⇌ Sa	(Hy, Ga, Qz, Si)	−ve
Cd + Sa + Hy ⇌ Ga	(Sp, Qz, Si)	−ve

Divariant Reactions Involving two Solid Solutions

Example:

$$\text{Cd} \rightleftharpoons \text{Ga} + \text{Si} + \text{Qz} \quad (\text{Hy, Sp, Sa}). \tag{1}$$

A reaction of this type for a cordierite of a particular composition can be represented graphically on a $P-T$ diagram as a band (Fig. 3). Within this band the four phases (Cd—Ga—Si—Qz) coexist. Both from experimental evidence and from natural rock data it is well established that Mg distributes in favour of cordierite with respect to garnet and therefore at any point within the divariant field of Fig. 3 we have $X_{Cd} > X_{Ga}$.

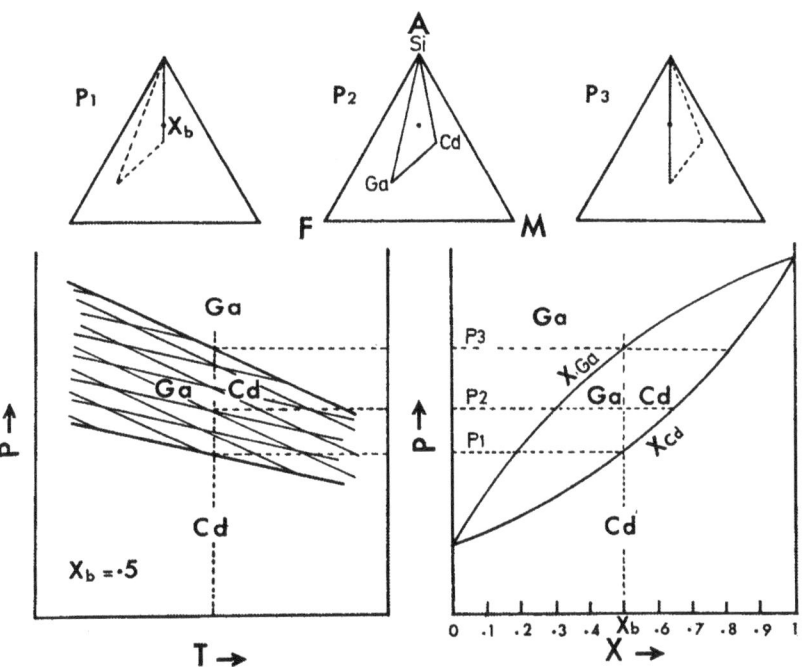

Fig. 3. Graphical representation of the divariant reaction $Cd \rightleftharpoons Ga + Si + Qz$ (1). Left: P—T diagram showing boundaries of divariant field and X_{Cd} and X_{Ga} contours. Right: P—X diagram (constant T). Top: AFM diagrams for P_1, P_2 and P_3 (constant T)

The compositions of the ferromagnesian minerals in divariant equilibrium are functions of pressure and temperature only (e. g. $X_{Cd} = f(P, T)$). X_{Cd} and X_{Ga} are also related by a distribution coefficient (K_D) [10] which is here treated as a function of temperature only, assuming ideal mixing and disregarding the effect of pressure [11]. The latter relation applies for any equilibrium situation, whether invariant, univariant or multivariant. This means that for a given K_D and composition of one phase, the composition of the other phase is determined. Thus the composition of the first garnet to appear at the low pressure boundary of the divariant band can be calculated for each chosen K_D and initial cordierite composition [12]. Similarly X_{Cd} at the upper boundary of the band is determined for a given K_D. As a result, the change in the composition of both minerals across the

---

10 The expression of the distribution coefficient K_D used is

$$K_{D(Cd-Ga)} = \frac{X_{Cd}(1-X_{Ga})}{X_{Ga}(1-X_{Cd})} = \frac{(Mg/Fe)\,Cd}{(Mg/Fe)\,Ga}$$

where $X = Mg/Mg + Fe^{2+}$. Because $X_{Cd} > X_{Ga}$, we have $K_D > 1$.

11 The effect of pressure on K_D is probably small. This effect is a function of the volume change resulting from the exchange of Fe and Mg between the two minerals. The ΔV is small but presently available thermodynamic data do not allow an accurate determination of its magnitude or sign.

12 $X_b = Mg/Mg + Fe^{2+}$ ratio of the bulk composition as used in Fig. 3.

divariant field is determined for a chosen K_D (i. e. fixed T)[13]. This information, however, does not tell us how X_{Cd} changes as a function of pressure or, in other words, what is the width of the divariant band. The width shown on the $P-T$ diagram (Fig. 3) has been chosen arbitrarily.

Also represented on the $P-T$ diagram are contours of constant X_{Cd} and X_{Ga}. From the foregoing it follows that lines of constant X_{Ga} will intersect the phase boundary for the incoming of garnet, giving rise to a different initial garnet composition for each temperature. The same applies for cordierite at the upper boundary marking the disappearance of that phase. Whether the contours for constant X_{Cd} and X_{Ga} are straight lines (as represented in Fig. 3) or curves, depends on the amount of reaction as a function of pressure and on the variation of K_D as a function of temperature[13].

The change in composition of cordierite and garnet for any bulk composition (X_b) as a function of pressure (at constant temperature) is conveniently represented in a $P-X$ diagram. This type of diagram is an isothermal section through P, T, X space, whereas a $P-T$ diagram is an iso-compositional section. The $P-X$ diagram presented here is a simplified case in that the reaction for the breakdown of cordierite involves the same phases (i. e. garnet, sillimanite and quartz) over the complete range of $Mg/Mg + Fe^{2+}$ ratios. In reality this does not occur because other reactions interfere near the iron and magnesian ends of the diagram.

As can be seen from the $P-X$ diagram, the pressure interval over which the four phases coexist (the divariant band) shifts to higher pressure from Fe-rich to Mg-rich compositions. The divariant pressure interval is largest for intermediate compositions. It can be determined from the $P-X$ diagram whether or not the divariant fields for two different bulk compositions will overlap on the $P-T$ diagram. This overlap is a function of the distribution coefficient (and thus of temperature), which is demonstrated graphically by a widening of the loop that limits the divariant field with increasing K_D (decreasing temperature)[14]. The Mg and Fe endpoints are also displaced with changing temperature, and may accentuate or counteract the widening effect.

As K_D increases with decreasing temperature, and the first garnet to appear becomes more Fe-rich and the last (disappearing) cordierite more Mg-rich, the width of the divariant band may increase with decreasing temperature. This is in harmony with experimental evidence and with the fact that the assemblage Cd—Ga—Si—Qz commonly occurs in natural metamorphic rocks.

The change in composition of garnet and cordierite with pressure (at constant temperature) for a bulk composition with $X_b = 0.5$ has also been shown in Fig. 3 in AFM triangular diagrams, which can be used conveniently for quartz-saturated assemblages when undersaturated phases like sapphirine and spinel do not occur.

Divariant Reactions Involving Three Solid Solutions

Example:

$$Cd + 4 Hy \rightleftharpoons 2 Ga + 3 Qz \quad (Sa, Sp, Si). \qquad (2)$$

---

[13] See Appendix 1 for mathematical treatment and graphical representation.

[14] It is theoretically possible that the change of K_D with temperature is insignificant. This would mean that the X_{Cd} and X_{Ga} contours on the $P-T$ plane become parallel.

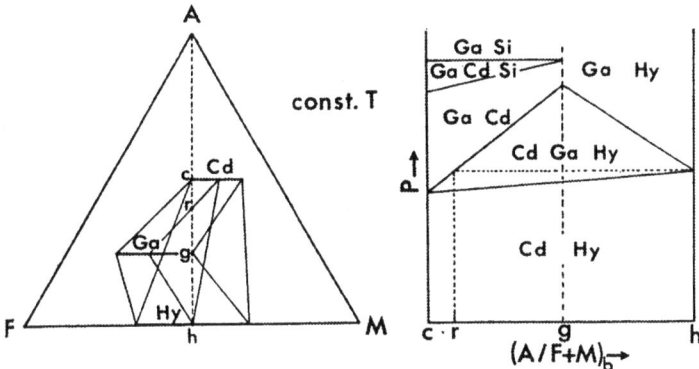

Fig. 4. AFM and $P-(A/F+M)_b$ diagrams (constant T) for the divariant reaction $Cd + Hy \rightleftharpoons Ga + Qz$ (2). The points c, g and h have a $(A/F+M)_b$ ratio of 1, 1/3 and 0 respectively

Reactions of this type, although basically similar to those involving two solid solutions, differ in some important aspects. The appearance of garnet from the reaction between cordierite and hypersthene is no longer a unique function of pressure, temperature and $Mg/Mg + Fe^{2+}$ ratio of the bulk composition. It is also determined by the relative amounts of cordierite and hypersthene present which are given by the $Al_2O_3/MgO + FeO$ ratio of the bulk composition $(A/F+M)_b$. This ratio ranges from 1 for cordierite to 1/3 for garnet and 0 for hypersthene. For bulk compositions with the same $Mg/Mg + Fe^{2+}$ ratio (X_b) but with different $A/F + M$ ratios, coexisting cordierite and hypersthene will have different compositions for the same temperature (or K_D), as shown in Figs. 4 and 5. Consequently, the composition of the first garnet (X_{Ga}) that can coexist with cordierite and hypersthene at a constant temperature and X_b, will be a function of $(A/F+M)_b$. Because the compositions of the three coexisting phases (at constant T) are unique functions of pressure, it follows that garnet must appear at different pressures depending on the $(A/F+M)_b$ ratio. This has been shown graphically in Fig. 4, where the $A/F + M$ ratio is plotted against pressure. The pressure for the incoming of garnet increases with decreasing $A/F + M$ ratio[15].

On a $P-T$ diagram the divariant reaction is represented by a band (Fig. 5). Only if cordierite and hypersthene are in the proper molecular proportions (i.e. 1:4) does garnet alone occur above the upper boundary of the divariant band. When $A/F + M < 1/3$, hypersthene persists to higher pressure and when $A/F + M > 1/3$ cordierite coexists with garnet above the divariant band. The pressure at which either cordierite or hypersthene (or both) disappear, is again determined by $(A/F+M)_b$.

As demonstrated by Fig. 4 (AFM) plot, the most magnesian garnet that can possibly coexist with hypersthene and cordierite, occurs in the composition with $(A/F+M)_b = 1/3$. In this composition X_{Ga} is equal to X_b at the upper boundary of the divariant field. The X_{Ga} is lower for all other values of $(A/F+M)_b$. Therefore

---

[15] For simplicity the effect of Al_2O_3 in the orthopyroxene has been disregarded. At high temperature (around 1000° C) hypersthene may have up to 10–15 mole % Al_2O_3 in solid solution.

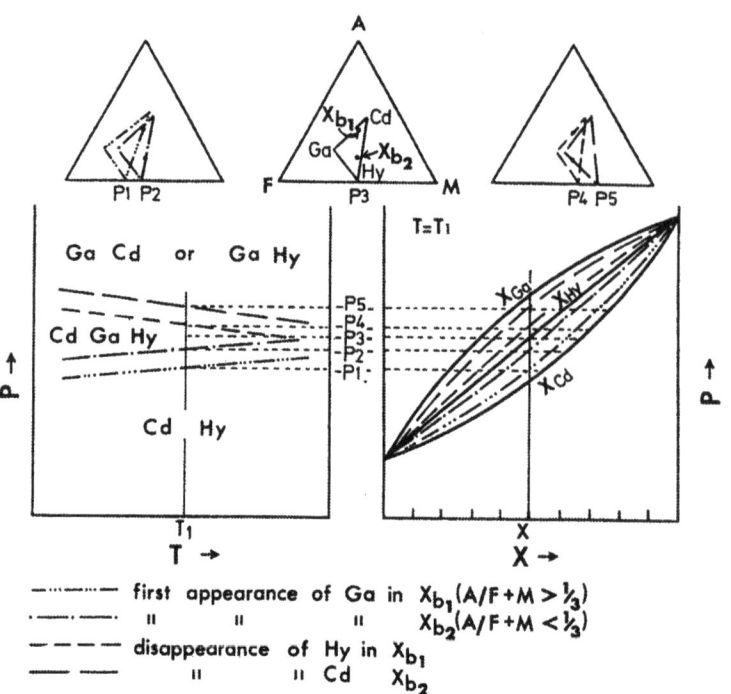

Fig. 5. Graphical representation of the divariant reaction $Cd + Hy \rightleftharpoons Ga + Qz$ (2). Left: P—T diagram. Right: P—X diagram. Top: AFM diagrams. Phase relations are shown for two bulk compositions X_{b_1} and X_{b_2} with the same $Mg/Mg + Fe^{2+}$ ratio but different $(A/F+M)_b$ ratios

the divariant band will extend to the highest pressure, and the pressure interval for the divariant reaction will be largest for the composition with the $A/F+M$ ratio of garnet. On the $(A/F+M)_b$—P diagram (Fig. 4) this is expressed by a maximum in the upper boundary. The position of this maximum is restricted by the point $(A/F+M)_b = r$. In a composition at point r, the position of which is fixed by K_D, hypersthene disappears at the same pressure as garnet (theoretically) appears in a composition at point h (see Fig. 4). Given the slope of the lower boundary as a function of pressure and $(A/F+M)_b$ ratio, the position of the maximum in the upper boundary at $(A/F+M)_b = 1/3$ can be constructed, assuming that the phase boundaries are represented by straight lines on this diagram.

A final point about the P—$(A/F+M)_b$ diagram is that the cordierite persisting above the upper phase boundary of the divariant band for compositions with $(A/F+M)_b > 1/3$ will eventually break down as shown in Fig. 4. This will be further explained in the next section where the interaction of divariant reactions will be considered. First, however, it is necessary to turn our attention to the P—X diagram for the type of divariant reaction under discussion, represented in Fig. 5.

On the P—X diagram a curve for the variation in X_{Hy} occurs inside the loop formed by the compositional curves (X_{Cd} and X_{Ga}) for the phases cordierite and garnet. The curves delineating the stability field of the divariant assemblage Cd—Hy—Ga—Qz, for a particular value of $(A/F+M)_b$, i. e. composition X_{b_1}[16], as a function of pressure, do not coincide with the X_{Cd} and X_{Ga} curves. The lenticular stability field for such a bulk composition is located inside the loop formed by X_{Cd} and X_{Ga}. The reason for this relationship is as follows: at the point where garnet appears in composition X_{b_1} (T_1, P_1) in Fig. 5, cordierite is more magnesian and hypersthene is more iron-rich than the bulk composition ($K_{D(Cd-Hy)} = 2$ was used for the construction of the diagram[17]. The garnet[18] that first appears in X_{b_1} will be more magnesian than the garnet in X_{b_2} (compare upper left hand AFM diagram in Fig. 5). Therefore the garnet in X_{b_2} has to occur at the higher pressure (see previous discussion).

At intermediate pressure (P_3) the divariant assemblage occurs in both X_{b_1} and X_{b_2}. Because X_{Cd}, X_{Ga} and X_{Hy} are uniquely fixed at (T_1, P_3) the compositions of the phases are the same in both bulk compositions though their relative amounts are different. With further increase in pressure, the upper boundary of the divariant interval for composition X_{b_1} is reached at P_4. At this pressure hypersthene disappears and we have the assemblage Ga—Cd—Qz. At P_5 cordierite breaks down in X_{b_2}. Ga—Hy—Qz occurs at high pressure. $P_5 > P_4$ because $X_{Ga_2} > X_{Ga_1}$, and garnet Ga_2 can therefore be stable only at higher pressure than P_4.

For a bulk composition with $A/F+M = 1/3$, garnet will appear at a pressure intermediate between those at which garnet appears in X_{b_1} and X_{b_2} (Figs. 4 and 5); while cordierite and hypersthene will simultaneously break down at the point where the line for constant X_b intersects the X_{Ga} curve. Only in this particular case does the X_{Ga} curve limit the divariant field for the assemblage Cd—Hy—Ga—Qz. For all other values of $(A/F+M)_b$ the upper boundary occurs below the X_{Ga} curve in Fig. 5.

Relationships between Divariant and Univariant Reactions

Example:

Univariant reaction

$$Cd + Ga \rightleftharpoons Hy + Si + Qz \quad (Sa, Sp).$$

The related divariant equilibria are:

Reactions	Missing phases		ΔV
Cd $\rightleftharpoons$ Ga + Si + Qz	(Hy, Sa, Sp)	(1)	—ve
Cd + Hy $\rightleftharpoons$ Ga + Qz	(Si, Sa, Sp)	(2)	—ve
Cd $\rightleftharpoons$ Hy + Si + Qz	(Ga, Sa, Sp)		—ve
Hy + Si $\rightleftharpoons$ Ga + Qz	(Cd, Sa, Sp)		—ve
Cd + Ga $\rightleftharpoons$ Hy + Si	(Qz, Sa, Sp)		—ve

[16] X_{b_1} stands for a *bulk composition* with a fixed value for X; the subscript 1 refers to a particular $A/F+M$ ratio.

[17] The diagram is not precisely constructed; K_D changes slightly from the middle of the diagram towards both ends.

[18] $K_{D(Cd-Ga)} = 4$ in the diagram.

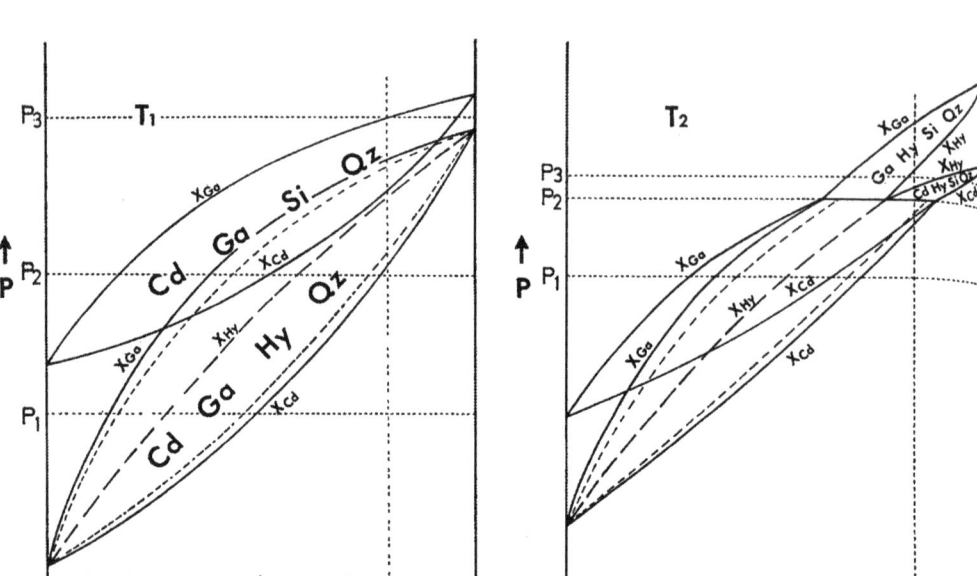

Fig. 6. P–X diagrams for reactions $Cd \rightleftharpoons Ga + Si + Qz$ (1) and $Cd + Hy \rightleftharpoons Ga + Qz$ (2). Left: T_1 reactions do not intersect (hypothetical). Right: reactions intersect on univariant boundary giving rise to two more reactions i. e. $Ga + Qz \rightleftharpoons Hy + Si$ and $Cd \rightleftharpoons Hy + Si + Qz$. Dashed lines mark the appearance of Ga and the disappearance of Hy in reaction (2)

Firstly, the relationship of divariant and univariant reactions will be illustrated on P–X diagrams. For arguments' sake it has been assumed that at low temperature a univariant curve does not occur[19]. The graphical representation of this situation is presented in Fig. 6 (left hand side, T_1). In this diagram reactions (1) and (2), discussed previously, have been represented. Three isothermal-isobaric sections (P_1, P_2, P_3) are given in the form of AFM diagrams in Fig. 7. These exhibit the phase assemblages occurring for all silica-saturated compositions at a specified pressure and temperature. Divariant assemblages are represented by triangles. For all compositions within such a triangle the $Mg/Mg + Fe^{2+}$ ratios of the coexisting phases are fixed. Trivariant fields have been contoured for constant K_D. The compositions of coexisting phases in these fields are determined by bulk composition and temperature (K_D). The distribution coefficients used for T_1 in Figs. 6 and 7 are $K_{D\,(Cd-Ga)} = 6$ and $K_{D\,(Cd-Hy)} = 2.5$.

For all compositions in which cordierite persists to pressures above the upper limit of the divariant band (i. e. compositions with $1/3 < (A/F+M)_b < 1$), the cordierite will eventually disappear with increasing pressure due to reaction (1). In the pressure interval between the upper boundary of reaction (2) and the

---

[19] In actual fact the reaction $Cd + Ga \rightleftharpoons Hy + Si + Qz$ is replaced at low temperature by the reaction $Cd + Ga \rightleftharpoons Hy + Ky + Qz$, due to the intersection of the former with the $Ky \rightleftharpoons Si$ phase boundary. As a result the stable coexistence of pyrope and Mg-cordierite in the presence of quartz is prohibited.

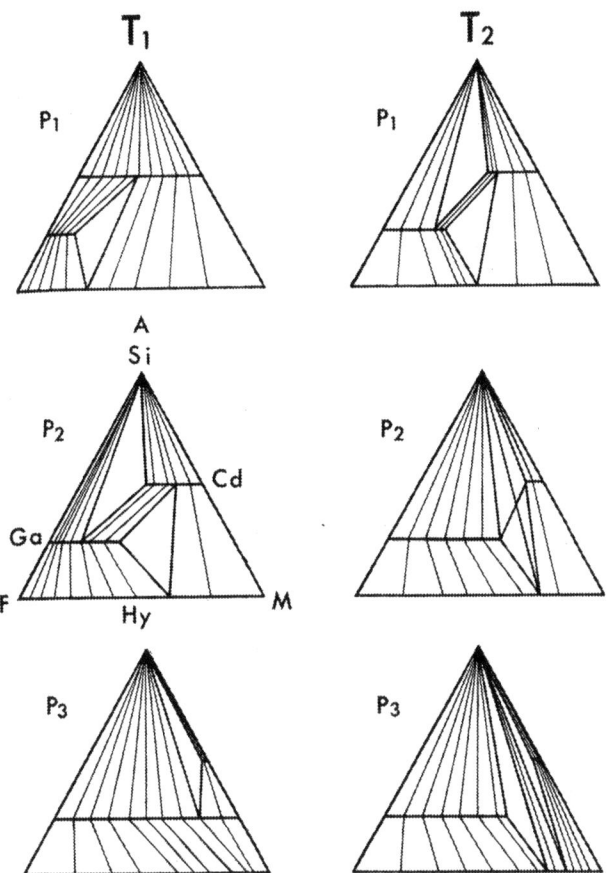

Fig. 7. AFM diagrams at $T_1(P_1, P_2, P_3)$ and $T_2(P_1, P_2, P_3)$ showing shifting tie-lines as a function of P at constant T. Compare Figs. 6 and 8

lower boundary of reaction (1) X_{Cd} and X_{Ga} remain unchanged. On the $P-X$ diagram X_{Cd} and X_{Ga} move vertically up from the loop of reaction (2) until both of them simultaneously reach the curves for reaction (1). X_{Cd} and X_{Ga} will reach their respective curves at the same pressure because the $K_{D\,(Cd-Ga)}$ is the same for both divariant reactions at constant temperature (T_1) and composition[20].

At higher temperature (T_2) reactions (1) and (2) intersect, which means that at one particular pressure the X_{Ga} and X_{Cd} in both reactions are identical. At that pressure (P_2 in Fig. 6), five phases coexist in univariant equilibrium. The intersection gives rise to two new quartz-saturated, divariant reactions which occur on the high pressure side of the univariant isobaric horizontal. These are the reactions

$$Cd \rightleftharpoons Hy + Si + Qz \quad \text{and} \quad Hy + Si \rightleftharpoons Ga + Qz.$$

AFM plots for three isobaric sections are shown in Fig. 7[21].

---

[20] Assuming K_D is independent of pressure.
[21] Distribution coefficients used for T_2 in Figs. 6 and 7 are: $K_{D(Cd-Ga)} = 4$ and $K_{D(Cd-Hy)} = 2$.

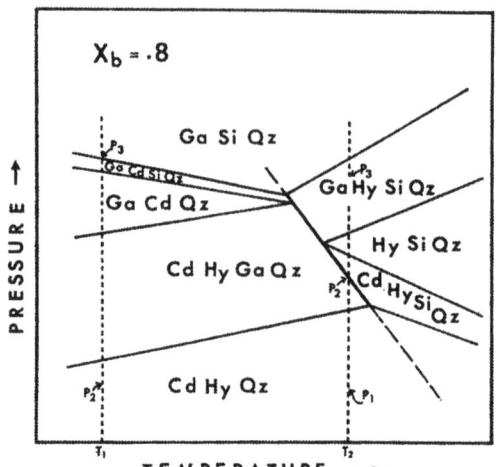

Fig. 8. P–T diagram illustrating phase relations for a composition with $X_b = 0.8$ and $1/3 < (A/F+M)_b < 1$. T_1 and T_2 refer to Figs. 6 and 7

A P–T diagram (Fig. 8) incorporating the T_1 and T_2 isothermal sections shows the phase relationships for a chosen bulk composition ($X_b = 0.8$ and $1/3 < (A/F+M)_b < 1$) as a function of pressure and temperature. Note the intervening trivariant assemblages (3 phases) between the divariant bands (4 phases).

Additional P–T diagrams could be constructed for a variety of bulk compositions (X_b) with the same $(A/F+M)_b$ ratio. If we change the $(A/F+M)_b$ ratio, the stippled lines delineating the divariant field for Cd–Hy–Ga–Qz in the P–X diagrams (Fig. 6) are displaced. The direction of displacement as a function of composition has been dealt with earlier (see Fig. 4).

For clarity, contours for constant compositions of the coexisting phases have been omitted from the P–T diagram (Fig. 8). Two sets of X_{Cd} and X_{Ga} curves and one set of X_{Hy} curves, related to reactions (1) and (2) can be represented on this diagram. These constitute the projections of sloping planes in P–T–X space onto the P–T diagram. The compositional curves of both reactions intersect on the univariant curve because the composition of all phases in univariant equilibrium is fixed (compare Fig. 7, T_2, P_2).

Stability Relations in Natural Rocks

The foregoing theoretical treatment cannot be directly applied to most natural mineral assemblages since additional chemical variables may play an important role in determining the relative stability of ferromagnesian minerals e. g. Ca and Mn (in garnet), oxygen fugacity, the occurrence of hydrous phases such as biotite and amphiboles. It will therefore be necessary to carefully evaluate the effect of such variables on the proposed equilibria in each particular case.

Further discussion of applications, in particular to experimental results on complex model pelitic compositions will be given elsewhere (Hensen and Green, in prep.).

Intersections of Reactions (1) and (2) with Dehydration Reactions Involving Biotite

Since in most natural metamorphic rocks containing cordierite-garnet assemblages, biotite is present as a major phase, it is worth considering the interactions between dehydration reactions involving biotite and the anhydrous solid-solid reactions discussed above. The dehydration of biotite produces ferromagnesian silicate(s) + orthoclase (Or) + vapour (H_2O). If we consider $Bi + Or + H_2O$ in addition to the phases cordierite, garnet, hypersthene, sillimanite and quartz in the 6 component system MgO—FeO—Al_2O_3—SiO_2—K_2O—H_2O, we can deduce five silica-saturated, univariant reactions, given the relative $Mg/Mg + Fe^{2+}$ ratios of the ferromagnesian phases (i. e. $X_{Cd} > X_{Bi} > X_{Hy} > X_{Ga}$). These reactions are:

	ΔV	
$Bi + Si + Qz \rightleftharpoons Cd + Ga + Or + H_2O$	+ve	(Hy)
$Bi + Ga + Qz \rightleftharpoons Cd + Hy + Or + H_2O$	+ve	(Si)
$Bi + Si + Qz \rightleftharpoons Cd + Hy + Or + H_2O$	+ve	(Ga)
$Bi + Ga + Qz \rightleftharpoons Hy + Si + Or + H_2O$	+ve	(Cd)
$Cd + Ga \rightleftharpoons Hy + Si + Qz$	+ve	(Bi, Or, H_2O)

Theoretically, these reactions would be expected to intersect and produce an invariant point. This invariant point and the sequence of univariant reactions around it have been described by Hess (1969).

The univariant reactions (Hy) and (Si) both involve cordierite and garnet, and are therefore of particular interest in relation to the anhydrous reactions (1) and (2) discussed previously.

The divariant reactions related to the univariant boundaries (Hy) and (Si) are given in Table 3. This table shows that reactions (1) and (2) intersect the univariant boundaries (Hy) and (Si) respectively. In this fashion, the anhydrous reactions are tied in with the dehydration reactions.

The volume change of the solids (ΔV_s) is negative in three of the dehydration reactions. Since the molar volume of water decreases rapidly with increasing pressure, the negative ΔV_s causes the total volume of reactions (Cd, Hy) and (Hy, Si) to change sign at relatively low water pressures. Calculations show that the reaction boundary (Cd, Hy) will backbend between 1 and 2 kb at both 700° C and 800° C, whereas the volume change of reaction (Hy, Si) will change sign around 7 or 8 kb depending on the temperature (Table 3). The calculated pressures constitute only an approximate estimate, since neither accurate volumes (except for end member minerals) nor thermal expansion data are available at the present time.

A hypothetical phase diagram (Fig. 9) for a bulk composition containing cordierite and hypersthene plus excess orthoclase, quartz and water, shows the relations among anhydrous and hydrous reactions on the P—T plane (compare Fig. 8).

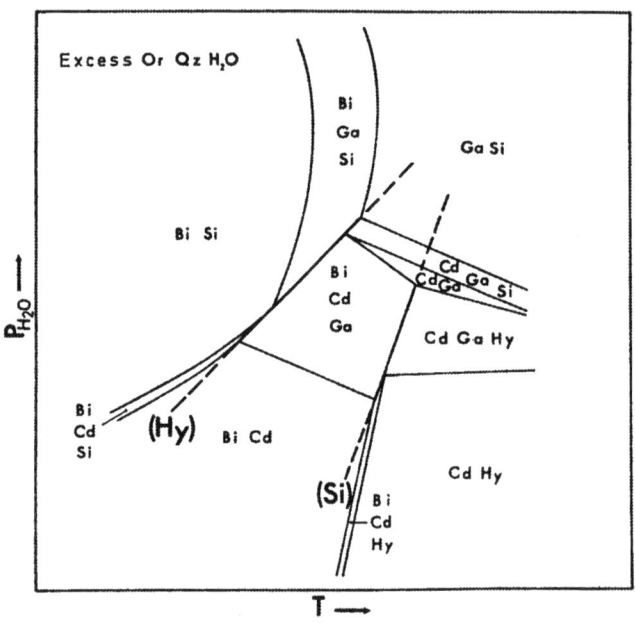

Fig. 9. Hypothetical diagram for a Cd—Hy—Qz bearing composition at temperatures of 700 to 900° C, showing relation of anhydrous equilibria to dehydration reactions. (Hy) and (Si) are univariant boundaries (see Table 3)

Table 3

Univariant reaction (Hy)		
Divariant reactions	ΔV_s [a]	
$2Bi + 6Si + 9Qz \rightleftharpoons 3Cd + 2Or + 2H_2O$	$+ 81.3$	(Ga, Hy)
$Bi + Si + 2Qz \rightleftharpoons Ga + Or + H_2O$	$- 22.3$	(Cd, Hy)
$4Bi + 3Cd + 3Qz \rightleftharpoons 6Ga + 4Or + 4H_2O$	-215	(Si, Hy)
$3Cd \rightleftharpoons 2Ga + 4Si + 4Qz$ (1)	-129.5	(Bi, Or, H_2O, Hy)

Univariant reaction (Si)		
Divariant reactions	ΔV_s	
$4Bi + 3Cd + 3Qz \rightleftharpoons 6Ga + 4Or + 4H_2O$	-215	(Hy, Si)
$Bi + 3Qz \rightleftharpoons 3Hy + Or + H_2O$	$- 14$	(Cd, Ga, Si)
$Cd + 4Hy \rightleftharpoons 2Ga + 3Qz$ (6)	$- 53$	(Bi, Or, H_2O, Si)

Backbending reactions	Change in sign of ΔV [b]	
Reaction	700° C	800° C
(Si, Hy)	1—2 kb	1—2 kb
(Cd, Hy)	6—7 kb	7—8 kb

[a] Volumes are the same as used by Hess (1969) i.e. $V_{Cd}=223$; $V_{Bi}=151$; $V_{Ga}=115$, $V_{Hy}=32$; $V_{Si}=49.9$.

[b] Using the volume data for water by Burnham et al. (1969).

The X_{Cd} and X_{Ga} contours (not shown in Fig. 9) for reactions (1) and (2) pass through the univariant curves (Hy) and (Si) uninterrupted. On the low temperature side of reaction (Si) both the divariant assemblages Cd—Ga—Hy—Qz—Bi—H_2O, and Cd—Ga—Hy—Qz—Bi—Or can occur (the former in a K_2O-deficient environment and the latter in a H_2O-deficient environment). In both these assemblages we are basically dealing with equilibrium (2), since biotite cannot react without orthoclase and water being produced. If biotite reacts the number of phases exceeds that possible for divariant equilibrium and either cordierite or garnet has to break down. The coexistence of Bi—Hy—Qz—Cd—Ga—Or—H_2O is only possible on the univariant phase boundary (Si). It follows from the above that the presence of biotite does not affect the (anhydrous) divariant equilibrium between the phases cordierite, hypersthene, garnet and quartz. However, because the biotite may have a $Mg/Mg + Fe^{2+}$ ratio different from that of the bulk composition, its presence can alter the *relative proportions* of the phases in divariant equilibrium and may also lead to the disappearance of one of the phases. The presence of biotite cannot change the *compositions* of the phases in divariant equilibrium (reaction Bi, Or, H_2O, Hy; Table 3), since these compositions are a unique function of P and T. The same also applies on the low temperature side of reaction (Hy) where we can have the assemblages Cd—Ga—Si—Qz—Bi—H_2O (K_2O-deficient) and Cd—Ga—Si—Qz—Bi—Or (H_2O-deficient).

The $Mg/Mg + Fe^{2+}$ ratio of each ferromagnesian phase increases with increasing pressure along the univariant boundaries in Fig. 9. With increasing temperature the $Mg/Mg + Fe^{2+}$ ratio of biotite decreases when it coexists with cordierite (reaction (Ga, Hy)), but increases when it occurs with garnet (reaction (Cd, Hy)). This is a result of the fact that $X_{Cd} > X_{Bi}$, whereas $X_{Ga} < X_{Bi}$, and follows from the general rule that when the $Mg/Mg + Fe^{2+}$ ratio is increased the stability field of the phase with the highest value of X is extended. This provides an explanation for the observation that cordierite becomes more iron-rich with increasing grade of metamorphism (Best and Weiss, 1964; Seifert, 1970).

On the other hand, garnet coexisting with biotite, orthoclase, sillimanite and quartz will become more magnesian with increasing grade (temperature).

Factors Bearing on the Role of Biotite in Natural Rocks

In the foregoing discussion it has been assumed that biotite is a pure annite-phlogopite solid solution. However, natural biotites from high grade metamorphic terranes contain (1) high Al_2O_3; Best and Weiss (1964) and Rutherford (1968). (2) high Ti; up to 4.5 weight % TiO_2, Kwak (1968); Barker (1962) and (3) may have considerable fluorine and chlorine substituting for hydroxyl group. In addition oxygen fugacity should also be taken into account. The incorporation of Al_2O_3 in the biotite would affect the divariant reactions extending the stability field of biotite but should not affect the univariant boundaries. Factors (2) and (3) could extend the stability of biotite to higher temperature and could also significantly influence the hypothetical univariant boundaries of Fig. 9. These boundaries could be displaced to higher temperatures and may become divariant or multi-variant.

210

Appendix 1

Mathematical and Graphical Analysis of Divariant Reactions

For each simple divariant equilibrium involving two solid solutions, an equation of the following type can be written [e. g. using reaction (1)]:

$$(Mg_{X_{Cd_1}}Fe_{1-X_{Cd_1}})_2Al_4Si_5O_{18} = (1-n)(Mg_{X_{Cd_2}}Fe_{1-X_{Cd_2}})_2Al_4Si_5O_{18} +$$
$$Cd_1 \quad\quad = \quad\quad Cd_2$$

$$\tfrac{2}{3}n(Mg_{X_{Ga}}Fe_{1-X_{Ga}})_3Al_2Si_3O_{12} + \tfrac{4}{3}n\ Al_2SiO_5 + \tfrac{5}{3}n\ SiO_2$$
$$Ga \quad\quad + \quad Si \quad + \quad Qz$$

where X_{Cd} = mole fraction of magnesian end member in cordierite ($= Mg/Mg+Fe^{2+}$ ratio).

X_{Ga} = mole fraction of pyrope in garnet ($= Mg/Mg+Fe^{2+}$ ratio).

n = proportion of cordierite broken down in the reaction.

$0 < X_{Cd} < 1$
$0 < X_{Ga} < 1$
$0 < n < 1$

In the case where cordierite is the only ferromagnesian phase present, Cd_1 has the same $Mg/Mg+Fe^{2+}$ ratio as the bulk composition. Therefore $X_{Cd_1} = X_b$ (where X_b is the X value for the bulk composition). Consequently we can relate the composition of the initial cordierite to that of the phases produced by the reaction. From element balance requirements for Mg in the equation for reaction (1) we can write

$$2(X_b) = (1-n)\ 2(X_{Cd}) + \tfrac{2}{3}\ 3(X_{Ga})$$

or

$$X_b = (1-n)\ X_{Cd} + n\ X_{Ga}.$$

Thus we can express the composition of one phase as a function of the composition of the other phase, the amount of reaction that has taken place and the bulk composition.

$$X_{Cd} = \frac{X_b - n\ X_{Ga}}{1-n}$$

$$X_{Ga} = \frac{X_b - X_{Cd}(1-n)}{n}$$

and

$$n = \frac{X_{Cd} - X_b}{X_{Cd} - X_{Ga}}.$$

The compositions of the coexisting phases are also related by a distribution coefficient (K_D)

$$K_{D(Cd-Ga)} = \frac{X_{Cd}(1-X_{Ga})}{X_{Ga}(1-X_{Cd})} = \frac{(Mg/Fe)\ Cd}{(Mg/Fe)\ Ga}$$

or

$$X_{Cd} = \frac{K_D \cdot X_{Ga}}{1 + (K_D - 1)\ X_{Ga}}$$

and

$$X_{Ga} = \frac{X_{Cd}}{K_D(1-X_{Cd}) + X_{Cd}}.$$

Therefore, it is possible to express the composition of a phase in terms of X_b, n and K_D. This relationship is best illustrated graphically (Fig. 10). In this diagram, having X_{Cd} and X_{Ga} as perpendicular axes, one can represent:

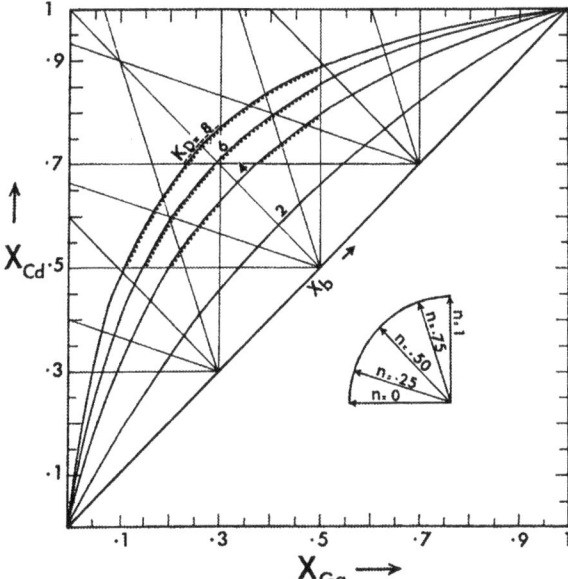

Fig. 10. Graphical representation of the relation between the amount of reaction (n), the composition of the ferromagnesian phases (X_{Cd} and X_{Ga}) and the distribution coefficient (K_D) for the reaction Cd $\rightleftharpoons$ Ga + Si + Qz (1). The overlap of n between different bulk compositions as a function of K_D is dotted (for $X_b = 0.3$ and 0.5, and for $X_b = 0.5$ and 0.7) and dashed (for $X_b = 0.3$ and 0.7)

1. the composition of two coexisting phases by a point on the diagram,
2. a constant distribution coefficient between two phases by a curve,
3. the amount of reaction that has taken place, by the slope of the straight line passing through the compositional point (1) and X_b, the initial composition of the cordierite.

If ideal mixing is assumed, implying that the K_D is independent of composition and if the effect of pressure is neglected K_D is a function of temperature only. In this case the change in composition of the coexisting phases with increasing pressure (and increasing n) at constant temperature (and K_D) can be illustrated as in Fig. 10. For each "bulk composition" (X_b) and chosen K_D (i. e. temperature) the compositional point giving X_{Cd} and X_{Ga} moves a corresponding distance along a constant K_D curve on the diagram. The intercept on the K_D curve is bounded by a horizontal line ($n = 0$; i. e. no garnet present) and a vertical line ($n = 1$; i. e. no cordierite present).

An interesting aspect to emerge from this treatment is the graphical demonstration of the effect of K_D on the overlapping of divariant assemblages for different bulk compositions. If we take, for instance, two initial cordierite compositions ($X_b = 0.5$ and $X_b = 0.7$), it follows from Fig. 10 that at $K_D = 2$ we cannot have the same cordierite + garnet assemblage (the same X_{Cd} and X_{Ga}) in both bulk compositions. When $K_D = 4$, however, there is a range of compositional points (overlap), where the same assemblage can occur in both bulk compositions. These points correspond to high values for n in the composition with $X_b = 0.5$ and to low values in the composition with $X_b = 0.7$. This overlap further increases, that is it involves a larger range of n-values, with rising K_D.

Because the compositions of the coexisting phases in divariant equilibrium are unique functions of pressure (at constant temperature), the amount of overlap in n-values corresponds to an overlap of divariant bands for different bulk compositions in a $P-T$ diagram. Fig. 10 also shows that, for low values of K_D, the divariant bands for certain bulk compositions cannot possibly overlap, whatever the value of n. Therefore this type of approach can provide criteria for equilibrium in the experimental study of divariant reactions where the Phase Rule is of restricted use. The amount of reaction (n) for each bulk composition (X_b) is some unknown function of pressure for each particular temperature.

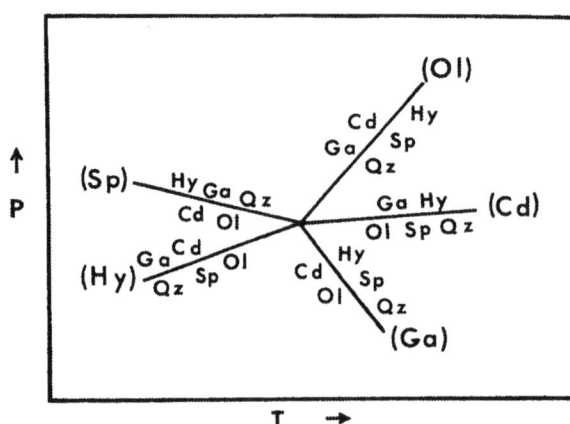

Fig. 11. Schematic P–T diagram of the invariant point involving the phases Cd, Ga, Hy, Ol, Sp and Qz assuming $X_{Cd} > X_{Hy} > X_{Ol} > X_{Sp} > X_{Ga}$

Appendix 2

Equilibria Involving Fe-Rich Olivine

In compositions with a very low $Mg/Mg+Fe^{2+}$ ratio Fe-rich olivine is stable with quartz. Thus the stability of hypersthene is restricted in Fe-rich compositions (at low pressures). The divariant reaction:

$$\text{olivine} + \text{quartz} \rightleftharpoons \text{hypersthene}$$

has recently been studied experimentally by Smith (1970).

The appearance of fayalite-rich olivine in quartz-bearing compositions gives rise to an additional invariant point in the system $MgO-FeO-Al_2O_3-SiO_2$. This point involves the phases cordierite, garnet, hypersthene, olivine, spinel and quartz. The relative values for the $Mg/Mg+Fe^{2+}$ ratios at this invariant point are:

$$X_{Cd} > X_{Hy} > X_{Ol} > X_{Sp} > X_{Ga}.$$

The univariant reactions and their sequence around the invariant point is given in Fig. 11. Reaction (Ol) in Fig. 11 is identical to reaction (Si, Sa) in Fig. 1 and Table 1. Reaction (Hy) terminates in an invariant point in the system $FeO-Al_2O_3-SiO_2$, involving the phases Fe-cordierite, almandine, fayalite, hercynite and quartz (Hsu, 1968).

Appendix 3

Alternative Solution for Univariant Curves Involving Garnet and Spinel with $X_{Sp} < X_{Ga}$

Although evidence has been found in natural rocks supporting the assumption that $X_{Sp} > X_{Ga}$ (Hensen and Green, in prep.), there remains some uncertainty about the relative composition of coexisting garnet and spinel (compare Hsu and Burnham, 1969).

If we take $X_{Sp} < X_{Ga}$, the following changes are required in two of the four univariant reactions simultaneously involving garnet and spinel. These are:

1. $Cd + Ga + Si \rightleftharpoons Sp + Qz$ $\Delta V = +\text{ve}$ (?)

   becomes   (Hy, Sa)

     $Ga + Si \rightleftharpoons Cd + Sp + Qz$ $\Delta V = +\text{ve}$.

2. $Ga + Sa + Si \rightleftharpoons Sp + Qz$ $\Delta V = +\text{ve}$ (Hy, Cd)

   becomes

     $Ga + Si \rightleftharpoons Sa + Sp + Qz$ $\Delta V = +\text{ve}$

However, the two remaining univariant reactions (Si, Sa) and (Si, Cd), do not necessarily change when the relative compositions of garnet and spinel are reversed. In fact no change can occur unless the $K_{D(Hy-Ga)}$ is very small and the $K_{D(Ga-Sp)}$ relatively large.

Considering the reaction:

$$Cd + Ga \rightleftharpoons Hy + Sp + Qz \qquad \text{(Si, Sa)}$$

and taking the extreme case where $X_{Hy} = 0.5$; $X_{Sp} \to 0$; $X_{Cd} = 0.66$, as an example, then this reaction will remain non-terminal unless $X_{Ga} > 0.33$ (i. e. if $K_{D(Hy-Ga)} < 2$), since only in that case will the garnet be situated inside the tetrahedron formed by the phases Cd—Hy—Sp—Qz. The reaction

$$Cd + Ga \rightleftharpoons Hy + Sp + Qz \qquad \Delta V = + \text{ve (?)} \qquad \text{(Si, Sa)}$$

will then change to

$$Ga \rightleftharpoons Cd + Hy + Sp + Qz \qquad \Delta V = + \text{ve}.$$

However, this alternative solution requires a relatively small difference between X_{Hy} and X_{Ga}, as well as a large difference between X_{Ga} and X_{Sp}. This is a very unlikely situation. Therefore the reactions (Si, Sa) and (Si, Cd) which theoretically could be terminal reactions representing the breakdown of a single phase (i. e. garnet), are most likely non-terminal and represent the breakdown of two joins (i. e. the Cd—Ga and Sa—Ga joins respectively).

Acknowledgements. In preparing this paper I greatly benefitted from discussions with Drs. D. H. Green and E. J. Essene at the Australian National University, Drs. A. L. Graham and A. J. Irving (A.N.U.) and Professor Schreyer (Ruhr-Universität) are thanked for critically reading the manuscript. The work has been carried out during a Ph. D. scholarship at the Australian National University and the final version of the manuscript has been prepared while holding a Fellowship at the Institute for Mineralogy at the Ruhr University, Bochum.

References

Barker, F.: Cordierite-garnet gneiss and associated microcline-rich pegmatite at Sturbridge, Massachusetts and Union, Connecticut. Am. Mineralogist **47**, 907–918 (1962).

Best, M. G., Weiss, L. E.: Mineralogical relations in some pelitic hornfelses from the southern Sierra Nevada. Am. Mineralogist **49**, 1240–1266 (1964).

Burnham, C. W., Holloway, J. R., Davis, N. F.: The specific volume of water in the range 1,000–8,900 bars, 20°–900° C. Am. J. Sci. Schairer Vol., 70–95 (1969).

Goldschmidt, V. M.: Die Kontaktmetamorphose im Kristianagebiet. Kristiana Vidensk Skr., Math.-Naturv. Kl. 11 (1911).

Grant, J. A.: Partial melting of common rocks as a possible source of cordierite-anthophyllite bearing assemblages. Am. J. Sci. **266**, 908–931 (1968).

Hensen, B. J., Essene, E. J.: Stability of Pyrope-Quartz in the system $MgO-Al_2O_3-SiO_2$. Contr. Mineral. and Petrol. **30**, 72–83 (1971).

— Green, D. H.: Experimental data on coexisting cordierite and garnet under high grade metamorphic conditions. Phys. Earth Planet. Int. **3**, 431–440 (1970).

Hess, P. C.: The metamorphic paragenesis of cordierite in pelitic rocks. Contr. Mineral. and Petrol. **24**, 191–207 (1969).

Hsu, L. C.: Selected phase relationships in the system Al—Mn—Fe—Si—O—H: A model for garnet equilibria. J. Petrol. **9**, 40–83 (1968).

— Burnham, C. W.: Phase relationships in the system $Fe_3Al_2Si_3O_{12}-Mg_3Al_2Si_3O_{12}-H_2O$ at 2.0 kilobars. Geol. Soc. Am. Bull. **80**, 2393–2408 (1969).

Khlestov, V. V.: Garnets from cordierite-bearing rocks of the sharyzhalgay complex (southern Transbaikal). Dokl. Akad. Nauk. SSSR **154**, 4, 842–845 (1964).

Kwak, T. A. P.: Ti in biotite and muscovite of metamorphic grade in almandine amphibolite facies rocks from Sudbury, Ontario. Geochim. Cosmochim. Acta **32**, 1222–1229 (1968).

Marakushev, A. A., Kudryavtsev, V. A.: Hypersthene-sillimanite paragenesis and its petrological implication. Dokl. Akad. Nauk SSSR **164**, 1, 179–182 (1965).

Ramberg, H.: Chemical thermodynamics in mineral studies. Physics and chemistry of the earth, 5 (ed. L. H. Ahrens, F. Press and S. K. Runcorn), p. 225–253. (1964).

Rutherford, M. J.: An experimental study of biotite phase equilibria. Johns Hopkins University, Ph. D. 1968.

Schreinemakers, F. A. H.: In-, mono-, and divariant equilibria. Penn. State Univ. Publ. **2**, 322 p. (1965).

Seifert, F.: Low-temperature compatibility relations of cordierite in haplopelites of the system K_2O—MgO—Al_2O_3—SiO_2—H_2O. J. Petrol. **11**, 73–99 (1970).

Smith, D.: Stability of iron-rich orthopyroxene. Carnegie Inst. Washington Year Book **68**, 229–231 (1970).

Zen, E.: Construction of pressure-temperature diagrams for the multi-component systems after the method of Schreinemakers, a geometric approach. U.S. Geol. Surv. Bull. **1225**, 56 (1966).

Dr. B. J. Hensen
Geophysical Laboratory
2801 Upton Street
N. W. Washington, D. C. 20008, U.S.A.

CHAPTER 9: THERMAL MODELS OF COLLISION BELTS

England, P.C. and Richardson, S.W. (1977) The influence of erosion upon the mineral facies of rocks from different metamorphic environments. *Journal of the Geological Society of London*, **134**, 201–213.

The new global tectonics of the 1960s prompted geophysicists to model temperatures and pressures in idealized sections through subduction zones (Isacks et al., 1968; Oxburgh and Turcotte, 1968, 1970, 1971; Bird et al., 1975). Oxburgh and Turcotte (1974) assumed that regional thrusting processes associated with collision in orogenic belts are related to plate tectonics and therefore share similar rates of movement. Taking a reasonable value for thermal diffusivity, they showed in a simple one-dimensional model that the rate at which thrust masses are emplaced is fast in comparison to the rate of heat transfer by conduction. As a result, crustal-scale overthrusting produces a major perturbation in the regional conductive thermal gradient that can persist for tens of millions of years. Thus, temperatures accompanying metamorphism in orogenic belts, contrary to long-held views, are likely not to be a monotonic function of depth. Anomalously low temperatures close to a major thrust may be expected to last for several million years and, if the thrust-nappe is thick enough, may be associated with an imprint of blueschist-facies metamorphism – something of a variation on the accepted Benioff-zone origin of blueschists (see Chapter 7). Relaxation of the geothermal perturbation may then superimpose greenschist- or amphibolite-facies metamorphism on the blueschists. Oxburgh and Turcotte applied their calculations to the metamorphic and structural history of the Tauern Window in the Eastern Alps of Austria, which Oxburgh had studied for his doctoral thesis.

England and Richardson (1977), both students of Oxburgh, examined the effect of erosion on the pressures and temperatures in crust whose total content of heat-producing radiogenic isotopes has been increased by thickening caused by orogenic collision, the results having major implications for metamorphic petrology. Their paper showed that erosion of tens of km of previously thickened crust concurrent with thermal relaxation over time scales of tens of millions of years is a factor of critical importance in determining the pressures and temperatures finally recorded by the metamorphic mineral assemblages. This was a challenging idea for metamorphic petrologists at the time.

Their model assumed that erosion commenced 20 m.y. after geologically instantaneous thrusting. As explained by England and Thompson (1984, p. 901), an initial densification of cold crust beneath the thrust (by means of reactions such as plagioclase → pyroxene + quartz) was believed to cancel out temporarily the immediate gain in surface elevation due to isostasy. This was a reasonable assumption for a model of instantaneous collision that involved an upper plate as thick as 30 km. Thus, after 20 m.y., pressure is unchanged, but the initial saw-tooth geotherm at $t = 0$ (implicit but not shown on Fig. 1 of England and Richardson) has already been smoothed out. With further passage of time, a rock such as A experiences thermal relaxation (T increase) while erosion at the top of the pile decreases the pressure on it. Rock A can be expected to recrystallize close to the T_{max} (or S_{max}) on its curving clockwise P-T-t path, which in this example is reached 33 m.y. after thrusting. The P-T locus of T_{max} for rocks at all structural levels defines what England and Richardson called a metamorphic geotherm. Because it is not strictly a geotherm, England and Thompson (1984) preferred to call it a "piezothermic array", or "P-T array". Three properties of the P-T array are of particular interest: (1) it is slightly concave to the temperature axis; (2) it is polychronic – structurally deeper rocks reach higher grades of metamorphism several m.y. later than lower-grade, shallower rocks; and (3) it is at a steep angle to the P-T-t paths of individual rocks.

The England and Richardson paper was a wake-up call for metamorphic petrologists. In earlier times, regional metamorphism was believed primarily to reflect a steady-state geotherm, convex to the T axis, and reaching no more than 500–600°C at the base of 35 km thick crust. Turner (1968, Fig. 8-4) already had compiled estimates of P-T field gradients from 12 different sources, which showed that many were concave to the T axis, and some reached >700°C. The curvature had much to do with the upgrade transition Ky → Sil, but it (and the higher temperatures) can also be taken as a field prediction of the results of the dynamic crust-thickening model of England and Richardson.

A distinction between space and time in the P-T trajectories of regional metamorphic rocks had hitherto not been part of our conceptual understanding. The working hypothesis of progressive metamorphism had always been taken to mean the growth of minerals in one zone from those in the immediately adjacent lower-grade zone. Furthermore, consistent with the outflow of heat from some poorly known source, the isograds of regional metamorphism, as in contact metamorphism, were thought to be broadly contemporaneous – not separated by millions of years. England and Richardson's paper provided P-T-t path criteria for distinguishing between contact (thermal) and regional metamorphism where field relations are otherwise ambiguous.

The additional supply of heat in tectonically thickened crust, after much time, leads to higher geothermal gradients than in undisturbed crust. Figure 1 of England and Richardson shows that temperatures in excess of 800°C should be reached deep in the crust after 60–80 m.y. Thus, partial melting can be expected to occur at depth several m.y. after T_{max} is reached at higher structural levels. Magma so produced will rise through the crust and create late regional-metamorphic intrusions, exactly as observed in numerous metamorphic belts. England and Richardson's model also showed that the combined effects of thermal relaxation and erosion in collisional belts are sufficient to eliminate blueschists over a time span of 500 m.y., an issue discussed in Chapter 7. Thus, some special circumstances would appear to be necessary to account for the survival of Precambrian-age blueschists.

Even before England and Richardson (1977), field investigations had begun to show that some terranes had experienced superposed metamorphisms of contrasting baric type within the same orogenic cycle. Prominent among these accounts were belts of Barrovian-style metamorphism showing vestiges of earlier blueschist- or eclogite-facies metamorphism, as in the Alps (Bearth, 1970; Dal Piaz et al., 1972; Frey et al., 1974; Miller, 1974). As the study of blueschist belts evolved in later years, it was realized that one of the common characteristics of Alpine (A-type) belts worldwide (Chapter 7) was a decompression path leading across Barrovian conditions.

In 1984, a pair of subsequently much-cited papers was devoted to a more thorough investigation of the same subject: the petrological consequences of a geologically instantaneous event of crustal-scale thickening, followed by isostatically driven erosion of the thickened amount. The first (England and Thompson, 1984) calculated P-T-t paths and geotherms — space limitations permitted only a selection of them to be shown — for combinations of what the authors considered to be the major controls. These were the modes (geometry) of crustal thickening (thrusting, homogeneous horizontal shortening, shortening of the entire lithosphere), and values (low, medium, and high) for thermal conductivity, heat fluxes and source distributions, thrust thickness, and erosion times. Their Fig. 1 is a good starting point for acquiring a graphical understanding of the consequences of the model, because it uses medium values of the controlling variables, and shows the transient geotherms, as well as the pre-thickening, immediate post-thickening, and (never attained) steady-state thickened crust geotherms, selected P-T-t paths, and the principal metamorphic facies series; it is a simple step to add the polychronic P-T array.

In the thrust-thickening case, their Fig. 1 shows two P-T-t paths, one traversing blueschist (Gln-Jd) and amphibolite facies (Ky-Sil) conditions, and another, deeper one emerging from eclogite into transitional amphibolite-granulite conditions, in principle much like the single path figured in England and Richardson (Fig. 1). Other figures in England and Thompson (1984) show major differences in P-T-t paths for the three selected heat-source distributions and extremes in thermal conductivity, although most of them intersect, at least for a time, the Gln-Jd facies series. In the thrust model, transient high-P/T conditions are encountered both below and above the thrust. Preservation of the Gln-Jd mineralogy in the course of uplift is favoured in cases of minimal heat supply and high conductivity (Fig. 3), and rapid erosion. After some tens of m.y., transient geotherms in most cases are steeper than the original undisturbed geotherm, because the new crust is thicker. And-Sil metamorphism is predicted to take place on the retrograde path in the collision models; when it is found in the field to be prograde, a heat contribution from intrusive activity is indicated. Eliminating the initial 20 m.y. delay in the onset of erosion lowers temperatures in the ensuing relaxation cycle, but not enough to miss Ky-Sil conditions altogether (England and Thompson, 1984, Figs 6, 7, 8, cf. Spear, 1993, p. 60).

In the second paper, Thompson and England (1984) expanded on the petrological implications of the results from the first paper, and examined the degree to which the metamorphic mineralogy can be expected to respond faithfully in nature to the uplift P-T-t path. Isograds of regional metamorphism (basically intersections of isotherms with the horizontal plane) are not addressed in the one-dimensional model, unless small-scale variations in surface relief are postulated. The common linear or concentric pattern of multiple isograds — corresponding to the metamorphic field gradient, which is the spatial expression of the P-T array — must then reflect a geographic variation in one or more of a number of additional factors. Variation in burial depth (wedge-shaped thrust packages) and spatially uneven heat supply are two possibilities modeled in two dimensions by Thompson and England (1984, pp. 950–952). Other possibilities include differential uplift and syn- to late-metamorphic folding deformation (e.g. Chamberlain 1986; Thompson and Ridley, 1987). Day (1987) and Day and Chamberlain (1989) examined the effect of spatial variations in burial depth and heat supply. Intersecting isograds and bathograds (as in northern New England or the Dalradian of N.E. Scotland) suggest spatial variations in thermal parameters, uplift history, or both (Day and Chamberlain, 1989). The thermal variable could, of course, include the magmatic contributions. Thompson and England (1984, pp. 944–946) also recognized that extensional thinning as a late phase in the development of a compressional terrane can cause fast exhumation and induce a steep P-T-t path of isothermal decompression to temperatures well above the steady-state geotherm. Extension will produce condensed isograd sequences (Thompson and Ridley, 1987). The end-stage of extension is one of largely isobaric cooling, which is not necessarily to be interpreted as related to magmatic involvement in the thermal budget.

Peacock (1989) provided a useful digest of the England, Richardson and Thompson papers, as well as help with the methods of numerical models (finite difference vs. finite element, etc.). A further account of the application of one-dimensional models to collisional orogenesis was provided by Spear (1993, pp. 56–72). He noted (p. 60) that, in models involving slow rather than instantaneous thrusting, Barrovian metamorphism was readily achieved without assuming an initial 20 m.y. time lag. He figured model P-T-t paths (p. 70-72) for crustal thickening (clockwise), magma intrusion, and magma intrusion combined with crustal thickening (counter-clockwise).

The landmark paper of England and Richardson, and the following two papers by England and Thompson, were a major part of the driving force behind efforts in the last two decades of the 20th century — in a marriage of mineral thermobarometry and geochronology — to determine the P-T-t paths experienced by individual rocks and by different parts of terranes (e.g. nappes) in regional metamorphism. Their papers provided the stimulus for important advances in techniques for extracting this P-T-t information from rocks (e.g. Spear and Selverstone, 1983; Spear et al., 1984; Spear, 1989a,b; Essene, 1989; Spear et al., 1991; see also Chapter 10 on Grt-Bt thermometry), and a better appreciation of their reliability. Our choice of mineral barometers (with their catchy acronyms) notably expanded during this time. The preservation of inclusions and compositional zoning in relatively non-reactive porphyroblasts of garnet (Spear and Rumble, 1986; St. Onge, 1987) played an important role in these efforts, and significant advances were made in radiometric methods in metamorphic minerals (e.g. Zeitler, 1989). More recently, low-temperature thermochronometers such as fission-track dating of apatite and zircon and U-Th/He dating of apatite have been widely used to determine denudation rates. In situ radiometric methods (SHRIMP) applied to metamorphic zircon and monazite have proven to be a remarkable addition to methods for exhumation rates (e.g. Gebauer et al., 1997). (The extensive literature on metamorphic pressure-temperature-time paths in the field that has accumulated in the last 2 decades can be readily accessed through GEOREF).

The assumption of geologically instantaneous thrusting or thickening in the modelling by England, Richardson and Thompson may be acceptable for orogenic time-scales on the order of 100 m.y. But if orogeny lasts only ~20 m.y., and metamorphism is broadly syn-kinematic, as is the case for the Taconic and Grampian belts of eastern N. America and Scotland, and many others, measured P-T-t paths are likely to reflect heat-flow patterns that develop locally with respect to the contacts of superimposed thrust sheets. Davy and Gillet (1986) pointed out that P-T-t paths will differ greatly in different thrust slices depending on the number of slices present, their thickness, position within the pile, and the time delays between each thrust. In a mountain belt such as the Alps, there will not be one set of nested paths for the orogen as a whole, cf. England and Thompson (1984). Karabinos and Ketcham (1988) used a one-dimensional, finite-difference conduction model to show that there were important differences in thermal profiles over a short term (1–10 m.y.) and over distances of up to 10 km from the surface of a thrust. Their model helped to explain some of the different polymetamorphic reaction histories recovered by mineral textures and zoning in different structural units of the Taconic in New England.

Realistic modelling of collision zones required consideration of heat and mass transfer in two dimensions, and this is better handled with finite-element solutions. Shi and Wang (1987) employed this numerical technique to show that the speed and duration of emplacement of a single 15 km thick thrust sheet are important controls on the P-T-t path of rocks in orogenic belts. Leading-edge and near-root zones showed major differences in P-T-t paths. In contrast to models with instantaneous thrusting, they found that saw-toothed or inverted T profiles at the end of thrusting were only possible at geologically unreasonable faulting velocities. Except for low-angle or thin thrust sheets, lateral heat transfer is important. It was becoming clear why different locations laterally would have different P-T histories, as found in field studies (e.g. Rubie, 1984; Cliff et al., 1985; Spear and Rumble, 1986; St. Onge, 1987, and many others).

The Ky-Sil (Barrovian) facies series typifies the product of continental collision. In the thermal models of collision, the mineral assemblages that survive are assumed to be those forming at equilibrium close to the T_{max} of a clockwise curving path involving many kbar of decompression. In this scenario, Barrovian-style metamorphism takes place during decompression while temperature continues to rise. We then have to consider if this is fundamentally true for all Barrovian terranes. For some terranes, particularly those that have undergone deep subduction, P-T-t paths determined for individual samples show it to be a valid model (e.g. Spear, 1993; Brown, 2001). By its very nature, details of the rising-temperature portion of a possible clockwise P-T-t path in a terrane undergoing decompression are difficult to recover using conventional mineral thermobarometric methods. Nonetheless, the paths that have been successfully extracted from high-pressure and ultrahigh pressure metamorphic (UHPM) terranes not uncommonly show that decompression was roughly isothermal or even attended by a gradual decline in temperature (see Chapters 2, 7 and 14). These paths are indicative of exhumation and slow cooling along imbricate thrusts while convergent forces were still operative. In the absence of convincing evidence of early-stage phase equilibria, the relative importance for the P-T-t path of thermal relaxation/decompression vs. magmatic heating remains controversial for some Barrovian terranes, including the type area (Chapter 2). If magmatic heat drives metamorphism in And-Sil, or Buchan-style, terranes, there is in principle no reason for magmatic heating not to contribute to the heat budget at deeper levels in the crust.

In the late 1980s, progress in the geodynamic modelling of convergent margins began to show that one-dimensional models are fundamentally limited by their inability to accommodate the two- or three-dimensional processes of deformation that accompany collision. Two-dimensional finite-element models are needed to follow rock-paths shaped by deformation following subduction-driven collision of an orogen with a buttress such as a rifted continental margin (Platt, 1986; Jamieson and Beaumont, 1989a). Two-dimensional modelling recognizes exhumation related to synmetamorphic deformation as well as to erosion. Critical-wedge theory, extended to the case where there are major changes in the slope of the basal decollement, was used by Jamieson and Beaumont (1989b) to define internal deformation in a one-sided orogen. They showed that the uplift rate is greatest where the basal decollement dips most steeply, at the site of the former rift. Clearly, we can expect that P-T-t paths will reflect these kinds of variations in uplift rate across the orogen. Uplift rates in general will be much faster than those dependent on erosional unroofing as in the Richardson-England-Thompson models.

Koons (1989, 1990) recognized the influence of climatic gradients on uplift rates in mountain ranges with prevailing high-moisture wind patterns. Collision zones such as the Southern Alps of New Zealand, the central Himalaya, and the western European Alps, have highly asymmetric erosion patterns, with the greatest exhumation rates coincident with high-grade metamorphic and igneous rocks. The feedback among climate, erosion, tectonic processes, and metamorphism has become a topic of great current interest.

The two-dimensional model of Willett *et al.* (1993) for doubly convergent orogens driven by mantle subduction accounts for the simultaneous operation of deformation and localized uplift, isostatic response, lower-crustal ductile deformation, gravitational extension, and surface erosion and sedimentation. These authors show that the intensity of metamorphic grade of rocks at the surface should increase smoothly in the direction of subduction; the location of maximum grade and uplift velocity will be where denudation rates are greatest, and that this may be determined by the polarity of orographic precipitation.

Beaumont *et al.* (1996) applied a mechanical model for Alpine-type orogens, with operation of single-vergent subduction of oceanic lithosphere, followed by a transition to double vergence with subduction of continental crust, and ending with double-vergent collision in response to the buoyancy of subducted continental crust. Their model successfully accounts for the key components of the large-scale tectonic evolution of the European Alps, and probably applies well to other collisional orogens. It also helps account for the complexity of *P-T-t* paths that we find in different parts of many mountain belts, including the observed inversion of Barrovian isograds (Jamieson *et al.*, 1996).

Jamieson *et al.* (1998) noted that orogen models using average values for crustal heat production and convergence and erosion rates tend to be too cool to reproduce the pressures and temperatures of the higher-grade parts of Barrovian sequences. Their two-dimensional thermal-mechanical model for collision driven by subduction showed that the attainment of Barrovian *P-T* conditions, unless there are magmatic influences or other sources of heat, requires accretion of a significant volume of radioactive sedimentary material prior to collision.

In a further study of the feedback and coupling of deformation, metamorphism, and exhumation in orogenic belts, Jamieson *et al.* (2002) found that different styles of thickening and exhumation do not necessarily give rise to fundamentally different *P-T-t* paths. All their models in fact showed clockwise loops with T_{max} later than P_{max}, a remarkable affirmation of the essence of England and Richardson.

Belts of ultra-high pressure metamorphism (Chapter 14) are extreme examples of collision. Fast initial exhumation rates (10–20 mm/y) and isothermal, adiabatic, or cooling *P-T-t* uplift paths have been determined for many of these terranes. These observations indicate tectonic processes of unroofing. At a later stage, there are opportunities for erosionally driven exhumation and thermal relaxation. I am sure that England, Richardson and Thompson had no illusions as to the quantitative applicability of their simple model to structurally complex metamorphic belts. Their papers nonetheless gave us powerful insights into the dynamic nature of metamorphism in an orogen, and in the ensuing years stimulated an enormous amount of research that greatly advanced our appreciation of metamorphic histories.

References

Bearth, P. (1970) Zur Eklogitbildung in den Westalpen. *Fortschritte der Mineralogie*, **47**, 27–33.Beaumont, C., Ellis, S., Hamilton, J. and Fullsack, P. (1996) Mechanical model for subduction-collision tectonics of Alpine-type compressional orogens. *Geology*, **24**, 675–678.

Bird, P., Toksoz, M.N. and Sleep, N.H. (1975) Thermal and mechanical models of continent-continent convergence zones. *Journal of Geophysical Research*, **80**, 4405–4416.

Brown, M. (2001) From microscope to mountain belts: 150 years of petrology and its contribution to understanding geodynamics, particularly the tectonics of orogens. *Journal of Geodynamics*, **32**, 115–164.

Chamberlain, C.P. (1986) Evidence for the repeated folding of isotherms during regional metamorphism. *Journal of Petrology*, **27**, 63–89.

Cliff, R.A., Droop, G.T.R. and Rex, D.C. (1985) Alpine metamorphism in the south-east Tauern Window, Austria: 2. Rates of heating, cooling and uplift. *Journal of Metamorphic Geology*, **3**, 403–416.

Dal Piaz, G.V., Hunziker, J.C. and Martinotti, G. (1972) La Zona Sesia-Lanzo e l'evoluzione tettonico-metamorfica delle Alpi noroccidentali interne. *Società Geologica Italiana*, **85**, 103–132.

Davy, P. and Gillet, P. (1986) The stacking of thrusts slices in collision zones and its thermal consequences. *Tectonics*, **5**, 913–929.

Day, H.W. (1987) Controls on the apparent thermal and baric structure of mountain belts. *Journal of Geology*, **95**, 807–824.

Day, H.W. and Chamberlain, C.P. (1989) Implications of thermal and baric structure for controls on metamorphism: northern New England. Pp. 215–222 in: *Evolution of Metamorphic Belts* (J.S. Daly, R.A. Cliff and B.W.D. Yardley, editors). Special Publication No. **43**, Geological Society, London.

England, P.C. and Richardson, S.W. (1977) The influence of erosion upon the mineral facies of rocks from different metamorphic environments. *Journal of the Geological Society of London*, **134**, 201–213.

England, P.C. and Thompson, A.B. (1984) Pressure-temperature-time paths of regional metamorphism I. Heat transfer during the evolution of regions of thickened continental crust. *Journal of Petrology*, **25**, 894–928.

Essene, E.J. (1989) The current status of thermobarometry in metamorphic rocks. Pp. 1–44 in: *Evolution of Metamorphic Belts* (J.S. Daly, R.A. Cliff and B.W.D. Yardley, editors). Special Publication No. **43**, Geological Society, London.

Frey, M., Hunziger, J.G., Frank, W., Bocquet, J., Dal Piaz, G.V., Jaeger, E. and Niggli, E. (1974) Alpine metamorphism of the Alps: a review. *Schweizerische Mineralogische und Petrographische Mitteilungen*, **54**, 247–290.

Gebauer, D., Schertl, H.-P., Brix, M. and Schreyer, W. (1997) 35 Ma old ultrahigh-pressure metamorphism and evidence for very rapid exhumation in the Dora Maira Massif, Western Alps, *Lithos*, **41**, 5–24.

Isacks, B., Oliver, J. and Sykes, L.R. (1968) Seismology and the new global tectonics. *Journal of Geophysical Research*, **73**, 5855–5899.

Jamieson, R.A. and Beaumont, C. (1989a) Deformation and metamorphism in convergent orogens: a model for uplift and exhumation of metamorphic terrains. Pp. 117–129 in: *Evolution of Metamorphic Belts* (J.S. Daly, R.A. Cliff and B.W.D. Yardley, editors). Special Publication No. **43**, Geological Society, London.

Jamieson, R.A. and Beaumont, C. (1989b) Orogeny and metamorphism: A model for deformation and pressure-temperature-time paths with application to the central and southern Appalachians. *Tectonics*, **7**, 417–445. Jamieson, R.A., Beaumont, C., Fullsack, P. and Lee, B. (1996) Tectonic assembly of inverted metamorphic sequences. *Geology*, **24**, 839–842.

Jamieson, R.A., Beaumont, C., Fullsack, P. and Lee, B. (1998) Barrovian regional metamorphism: where's the heat? Pp. 23–51 in: *What Drives Metamorphism and Metamorphic Reactions?* (P.J. Treloar and P.J. O'Brien, editors). Special Publication No. **138**, Geological Society, London.

Jamieson, R.A., Beaumont, C., Nguyen, M.H. and Lee, B. (2002) Interaction of metamorphsim, deformation and exhumation in large convergent orogens. *Journal of Metamorphic Geology*, **20**, 9–24.

Karabinos, P. and Ketcham, R. (1988) Thermal structure of active thrust belts. *Journal of Metamorphic Geology*, **6**, 559–570.

Koons, P.O. (1989) The topographic evolution of collisional mountain belts: A numerical look at the Southern Alps, New Zealand. *American Journal of Science*, **289**, 1041–1069.

Koons, P.O. (1990) Two-sided orogen: collision and erosion from the sandbox to the Southern Alps, New Zealand. *Geology*, **18**, 679–682.

Miller, C. (1974) On the metamorphism of the eclogites and high-grade blueschists from the Penninic terrain of the Tauern Window, Austria. *Schweizerische Mineralogische und Petrographische Mitteilungen*, **54**, 371–384.

Oxburgh, E.R. and Turcotte, D.L. (1968) Problems of high heat flow and volcanism associated with zones of descending mantle convective flow. *Nature*, **218**, 1041–1043.

Oxburgh, E.R. and Turcotte, D.L. (1970) Thermal structure of island arcs. *Geological Society of America Bulletin*, **81**, 1665–1688.

Oxburgh, E.R. and Turcotte, D.L. (1971) Origin of paired metamorphic belts and crustal dilation in island arc regions. *Journal of Geophysical Research*, **76**, 1315–1327.

Oxburgh, E.R. and Turcotte, D.L. (1974) Thermal gradients and regional metamorphism in overthrust terrains with special reference to the Eastern Alps. *Schweizerische Mineralogische und Petrographische Mitteilungen*, **54**, 641–662.

Peacock, S.M. (1989) Thermal modeling of metamorphic pressure-temperature time paths: A forward approach. Pp. 57–102 in: *Metamorphic Pressure-Temperature-Time Paths* (F.S. Spear and S.M. Peacock, editors). Short Course in Geology, American Geophysical Union, Washington, D.C.

Platt, J.P. (1986) Dynamics of orogenic wedges and the uplift of high-pressure metamorphic rocks. *Geological Society of America Bulletin*, **97**, 1037–1053.

Rubie, D.C. (1984) Thermal-tectonic model for H-P metamorphism and deformation in the Sesia-Lanzo zone, western Alps. *Journal of Geology*, **92**, 21–36.

Shi, Y. and Wang, C.Y. (1987) Two-dimensional model of the *P-T-t* paths of regional metamorphism in simple overthrust terranes. *Geology*, **15**, 1048–1051.

Spear, F.S. (1989a) Relative thermometry and metamorphic *P-T* paths. Pp. 63–81 in: *Evolution of Metamorphic Belts* (J.S. Daly, R.A. Cliff and B.W.D. Yardley, editors). Special Publication No. **43**, Geological Society, London.

Spear, F.S. (1989b) Petrologic determination of metamorphic pressure-temperature-time paths. in: *Metamorphic Pressure-Temperature-Time Paths* (F.S. Spear and S.M. Peacock, editors). Short Course in Geology, American Geophysical Union, Washington, D.C.

Spear, F.S. (1993) *Metamorphic Phase Equilibria and Pressure-Temperature-Time Paths*. Mineralogical Society of America Monograph, 799 pp.

Spear, F.S. and Rumble, D. III (1986) Pressure, temperature and structural evolution of the Orfordville Belt, west-central New Hampshire. *Journal of Petrology*, **27**, 1071–1093.

Spear, F.S. and Selverstone, J. (1983) Quantitative P-T paths from zoned minerals: Theory and tectonic applications. *Contributions to Mineralogy and Petrology*, **83**, 348–357.

Spear, F.S., Selverstone, J., Hickmont, D., Crowley, P. and Hodges, K.V. (1984) *P-T* paths from garnet zoning: A new technique for deciphering tectonic processes in crystalline terranes. *Geology*, **12**, 87–90.

Spear, F.S., Peacock, S.M., Kohn, M.J., Florence, F.P. and Menard, T. (1991) Computer programs for petrologic *P-T-t* path calculations. *American Mineralogist*, **76**, 2009–2012.

St-Onge, M.R. (1987) Zoned poikiloblastic garnets: *P-T* paths and syn-metamorphic uplift through 30 km of structural depth, Wopmay Orogen, Canada. *Journal of Petrology*, **28**, 1–22.

Thompson, A.B. and England, P.C. (1984) Pressure-temperature-time paths of regional metamorphism II. Their influence and interpretation using mineral assemblages in metamorphic rocks. *Journal of Petrology*, **25**, 929–955.

Thompson, A.B. and Ridley, J.R. (1987) Pressure-temperature-time (*P-T-t*) histories of orogenic belts. *Philosophical Transactions of the Royal Society of London*, **A321**, 27–45.

Turner, F.J. (1968) *Metamorphic Petrology*. McGraw Hill, New York, 403 pp. Willett, S., Beaumont, C. and Fullsack, P. (1993) Mechanical models for the tectonics of doubly convergent compressional orogens. *Geology*, **21**, 371–374.

Zeitler, P.K. (1989) The geochronology of metamorphic process. Pp. 131–147 in: *Metamorphic Pressure-Temperature-Time Paths* (F.S. Spear and S.M. Peacock, editors). Short Course in Geology, American Geophysical Union, Washington, D.C.

The influence of erosion upon the mineral facies of rocks from different metamorphic environments

P. C. ENGLAND & S. W. RICHARDSON

SUMMARY

Metamorphism of tectonically thickened continental crust or subducted sediment wedges is likely to take place in a thermal regime where temperature increases by conductive relaxation whilst concurrently pressure decreases by erosion of the pile. The mineral facies of rocks reaching the surface do not reflect any one geotherm through the pile but lie on a locus of P–T conditions, the metamorphic geotherm, which will generally be concave towards the temperature axis. Maximum pressures on the metamorphic geotherm are significantly less than maximum pressures experienced by rocks during the early stages of recrystallization. The metamorphic geotherm is polychronic, points at lower temperatures reflecting conditions earlier in the development than those at higher temperature; crustal melts are developed after low-medium temperature metamorphism and the amount of such melts could be significant.

Blueschists develop on the low temperature end of the metamorphic geotherm and are succeeded in exposure at the surface by greenschist- or amphibolite-facies rocks; the timescale for this process is consistent with the virtual absence of Precambrian blueschists. Crust thickened by addition of hot magma is likely to yield a metamorphic geotherm convex towards the temperature axis. Recognition of differently curving metamorphic geotherms can be used to assess the part played by magmatic activity in older metamorphic terrains.

1. Introduction

STABILIZED OR CRATONIZED AREAS of continental crust are made up largely of deformed and recrystallized igneous and sedimentary rocks, often covered by a layer of unmetamorphosed sediment. The highest grade metamorphic rocks crystallized at pressures equivalent to depths of 10, 20 or even 50 km and their presence at the surface today implies considerable erosion subsequent to crystallization (cf. Watson 1976); that the average metamorphic rock is today underlain by crust of (by definition) average thickness implies an initial tectonic anomaly which either thickened pre-existing crust or created pristine over-thick crust; the relaxation of the anomalous conditions is inevitably followed by erosion.

It is the purpose of this paper to show that erosion has significant effects upon the thermal structure of thick crust and in many instances may be the most important factor determining the pressure and temperature which are recorded by mineral assemblages.

2. The erosional process

Following Schumm (1963), Ahnert (1970) and Carson & Kirby (1972, p. 414), we take the erosion rate dE/dt to be proportional to surface height above sea level, H. Thus:

$$dE/dt = cH \qquad (1)$$

where E is the amount eroded after time t and c is a constant. Clearly such a simplification cannot be held to represent the state of affairs at any specific locality but its validity as a general description is borne out by the studies of present day large drainage basins cited above.

If the change in surface height is governed by isostasy:

$$dH/dE = -F \text{ or } H = H_0 - FE \qquad (2)$$

where $F = (1 - \rho_c/\rho_m)$
and ρ_c and ρ_m are the densities of eroded material and that at the compensation depth respectively. H_0 is the surface height at time $t = 0$. Combining (1) and (2):

$$dE/dt = c(H_0 - FE) \qquad (3)$$

whence
$$E = H_0/F(1 - \exp[-Fct]) \qquad (4a)$$

or
$$E = E_{max}(1 - \exp[-t/\lambda]) \qquad (4b)$$

Equation 4b is in the form used by Richardson (1975) where λ is the time constant for the erosional process and E_{max} is the total amount eroded at infinite time.

The studies of Ahnert and Carson & Kirby (*op. cit.*) indicate values for c of about 0·1 (km/Ma.km of surface height) which yields a value for the erosional time constant of about 60 Ma. As we are dealing with situations influenced by a tectonic control which yields thickened crust and as this control may continue for times characteristic of the development of mobile belts we adopt as likely values of λ the range 60–300 Ma depending upon the type of mobile belt (see below).

To investigate the thermal effects of erosion on temperature profiles in various types of mobile belt we use a numerical solution to the heat conduction equation in one dimension. The properties chosen to represent rocks and the method of solution are detailed in the appendix. A variety of initial conditions is chosen to represent differing metamorphic situations and in addition to producing a temperature-depth profile at chosen time intervals, the solution also permits the labelling of rocks in the eroding pile with the time, depth and temperature at which their maximum temperature is reached. Thus it is possible to discover, for a given choice of initial conditions, erosional time constant and E_{max}, what 'metamorphic conditions' are recorded by a rock which has been exposed at the surface after t Ma of erosion.

It is clearly a simplification to regard the metamorphic mineral equilibria as being established at the maximum temperature experienced by a rock. In most circumstances it would be more reasonable to expect rocks to record conditions at which their mineral assemblages reached maximum entropy, S_{max}. Only in the special case where the volume of the assemblage at equilibrium has pressure and temperature derivatives equal to zero will S_{max} necessarily correspond to T_{max}. Most of the reactions important in regional metamorphism (both continuous and discontinuous equilibria of Thompson 1957) have positive dP/dT so that we may expect T_{max} to be greater than, or equal to, $T_{S_{max}}$ and the pressure condition at which T_{max} is reached to be likewise greater than, or equal to, P at which S_{max} is achieved. This source of approximation is discussed further in the following sections.

3. Erosion of overthrust continental crust

One way in which continental crust may be thickened is by overthrusting or obduction (Coleman 1971, p. 1216, Oxburgh 1972, Armstrong & Dick 1974). If the continental mass below the thrust has a normal thermal structure, that is, the temperature is in equilibrium with heat production in the crust and a normal heat flow from the mantle, the temperature at any depth below the thrust will rise as described by Oxburgh & Turcotte (1974). The erosion of this thickened crust will effect instantaneous pressure changes on a given rock but at any depth below a few kilometres the temperature will materially increase for some time. The deeper the rock within the pile, the longer will be the period for which its temperature increases.

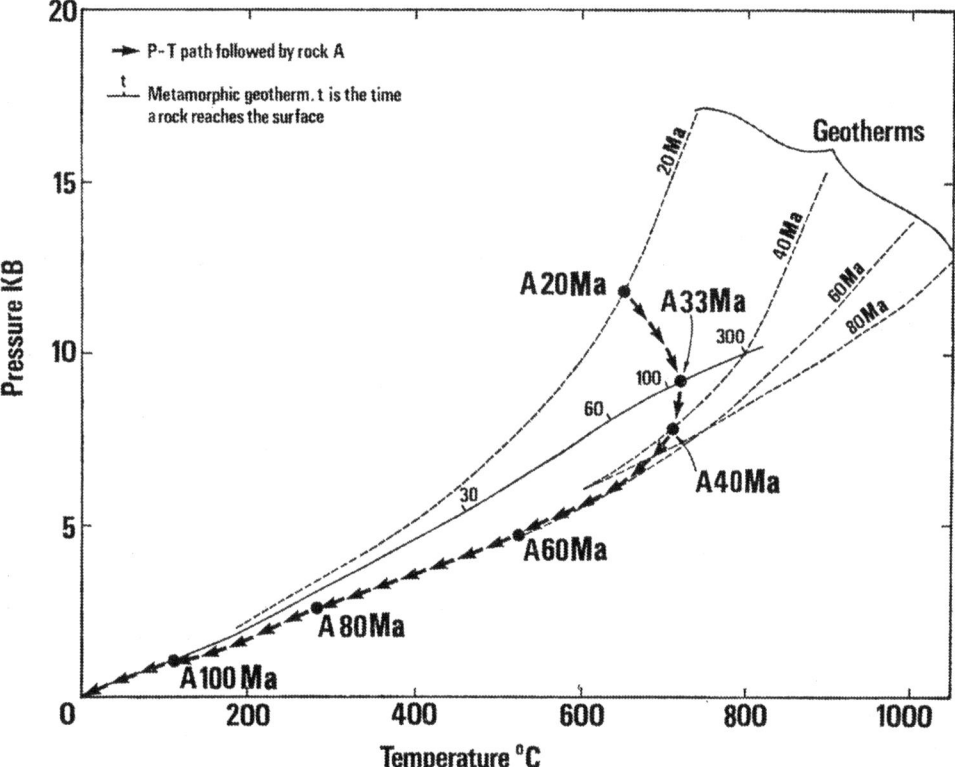

FIG. 1. Geothermal gradients (dashed) in continental terrain at times of 20, 40, 60 and 80 Ma after overthrusting. Erosion of the pile above rock A lowers the pressure whilst heating is in progress (arrows) so that maximum temperature is reached at 33 Ma. The array of $T_{max} - P_{T_{max}}$ points achieved by rocks which subsequently reach the surface is shown as the metamorphic geotherm (solid line) which is labelled with the times at which the rocks appear on surface.

Initial conditions: above thrust — 30 km of crust with initial gradient 25K/km; below thrust — temperature in equilibrium with heat production $A_x = A_{30}\exp(-[x-30]/D)$ where $A_{30} = 8$ HGU, $D = 10$ km (Sclater & Franchetau 1970, Richardson 1975) and heat flux at the lower boundary of 1·0 HFU. Erosion commences at 20 Ma with the parameters of Equation 4, $\lambda = 60$ Ma, $E_{max} = 60$ km.

For the specific conditions stated in the legend, Fig. 1 shows the effects of this process on particular rocks. As an example we trace the development of P–T conditions in rock A. At the onset of erosion, 20 Ma after the overthrusting event, A is at 12 kbar, 650°C. From 20–33 Ma pressure decreases but temperature increases until the maximum temperature, 720°C is reached at a pressure of 9 kbar. At 40 Ma the temperature has hardly decreased, being over 700°C at a pressure of 7·5 kbar, but thereafter the temperature declines rapidly, reaching 500°C shortly after 60 Ma and 300°C around 80 Ma. At 115 Ma the rock reaches the surface.

The conditions of metamorphism recorded in the mineralogy of A will clearly depend upon the bulk composition; but in general, it is most likely to have established its mineral facies by discontinuous reactions occurring during its negatively-sloping transect of the P–T diagram (cf. Schuiling 1963). Continuous reactions are likely to record temperatures above 700°C although the pressures could be anywhere from 7–9 kbar.

The condition of maximum temperature for all rocks ultimately arriving at the surface is shown in Fig. 1 as the *metamorphic geotherm*, part of which might be constructed from field studies of an area with local slight variations in structural relief. It is clear that the upper part of this line has no resemblance to any true geotherm, all of which are convex towards the temperature axis: the shape of the polychronic metamorphic geotherm is controlled by the erosional history. The common occurrence of regional metamorphic facies series in which kyanite-bearing rocks at lower temperatures are succeeded by sillimanite-bearing ones at higher temperatures is evidence for a common occurrence of metamorphic geotherms concave towards the temperature axis or with low dP/dT. Indeed, most equilibria amongst solid silicates have dP/dT around 24 bar/K which would be nearly tangential to the true geotherm and could lead to assemblages of lower entropy being established at higher temperature if metamorphism took place on any single geotherm.

Fig. 1 also shows that the maximum pressures which could be recorded by continous reactions in rocks reaching the surface are only about 10 kbar; quite possibly the actual pressures recorded would be only about 8 kbar as explained above for the rock A. The maximum depth of burial to which such rocks were subjected is equivalent to about 15 kbar. We note the general absence of evidence from high-grade amphibolite and granulite terrains for metamorphic conditions much in excess of 10 kbar (Touret 1970, Newton *et al.* 1974, Wells 1976); also the common occurrence of prograde (that is higher entropy) reaction products of the highest pressure assemblages (Chinner & Sweatman 1968, Wells 1976, Yardley 1976) such as might be expected to form on the almost isothermal segment of the cooling path.

Because erosion transposes hot material upwards in the crust, there will be a tendency for rocks deep within the section to melt. We have not tried to incorporate the possible thermal consequences of melting into the solutions described here but some geological consequences can be seen qualitatively from Fig. 1. As maximum temperatures are reached later at successively greater depths, melts produced in response to the maximum conditions and migrating upwards will come into contact with rocks which have already started to cool from their maximum temperature: intrusions will be post-regional-metamorphic. This is exactly the case observed in many low and medium pressure amphibolite-facies terrains (see Miyashiro 1973, p. 96).

The amount of melt to be expected from this type of metamorphic sequence is difficult to estimate both on thermal grounds, as its upward migration carries heat away from the zone of melt formation, and because the composition of the pile must be specified. A rough idea of the melting potential may be gained by examining the excess heat carried by all crustal rocks over their likely melting temperature on a particular geotherm. As an example we take the situation of Fig. 1 and assume that at infinite time, the crust is 35 km thick. At 80 Ma, 20 km of the crust is above a likely melting temperature of 800 °C and the base of the crust is at 1050 °C. The excess heat carried by this section is therefore $1\cdot6 \times 10^8$ cal/cm^2; if this heat were used in the production of liquid at 800 °C involving an enthalpy change of 250 cal/cm^3, a 6 km column of liquid could be produced (this calculation is appropriate to a vapour-absent melting equilibrium involving a 4·5 per cent volume change and with a slope of 200 bar/K (see Huang & Wyllie 1974)). The exact quantity is not to be taken seriously: what matters is that the result is in kilometres of liquid rather than centimetres and may be expected to have a bulk sufficient to form intrusive or extrusive magmas.

By slightly varying the initial conditions, E_{max} and λ, it is possible to model the P–T conditions and facies series of many amphibolite- and granulite-facies terrains. The essential features of such terrains can be reproduced, as demonstrated by Richardson & Powell (1976) and England (1976 and unpublished work) without recourse to extreme values of mantle heat flow (cf. Bird *et al.* 1975, p. 4415) and we would expect these features to be characteristic of orogenesis along sites of continental collision, sites of obduction over continental margins and sites of thickening by massive shear failure within continental crust (Miyashiro 1973, ch. 17, Oxburgh 1972, Bridgwater *et al.* 1973, Bridgwater *et al.* 1974).

4. Erosion of a subducted sediment wedge

In this section we consider the thermal development of the kind of terrain which may have produced blueschist- to zoisite-eclogite-facies metamorphism (Ernst *et al.* 1970, Oxburgh & Turcotte 1970, 1971, Ernst 1975). The situation envisaged is one in which a thick wedge of cold sediment overlies oceanic lithosphere or perhaps thin continental crust; part of the load upon the sediment may be due to superincumbent lithosphere or it could be self-loaded only. In either case we assign no heat production to the pile and take a very low initial thermal gradient of 10 K/km to a depth of E_{max}; below this level we assume lithosphere in equilibrium with a heat-flux of 0·8 HFU. These conditions are chosen to produce blueschist-facies conditions in the first place and severely to limit the rate of conductive relaxation in the pile; we wish to err on the side of preserving blueschist conditions for as long as possible.

Two particular problems associated with blueschist belts are:

(1) The close correlation between age and intensity of metamorphism and

the systematic decrease of both with distance from the inferred line of suture (Ernst 1975).

(2) The increasing scarcity of blueschist belts in older parts of the Phanerozoic record and their virtual absence from Precambrian metamorphic suites (de Roever 1956, Ernst 1972).

The first of these problems has been discussed by Graham & England (1976) who pointed out the importance of shear strain as an energy source in underthrust masses. The second observation was explained by Ernst as reflecting the secular decline in the Earth's energy budget, manifesting itself in growing lithosphere thicknesses. Whilst accepting that world heat production rates are likely to influence plate motions and lithosphere dimensions in ways which are beyond the scope of this paper, we doubt the magnitude of the effects proposed over the past 500 Ma. The heat production half-life of a chondritic model Earth is roughly 2000 Ma so that heat production 500 Ma ago was only 1·2 times its present value. Instead of appealing to changes in the operation of plate tectonics we wish with Zwart (1967, p. 305) to point out that both thermal relaxation and erosion will tend to increase temperature within blueschist terrains and this could result in the production of higher entropy assemblages characteristic of greenschist or amphibolite facies, a process which is manifestly occurring within the Alpine chain (Miller 1974, Frey et al. 1974). Fig. 2 shows the P–T field chosen to model the blueschist facies and its boundary with greenschists and amphibolites (called, for convenience, greenschists). Using ranges of the erosional time constant λ of 50–500 Ma and the ultimate amount eroded E_{max} of 30–90 km, ranges which are both almost certainly more extensive than physical realism would permit, we show in Fig. 3 the mineral facies to be expected at the surface 50, 100, 300 and 500 Ma after cessation of the tectonic control which creates subducted sediment wedges.

For reasonable values of λ up to 200 Ma (see above) we would expect a maxi-

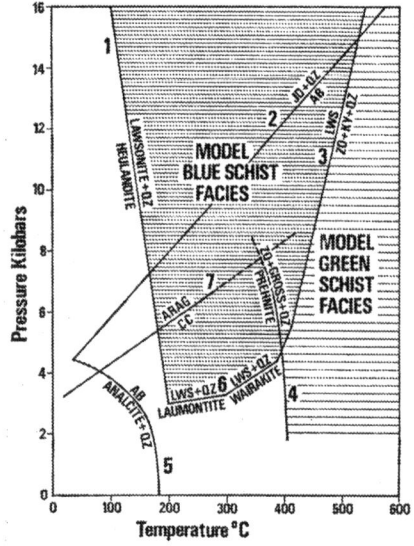

FIG. 2

P–T conditions used to describe production of blueschists and higher grade rocks (here called greenschists). In the absence of a well determined stability field for the diagnostic blueschist assemblages, glaucophane + Ca–Al silicate, a guide to blueschist-facies conditions is provided by the stability field of lawsonite + quartz.

Sources of experimental information:

1—Nitsch (1968); 2—Newton & Smith (1967); 3—Newton & Kennedy (1963); 4 Liou (1971a); 5—Campbell & Fyfe (1965); 6—Liou (1971b); 7—Boettcher & Wyllie (1968).

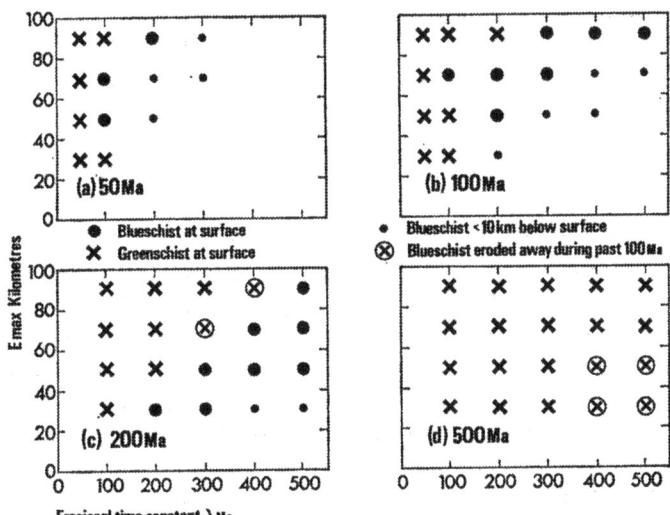

FIG. 3. Blueschist exposures at the surface are underlain by rocks relaxing upgrade into greenschist (or amphibolite) facies conditions as defined on Fig. 2. Continued erosion at any reasonable choice of E_{max} and λ will exhume greenschist (or amphibolite) by the time 500 Ma have elapsed.

mum in blueschist occurrence at 50–100 Ma, precisely the age range of most jadeite-bearing blueschist belts (Ernst 1972): this correlation indicates that our choice of initial conditions is reasonable. Continued erosion of such belts for 200 or 500 Ma reduces the probability of blueschist preservation; for erosion which exactly obeys equation (4) the probability goes to zero by 500 Ma. Progradation of blueschist as a means of destruction is well documented in older occurrences of these rocks (Bloxam & Allen 1959, p. 8, Carpenter & Civetta 1976, p. 277) and may be regarded as internal evidence for the combined effect of thermal relaxation and erosion which tends to carry individual rocks to higher temperature whilst decreasing pressure; any heat production within the pile, which we have here neglected, will of course reinforce the effect.

5. Erosion of crust thickened by magma addition

It has been proposed that the heat flow anomaly over island arcs is caused by additions of magma from subducted oceanic lithosphere or the superjacent upper mantle (Oxburgh & Turcotte 1971, Ringwood 1974). Likewise, the Andean mountain chain in Peru is underlain by thickened crust which may be characterized by extensional tectonics (James 1971, Knox, in discussion of Cobbing & Pitcher 1974). The most plausible source of crustal thickening in such environments is not compressive strain but the bulk of magmatic bodies whose buoyant uprise is halted within the crust by loss of heat. In this situation, conductive relaxation is towards lower temperatures and upon cessation of magma injection, erosion might be expected to expose a rock suite whose maximum temperatures were attained along something like a single true geotherm which will be convex towards the temperature axis in a P-T diagram. The exact form of the metamorphic geotherm will clearly be controlled by the exact distribution of magmatic heat sources in space and time: we imply only that the *average* metamorphic geotherm could be expected to assume the shape of a true geotherm with d^2P/dT^2 positive, in contrast to the situation within tectonically thickened crust discussed above.

The problem of how far the conditions T_{max} and $T_{S_{max}}$ diverge is particularly acute in these situations and might substantially modify the expected shape of the metamorphic geotherm. If erosion were very rapid a rock deep within the pile would reach shallow depths essentially along an isotherm, crossing as it did so reactions with positive dP/dT. If on the other hand, erosion were very slow the conductive relaxation would allow cooling essentially on an isobar and the possibility of prograde reaction during return to the surface would be minimized. In the following three paragraphs we examine a model of this situation and conclude that cooling during the early part of the return towards surface will be nearly isobaric and in consequence the array of preserved metamorphic conditions is likely to follow closely the array of initial conditions.

We attempt to assess the relative effects of cooling due to relaxation and of heating at a particular depth below surface due to erosion by examining their effects in the most realistic situation for which an analytical solution is available: constant erosion rate dE/dt and initially-constant temperature gradient β along which the temperature is everywhere T_0 above equilibrium. The solution given by Carslaw & Jaeger (1959, p. 388) may be differentiated to give the separate effects as they are seen at the surface, which may be shown to reflect the relative effects on the gradient at any other depth.

The change in surface gradient due to relaxation, $\Delta\alpha_r$, is:

$$\Delta\alpha_r/T_0 = \frac{1}{y}\exp(-x^2) + \frac{x\sqrt{\pi}}{y}(2 - \text{erfc}(x)) \qquad (5)$$

and the change due to erosion, $\Delta\alpha_e$, is

$$\Delta\alpha_e/\beta = 1 + 4x^2 - 4i^2\text{erfc}(x) \qquad (6)$$

where $x = \dfrac{dE}{dt}\sqrt{\dfrac{t}{4\kappa}} \quad y = \sqrt{\pi\kappa t}$

and κ is the diffusivity.

For erosion rates corresponding to erosional time constants (λ, see Equation 4) of 60–200 Ma there is little variation in the dimensionless erosion rate parameter, x, which takes values between $10^{-0.7}$ and $10^{-1.7}$ for times of 1–50 Ma.

The relative size of $\Delta\alpha_e$ and $\Delta\alpha_r$ depends upon a choice of β and T_0; reasonable quantities for these parameters in the situation we consider are tens of K/km and hundreds of kelvins respectively, corresponding for example to a crustal section with initial temperature gradient 25 K/km initially everywhere 250 K hotter than equilibrium. Fig. 4 shows values of the ratio $\Delta\alpha_r/\Delta\alpha_e$ when β/T_0 assumes a value of 0·1/km; during the critical early stages of relaxation and erosion the effect of the

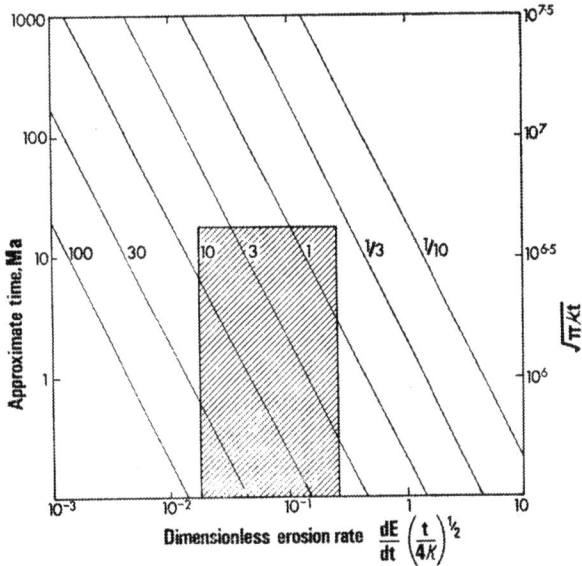

FIG. 4. Contours showing the relative effects of cooling an intrusion-thickened metamorphic belt due to conductive relaxation, and maintenance of high temperature at a particular depth due to erosion. For conditions in the lower left hand corner of the diagram individual rocks will cool almost isobarically; for conditions in the upper right hand corner pressure will decrease almost isothermally. The box encloses a reasonable parametral range influencing the cooling path soon after metamorphism: cooling is likely to be along a path which will produce lower entropy retrograde assemblages. The contours are numbered in terms of the ratio $\Delta\alpha_r.\beta/\Delta\alpha_e.T_0$ as defined in the text, giving values of $\Delta\alpha_r/\Delta\alpha_e$ when $\beta/T_0 = 0 \cdot 1/\text{km}$.

former is ten times to approximately equal the effect of the latter. We conclude that the form of the metamorphic geotherm will not be very different from that expected in the case of extremely slow erosion and is indeed likely to be convex towards the temperature axis.

In addition to the shape of the metamorphic geotherm, the identification of conditions extreme in both pressure and temperature is to be expected from this sort of metamorphic terrain. Thus the sapphirine granulites of Morse & Talley (1971, see also Newton *et al.* 1974, p. 309) could, on these grounds have formed at the high temperature end of a metamorphic geotherm within magma-thickened crust, the essentially isobaric retrograde reactions observed by them reflecting the local dominance of the conductive relaxation term in the cooling history.

6. Conclusions

We propose a test, based upon the array of P–T conditions indicated by metamorphic rocks at the surface, between two types of mechanism whereby the continental crust can be thickened and metamorphism take place. We do not mean to imply that these mechanisms, tectonic thickening and the addition of magma, are mutually exclusive; but it may be possible to assess their regional importance within a particular metamorphic belt by examination of facies series.

For the northeast Dalradian metamorphism Harte (1975), following Chinner (1966), has recognized local variations in facies series. Over the greater part of the area, facies series define P–T arrays which are of shallow slope, not necessarily concave towards the temperature axis but certainly not markedly convex (*op. cit.*, fig. 26, A, B, C & D). We would associate these metamorphic geotherms with the type of situation illustrated in Fig. 1 of this paper. Harte's type-E facies series is decidedly convex towards the temperature axis and we attribute the Buchan E series to the local influence of a convective heat source, although this is not identified with the Newer Basic bodies of Ashworth (1975).

On a larger scale, we have drawn attention to the powerful effects of erosion upon temperatures within thick continental crust. Convection of heat by uplift and erosion is as potent a means of energy supply to the upper lithosphere in this situation as are radiogenic heating or transfer of heat from beneath the lithosphere. Consequently, rocks in eroded orogenic regions will generally reflect temperatures higher than those predicted from the latter effects alone.

Finally, we note the essentially 3-dimensional reality of metamorphic terrains which we have here discussed in terms of one dimension. The geographic variation of initial conditions, E_{max} and λ could give rise to facies series which are unrelated both to true geotherms and also to the metamorphic geotherm of this paper (see Turner 1968, ch. 8, Richardson 1970). It is only small areas showing variation in structural relief which we would take to be capable of revealing a metamorphic geotherm of the type discussed here.

APPENDIX

To investigate the thermal effects of erosion on the temperatures in the continental crust we use a finite difference approximation to the equation:

$$\frac{\partial v}{\partial t} = \frac{\partial}{\partial x}\left(\kappa \frac{\partial v}{\partial x}\right) + \frac{A(x, v, t)}{\rho C_p} + u_x \frac{\partial v}{\partial x}$$

where v is temperature, t is time, κ is diffusivity, A is heat production and C_p is specific heat at constant pressure. It is assumed throughout that the systems discussed have significant temperature derivatives only perpendicular to the surface, that the diffusivity is a function only of temperature (see below) and that u_x, representing the upward movement of the system, is a function of time of the kind given in Equation 4. The equation is solved using the Crank-Nicolson implicit method.

The program was checked against the analytical solutions of Carslaw & Jaeger (1959, p. 388) for constant erosion/linear initial thermal gradient situations; in all cases the solutions were consistent to better than 2 per cent with the analytical ones, and when the model is extended to more complicated situations, the solutions obtained with a time-step of 0·25 Ma and distance steps of 2 km were consistent with those obtained using steps of half these sizes to better than 2 per cent.

The temperature dependence of diffusivity used in the program is made up as follows. Conductivity has the form:

$$K = K_L + K_R$$

where the lattice component of conductivity has the form:

$$K_L = \frac{1}{A + BT} \text{ cal/cm s K}$$

$$\text{T in kelvins, } A = 122, B = 0.078$$

(see, for example, Clark 1966)

and
$$K_R = C(T - T_C); C = 4 \times 10^{-6}, T_C = 773K$$

(see, for example, Schatz & Simmons 1972).

The specific heat function used, with T in kelvins, is

$$C_p = 0.18(1 + 6.14 \times 10^{-4}T - 1.928 \times 10^4/T^2) \text{ cal/gm K}$$

The boundary conditions on the model are a constant temperature of 0°C on the upper surface and a constant heat flux across the lower boundary.

ACKNOWLEDGEMENTS. We acknowledge support from the NERC through a studentship to PCE and research grant GR3/1340. We have enjoyed the benefits of discussing a draft manuscript with E. R. Oxburgh, E. T. C. Spooner and P. R. A. Wells.

References

AHNERT, F. 1970. Functional relationships between denudation, relief, and uplift in large mid-latitude drainage basins. *Am. J. Sci.* **268**, 243–63.

ARMSTRONG, R. L. & DICK, H. J. B. 1974. A model for the development of thin overthrust sheets of crystalline rock. *Geology* **2**, 35–40.

ASHWORTH, J. G. 1975. The sillimanite zones of the Huntly-Portsoy area in the northeast Dalradian, Scotland. *Geol. Mag.* **112**, 113–24.

BIRD, P., TOKSOZ, M. N. & SLEEP, N. H. 1975. Thermal and mechanical models of continent-continent convergence zones. *J. geophys. Res.* **80**, 4405–16.

BLOXAM, T. W. & ALLEN, J. B. 1959. Glaucophane-schist, eclogite and associated rocks from Knockormal in the Girvan-Ballantrae complex, South Ayrshire. *Trans. R. Soc. Edinb.* **64**, 1–27.

BOETTCHER, A. L. & WYLLIE, P. J. 1968. The calcite-aragonite transition measured in the system $CaO-CO_2-H_2O$. *J. Geol.* **76**, 314–30.

BRIDGWATER, D., ESCHER, A. & WATTERSON, J. 1973. Tectonic displacements and thermal activity in two contrasting Proterozoic mobile belts from Greenland. *Phil. Trans. R. Soc.* **273A**, 513–33.

——, MCGREGOR, V. R. & MYERS, J. S. 1974. A horizontal tectonic regime in the Archaean of Greenland and its implications for early crustal thickening. *Precamb. Res.* **1**, 179–97.

CAMPBELL, A. S. & FYFE, W. S. 1965. Analcime-albite equilibria. *Am. J. Sci.* **263**, 807–16.

CARPENTER, M. S. N. & CIVETTA, L. 1976. Hercynian high pressure/low temperature metamorphism in the Île de Groix bluechists. *Nature, Lond.* **262**, 276–7.

CARSLAW, H. S. & JAEGER, J. C. 1959. *Conduction of heat in solids* (2nd ed.). Oxford Univ. Press, 510 pp.

CARSON, M. A. & KIRBY, M. J. 1972. *Hillslope form and process*. Cambridge Univ. Press, 475 pp.

CHINNER, G. A. 1966. The distribution of pressure and temperature during Dalradian metamorphism. *Q. Jl geol. Soc. Lond.* **122**, 158–86.

—— & SWEATMAN, T. R. 1968. A former association of enstatite and kyanite. *Mineralog. Mag.* **36**, 1052–60.

CLARK, S. P. 1966. Handbook of Physical Constants. *Mem. geol. Soc. Am.* **97**.

COBBING, E. J. & PITCHER, W. S. 1974. The coastal Batholith of central Peru. *Jl geol. Soc. Lond.* **128**, 421–60.

COLEMAN, R. G. 1971. Plate tectonic emplacement of upper mantle peridotites along continental edges. *J. geophys. Res.* **76**, 1212–22.

ENGLAND, P. C. 1976. *Thermal modelling from heat flow and velocity modelling from seismological data, with special reference to the Eastern Alps.* Thesis, D.Phil., Oxford Univ. (unpubl.).

ERNST, W. G. 1972. Occurrence and mineralogic evolution of blueschist belts with time. *Am. J. Sci.* **272**, 657–68.

—— 1975. Systematics of large-scale tectonics and age progressions in Alpine and circum-Pacific blueschist belts. *Tectonophys.* **26**, 229–46.

——, Seki, Y., Onuki & Gilbert, M. C. 1970. Comparative study of low-grade metamorphism in the California Coast Ranges and the outer metamorphic belt of Japan. *Mem. geol. Soc. Am.* **124**.

Frey, M., Hunziker, J. G., Frank, W., Bocquet, J., Dal Piaz, G. V., Jäger, E. & Niggli, E. 1974. Alpine metamorphism of the Alps: a review. *Schweiz. miner. petrogr. Mitt.* **54**, 247–290.

Graham, C. M. & England, P. C. 1976. Thermal regimes and regional metamorphism in the vicinity of overthrust faults: an example of shear heating and inverted metamorphic zonation from Southern California. *Earth Planet. Sci. Lett.* **31**, 142–52.

Harte, B. 1975. Determination of a pelite petrogenetic grid for the Eastern Scottish Dalradian. *Yb. Carnegie Instn. Wash.* **74**, 438–46.

Huang, W. L. & Wyllie, P. J. 1974. Melting relations of muscovite with quartz and sanidine. *Am. J. Sci.* **274**, 378–95.

James, D. E. 1971. Plate tectonic model for the evolution of the central Andes. *Bull. geol. Soc. Am.* **82**, 3325–46.

Liou, J. G. 1971a. Synthesis and stability relations of prehnite, $Ca_2Al_2Si_3O_{10}(OH)_2$. *Am. Miner.* **56**, 507–31.

—— 1971b. P–T stabilities of Laumontite, Wairakite, Lawsonite and related minerals in the system $CaAl_2Si_2O_8$–SiO_2–H_2O. *J. Petrology* **12**, 379–411.

Miller, C. 1974. On the metamorphism of the eclogites and high grade blueschists from the Penninic terrain of the Tauern Window, Austria. *Schweiz. miner. petrogr. Mitt.* **54**, 371–84.

Miyashiro, A. 1973. *Metamorphism and metamorphic belts*. Allen & Unwin, London, 492 pp.

Morse, S. A. & Talley, J. H. 1971. Sapphirine reactions in deep-seated granulite near Wilson Lake, Central Labrador, Canada. *Earth Planet. Sci. Lett.* **10**, 325–8.

Newton, R. C., Charlu, T. V. & Kleppa, O. J. 1974. A calorimetric investigation of the stability of anhydrous Mg-Cordierite with application to granulite facies metamorphism. *Contrib. Mineral. Petrol.* **44**, 295–311.

—— & Kennedy, G. C. 1963. Some equilibrium reactions in the join $CaAl_2Si_2O_8$–H_2O. *J. geophys. Res.* **68**, 2967–83.

—— & Smith, J. V. 1967. Investigations concerning the breakdown of albite at depth in the Earth. *J. Geol.* **75**, 268–86.

Nitsch, K. H. 1968. Die stabilität von Lawsonit. *Naturwissenschaften* **55**, 388.

Oxburgh, E. R. 1972. Flake tectonics and continental collision. *Nature, Lond.* **239**, 202–4.

—— & Turcotte, D. L. 1970. Thermal structure of island arcs. *Bull. geol. Soc. Am.* **81**, 1665–88.

—— & —— 1971. Origin of paired metamorphic belts and crustal dilation in island arc regions. *J. geophys. Res.* **76**, 1315–27.

—— & —— 1974. Thermal gradients and regional metamorphism in overthrust terrains with special reference to the Eastern Alps. *Schweiz. miner. petrogr. Mitt.* **54**, 641–62.

Richardson, S. W. 1970. The relation between a petrogenetic grid, facies series and the geothermal gradient in metamorphism. *Fortschr. Miner.* **47**, 65–76.

—— 1975. Heat flow processes. In: *Geodynamics Today*. Royal Society, pp. 123–32.

—— & Powell, R. 1976. Thermal causes of the Dalradian metamorphism in the central Highlands of Scotland. *Scott. J. Geol.* **12**, 237–68.

Ringwood, A. E. 1974. The petrological evolution of island arc systems. *Jl geol. Soc. Lond.* **130**, 183–204.

de Roever, W. P. 1956. Some differences between post-Palaeozoic and older regional metamorphism. *Geologie Mijnb.* **18**, 123–7.

Schatz, J. F. & Simmons, G. 1972. Thermal conductivity of earth materials at high temperatures. *J. geophys. Res.* **77**, 6966–83.

Schuiling, R. D. 1963. Some remarks concerning the scarcity of retrograde vs. progressive metamorphism. *Geologie Mijnb.* **42**, 177–9.

Schumm, S. A. 1963. Denudation and Orogeny. *Prof. Pap. U.S. geol. Surv.* **454**-H.

Sclater, J. G. & Franchetau, J. 1970. The implications of terrestrial heat flow observations on current tectonic and geochemical models of the crust and upper mantle of the Earth. *Geophys. J. R. astr. Soc.* **20**, 509–42.

THOMPSON, J. B. 1957. The graphical analysis of mineral assemblages in pelitic schists. *Am. Miner.* **42,** 842–58.

TOURET, M. J. 1970. Le facies granulite, metamorphisme en milieu carbonique. *Compt. Rend.* **271,** 2228–31.

TURNER, F. J. 1968. *Metamorphic Petrology.* McGraw-Hill, 403 pp.

WATSON, J. V. 1976. Vertical movements in Proterozoic structural provinces. *Phil. Trans. R. Soc.* **280A,** 629–40.

WELLS, P. R. A. 1976. Late Archaean metamorphism in the Buksefjorden region, Southwest Greenland. *Contrib. Mineral. Petrol.* **56,** 229–42.

YARDLEY, B. W. D. 1976. Deformation and metamorphism of Dalradian rocks and the evolution of the Connemara cordillera. *Jl geol. Soc. Lond.* **132,** 521–42.

ZWART, H. J. 1967. The duality of orogenic belts. *Geologie Mijnb.* **46,** 283–309.

Received 26 November 1976; revised typescript received 12 February 1977.

PHILIP C. ENGLAND, Dept. of Geodesy and Geophysics, University of Cambridge, Madingley Rise, Cambridge.

STEPHEN W. RICHARDSON, Dept. of Geology and Mineralogy, University of Oxford, Parks Road, Oxford.

Requests for offprints should be sent to Dr. Richardson.

CHAPTER 10: MINERAL THERMOMETRY

Ferry, J.M. and Spear, F.S. (1978) Experimental calibration of the partitioning of Fe and Mg between biotite and garnet. *Contributions to Mineralogy and Petrology*, 66, 113–117.

Modern petrogenetic grids (Chapters 4 and 8) provide precise constraints on the possible *P-T* conditions for metamorphic parageneses, although the loci of many of the curves on them are also a function of the pressure of a volatile component, such as H_2O or CO_2. Element-exchange reactions between mineral solid solutions are theoretically superior indicators of temperature in that: (1) they are independent of fluid pressure, (2) they have a small dependence on total pressure, and (3) all of them are continuous functions of temperature and thus potentially offer high precision. However, they do share some problems with the petrogenetic grid in compositionally complex systems when, as is usually the case, non-ideal mineral solutions are involved. Also, exchange reactions have a tendency to reset on cooling, whereas the progress of retrograde net-transfer reactions is in many cases inhibited by the absence of the needed fluid components.

The partition of ferrous iron and magnesium is one of the commonest element exchanges we encounter in mineral pairs in metamorphic rocks. When the partition is strong, lnK_D is large and so is the standard free energy of the exchange reaction. It then generally follows that the standard exchange enthalpy is also large, which means that the reaction has strong temperature dependence. Into this category fall mineral pairs involving garnet, several of which have been found useful in thermometry, for example: garnet with biotite, clinopyroxene, orthopyroxene, olivine, hornblende, chlorite and phengite. The electron microprobe, indispensable as the analytical tool of choice for minerals *in situ*, became widely available in the late 1960s, when mineral thermometry was in its infancy. Its inability to distinguish in routine analysis between ferrous and ferric iron has, however, always been a problem.

The potential of garnet-biotite thermometry was recognized long ago by many authors (Kretz, 1959, 1964; Mueller, 1961; Frost, 1962; Albee, 1965; Perchuk, 1967; Sen and Chakraborty, 1968; Hietanen, 1969; Saxena, 1969). Already at that time a decrease in the lnK_D of the reaction (as written here) due to the presence of Ca and Mn in the garnet was noted. Empirical calibrations of lnK_D vs. estimated temperature based on petrological field studies were attempted by Saxena (1969), Thompson (1976), Perchuk (1977), Holdaway and Lee (1977), and, with reference to $^{18}O/^{16}O$ fractionation between magnetite and quartz, by Goldman and Albee (1977). The careful laboratory investigation of the pure-system exchange by Ferry and Spear (1978) that we highlight here was thus the greatly anticipated controlled calibration of the garnet-biotite thermometer. Some workers continued to use the empirical calibrations in the belief that they applied better to complex-system compositions, but this practice has now ceased.

In terms of mineral-components the exchange reaction is:

phlogopite + almandine = annite + pyrope

Aware of the sluggish nature of garnet equilibration on the laboratory time scale, Ferry and Spear chose a strategy that minimized the need for garnet to change composition during the experiment and maximized the requirement for change in the composition of biotite. Thus, charges contained 98% garnet (Alm_{90} and Alm_{80}) and 2% biotite, so that the pivot of tie-line rotation toward equilibrium during the experiments was very close to garnet. The success of this approach in the present case was an encouraging sign for future work on other mineral pairs where only one mineral is reactive. The laboratory techniques used were by then standard: cold-seal pressure vessels, a graphite-methane buffer to keep Fe largely ferrous, and analysis of run products by electron microprobe. Starting biotites and garnets were synthesized in the system KFMASH; the biotites were inferred to be on the phlogopite–annite join, i.e. free of octahedral Al. The experimental calibration of lnK_D was expressed as a straight-line fit to inverse temperature, with an added term for pressure. This made the reasonable assumption of constant ΔH for the reaction ($\Delta Cp \approx 0$) over the range 550 to 800°C. Ferry and Spear recommended that the thermometer not be used if the combined content of grossular and spessartine components in the garnet exceeded 20% and if much Ti and octahedral Al were present in the biotite. The uptake of Al on the octahedral site of ferromagnesian silicate minerals was known to increase their Fe/Mg ratio. Later, it became clear that these limitations were not stringent enough. Their estimate of uncertainty (±50°C) referred to the chemical analysis of unknowns, and did not include the accuracy of the calibration, or possible equilibrium shifts caused by non-ideal solution interactions with extra components.

Through the years, the garnet-biotite exchange has remained a favoured choice of thermometer for medium-grade rocks, in most cases pelitic schists. In many of the earlier applications of the thermometer, the Ferry and Spear calibration for the simple system KFMASH was used as is, in the hope that the chemical potentials of the phase components in the exchange reaction were not greatly influenced by non-ideal solution interactions among the four garnet components and by the extra components in biotite. The differences between the Ferry and Spear (1978) pure-system calibration and the empirical calibrations (Fig. 3) reflect some of these compositional effects. The Goldman and Albee thermometer used an isotope calibration that has since been revised; furthermore, the isotopes may have been partially reset during cooling. There has never been any question that the uncertainties introduced by non-ideal interactions need to be quantitatively resolved in order for the thermometer to be used with confidence. Fortunately, the contents of F and Cl in the biotite of pelitic schists are small except at high metamorphic grades. An issue of concern has been the presence of unanalysed Fe^{3+}, usually minor in almandine-rich garnet, but in some cases major in biotite. Finally, we had to recognize that partial resetting of garnet and biotite compositions takes place in rocks during cooling from granulite and upper-amphibolite facies conditions. A strategy that has evolved to accommodate this, as well as the presence of primary growth zoning in garnet, is to obtain maximum metamorphic temperatures by matching analyses of biotite (commonly homogenous) with those of spots on garnet just inside the crystal margin where retrograde compositional effects (reverse zoning) are usually manifest. For those starting to do thermobarometry, I recommend the paper by Eric Essene (1989), our thermobarometric guru, for critical information and advice.

Major strides in refining the thermometer have been made in the past 20–30 y. These efforts have followed three lines of enquiry: (1) testing the thermometer calibration against independent thermometry based on alternative mineral assemblages present in the rocks, (2) further calibration experiments, and (3) experimental and empirical measurement of activity-composition relations among non-ideal solution components in garnet and biotite, expressed as Margules parameters.

In pelitic schists in south-central Maine, Ferry (1980) found that Grt-Bt temperatures obtained without modification from the experimental calibration by Ferry and Spear were in agreement with those from other fluid-independent mineral thermometers. Temperatures from fluid-dependent thermometers assuming $P_{H_2O} = P_{total}$ were greater; this difference, found repeatedly in subsequent studies, has usually been attributed to low H_2O activities attending metamorphism. We should not forget, however, that metamorphism in south-central Maine was the first of many cases shown by Ferry (Chapter 12) to have been accompanied by major infiltration of aqueous fluid.

In another test of the thermometer, Hodges and Spear (1982) showed that the Ferry and Spear calibration apparently worked well for sillimanite-andalusite metapelites at Mt. Moosilauke, New Hampshire, thought to have crystallized close to 501°C, 3.8 kbar, the aluminosilicate triple point of Holdaway (1971). The results agreed with other semi-empirical fluid-independent thermobarometers, although the suggestion was made that a correction for non-ideal Ca-Mg interaction in garnet was appropriate. In hindsight, this agreement appears to have been a sort of false positive. More recent petrogenetic grids (e.g. Spear, 1993; Xu et al., 1994) show that, as emphasized by Pattison (2001), higher temperatures, namely >550°C, are necessary for aluminosilicate + biotite to form in ordinary prograde muscovite-quartz pelites, unless we assume substantially reduced H_2O activities. Use of the most recent refinement of the thermometer by Holdaway (2000, see below) gives temperatures in the range 551 to 565°C for most of the Mt. Moosilauke samples (M.J. Holdaway, pers. comm.). On the other hand, Ferry-Spear temperatures from low-Mn garnet-biotite pairs from Sta-Ky zone rocks (555 to 615°C) at Mica Creek, British Columbia (Ghent et al., 1979) seemed about right.

Perchuk and Lavrent'eva (1983) performed a large number of experiments at 6 kbar over a range of temperatures re-equilibrating natural garnets, high Al-biotites, and cordierites with a wide spread of Fe-Mg compositions. Unfortunately, their chemical characterization of run products was incomplete, although it has been possible to reconstruct some of the data. Compositionally complex natural-system experiments, which include those of Patino-Douce and Johnston (1991), minimize the problems of non-ideality in garnet and biotite, and serve as important, independent checks on optimized thermodynamic models of the thermometer.

Based on (un-reset) garnet-biotite compositions in granulite facies rocks, Indares and Martignole (1985) proposed an empirical modification to the Ferry-Spear calibration to accommodate the effect of non-ideal interactions involving Ti and ^{VI}Al in biotite. These constituents tend to be quite high in granulite-facies biotite. This correction appeared to reconcile Thompson's (1976) empirical calibration with Ferry and Spear's experimental data in the KFMASH system. Empirical or semi-empirical modifications were also proposed by a number of other authors: Perchuk and Aranovich (1984) for F in biotite; Hoinkes (1986) for Ca in garnet; Bhattacharya and Raith (1987) for non-ideal mixing in garnet and biotite; Williams and Grambling (1990) for Mn in garnet, and Fe^{3+} in biotite; and Hoisch (1991) for non-ideal mixing on the octahedral site of biotite.

McMullin et al. (1991) applied linear programming to a revised thermometer calibration using the Ferry and Spear data, a ternary solution model for garnet from Berman (1990), and field and experimental constraints on biotite solution properties. Dasgupta et al. (1991) presented a new formulation of the thermometer using the Ferry and Spear (F & S) experiments but not those of Perchuk and Lavrent'eva (P & L). They used the available data on non-ideal Fe-Mg-Ca-Mn mixing in garnet and estimates of Ti and Al mixing based on natural biotites, and then derived the Fe-Mg mixing properties of biotite from the F & S data. Bhattacharya et al. (1992) reconciled the low-Al biotite F & S and high-Al biotite P & L datasets by adopting experimental data on Fe-Mg-Ca mixing in garnet coupled with a non-ideal Mg-Fe mixing model for biotite. Patino-Douce et al. (1993) used the results of their experiments on the dehydration melting of a natural metapelite sample to derive Margules parameters for biotite based on a model that limits non-ideality to the mixing of Al and Ti on the two M2 octahedral sites. Their temperatures for Mt. Moosilauke were greater than those of Hodges and Spear (1982), but subject to uncertainty about the role of Fe^{3+}. Kleeman and Reinhardt (1994) also reconciled the F & S and P & L data using revised expressions for Al and Ti solution in biotite. Altogether, it was an intense few years of fine-tuning of the thermometer.

Rathmell et al. (1999) presented a table comparing garnet-biotite temperatures for 24 samples from the Grenville Orogen, Ontario, calculated from 10 different formulations of the thermometer published between 1978 and 1993. Even ignoring the worst 'offender', there is no one sample where the estimates did not spread over a range of 100°C, although some exceeded it only slightly. I think that this spread should not be construed as an indictment of the thermometer's reliability. It reflects the problem of piecemeal attempts to refine the thermometer, and the inability of empirical adjustments to it to recover accurate temperatures for rocks from PTX environments beyond the confines of the source data.

Garnet is supreme in its importance to thermobarometry. For this reason, much effort has been expended over many years to determine by phase-equilibrium and other methods the thermodynamic mixing properties of garnet along binary joins in quaternary Ca-Mg-Fe-Mn garnet solid solutions. Mukhopadhyay et al. (1997) performed a great service for thermobarometry with their thorough statistical analysis of the then available volume, experimental, and calorimetric data for solution of Fe, Mg and Ca in garnet, and derivation of an optimized set of activity-composition relationships expressed in terms of T and P dependent Margules parameters. The thermodynamic end-member and solution properties of garnet are, of course, relevant to garnet participating in any thermometer, barometer, or grid calculation.

Holdaway et al. (1997) provided what was desperately needed, namely, a serious effort to optimize all the diverse experimental data relevant to the thermometer. Their recalibration of the garnet-biotite thermometer used: (1) revised Fe-Mg-Ca Margules parameters for garnet from Mukhopadhyay et al. (1997); (2) Margules parameters for Mn in garnet; (3) newly retrieved parameters for non-ideal mixing of Fe, Mg and Al in biotite, based on the F & S and P & L experiments; and (4) refined estimates of the oxidation state of Fe in both minerals. Their new calibration, when applied to rocks from Maine gave an average of 574°C for staurolite-zone zone rocks, 44°C higher than an earlier estimate (Holdaway et al., 1988) based on Ganguly and Saxena (1984), and 66°C higher than that of Ferry (1980) based on the raw F & S calibration.

Gessman et al. (1997) conducted experiments at 2 kbar on the QFM oxygen buffer with synthetic Ca- and Mn-free garnet and biotite, following the design of Ferry and Spear (1978). They found that $\ln K_D$ decreased with decreasing X_{alm} in the garnet, indicating non-ideal thermodynamic mixing behaviour on the FeMg join of garnet.

Holdaway (2000, 2004) incorporated constraints from the Gessman experiments, and new Margules data for garnet published by Berman and Aranovich (1996) and Ganguly et al. (1996), to produce an updated calibration of the garnet-biotite thermometer, and one that will survive, let us hope, unchanged for a while. He took 20 different combinations of all the relevant experimental data, and regressed calculated vs. experimental temperatures constrained to a slope of unity, maximizing the fit through stepwise adjustment of six variable thermodynamic parameters. The best of all fits was obtained using an average of the garnet Margules parameters; from this he extracted revised values of the enthalpy and entropy of the exchange reaction and four biotite Margules parameters. Temperatures for an analyzed pair are calculated using equations 1 and 2 in Holdaway (2000), and activities of alm and py in garnet and ann and phl in biotite are calculated from expressions in Mukhopadhyay et al. (1993). For the staurolite zone in Maine, Holdaway found the average temperature to be 571°C, practically unchanged from 1997. The error of the thermometer under optimum conditions was estimated to be ±15°C relative and ±25°C absolute. This is about the best you can do in mineral thermometry; it is hard to believe that further fine-tuning could reduce these numbers. In practice, the ultimate limitation derives from the need to exercise good petrographic judgment in selecting compositions in polished sections that most likely represent frozen exchange equilibrium.

References

Albee, A.L. (1965) Distribution of Fe, Mg and Mn between garnet and biotite in natural mineral assemblages. *Journal of Geology*, **73**, 155–164.

Battacharya, A. and Raith, M. (1987) An updated calibration of Mg-Fe partitioning between garnet-biotite and orthopyroxene-biotite based on natural assemblages. *Fortschritte in Mineralogie*, **65**, Beiheft 1, 25.

Battacharya, A., Mohan, L., Maji, A., Sen, K. and Raith, M. (1992) Non-ideal mixing in the phlogopite-annite binary: constraints from experimental data on Mg-Fe partitioning and a reformulation of the biotite-garnet thermometer. *Contributions to Mineralogy and Petrology*, **111**, 87–93.

Berman, R.G. (1990) Mixing properties of Ca-Mg-Fe-Mn garnets. *American Mineralogist*, **75**, 328–344.

Berman, R.G. and Aranovich, L.Y. (1996) Optimized standard state and solution properties of minerals. I. Model calibration for olivine, orthopyroxene, cordierite, garnet, and ilmenite in the system FeO-MgO-CaO-Al_2O_3-TiO_2-SiO_2. *Contributions to Mineralogy and Petrology*, **126**, 1–24.

Dasgupta, S., Sengupta, P., Guha, D. and Fukuoka, M. (1991) A refined garnet-biotite Fe-Mg exchange geothermometer and its application in amphibolites and granulites. *Contributions to Mineralogy and Petrology*, **109**, 130–137.

Essene, E.J. (1989) The current status of thermobarometry in metamorphic rocks. Pp. 1–44 in: *Evolution of Metamorphic Belts* (J.S. Daly et al. editors). Special Publication, **43**, Geological Society, London.

Ferry, J.M. (1980) A comparative study of geothermometers and geobarometers in pelitic schists from south-central Maine. *American Mineralogist*, **65**, 720–732.

Ferry, J.M. and Spear, F.S. (1978) Experimental calibration of the partitioning of Fe and Mg between biotite and garnet. *Contributions to Mineralogy and Petrology*, **66**, 113–117.

Frost, M.J. (1962) Metamorphic grade and iron-magnesium distribution between coexisting garnet-biotite and garnet-hornblende. *Geological Magazine*, **99**, 427–438.

Ganguly, J. and Saxena, S., (1984) Mixing properties of aluminosilicate garnets: constraints from natural and experimental data, and applications to geothermo-barometry. *American Mineralogist*, **69**, 88–97.

Ganguly, J., Cheng, W. and Tirone, M. (1996) Thermodynamics of aluminosilicate garnet solid solution: new experimental data, an optimized model, and thermometric applications. *Contributions to Mineralogy and Petrology*, **126**, 137–151.

Gessman, C.K., Spiering, B. and Raith, M. (1997) Experimental study of the Fe-Mg exchange between garnet and biotite: Constraints on the mixing behavior and analysis of the cation-exchange mechanism. *American Mineralogist*, **82**, 1225–1240.

Ghent, E.D., Robbins, D.B. and Stout, M.Z. (1979) Geothermometry, geobarometry, and fluid compositions of metamorphosed calc-silicates and pelites, Mica Creek, British Columbia. *American Mineralogist*, **64**, 874–885.

Goldman, D.S. and Albee, A.L. (1977) Correlation of Mg/Fe partitioning between garnet and biotite with O^{18}/O^{16} partitioning between quartz and magnetite. *American Journal of Science*, **277**, 750–761.

Hietanen, A. (1969) Distribution of Fe and Mg between garnet, staurolite, and biotite in an aluminum-rich schist in various metamorphic zones north of the Idaho batholith. *American Journal of Science*, **267**, 422–456.

Hodges, K.V. and Spear, F.S. (1982) Geothermometry, geobarometry and the Al$_2$SiO$_5$ triple point at Mt. Moosilauke, New Hampshire. *American Mineralogist*, **67**, 1118–1134.

Hoinkes, G. (1986) Effect of grossular content on the partitioning of Fe and Mg between garnet and biotite. An empirical investigation on staurolite zone samples from the Austroalpine Schneeberg complex. *Contributions to Mineralogy and Petrology*, **92**, 393–399.

Hoisch, T.D. (1991) Equilibria within the mineral assemblage quartz + muscovite + biotite + garnet + plagioclase, and implications for the mixing properties of octahedrally-coordinated cations in muscovite and biotite. *Contributions to Mineralogy and Petrology*, **108**, 43–54.

Holdaway, M.J. (1971) The stability of andalusite and the aluminum silicate phase diagram. *American Journal of Science*, **271**, 97–131.

Holdaway, M.J. (2000) Application of new experimental and garnet Margules data to the garnet-biotite geothermometer. *American Mineralogist*, **85**, 881–892.

Holdaway, M.J. (2004) Optimization of some key geothermobarometers for pelitic assemblages. *Mineralogical Magazine*, **68**, 1–14.

Holdaway, M.J. and Lee, S.M. (1977) Fe-Mg cordierite stability in high-grade pelitic rocks based on experimental, theoretical, and natural observations. *Contributions to Mineralogy and Petrology*, **63**, 175–198.

Holdaway, M.J., Dutrow, B.L. and Hinton, R.W. (1988) Devonian and Carboniferous metamorphism in west-central Maine: the muscovite-almandine geobarometer; the staurolite problem revisited. *American Mineralogist*, **73**, 20–47.

Holdaway, M.J., Mukhopadhyay, B., Dyar, M.D., Guidotti, C.V. and Dutrow, B.L. (1997) Garnet-biotite geothermometry revised: New Margules parameters and a natural specimen data set from Maine. *American Mineralogist*, **82**, 582–595.

Indares, A. and Martignole, J. (1985) Biotite-garnet thermometry in the granulite facies: the influence of Ti and Al in biotite. *American Mineralogist*, **70**, 272–278.

Kleeman, U. and Reinhardt, J. (1994) Garnet-biotite thermometry revisited: The effect of Ti and Al in biotite. *American Mineralogist*, **70**, 272–278.

Kretz, R. (1959) Chemical study of garnet, biotite, and hornblende from gneisses from S.W. Quebec, with emphasis on the distribution of elements in coexisting minerals. *Journal of Geology*, **67**, 371–402.

Kretz, R. (1964) Analysis of equilibrium in garnet-biotite-sillimanite gneisses from Quebec. *Journal of Petrology*, **5**, 1–20.

McMullin, D.W.A., Berman, R.G. and Greenwood H.J. (1991) Calibration of the SGAM thermobarometer for pelitic rocks using data from phase equilibrium experiments and natural assemblages. *The Canadian Mineralogist*, **29**, 889–908.

Mueller, R.F. (1961) Analysis of relations among Mg, Fe, Mn in certain metamorphic minerals. *Geochimica et Cosmochimica Acta*, **25**, 267–296.

Mukhopadhyay, B., Holdaway, M.J. and Koziol, A.M. (1997) A statistical model of thermodynamic mixing properties of Ca-Mg-Fe^{2+} garnets. *American Mineralogist*, **82**, 165–181.

Mukhopadhyay, B., Sabyasachi, B. and Holdaway, M.J. (1993) A review of Margules-type formulations for multicomponent solutions with a generalized approach. *Geochimica et Cosmochimica Acta*, **57**, 277–283.

Patino-Douce, A.E., Johnston, A.D. and Rice, J.M. (1993) Octahedral excess mixing properties in biotite: a working model with applications to geobarometry and geothermometry. *American Mineralogist*, **78**, 113–131.

Patino-Douce, A.E. and Johnston, A.D. (1991) Phase equilibria and melt productivity in the pelitic system: Implications for the origin of peraluminous granitoids and aluminous granulites. *Contributions to Mineralogy and Petrology*, **107**, 202–218.

Pattison, D.R.M. (2001) Instability of Al$_2$SiO$_5$ "triple-point" assemblages in muscovite+biotite+quartz-bearing metapelites, with implications. *American Mineralogist*, **86**, 1414–1422.

Perchuk, L.L. (1967) Biotite-garnet thermometer. *Doklady Akademie Nauk SSSR*, **177**, 411–414.

Perchuk, L.L. (1977) Thermodynamic control of metamorphic processes. Pp. 110–129 in: *Advances in Physical Geochemistry*, **1** (S.K. Saxena, A. Navrotsky and B.J. Wood, editors). Springer-Verlag, New York.

Perchuk, L.L. and Aranovich, L. (1984) Improvement of biotite-garnet thermometer: correction for fluorine content of biotite. *Doklady Akademie Nauk SSSR*, **277**, 131–135.

Perchuk, L.L. and Lavrent'eva, I.V. (1983) Experimental investigation of exchange equilibria in the system cordierite-garnet-biotite. Pp. 199–239 in: *Kinetics and Equilibrium in Mineral Reactions* (S.K. Saxena, editor). *Advances in Physical Geochemistry*, **3**, Springer, New York.

Rathmell, M.A., Streepey, M.M., Essene, E.J. and van der Pluijm, B.A. (1999) Comparison of garnet-biotite, calcite-graphite, and calcite dolomite thermometry in the Grenville Orogen, Ontario, Canada. *Contributions to Mineralogy and Petrology*, **134**, 217–231.

Saxena, S.K (1969) Silicate solid solution and geothermometry. 3. Distribution of Fe an Mg between coexisting garnet and biotite. *Contributions to Mineralogy and Petrology*, **22**, 259–267.

Sen, S.K. and Chakraborty, K.R. (1968) Magnesium-iron exchange equilibrium in garnet-biotite and metamorphic grade. *Neues Jahrbuch für Mineralogie Abhandlungen*, **108**, 181–207.

Spear, F.S. (1993) *Metamorphic Phase Equilibria and Pressure-Temperature-Time Paths*. Mineralogical Society of America Monograph, 799 pp.

Thompson, A.B. (1976) Mineral reactions in pelitic rocks II: Calculations of some P-T-X (Fe-Mg) phase relations. *American Journal of Science*, **276**, 425–454.

Williams, M.L. and Grambling, J.A. (1990) Manganese, ferric iron, and the equilibrium between garnet and biotite. *American Mineralogist*, **75**, 886–908.

Xu, G., Will, T.M. and Powell, R. (1994) A calculated petrogenetic grid for the system K$_2$O-FeO-MgO-Al$_2$O$_3$-SiO$_2$-H$_2$O, with particular reference to contact-metamorphosed pelites. *Journal of Metamorphic Geology*, **12**, 99–119.

Experimental Calibration of the Partitioning of Fe and Mg Between Biotite and Garnet

J.M. Ferry* and F.S. Spear

Geophysical Laboratory, 2801 Upton St., N.W., Washington, D.C. 20008, USA

Abstract. The cation exchange reaction $Fe_3Al_2Si_3O_{12} + KMg_3AlSi_3O_{10}(OH)_2 = Mg_3Al_2Si_3O_{12} + KFe_3AlSi_3O_{10}(OH)_2$ has been investigated by determining the partitioning of Fe and Mg between synthetic garnet, $(Fe, Mg)_3Al_2Si_3O_{12}$, and synthetic biotite, $K(Fe, Mg)_3AlSi_3O_{10}(OH)_2$. Experimental results at 2.07 $kbar$ and 550°–800° C are consistent with $\ln[(Mg/Fe)\ garnet/(Mg/Fe)\ biotite] = -2109/T(°K) + 0.782$. The preferred estimates for $\Delta \bar{H}$ and $\Delta \bar{S}$ of the exchange reaction are 12,454 cal and 4.662 e.u., respectively. Mixtures of garnet and biotite in which the ratio garnet/biotite = 49/1 were used in the cation exchange experiments. Consequently the composition of garnet-biotite pairs could approach equilibrium values in the experiments with minimal change in garnet composition (few tenths of a mole percent). Equilibrium was demonstrated at each temperature by reversal of the exchange reaction. Numerical analysis of the experimental data yields a geothermometer for rocks containing biotite and garnet that are close to binary Fe—Mg compounds.

Introduction

One of the major goals of experimental petrologists is to provide methods that can be used to estimate the temperature at which minerals in rocks crystallize. Laboratory calibration of element partitioning between mineral phases is a particularly useful method because it is a temperature-dependent phenomenon that is independent of the activity of volatile components (e.g., CO_2 and H_2O) in the geologic environment. The fugacity of volatile components may determine the stability of a phase but not the partitioning of elements between phases. This report presents data on the partitioning of Fe^{2+} and Mg between garnet, $Fe_3Al_2Si_3O_{12}$ (alm)—$Mg_3Al_2Si_3O_{12}$ (py), and biotite, $KFe_3AlSi_3O_{10}(OH)_2$(ann)—$KMg_3AlSi_3O_{10}(OH)_2$ (phl). An experimental procedure is presented by which the determination of Fe—Mg partitioning between two phases can be achieved below 800° C. The procedure is significant because determination of Fe—Mg partitioning between phases below 800° C has in the past been unsuccessful (Medaris, 1969; Gunter, 1974).

Experimental Procedures

Garnets ($alm_{80}py_{20}$ and $alm_{90}py_{10}$) were synthesized dry in graphite capsules in a solid medium, high-pressure apparatus at 1100° C, 26 $kbar$ (piston-out, no friction correction) for 18 h. Biotites ($ann_{25}phl_{75}$, $ann_{50}phl_{50}$, $ann_{75}phl_{25}$, and $ann_{100}phl_{00}$) were synthesized hydrothermally in $Ag_{70}Pd_{30}$ capsules in cold-seal pressure vessels at 750°–780° C, 2.07 $kbar$, at f_{O_2} defined by the iron-wustite buffer, for 124–206 h. Both minerals were prepared from appropriate mixtures of fayalite, γ-Al_2O_3, cristobalite, MgO, and crystalline $K_2Si_2O_5$, ground for one hour under acetone. Fayalite was synthesized from a mixture of cristobalite and Fisher reagent-grade Fe_2O_3 held at 1000° C, $\log_{10} f_{O_2} = -14$ in a one atmosphere CO—CO_2 gas-mixing furnace for 24 h. The mixture was quenched and crushed at six hour intervals. Cristobalite, γ-Al_2O_3, and MgO were synthesized by firing Fisher reagent-grade silicic acid, $Al(OH)_3$, and MgO, respectively, at 1400° C, one atmosphere, for 2 h. The $Al(OH)_3$ was heated gently over a Bunsen burner for 5 min before firing at 1400° C. The crystalline $K_2Si_2O_5$ was synthesized from Fisher reagent-grade K_2CO_3 and cristobalite by the method of Schairer and Bowen (1955).

Synthetic products were >99% biotite or garnet, as estimated optically. Powder X-ray diffraction patterns of synthetic products contained only biotite or garnet peaks. The composition of synthetic $alm_{80}py_{20}$ was checked by a four-element chemical analysis with the automated MAC electron microprobe at the Geophysical Laboratory on a flat, polished specimen using various natural silicate minerals as standards, and its composition was found to be as planned within the error of measurement ($\pm 0.01\ X_{alm}$). The intended composition of $alm_{90}py_{10}$ was verified by a partial chemical analysis by electron microprobe for Fe and Mg using synthetic $alm_{80}py_{20}$ as a standard. Approximately ten analyses were made of

---

* *Present address:* Department of Geology, Arizona State University, Tempe, Arizona 85281, USA

Table 1. Unit cell parameters and refractive indices of synthetic biotites, $K(Fe, Mg)_3AlSi_3O_{10}(OH)_2$

Composition	a (Å)	b (Å)	c (Å)	β (°)	V (Å³)	d_{331}^{ber} (calc)	n_γ
$ann_{00}phl_{100}$	5.3138(8)	9.204(1)	10.311(2)	99.86(2)	469.9(2)	1.5340	—
$ann_{25}phl_{75}$	5.341(4)	9.236(5)	10.315(5)	99.93(6)	501.2(5)	1.5412	—
$ann_{50}phl_{50}$	5.361(2)	9.288(3)	10.330(4)	99.91(5)	506.7(4)	1.5478	—
$ann_{75}phl_{25}$	5.382(1)	9.309(5)	10.337(2)	100.13(3)	509.9(3)	1.5532	1.659(3)
$ann_{100}phl_{00}$	5.391(3)	9.337(4)	10.324(6)	100.02(6)	511.8(5)	1.5562	1.684(3)

Figures in parentheses represent the standard deviation in terms of least units cited for the value to their immediate left; these were calculated by the program used to refine the parameters and represent precision only.

garnets $alm_{90}py_{10}$ and $alm_{80}py_{20}$, randomly spaced over the sample mounts, and no chemical zonation was observed in the synthetic garnets (within $\pm 0.01\ X_{alm}$).

The unit cell dimensions of biotites synthesized in this study are listed in Table 1; in addition, refractive index data were collected for some of the synthetic biotites. The unit cell dimensions for synthetic biotites were calculated by refining powder patterns obtained with an X-ray diffractometer, CuK_α radiation, a silicon internal standard (National Bureau of Standards, SRM 640), and the computer program LCLSQ written by Burnham (1962). Refractive index, n, was measured at 25° C using sodium light ($\lambda = 589$ nm).

The unit cell dimensions and refractive indices of biotites listed in Table 1 agree very well (in most cases within experimental error) with the values listed for corresponding biotite compositions in Hewitt and Wones (1975), with the exception of the biotite with composition $ann_{100}phl_{00}$. The discrepancy between the unit cell dimensions of $ann_{100}phl_{00}$ from this study and those of Hewitt and Wones may be explained by the presence of a small amount (approximately 6%) of Al in the octahedral sites of our biotite synthesized from the $ann_{100}phl_{00}$ mix. The consistency between the results of this study in Table 1 and the results of Hewitt and Wones (1975) is a strong argument that the synthetic biotites of intermediate Fe—Mg composition had compositions as planned.

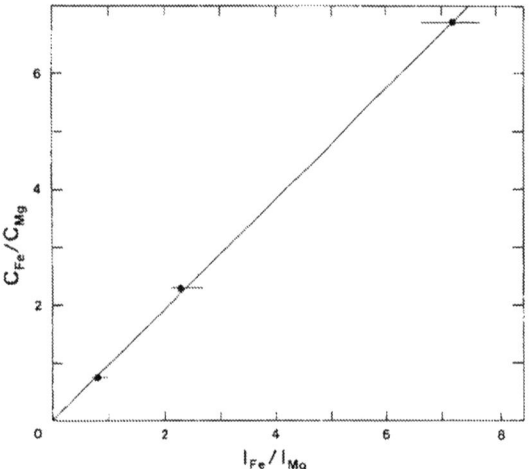

Fig. 1. Measured relation between I_{Fe}/I_{Mg} and C_{Fe}/C_{Mg} for synthetic biotites ($ann_{25}phl_{75}$, $ann_{50}phl_{50}$, $ann_{75}phl_{25}$). Circles and light horizontal lines are the average and range, respectively, of the population. The C_i and I_i are defined in the text. Operating conditions are 15 kv, 50 na; 100 second counting times. Grains in the size range 3–8 microns were analyzed

Further evidence for this conclusion is presented in Figure 1. Synthetic biotites were dispersed on flat, polished diamond surfaces, and the intensities of background-corrected X-ray peaks for Fe and Mg, I_{Fe} and I_{Mg}, respectively, were measured. For $ann_{25}phl_{75}$, $ann_{50}phl_{50}$, and $ann_{75}phl_{25}$ measured I_{Fe}/I_{Mg} is a linear function, passing through (0,0), of C_{Fe}/C_{Mg}, where C_i is the weight fraction of element i. The internal consistency of the I_{Fe}/I_{Mg} vs. C_{Fe}/C_{Mg} data would be a necessary consequence of synthetic biotites having compositions as planned.

The Fe—Mg partitioning experiments consisted of combination of garnet with biotite in a mixture in which molar garnet/biotite = 98/2 (on a 12 oxygen atom basis). This mixture was ground under acetone to an average particle size of approximately one micron, sealed with H_2O in $Ag_{70}Pd_{30}$ capsules, and held at temperatures of 550°–800° C and 2.07 kbar for 3–9 weeks. Temperature was measured by unsheathed chromel-alumel thermocouples, newly prepared from unoxidized wire before each experiment and calibrated in situ (1 atm pressure) against the melting point of NaCl (800.4° C). Temperature in each experiment was maintained within a $\pm 3°$ C cycle by a Honeywell-Brown controller. Reported temperatures are believed to be accurate to $\pm 6°$. Pressure was measured with a bourdon tube gauge calibrated against a factory-calibrated Heise gauge. Reported pressures are believed to be accurate to ± 50 bars. The fugacity of oxygen was controlled in the partitioning experiments by the graphite-methane buffer (Eugster and Skippen, 1967). The high H_2O content of the experimental samples (Table 2), intended to facilitate reaction, apparently did not significantly change the bulk composition of the solids because garnet composition remained unchanged, within uncertainty of electron microprobe analysis, after the experiments. The refractive index of the garnet after the experiments was ~1.82, the value appropriate for $alm_{90}py_{10}$ (Chinner et al., 1960), and hydrogarnet, therefore, could not have formed during the partitioning experiments (H.S. Yoder, 1977, personal commun.). At the conclusion of the experiments, the capsules were opened and the contents were dispersed on flat, polished diamond surfaces. The compositions of biotite grains were subsequently determined by electron microprobe using a linear relation between I_{Fe}/I_{Mg} and C_{Fe}/C_{Mg} (Fig. 1 is a sample calibration curve). Five to fifteen grains in a population of unknowns were analyzed and their compositions averaged. This method was developed by White (1964) and has been successfully exploited by Eugster et al. (1972) and Spear (1976). Repeated analysis of synthetic biotites of known compositions, with the procedures described above, indicated that measured biotite compositions are accurate to $\pm 0.01\ X_{ann}$.

Equilibrium was demonstrated at each temperature by placing two capsules in the same pressure vessel: one capsule contained garnet and a biotite more magnesian and the other capsule garnet and a biotite more iron-rich than the composition of the biotite in equilibrium with the garnet at a given

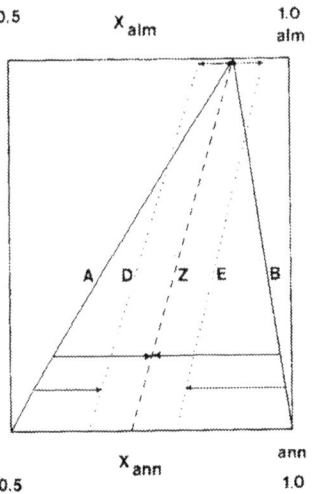

Fig. 2. Compositions of coexisting biotites and garnets at some arbitrary temperature and pressure. Tie lines D, E, and Z are equilibrium tie lines. All tie lines are explained further in the text

Fig. 3. Plot of $\ln K = \ln\{(Mg/Fe)_{garnet}/(Mg/Fe)_{biotite}\}$ vs. $1/T(°K)$ for the data in Table 1A. The solid line is calculated from a least-squares fit to the data points (see inset). Size of boxes corresponds to $\pm 5°$ uncertainty in temperature and ± 0.01 uncertainty in X_{ann}. Dashed line is from Thompson (1975, Fig. 1 – B). Dotted line is from Goldman and Albee (1977, Equation (8))

temperature. In experiments with garnet $alm_{90}py_{10}$, the compositions of biotites at the end of the two different experiments did not differ by more than 0.04 mole fraction *ann*.

The advantage of the 98-2 garnet-biotite ratio in the experimental samples is that garnet, which is sluggish to react, does not have to change significantly in composition during the experiments. For example, two mixtures, $alm_{90}py_{10} + ann_{50}phl_{50}$ and $alm_{90}py_{10} + ann_{100}phl_{00}$ (solid circles on tie lines A and B, respectively, Fig. 2), have very nearly the same bulk composition. If the two mixtures are held at the arbitrary temperature and pressure of Figure 2, tie lines A and B will rotate to the equilibrium tie line, Z. Biotite composition in the two mixes changes considerably as equilibrium is approached, but the composition of garnet changes negligibly. In the 740°C experiment (Table 2) a garnet $alm_{90}py_{10}$ was mixed with $ann_{100}phl_{00}$. At the conclusion of the experiment, the measured composition of the biotite was $ann_{73}phl_{27}$, requiring, by mass balance, a change in the garnet composition of only $+0.0055$ mole fraction *alm*. The largest change in garnet composition occurred in the 740°C experiment; garnets in the other experiments changed even less, usually by about 0.002 mole fraction. This predicted small change in the garnet composition was verified experimentally by determining that the composition of the garnet from selected experiments remained unchanged within the uncertainty of electron microprobe analysis (± 0.01 X_{alm}). This technique should be applicable to determination of element partitioning between any mineral pair provided at least one mineral (such as biotite in this study) is sufficiently reactive.

The narrow reversal brackets in Figure 3 indicate that biotite did not equilibrate with rims of garnets that were significantly different in composition from $alm_{90}py_{10}$. If biotite reacted only with garnet rims in the experiments, then the effective bulk composition of the biotite-garnet couple would lie along tie lines A or B (Fig. 2) but at garnet-biotite ratios less than 98-2. Starting tie lines A and B consequently would rotate, for example, to tie lines D and E, respectively, which differ greatly in biotite composition. Because starting tie lines A and B converged very nearly to the same tie line, Z, at each temperature investigated, biotite must have effectively equilibrated with the bulk of the garnet in the samples.

Results

Results of experiments with garnet, $alm_{90}py_{10}$, and biotite are presented in Table 2 and Figure 3. The data in Table 2 and Figure 3 are consistent with the expression (at 2.07 *kbar*):

$$\ln K = -2109/T(°K) + 0.782, \qquad (1)$$

Table 2. Compositions of biotite equilibrated at various temperatures and 2.07 kbar with garnet $alm_{90}py_{10}$ or $alm_{80}py_{20}$. Garnet and biotite were mixed in a molar garnet/biotite = 98/2

Run #	T(°C) (±6°)	Starting Material (X_{ann})	Time (h)	Final X_{ann} (±0.01)	Number of Grains Analyzed	ln K
A. Experiments with garnet $alm_{90}py_{10}$						
112	799	1.00	551	0.750	8	−1.155
116	799	0.50	551	0.710	12	−1.258
89	738	1.00	316	0.730	4	−1.271
123	749	0.50	818[a]	0.695	12	−1.330
128	698	0.75	672	0.704	15	−1.342
125	698	0.50	1151[b]	0.690	10	−1.353
138	651	0.75	523	0.679	11	−1.459
135	651	0.50	1002[c]	0.661	11	−1.497
139	599	0.75	498	0.645	8	−1.623
137	599	0.50	1344[d]	0.608	4	−1.728
126	550	0.75	911	0.620	5	−1.741
118	550	0.50	911	0.587	3	−1.811
B. Experiments with garnet $alm_{80}py_{20}$						
153	697	0.75	869	0.574	15	−1.113
149	697	0.25	869	0.468	11	−1.486
152	601	0.50	869	0.500	6	−1.386
150	601	0.25	869	0.392	2	−1.806

[a] (316/502) = Sample run for 316 h, opened, ground under acetone for 1 h, reloaded, and run for an additional 502 h
[b] (479/672)
[c] (523/479)
[d] (846/498)

where $K = (Mg/Fe)_{garnet}/(Mg/Fe)_{biotite}$ (either on a weight or atomic metal basis) and the coefficients were determined by a linear least-squares fit of the ln K values vs. 1/T (Table 2A; data were unweighted in the regression). At equilibrium for the reaction

$$Fe_3Al_2Si_3O_{12} + KMg_3AlSi_3O_{10}(OH)_2$$
$$= KFe_3AlSi_3O_{10}(OH)_2 + Mg_3Al_2Si_3O_{12}, \quad (2)$$

$$\Delta \bar{G} = \Delta \bar{H} - T\Delta \bar{S} + P\Delta \bar{V} + 3RT\ln K = 0. \quad (3)$$

If $\Delta \bar{H}$, $\Delta \bar{S}$, and $\Delta \bar{V}$ for reaction (2) are independent of pressure and temperature in the range of conditions experimentally investigated, by comparing Equations (1) and (3),

$$\Delta \bar{S} = 3R(0.782) = 4.662 \text{ e.u.}; \quad (4)$$

$$\Delta \bar{H} + 2070\Delta \bar{V} = 3R(2109). \quad (5)$$

For reaction (2), $\Delta \bar{V} = +0.057$ cal/bar (Robie et al., 1967), and

$$\Delta \bar{H} = 3R(2109) - 2070(0.057) = 12{,}454 \text{ cal.} \quad (6)$$

The preferred estimates of $\Delta \bar{S}$ and $\Delta \bar{H}$ for reaction (2) are 4.662 e.u. and 12,454 cal. respectively. The experimental data in Table 1 are consistent (considering errors in measurement of T and K), however, with values of $\Delta \bar{S}$ in the range 0.143 – 9.227 e.u. and with values of $\Delta \bar{H}$ in the range 8166 – 16,784 cal. A polybaric, polythermal expression for Fe—Mg partitioning between biotite and garnet may be formulated by substituting values of $\Delta \bar{H}$, $\Delta \bar{S}$, and $\Delta \bar{V}$ (from Equations (4) and (6) above) into Equation (3):

$$12{,}454 - 4.662T(°K) + 0.057P(bars) + 3RT\ln K = 0. \quad (7)$$

Experiments were conducted at 600° and 700°C, 2.07 kbar, with garnet $alm_{80}py_{20}$ starting material, to test the compositional dependence of ln K within the system alm-py-ann-phl. The brackets of ln K as a function of 1/T from experiments using $alm_{80}py_{20}$ are much larger than the brackets using $alm_{90}py_{10}$ (Table 2). The results of experiments with $alm_{80}py_{20}$, however, are consistent with Equation (1) and suggest that Fe and Mg mix ideally in biotite and garnet solid solutions at least in the composition interval $0.80 \leq Fe/(Fe+Mg) \leq 1.00$.

Applications

Figure 3 and Equation (7) represent a geothermometer for rocks containing biotite and garnet that are close to binary Fe—Mg compounds. It is clear, from analyses of biotite and garnet of different compositions that crystallized at the same metamorphic grade, that K is also a function of Ca and Mn content of the garnet and the Ti and Al^{VI} content of the biotite (Dallmeyer, 1974; Saxena, 1969). Caution should be exercised in applying the data in Figure 3 to systems containing significant amounts of Ca, Mn, or Ti. The experiments of this study, nevertheless, demonstrate a procedure by which the dependence of K on Ca, Mn, and Ti in garnet and biotite could be experimentally determined.

The partitioning of Fe and Mg between biotite and garnet has been calibrated in two studies by measuring the compositions of biotite and garnet from natural rock specimens for which temperatures of crystallization were estimated by independent means (Goldman and Albee, 1977: temperatures from $^{18}O/^{16}O$ fractionation between quartz and magnetite; Thompson, 1975: temperatures mainly from comparison of mineral assemblages with results of phase equilibrium experiments). Both calibrations assumed no pressure affect on Fe—Mg fractionation between biotite and garnet. The experimental results of this study and an extrapolation of them to lower temperatures agree with the calibration of Thompson (Fig. 3, dashed line) in the temperature range 400°–600°C and lie approximately 50° higher than the

Goldman and Albee calibration (dotted line). The agreement between the laboratory results of this study and the two calibrations based on natural samples suggests that Figure 3 and equation (7) may be a useful geothermometer without correction for components (up to ~ 0.2 (Ca+Mn)/(Ca+Mn+Fe+Mg) in garnet and up to ~ 0.15 (Al^{VI}+Ti)/(Al^{VI}+Ti+Fe+Mg) in biotite) other than Fe and Mg. The geothermometer in Figure 3 has a maximum practical resolution of approximately $\pm 50°$, which corresponds to the error in temperature that results when ± 0.01 errors in X_{ann}, X_{phl}, X_{alm}, and X_{py} are propagated through Equation (1).

Acknowledgements. The study was completed while both authors were postdoctoral fellows at the Geophysical Laboratory. A.A. Finnerty, R.K. Popp, D. Rumble, E.B. Watson, D.R. Wones and H.S. Yoder, Jr. assisted through discussion and critical review.

References

Burnham, C.W.: Lattice constant refinement. Carnegie Inst. Washington Yearbook **61**, 132–135 (1963)

Chinner, G.A., Boyd, F.R., England, J.L.: Physical properties of garnet solid solutions. Carnegie Inst. Washington Yearbook **59**, 76–78 (1960)

Dallmeyer, R.D.: The role of crystal structure in controlling the partitioning of Mg and Fe^{2+} between coexisting garnet and biotite. Am. Mineralogist **59**, 201–203 (1974)

Eugster, H.P., Albee, A.L., Bence, A.E., Thompson, J.B., Jr., Waldbaum, D.R.: The two-phase region and excess mixing properties of paragonite-muscovite crystalline solutions. J. Petrol. **13**, 147–179 (1972)

Eugster, H.P., Skippen, G.B.: Igneous and metamorphic reactions involving gas equilibria. In: Researches in geochemistry (P.H. Abelson, ed.), Vol. 2, pp. 377–396. New York: John Wiley 1967

Goldman, D.S., Albee, A.L.: Correlation of Mg/Fe partitioning between garnet and biotite with O^{18}/O^{16} partitioning between quartz and magnetite. Am. J. Sci. **277**, 750–761 (1977)

Gunter, A.E.: An experimental study of iron-magnesium exchange between biotite and clinopyroxene. Can. Mineralogist **12**, 258–261 (1974)

Hewitt, D.A., Wones, D.R.: Physical properties of some synthetic Fe–Mg–Al trioctahedral biotites. Am. Mineralogist **60**, 854–862 (1975)

Medaris, L.G.: Partitioning of Fe^{2+} and Mg^{2+} between coexisting synthetic olivine and orthopyroxene. Am. J. Sci. **267**, 945–968 (1969)

Robie, R.A., Bethke, P.M., Beardsley, K.M.: Selected x-ray crystallographic data, molar volumes, and densities of minerals and related substances. U.S. Geol. Surv. Bull. **1248** (1967)

Saxena, S.K.: Silicate solid solution and geothermometry: 3. Distribution of Fe and Mg between coexisting garnet and biotite. Contrib. Mineral. Petrol. **22**, 259–267 (1969)

Schairer, J.F., Bowen, N.L.: The system K_2O–Al_2O_3–SiO_2. Am. J. Sci. **253**, 681–746 (1955)

Spear, F.S.: Ca-amphibole composition as a function of temperature, fluid pressure, and oxygen fugacity in a basaltic system. Carnegie Inst. Washington Yearbook **75**, 775–779 (1976)

Thompson, A.B.: Mineral reactions in pelitic rocks: II. Calculation of some P–T–X(Fe–Mg) phase relations. Am. J. Sci. **276**, 425–454 (1975)

White, E.W.: Microprobe technique for the analysis of multiphase microcrystalline powders. Am. Mineralogist **49**, 196–197 (1964)

Received July 5, 1977/Accepted December 10, 1977

CHAPTER 11: THERMODYNAMIC DATABASE OF MINERALS

(a) Helgeson, H.C., Delaney, J.M., Nesbitt, H.W. and Bird, D.K. (1978) Summary and critique of the thermodynamic properties of rock-forming minerals. *American Journal of Science*, **278-A**, 1–229.

(b) Berman, R.G. (1988) Internally-consistent thermodynamic data for minerals in the system Na_2O-K_2O-CaO-MgO-FeO-Fe_2O_3-Al_2O_3-SiO_2-TiO_2-H_2O-CO_2. *Journal of Petrology*, **29**, 445–522.

(c) Holland, T.J.B and Powell, R. (1990) An enlarged and updated internally consistent thermodynamic dataset with uncertainties and correlations: the system K_2O-Na_2O-CaO-MgO-MnO-FeO-Fe_2O_3-Al_2O_3-TiO_2-SiO_2-C-H_2-O_2. *Journal of Metamorphic Geology*, **8**, 89–124.

Aside from Goldschmidt's classic calculation in 1912 of the wollastonite-forming reaction (Chapter 1), there were few attempts to calculate the PT conditions of metamorphic reactions from thermodynamic data in the early decades of the 20th century. Among the first efforts were those of Kracek *et al.* (1951) and Adams (1953) on the stability of jadeite, Danielsson (1950) and Ellis and Fyfe (1956) on the wollastonite reaction, and McDonald (1955) on the brucite-periclase equilibrium.

There are two likely explanations for this relatively slow start: (1) thermodynamic reference data were available for only a small number of compounds that were also metamorphic minerals (Birch *et al.*, 1942), and the coverage of P-V-T data for gases was inadequate; (2) many petrologists believed that factors in addition to P, T and composition were responsible for what minerals developed in metamorphic rocks (as in dynamothermal metamorphism, for example). Some thought that metamorphism was not necessarily "progressive" (proceeding via equilibrium steps), and that disequilibrium was prevalent. These views discouraged a thermodynamic approach.

A major effort during and after World War II by chemists of the U.S. Bureau of Mines and the U.S. Geological Survey produced a large amount of high-quality, low- and high-temperature calorimetric data for elements, simple compounds, and minerals. These data were summarized at various times by K.K. Kelley and coworkers in Bulletins of the U.S.B.M. and R.A. Robie and coworkers in Bulletins of the U.S.G.S. Their heat capacity and entropy data were extremely valuable; unfortunately, the magnitude of uncertainties in the enthalpies of formation of minerals derived from solution calorimetry (typically 2–3 kJ/mol) in most cases were too large for satisfactory calculation of PT phase equilibria.

Brackets on reactions defining the stability fields of many end-member metamorphic minerals were obtained in the 1950s, 1960s and 1970s in laboratories worldwide, using high-PT pressure vessels such as the Tuttle cold-seal rod bomb and the Boyd and England piston-cylinder. These did not immediately find their way into a comprehensive and accurate thermodynamic reference dataset for minerals because the thermodynamic properties of so many of the mineral products were poorly known. It was found useful at that time to fit experimental brackets on devolatilization reactions to an expression for the equilibrium constant:

$\ln K = -\Delta H^\circ/RT + \Delta S^\circ/R - \Delta V^\circ_{solids}(P-1)/RT$, where T (in Kelvin) and P (in bars) is the standard state for the solids and T and 1 bar the standard state for the gases (Eugster and Wones, 1962; Orville and Greenwood, 1965; Eugster and Skippen, 1967; Skippen 1971, 1974; Wood and Fraser, 1977). The equilibrium constant K is simply the gas fugacity, or gas fugacity product, when pure solids were used in the experiments. Over a modest range of T, within which $\Delta C_p \approx 0$, a linear, isobaric plot of $\ln K$ vs. inverse temperature yielded values of ΔH° and ΔS° for the reaction. This allowed the results of experiments on metamorphic reactions to be numerically manipulated in two important ways: (1) adjustment of gas fugacities for equilibria involving impure solids using an activity product term $\ln K_{solids}$, and (2) derivation of $\ln K$ expressions for linearly dependent reactions by algebraic combination of known $\ln K$ expressions (e.g. Skippen, 1971, 1974). LnK expressions were widely used by metamorphic petrologists in the 1970s and 1980s (for example, Chatterjee and Johannes, 1974; Joesten, 1974; Ferry, 1976, 1980, 1987; Evans and Trommsdorff, 1974; Evans *et al.*, 1976; Black, 1977; Ghent *et al.*, 1979; Franz and Spear, 1983).

At this time, a number of helpful books and reports on thermodynamic methods in metamorphic petrology appeared: Kern and Weisbrod (1967), Froese (1976), Wood and Fraser (1977), Greenwood (1977), and Powell (1978). Chapter 7 of Wood and Fraser (1977) gives a very clear account of how thermodynamic data may be derived from reversed phase-equilibrium experiments. In his Preface to the English edition of Kern and Weisbrod, R.M. Garrels eloquently wrote: "Thermodynamics is perhaps the single great tool of Science that never ages: it flowers in each science in succession, and then returns to flower again. First the applications of thermodynamics were to physics, then to chemistry, then to engineering; today (1967) its widest use is in biology and geology. In each instance, it simply awaits the availability of data good enough to fit its elegant structure, and then permits rapid progress in solving problems of the particular discipline".

It became clear over time that the uncertainty in a carefully determined experimentally bracketed metamorphic reaction (for example ±20°C), when expressed in terms of uncertainty in the Gibbs free energy or enthalpy of reaction, is typically much smaller than the calorimetric measurement of the same properties. This fact was exploited in the monumental effort by H.C. Helgeson and coworkers in 1978 to derive a comprehensive set of standard-state thermodynamic data for some 70 rock-forming minerals. The publication of this volume made it possible to start our calculations of phase equilibria with the properties of the phases and chemical species, rather than operating on numerical values for reactions in PT expressions. It opened the door to calculating the conditions for reactions which, because of their sluggish kinetics, could not be investigated experimentally. Could it be that the "tyranny" of numbers, the alarming phrase used by E-an Zen (1977) only one year before in his Presidential Address to the Mineralogical Society of America for what confronted the user of thermodynamic data, was about to be overcome? The Helgeson paper gave geochemists and petrologists access to the P, T and activity coordinates of a countless number of balanced reactions. The paper filled a single, 229-page issue (volume 278-A) of the *American Journal of Science*. Unfortunately we have room here only for the Preface, Abstract, and introductory remarks.

I believe that two later efforts also deserve inclusion in this book: Berman (1988) and Holland and Powell (1990), even if we again have room for only their Abstracts and Introductions. Citations of all three of these works number in the thousands. The main difference introduced in the latter two studies was that mathematical techniques were used to derive simultaneously the thermodynamic properties of all the minerals, so that any properties that were established early on did not skew properties derived later.

All three studies assembled their databases from a mix of published calorimetric, crystallographic, solubility, and phase-equilibrium data. Not surprisingly, there is good agreement among them for simple stoichiometric minerals in well determined systems. All three make appropriate provision for the energetic consequences of λ transitions and order-disorder effects. All were accompanied by matching computer routines for calculating phase diagrams. All employ expressions for the "apparent" standard Gibbs free energy and enthalpies of reaction at P and T; that is, calculations exclude changes in the thermodynamic properties of the elements above reference-state conditions, because these cancel out in balanced reactions. Unlike Helgeson *et al.*, Berman and Holland and Powell use an apparent free energy function that also excludes the entropies of the elements at 298 K and 1 bar, except in the definition of the enthalpy of formation in the reference state.

Berman's goal (1988) was to provide a thermodynamic database of minerals only where there were sufficient independent sources of input constraints to satisfy the requirements of internal consistency and ensure confidence in the final numbers. Thus, the total number of minerals evaluated did not expand beyond those covered by Helgeson *et al.* (1978). Internal consistency applied to the full 11-component system. Experimental half-brackets on P, T and aqueous species concentrations – expanded outwards to accommodate possible experimental error – were used to compile a large set of inequalities in $\Delta G_r(P,T)$ for each experiment. These were fit simultaneously in combination with thermophysical and volume data in the derivation of optimized mineral properties by a combination of linear programming (LIP) and mathematical programming (MAP) techniques. Some newer calorimetric measurements and phase-equilibrium constraints were included. Heat capacities of minerals were parameterized by a function permitting reasonable extrapolation, without inflections above 298 K, to temperatures greater than calorimetric data on them. In contrast to Helgeson *et al.* (1978), measured or estimated expansivities and compressibilities were included. Since 1988, Berman's work has focused on expanding internal consistency to include the solution properties of minerals, e.g. for five minerals relevant to thermometry in the granulite facies (Berman and Aranovich, 1996). These properties among others can be accessed with the Web-accessible TWQ software (*http://gsc.nrcan.gc.ca/sw/twq_e.php*) which automates the process of calculating simple phase diagrams and plotting the results of thermobarometric measurements.

The first thermodynamic dataset for minerals compiled by Holland and Powell was published in 1985(a,b), where the logic for their dataset generation was spelled out. The 1990 paper, of which we include a fraction here, expanded the dataset to 123 minerals and fluid end-members. The minerals were separated into three reliability levels. The addition of data for Fe-endmember and Tschermaks-substituted minerals was aimed at greatly increasing our capacity to calculate total-system phase diagrams for metamorphic mineral solid solutions in PT-dependent invariant, univariant and higher-variance assemblages (see pseudosections, Chapter 8). Improved fugacity polynomials for CO_2 and H_2O extended the range satisfactorily covered up to 40 kbar and higher temperatures. The new algorithm of Holland (1989) was used for third-law entropies where no measured values were available. Landau theory was employed to model the C_P of λ transitions and order-disorder relations. Natural and experimental partitioning data for Mg/Fe^{2+} and Tschermaks substitution enabled data to be presented for many end-members not figuring in earlier databases. However, users are cautioned to use these (reliability level 3 entries) only in the calculation of net-transfer reactions and not for exchange thermometry. Holland and Powell convert experimental brackets into reaction enthalpy brackets, fit the data by least-squares methods, and extract preferred values of reaction enthalpy with their standard deviations. They argue strongly (Powell and Holland, 1993) that there are advantages in this procedure over the LIP method, in which each experimental half-bracket is treated separately (Gordon, 1973; Zen, 1977; Berman *et al.*, 1986). Berman asserts that the latter method provides a better opportunity to eliminate bad experiments, and notes that it is in fact capable of yielding uncertainties (Chatterjee *et al.*, 1998).

The development of internally consistent thermodynamic databases has provided petrologists and geochemists with an enormously powerful tool for researching phase equilibria in natural rocks. At the same time, it has created the demand for coverage of an expanded number of species as well as improvements in the quality of end-member data and especially in solid solution models. Gottschalk (1997) published an internally consistent thermodynamic dataset that, like Holland and Powell (1990), uses least-squares regression, but differs in that it is based on iterative fits in plots of $\ln K$ *vs.* $1/T$, which permit extraction of optimized data for entropy as well as enthalpy of formation. We can be sure that the recent database update by Holland and Powell (1998) will not be the last of its kind. This paper lists reference-state thermodynamic properties for 154 minerals, and in addition, includes 13 silicate liquid end-members and 22 aqueous fluid species. Some simplified solid-solution models are included. Ultra-high pressure calculations have been improved through the use of temperature-dependent thermal expansion and bulk modulus, and a pressure-dependent Landau model for order-disorder. The minerals are no longer separated into reliability levels. Some Fe end-members have fictive (i.e. projected) properties (e.g. Fe-talc, Fe-anthophyllite, Al-free chlorite); these should be used with caution. Examples of phase diagrams involving solid solutions calculated with THERMOCALC were given in Powell et al. (1998). Information on the database and calculation methods of THERMOCALC can be found at www.thermocalc.com.

Sadly, there are few laboratories worldwide where experimental phase equilibrium and calorimetric work is currently being performed. Some refinements to the datasets are likely to come from the study of equilibrium phase assemblages in the field. These constitute very good critical tests of the quality of the database (e.g. Pattison et al., 2002). But they are not a substitute for what we really need: more studies aimed at defining solid-solution properties (critical for accurate thermometry, see Chapter 12), and better experimental brackets, especially for UHPM reactions (Chapter 14), which can be accessed with the multi-anvil press.

References

Adams L.H. (1953) A note on the stability of jadeite. *American Journal of Science*, **251**, 299−308.

Berman, R.G. (1988) Internally-consistent thermodynamic data for minerals in the system Na_2O-K_2O-CaO-MgO-FeO-Fe_2O_3-Al_2O_3-SiO_2-TiO_2-H_2O-CO_2. *Journal of Petrology*, **29**, 445−522.

Berman, R.G. and Aranovich, L.Y. (1996) Optimized standard state and solution properties of minerals. *Contributions to Mineralogy and Petrology*, **126**, 1−24.

Berman, R.G., Engi, M., Greenwood, H.J. and Brown, T.H. (1986) Derivation of internally-consistent thermodynamic data by the technique of mathematical programming: a review with application to the system MgO-SiO_2-H_2O. *Journal of Petrology*, **27**, 1331−1364.

Birch, F., Schairer, J.F. and Spicer, H.C. (1942) Handbook of physical constants. *Geological Society of America Special Paper*, **36**, 325 pp.

Black, P.M. (1977) Regional high-pressure metamorphism in New Caledonia: phase equilibria in the Ouegoa district. *Tectonophysics*, **43**, 89−107.

Chatterjee, N.D. and Johannes, W. (1974) Thermal stability and standard thermodynamic properties of synthetic $2M_1$-muscovite, $KAl_2[AlSi_3O_{10}(OH)_2]$. *Contributions to Mineralogy and Petrology*, **48**, 89−114.

Chatterjee, N.J., Krueger, R., Haller, G. and Olbricht, W. (1998) The Bayesian approach to an internally consistent thermodynamic database; theory, database, and generation of phase diagrams. *Contributions to Mineralogy and Petrology*, **133**, 149−168.

Danielsson, A. (1950) Das Calcit-Wollastonitgleichgewicht. *Geochimica et Cosmochimica Acta*, **1**, 55−69.

Ellis, A.J. and Fyfe, W.S. (1956) A note on the calcite-wollastonite equilibrium. *American Mineralogist*, **41**, 805−807.

Eugster, H.P. and Skippen, G.B. (1967) Igneous and metamorphic reactions involving gas equilibria. Pp. 492−520 in: *Researches in Geochemistry*, **2** (P.H. Abelson, editor). John Wiley, New York.

Eugster, H.P. and Wones, D.R. (1962) Stability relations of the ferruginous biotite, annite. *Journal of Petrology*, **3**, 82−125.

Evans, B.W. and Trommsdorff, V. (1974) Stability of enstatite, talc and CO_2 metasomatism of metaperidotite, Val d'Efra, Lepontine Alps. *American Journal of Science*, **274**, 274−296.

Evans, B.W., Johannes, W., Oterdoom, H. and Trommsdorff, V. (1976) Stability of chrysotile and antigorite in the serpentinite multisystem. *Schweizerische Mineralogische und Petrographische Mitteilungen*, **56**, 79−93.

Ferry, J.M. (1976) P, T, f_{CO2}, and f_{H2O} during metamorphism of calcareous sediments in the Waterville-Vassalboro area, south-central Maine. *Contributions to Mineralogy and Petrology*, **57**, 119−143.

Ferry, J.M. (1980) A comparative study of geothermometers and geobarometers in pelitic schists from south-central Maine. *American Mineralogist*, **65**, 720−732.

Ferry, J.M. (1987) Metamorphic hydrology at 13-km depth and 400−550°C. *The American Mineralogist*, **72**, 39−58.

Franz, G. and Spear, F.S. (1983) High pressure metamorphism of siliceous dolomites from the central Tauern window, Austria. *American Journal of Science*, **283-A** (Orville Volume), 396−413.

Froese, E. (1976) Application of thermodynamics in metamorphic petrology. *Geological Survey of Canada*, Paper 75-43.

Ghent, E.D., Robbins, D.B. and Stout, M.Z. (1979) Geothermometry, geobarometry, and fluid compositions of metamorphosed calc-silicates and pelites, Mica Creek, British Columbia. *American Mineralogist*, **64**, 874−885.

Gordon, T.M. (1973) Derivation of internally consistent thermodynamic data from phase equilibrium experiments. *Journal of Geology*, **81**, 199−208.

Gottschalk, M. (1997) Internally consistent thermodynamic data for rock-forming minerals in the system SiO_2-TiO_2-Al_2O_3-Fe_2O_3-CaO-MgO-FeO-K_2O-Na_2O-H_2O-CO_2. *European Journal of Mineralogy*, **9**, 175−223.

Greenwood, H.J., editor (1977) *Short Course in Application of Thermodynamics to Petrology and Ore Deposits*. Mineralogical Association of Canada, 230 pp. (reprinted in 1978, 1981 and 1983).

Helgeson, H.C., Delaney, J.M., Nesbitt, H.W. and Bird, D.K. (1978) Summary and critique of the thermodynamic properties of rock-forming minerals. *American Journal of Science*, **278-A**, 1−229.

Holland, T.J.B and Powell, R. (1990) An enlarged and updated internally consistent thermodynamic dataset with uncertainties and correlations: the system K_2O-Na_2O-CaO-MgO-MnO-FeO-Fe_2O_3-Al_2O_3-TiO_2-SiO_2-C-H_2O-O_2. *Journal of Metamorphic Geology*, **8**, 89−124.

Holland, T.J.B. (1989) The dependence of entropy on volume for silicate and oxide minerals: a review and a predictive model. *American Mineralogist*, **74**, 134−142.

Holland, T.J.B. and Powell, R. (1985a) An internally consistent thermodynamic dataset with uncertainties and correlations: 1. Methods and a worked example. *Journal of Metamorphic Geology*, **3**, 327−342.

Holland, T.J.B. and Powell, R. (1985b) An internally consistent thermodynamic dataset with uncertainties and correlations: 2. Data and results. *Journal of Metamorphic Geology*, **3**, 343−370.

Holland, T.J.B and Powell, R. (1998) An internally consistent thermodynamic data set for phases of petrological interest. *Journal of Metamorphic Geology*, **16**, 309−343.

Joesten, R. (1974) Local equilibrium and metasomatic growth of zoned calc-silicate nodules from a contact aureole, Christmas mountains, Big Bend Region, Texas. *American Journal of Science*, **274**, 876−901.

Kern, R. and Weisbrod, A. (1967) *Thermodynamics for Geologists*. Freeman, Cooper and Company, San Francisco, 304 pp.

Kracek, F.C., Neuvonen, K.J. and Burley, G. (1951) Thermochemistry of mineral substances. Part I. A thermodynamic study of the stability of jadeite. *Washington Academy of Sciences Journal*, **41**, 373−383.

McDonald, G.J.F. (1955) Gibbs free energy of water at elevated temperatures and pressures with application to the brucite-periclase equilibrium. *Journal of Geology*, **63**, 244−252.

Orville, P.M. and Greenwood, H.J. (1965) Determination of ΔH of reaction from experimental pressure-temperature curves. *American Journal of Science*, **263**, 678−683.

Pattison, D.M., Spear, F.S., Debuhr, C.L., Cheney, J.T. and Guidotti, C.V. (2002) Thermodynamic modeling of the reaction muscovite +

cordierite → Al_2SiO_5 + biotite + quartz + H_2O: constraints from natural assemblages and implications for the metapelitic petrogenetic grid. *Journal of Metamorphic Geology*, **20**, 99–118.

Powell, R. (1978) *Equilibrium Thermodynamics in Petrology*. Harper and Row, London, 284 pp.

Powell, R. and Holland, T.J.B. (1993) The applicability of least squares in the extraction of thermodynamic data from experimentally bracketed mineral equilibria. *American Mineralogist*, **78**, 107–112.

Powell, R., Holland, T.J.B. and Worley, B. (1998) Calculating phase diagrams involving solid solutions via nonlinear equations, with examples from THERMOCALC. *Journal of Metamorphic Geology*, **16**, 577–588.

Skippen, G.B. (1971) Experimental data for reactions in siliceous marbles. *Journal of Geology*, **79**, 451–481.

Skippen, G.B. (1974) An experimental model for low-pressure metamorphism of siliceous dolomitic marble. *American Journal of Science*, **274**, 487–509.

Wood, B.J. and Fraser, D.G. (1977) *Elementary Thermodynamics for Geologists*. Oxford University Press, Oxford, UK, 303 pp.

Zen E-an (1977) The phase equilibrium calorimeter, the petrogenetic grid, and a tyranny of numbers. *American Mineralogist*, **62**, 189–204.

SUMMARY AND CRITIQUE OF THE THERMODYNAMIC PROPERTIES OF ROCK-FORMING MINERALS

HAROLD C. HELGESON, JOAN M. DELANY[*],
H. WAYNE NESBITT[**], and DENNIS K. BIRD[***]

Department of Geology and Geophysics,
University of California,
Berkeley, California 94720

[*] Present address: Department of Earth and Space Sciences, University of California, Los Angeles, California 90024
[**] Present address: Department of Geology, La Trobe University, Bundoora, Victoria, Australia 3083
[***] Present address: Department of Geosciences, University of Arizona, Tucson, Arizona 85721

PREFACE

The research reported in the following pages was carried out in recognition of the need for a well documented, comprehensive, and critical compilation of thermodynamic data for minerals which can be used with confidence to characterize chemical equilibrium in geologic systems. The credibility of such a compilation is largely a function of the extent to which the authors demonstrate that the thermodynamic data adopted for a given mineral are consistent with those derived from experimental data for all other minerals, and that none contravenes reliable geologic observations. We make no pretense at having achieved these goals with the present contribution, which represents little more than a beginning step in this direction. Nevertheless, we consider it an important step and hope that it will serve as a comprehensive foundation for future refinement of the values adopted for the thermodynamic properties of minerals. At the same time we trust that the near-coincident appearance of the present communication and the revision of U.S. Geological Survey Bulletin 1259 (Robie and Waldbaum, 1968) by Robie, Hemingway, and Fisher (1978), which is based almost entirely on calorimetric data, will in no way polarize the geologic community. The two publications serve different needs and both should enhance the geologist's ability to understand and meet the many challenges inherent in applying thermodynamic analysis to the interpretation of phase relations in natural systems.

Most compilations of thermodynamic data contain few if any illustrations to facilitate assessment of the reliability of the values given for the thermodynamic properties of the species considered. Furthermore, as a rule they contain little or no comparative discussion or documentation of the basis for choosing the values adopted and the extent to which they agree or disagree with other values reported in the literature. Perhaps more importantly, attention is rarely devoted in such compilations to the chemical or geologic consequences of choosing one value over another. We consider these serious shortcomings which we have endeavored to overcome in the present communication.

The calculations summarized below were carried out with the firm conviction that there are no "correct" thermodynamic properties of minerals, just as there are no "correct" physical, compositional, and crystallochemical states of minerals. That this is indeed true can be easily verified by little more than a cursory appraisal of the spectrum of calorimetric enthalpies of formation that have been reported in recent years for different natural and synthetic samples of the same mineral.

The discussion in the following pages combines many elements of a textbook with advanced concepts and numerical results of recent research. We make no apology for this but instead submit that the present state of understanding in the application of thermodynamics to the interpretation of geologic systems requires careful documentation of the equations used in the calculations and continued reappraisal of the factors contributing to the thermodynamic behavior of minerals.

H. C. Helgeson
Berkeley, California
January 15, 1978

SUMMARY AND CRITIQUE OF THE THERMODYNAMIC PROPERTIES OF ROCK-FORMING MINERALS

ABSTRACT. Critical analysis of experimental high-pressure/temperature solubility data and univariant/divariant phase relations reported in the literature with equations representing the thermodynamic properties of ionic aqueous species, $SiO_{2(aq)}$, H_2O, CO_2, O_2, and the temperature dependence of the standard molal heat capacities of minerals affords an internally consistent set of thermodynamic data for the bulk of the abundant rock-forming silicates, carbonates, and oxides in the crust and upper mantle. Standard molal Gibbs free energies and enthalpies of formation were obtained in this manner for $\sim$ 70 minerals in the system $Na_2O-K_2O-CaO-MgO-FeO-Fe_2O_3-Al_2O_3-SiO_2-CO_2-H_2O$. The calculations were carried out using standard molal entropies, heat capacities, and volumes derived from calorimetric, crystallographic, and density data or estimated from correlation algorithms and Clapeyron slope constraints. Where necessary and appropriate, provision was included for configurational entropy contributions and enthalpy changes arising from the temperature dependence of substitutional and displacive order/disorder in minerals. In most cases, simultaneous consideration of multiple equilibria reduced relative uncertainties in the standard molal enthalpies and Gibbs free energies of formation generated from the high pressure/temperature data by more than an order of magnitude from those inherent in corresponding values derived from calorimetric measurements. Apparent inconsistencies among experimental data and phase relations in nature were resolved by taking account of geologic observations, the thermodynamic consequences of metastable equilibria, and compositional variation in minerals at both high and low temperatures and pressures. Experimental solution compositions and/or reversal temperatures for more than 130 reactions at various pressures are depicted in phase diagrams, where they can be compared with calculated equilibrium constants and univariant/divariant curves generated from the thermodynamic data summarized in the tables and discussed in the text. These data, together with the equations of state employed in the calculations, permit comprehensive prediction of the thermodynamic consequences of equilibrium and mass transfer among minerals and aqueous solutions in geochemical processes at high pressures and temperatures.

1

INTRODUCTION

In the years following the appearance of Robie and Waldbaum's (1968) and Naumov, Ryzenko, and Khodakovskii's (1971) compilations of thermodynamic data for minerals, numerous attempts have been made to resolve inconsistencies among reported values of the standard molal enthalpies and Gibbs free energies of formation of various oxides and silicates (for example, Helgeson, 1969; Anderson, 1970; Zen, 1969a and b, 1971, 1972, 1973, 1977; Fisher and Zen, 1971; Zen and Chernosky, 1976; Parks, 1972; Thompson, 1973a and b, 1974a and b; Bird and Anderson, 1973; Ulbrich and Merino, 1974; Chatterjee, 1977; Chatterjee and Johannes, 1974; Haas and Fisher, 1976; Hemingway and Robie, 1977; Robie, Hemingway, and Fisher, 1978). Unfortunately, most of these are restricted to one or another mineral or system, some are based on invalid assumptions, others fail to take adequate account of thermodynamic constraints on phase relations, and none is compatible with both geologic observations and recent experimental data. These deficiencies can be avoided by simultaneous consideration of multiple reactions in experimental as well as natural systems with equations incorporating rigorous and explicit provision for the thermodynamic behavior of minerals and aqueous species. The object of the present communication is to derive with the aid of such equations a comprehensive and reliable set of internally consistent thermodynamic data for all the abundant minerals in the Earth's crust which will, (1) reproduce accurately geologic phase relations and the many experimental observations of high pressure/temperature phase equilibria that have accumulated over the past twenty years, and (2) afford realistic prediction of the chemical consequences of reversible reactions among minerals and aqueous solutions in multicomponent systems at both high and low temperatures and pressures.

Although many calorimetric investigations of minerals have been carried out over the past several decades (for example, Kracek, Neuvonen, and Burley, 1951; Neuvonen, 1952; Kelley and others, 1953; Kelley, 1960, 1962; Barany and Kelley, 1961; Barany, 1962, 1963, 1964; King and Weller, 1961a and b; Pankratz, 1964a and b; Hlabse and Kleppa, 1968; Holm and Kleppa, 1966, 1968; Navrotsky, 1971; Waldbaum, ms; Waldbaum and Robie, 1971; Robie and Hemingway, 1973; Shearer, ms; Charlu, Newton, and Kleppa, 1975; Newton, Charlu, and Kleppa, 1977; Hemingway and Robie, 1977a), few have resulted in sufficiently accurate standard molal enthalpies of formation to afford reliable prediction of equilibrium constants at high pressures and temperatures. Uncertainties in enthalpies of formation at 25°C and 1 bar derived from high temperature molten salt calorimetry are of the order of 500 cal mole^{-1}, and those resulting from measurements of low-temperature heats of solution in hydrofluoric acid solutions commonly exceed a kcal mole^{-1}. Uncertainties of this order of magnitude may lead to errors of more than 100°C in computed equilibrium temperatures. In contrast, *relative* uncertainties in standard molal enthalpies and Gibbs free energies of formation of minerals derived from compositional data and experimental observations of equilibrium temperatures and pressures for a series of independent equilibria involving minerals and a fluid phase are commonly of the order of 50 cal mole^{-1} or less, which permits close approximation of the thermodynamics of reactions among minerals and aqueous solutions at high pressures and temperatures.

Many outstanding studies of reversible reactions among minerals, Na^+, K^+, $SiO_{2(aq)}$, O_2, CO_2, and/or H_2O have been reported in the literature (for example, Hemley, 1959; Hemley, Meyer, and Richter, 1961; Hemley and others, 1971, 1977a and b; Greenwood, 1962, 1967a; Orville, 1963, 1972; Eugster and Skippen, 1967; Skippen, 1971, 1974; Chatterjee and Johannes, 1974) which are now sufficient in number and cover a wide enough spectrum of equilibrium states to permit reliable calculation of the standard molal enthalpies and Gibbs free energies of formation of the bulk of the rock-forming minerals in the crust and upper mantle. Calculations of this kind were carried out by combining data reported in these and similar studies with equations of state for aqueous species (Helgeson and Kirkham, 1974a, 1976, and in press; Walther and Helgeson, 1977; Delany and Helgeson, 1978) and fugacity coefficients for gases (Holloway and Reese, 1974; Holloway, 1977) to evaluate simultaneously equilibrium constants for sets of independent reversible reactions. Calculation of the standard molal Gibbs free energies of the reactions at 25°C and 1 bar were carried out using independently derived volumes and calorimetric and/or estimated entropies and heat capacities of the minerals. The standard molal Gibbs free energies of formation of the minerals were then computed from the standard molal Gibbs free energies of the reactions at 25°C and 1 bar by simultaneous solution of the set of equations. Experimental uncertainties were optimized by carrying out comparative calculations and taking account of Clapeyron slope constraints, which were used to determine the extent to which third law entropies should be adjusted to take account of configurational contributions.

The reliability of standard molal Gibbs free energies of formation of minerals computed from high pressure/temperature phase equilibrium data can be assessed by comparing the results of the calculations with geologic observations and/or calorimetric measurements, which often afford corroboration in spite of (but more commonly because of) the relatively large uncertainties inherent in these data. Contradictions in values of the thermodynamic properties of a given mineral generated from different experimental data may be resolved by comparing the geologic consequences of the discrepancies with phase relations in natural systems. Such comparisons are particularly instructive if the pressure, temperature, and the system considered differ from those in the experimental studies from which the data were obtained. Activity diagrams facilitate correlation of predicted and observed equilibrium states with the compositions of fluid inclusions and interstitial waters in geologic systems. Correlations of this kind commonly reveal inconsistencies arising from metastability and differences in crystallinity and order/disorder in minerals formed under different conditions in different geologic environments. Because the latter factors affect considerably the extent to which thermodynamic calculations represent geologic reality, the effects of various physical and crystallochemical properties of minerals on their thermodynamic behavior in geologic systems are reviewed briefly below.

THERMODYNAMIC CONSEQUENCES OF THE PHYSICAL AND CRYSTALLOCHEMICAL PROPERTIES OF MINERALS

The crystallochemical properties of anisotropic minerals are vector quantities which may differentially affect the extent to which the minerals react in geochemical processes.[1] Furthermore, a mineral with a given name and composition occurring under the same pressure/temperature conditions in one or another laboratory experiment or different parts of the Earth may have significantly different thermodynamic properties (see below). These differences are manifested in part by differences in the thermodynamic behavior of natural minerals and their synthetic analogs.

In addition to their dependence on pressure, temperature, and composition, the thermodynamic properties of a mineral are sensitive to perturbations in energetic, configurational, and crystallochemical contributions to its stability arising from the cooling rate and strain history of the mineral in different environments. Perturbations of this kind are responsible for differences in crystallinity, substitutional and displacive order/disorder, and the number and kinds of defects and dislocations in the mineral, all of which contribute to metastable equilibrium states both in nature and in laboratory experiments. Metastable states also occur as a consequence of incomplete reaction resulting from kinetic constraints. It thus follows that experimental observations of the behavior of minerals cannot be related precisely to the behavior of minerals in geochemical processes without correlating the measurements with the physical and crystallochemical properties of the minerals. Unfortunately, few experimental investigations reported in the literature are comprehensive enough to permit such correlation. Until recently, the thermodynamic consequences of order/disorder in minerals received little attention in experimental investigations of phase equilibria, and the crystallographic properties of the minerals used in calorimetric studies were rarely determined. Even compositional data reported in many early investigations are inadequate to permit unambiguous interpretation of the experimental results.

Despite the fact that thermodynamic calculations of mineral stabilities using data derived from experimental studies afford little more than

---

[1] In this respect, the thermodynamic properties of anisotropic crystals are somewhat analogous to the electrostatic properties of polar molecules. Although in both cases overall electrical neutrality and conservation of mass are maintained, the magnitudes of the properties are directional.

approximations of geologic reality, comparison of predicted and observed compatibilities and compositions indicates that many of these are close approximations. Furthermore, experimental results obtained using a variety of starting materials, equipment, and techniques in different laboratory studies of a given equilibrium mineral assemblage are commonly compatible with one another. It thus seems reasonable to conclude that in many reactions, physical and crystallochemical factors have a secondary effect on the thermodynamic behavior of minerals. In others, such as solid/solid phase transitions, they are clearly of primary importance.

Solid state phase transitions.—Phase transitions in minerals range from those manifested by abrupt and obvious changes in morphology (such as clinoenstatite/enstatite) to those caused by gradual and nearly imperceptible changes in bond angles or substitutional order/disorder among atoms on energetically distinct lattice sites. Although chemical equilibrium at constant pressure and temperature requires the Gibbs free energy of transformation to be zero for all phase transitions, the partial derivatives of the Gibbs free energies of minerals may exhibit either abrupt or gradual changes with increasing pressure and/or temperature, depending on the nature of the transition taking place.

Despite a multitude of experimental studies, the nature of many phase transitions exhibited by oxides and silicates remains ambiguous. Some of these are thought to be first-order transitions, which are characterized by discontinuities in the temperature dependence of the entropy, enthalpy, volume, heat capacity, expansibility, and compressibility of the compound. Others are almost certainly lambda transitions[2] with no discontinuity in the temperature dependence of these properties. A lambda transition can be thought of as a continuum of infinitesimal changes in the internal structure and/or distribution of atoms in a mineral in response to changes in pressure and/or temperature. The heat capacity, expansibility, and compressibility of a mineral undergoing a lambda transition change dramatically and (in contrast to their first-order counterparts) their dependence on temperature before and after the transition is strikingly different. Typical perturbations of the heat capacities of minerals by structural changes accompanying first-order and lambda transitions are depicted schematically in figure 1. The heat capacity curve in figure 1B is similar to that of β-brass, which exhibits a lambda transition caused by increasing long-range disorder with increasing temperature.

The nature of a phase transition cannot always be deduced unambiguously from the behavior of the heat capacity of a mineral with increasing temperature. In some cases measurements of the heat capacities of minerals as a function of temperature must be carried out in conjunction with annealing experiments to detect lambda transitions.

---

[2] All transitions other than first-order are referred to as lambda transitions to avoid theoretical implications inherent in the designations, second-order, third-order, et cetera (Denbigh, 1971).

Internally-Consistent Thermodynamic Data for Minerals in the System Na$_2$O–K$_2$O–CaO–MgO–FeO–Fe$_2$O$_3$–Al$_2$O$_3$–SiO$_2$–TiO$_2$–H$_2$O–CO$_2$

by R. G. BERMAN*

Department of Geological Sciences, University of British Columbia, Vancouver, B. C. Canada V6T 2B4

(Received 16 March 1987; revised typescript accepted 7 October 1987)

ABSTRACT

Internally consistent standard state thermodynamic data are presented for 67 minerals in the system Na$_2$O–K$_2$O–CaO–MgO–FeO–Fe$_2$O$_3$–Al$_2$O$_3$–SiO$_2$–TiO$_2$–H$_2$O–CO$_2$. The method of mathematical programming was used to achieve consistency of derived properties with phase equilibrium, calorimetric, and volumetric data, utilizing equations that account for the thermodynamic consequences of first and second order phase transitions, and temperature-dependent disorder. Tabulated properties are in good agreement with thermophysical data, as well as being consistent with the bulk of phase equilibrium data obtained in solubility studies, weight change experiments, and reversals involving both single and mixed volatile species. The reliability of the thermodynamic data set is documented by extensive comparisons (Figs. 4–45) between computed equilibria and phase equilibrium data. The high degree of consistency obtained with these diverse experimental data gives confidence that the refined thermodynamic properties should allow accurate prediction of phase relationships among stoichiometric minerals in complex chemical systems, and provide a reasonable basis from which activity models for minerals may be derived.

INTRODUCTION

Thermodynamics provides a fundamental tool for gaining insight into and quantifying diverse processes which span most geologic disciplines. One of the most important applications of thermodynamics is the prediction of stable phase relationships, and software is now available for quick and accurate calculation of stable phase relationships of any complexity among phases of geologic interest (e.g., Ghiorso *et al.*, 1983; Perkins *et al.*, 1986; Berman *et al.*, 1987). The only limitation on the accuracy of calculated phase relationships is that imposed by our knowledge of the thermodynamic parameters describing the state functions of each phase, component, or species in the databases utilized by these software.

Over the past several decades, considerable effort has been directed towards formulating thermodynamic databases, starting with compilations which drew heavily upon calorimetric data (e.g., Kelley, 1960; Kelley & King, 1961; Robie *et al.*, 1979). More recent efforts have made use of phase equilibrium data to verify and refine these calorimetric data, an important step in the light of conclusions that enthalpies of formation have not in general been determined calorimetrically with the accuracy necessary to reproduce phase equilibrium data (Helgeson *et al.*, 1978; Berman *et al.*, 1985). The most extensive compilation of refined thermodynamic data remains that of Helgeson et al. (1978), who presented standard state properties for some 70 common rock-forming minerals, including many that exhibit considerable non-stoichiometry and complications due, for example, to temperature-dependent disordering. Since completion of this pioneering work, many additional experimental data have been gathered, but recent studies (e.g., Robinson et al., 1982; Halbach & Chatterjee, 1984; Holland & Powell, 1985; Berman et al., 1986) have focussed on stoichiometric minerals in smaller chemical subsystems, and can be viewed primarily as efforts to explore the results of using different mathematical techniques for extracting thermodynamic parameters from diverse sets of experimental observations (see below).

In this paper internally consistent thermodynamic data are presented for 67 minerals (Table 1) in the system $Na_2O-K_2O-CaO-MgO-FeO-Fe_2O_3-Al_2O_3-SiO_2-TiO_2-H_2O-CO_2$. This database has been derived through consideration of the most recent experimental data, using equations that take into account important effects such as temperature-dependent disordering and lambda transitions, without which it is impossible to reproduce the experimental data faithfully, or to apply them confidently to geological problems. For the most part, this work considers minerals that are stoichiometric in restricted portions of this chemical system. Solid solution effects have been evaluated only when sufficient data are available to allow reasonable confidence in derived activity models *and* when treatment of such data allows access to experimental data which contribute to significant refinement in the thermodynamic properties of other phases in the database (e.g., hydrous cordierite and aluminous enstatite). By refining thermodynamic parameters with experimental data involving stoichiometric phases, analysis of solution properties and more complex experimental data is avoided and greater accuracy is achieved in derived standard state properties. Although it is anticipated that some adjustments may be necessary, particularly in enthalpies of formation that are used as reference values (see discussion of reference values below), these standard state properties can then be used as a starting point to constrain activity models for the same minerals in chemically more complex systems.

The main purpose of this paper is to document fully the results of this thermodynamic analysis, showing which experimental data are reproduced well, which are not, and discussing sources of discrepancies. This detailed documentation is important not only to help define a reliable set of experimental data, but also because it offers the only means to assess the overall accuracy of derived thermodynamic parameters. The underlying premise of this work is that it is only possible to evaluate the validity of thermodynamic properties by comparison with all available data, because subsets of preferred experimental data are never sufficiently precise to define fully all thermodynamic parameters. This conclusion stems from the nature of phase equilibrium data which bracket the position of, but do not indicate the exact location of an equilibrium (see discussion of phase equilibrium data below).

THERMODYNAMIC RELATIONS

At a pressure P (b) and temperature T (K), the apparent free energy of formation of a pure phase from the elements*, $\Delta_a G^{P,T}$, is defined by

$$\Delta_a G^{P,T} = \Delta_a H^{P,T} - T \cdot S^{P,T} \qquad (1)$$

---

* In this paper, the IUPAC recommendation (Cox, 1979) is followed whereby a subscript modifying the Δ is used to denote a specific process. For simplicity, the convention of denoting standard state thermodynamic properties with a ° has not been adopted because this paper deals in entirety with standard state properties.

where

$$\Delta_a H^{P,T} = \Delta_f H^{P_r,T_r} + \int_{T_r}^{T} Cp \, dT + \int_{P_r}^{P} \left\{ V^{P_r,T_r} - T\left(\frac{\partial V}{\partial T}\right)_P \right\} dP \qquad (2)$$

and

$$S^{P,T} = S^{P_r,T_r} + \int_{T_r}^{T} \left(\frac{Cp}{T}\right) dT - \int_{P_r}^{P} \left(\frac{\partial V}{\partial T}\right)_P dP. \qquad (3)$$

$\Delta_a H^{P,T}$ is the apparent enthalpy of formation of the pure phase from the elements at P and T, $\Delta_f H^{P_r,T_r}$, S^{P_r,T_r}, and V^{P_r,T_r} are the enthalpy of formation from the elements, third law entropy, and molar volume, respectively, at the reference pressure ($P_r = 1$ b $= 10^5$ Pa) and temperature ($T_r = 298 \cdot 15$ K), and Cp is the heat capacity at constant pressure. Note that in eqns (1–3) the properties of the elements are used only to define the standard state enthalpy of formation $\Delta_f H^{P_r,T_r}$. In this regard, the apparent free energy function expressed through eqns (1–3) differs from that defined by Benson (1968) and used by Helgeson *et al.* (1978) which incorporates the entropies of the elements at 298·15 K and 1 b.

Heat capacity of minerals that do not undergo lambda transitions is parameterized with the function (Berman & Brown, 1985)

$$Cp = k_0 + k_1 T^{-0.5} + k_2 T^{-2} + k_3 T^{-3} \qquad (4)$$

where k_1 and k_2 are constrained to be less than or equal to zero. These constraints ensure that no inflections occur in the $Cp(T)$ function above 298·15 K, and it has been demonstrated (Berman & Brown, 1985) that the heat capacity of minerals can be reliably extrapolated to higher temperatures than available calorimetric data. This latter point is extremely important in that many important rock-forming minerals have stabilities far in excess of the temperatures of present calorimetric measurements. Because the k_3 term can be positive, inflections can occur at temperatures below the lowest temperature data point used in deriving the thermal functions (usually 250 K), and caution should be exercised not to use the Cp functions below these temperatures (Table 3).

In a previous study (Berman *et al.*, 1985) it was found that adequate representation of phase equilibrium data was possible using the assumption that mineral volumes are independent of pressure and temperature. In spite of the fact that the requisite data are incomplete with respect to many important minerals, this simplication has not been made in this work because expansivity and compressibility terms make significant contributions to low entropy (notably solid–solid) reactions and to reactions in which the reactant and product mineral assemblages have highly contrasting volumetric properties (e.g., muscovite dehydration equilibria). The pressure and temperature dependence of mineral volumes is represented with the following equation:

$$\frac{V^{P,T}}{V^{P_r,T_r}} = 1 + v_1(P - P_r) + v_2(P - P_r)^2 + v_3(T - T_r) + v_4(T - T_r)^2. \qquad (5)$$

This equation was chosen over alternate representations for two reasons. First, it is linear in the fit parameters v_1–v_4, a feature which greatly simplifies the refinement process of these terms (see below). Second, use of a more complex equation that allows accurate extrapolations to high pressures is unwarranted because the emphasis in this paper is on deriving standard state thermodynamic properties for minerals from phase equilibrium data that for the most part are restricted to pressures less than 10 kb. Although the parameterization of eqn (5) gives smooth extrapolations to high pressure and temperature, this equation has no theoretical basis and caution should therefore be exercised in using it at conditions beyond those of the volumetric data listed in Table 4 or the phase equilibrium data shown in Figs. 4–45.

Substitution of eqns (2–5) into (1), and rearrangement, yields:

$$\Delta_a G^{P,T} = \Delta_f H^{P_r,T_r} - TS^{P_r,T_r} + k_0\{(T-T_r) - T(\ln T - \ln T_r)\}$$
$$+ 2k_1\{(T^{0.5} - T_r^{0.5}) + T(T^{-0.5} - T_r^{-0.5})\}$$
$$- k_2\{(T^{-1} - T_r^{-1}) - T/2(T^{-2} - T_r^{-2})\}$$
$$- k_3\{(T^{-2} - T_r^{-2})/2 - T/3(T^{-3} - T_r^{-3})\}$$
$$+ V^{P_r,T_r}[(v_1/2 - v_2)(P^2 - P_r^2) + v_2/3(P^3 - P_r^3)$$
$$+ \{1 - v_1 + v_2 + v_3(T - T_r) + v_4(T - T_r)^2\}(P - P_r)]. \qquad (6)$$

For gases, the explicit volume integrations in eqn (6) are replaced by:

$$\int_{P_r}^{P} V_{\text{gas}} \, dP = G^{P,T} - G^{P_r,T} \qquad (7)$$

where the right hand side of eqn (7) has been evaluated using the equations of state of Haar *et al.* (1984) for H_2O below 10 kb, Delany & Helgeson (1978) for H_2O above 10 kb, and Kerrick & Jacobs (1981) for CO_2.

Polymorphic transitions

For accurate representation of the thermodynamic properties of minerals it is necessary to account for the consequences of both first- and second-order (lambda) polymorphic transitions, although in practice the exact nature of a phase transition may be ambiguous. Treatment of the former is the easier task because the temperature dependence of thermodynamic properties is quite similar above and below the transition temperature. Heat content measurements spanning the transition are sufficient to define the enthalpy of transition, $\Delta_t H$, and because the change in Gibbs free energy of the reaction $\Delta_r G = 0$ at the transition temperature (T_t), these data also define the entropy of transition, $\Delta_t S(\Delta_t S = \Delta_t H/T_t)$. Volumetric data are needed to define the volume of transition $\Delta_t V$, as well as changes in expansivity and compressibility. The internal consistency of these various data may be assessed (and refined) with use of the Clapeyron equation, $(dP/dT)_{\Delta_r G = 0} = (\Delta_r S/\Delta_r V)$, if the P–T slope of the transition is known.

Treatment of lambda transitions is more complicated because of the marked difference in the temperature dependence of thermodynamic properties above and below the lambda point (T_λ). In this paper, lambda transitions are modelled with the function proposed by Berman & Brown (1985), wherein heat capacity is separated into two components: a 'lattice' heat capacity that is represented by the same $k_0 - k_3$ coefficients (eqn 4) for both polymorphs, and a 'lambda' component (Cp_λ) that, for $T_{\text{ref}} < T < T_\lambda$ is computed at 1 b pressure by:

$$Cp_\lambda^{1\text{bar}} = T(l_1 + l_2 T)^2. \qquad (8)$$

The parameters l_1, l_2, T_{ref} (usually 298.15 K), and T_λ are defined by Cp or heat content data at 1 b, and for the most part have been taken from Berman & Brown (1985). Equation (8) allows reasonable representation of calorimetric data usually to within about 30° of the lambda point. The simple form of the equation does not allow more accurate representation closer to the lambda point, nor does it account for the decrease in Cp (relative to the 'lattice' Cp) commonly observed just above the lambda point.

Heat capacity measurements at elevated pressure are not presently available, necessitating an assumption regarding the pressure dependence of Cp_λ. Here it is assumed that the magnitude of Cp_λ is independent of P, and that the position of onset and width of the lambda transition remains fixed in relation to the lambda point at all pressures (Fig. 1b and c). Most phase equilibrium data are broadly consistent with modelling polymorphic transitions as straight lines in P–T, such that:

$$T_\lambda^P = T_\lambda^{1\,\text{bar}} + k(P-1). \tag{9}$$

By defining $T_d = T_\lambda^{1\,\text{bar}} - T_\lambda^P$, and $T' = T + T_d$, $C_{P\lambda}^P$ is computed at any pressure for $T_{\text{ref}} < T' < T_\lambda^{1\,\text{bar}}$ by:

$$Cp_\lambda^P = T'(l_1 + l_2 T')^2. \tag{8a}$$

In order to calculate the effect of the lambda transition on other thermodynamic variables, Cp_λ^P is integrated between a reference temperature defined by $t_r = T_{\text{ref}} - T_d$ and $T(T \leqslant T_\lambda^P)$, giving:

$$\Delta_\lambda H^{P,T} = \int_{t_r}^{T} Cp_\lambda \, dT = x_1(T - t_r) + \frac{x_2}{2}(T^2 - t_r^2) + \frac{x_3}{3}(T^3 - t_r^3) + \frac{x_4}{4}(T^4 - t_r^4) \tag{10}$$

$$\Delta_\lambda S^{P,T} = \int_{t_r}^{T} (Cp_\lambda / T) \, dT = x_1(\ln T - \ln t_r) + x_2(T - t_r)$$
$$+ \frac{x_3}{2}(T^2 - t_r^2) + \frac{x_4}{3}(T^3 - t_r^3) \tag{11}$$

where

$$x_1 = l_1^2 T_d + 2 l_1 l_2 T_d^2 + l_2^2 T_d^3$$
$$x_2 = l_1^2 + 4 l_1 l_2 T_d + 3 l_2^2 T_d^2$$
$$x_3 = 2 l_1 l_2 + 3 l_2^2 T_d$$
$$x_4 = l_2^2.$$

By performing the integrations in this manner, $\Delta_\lambda H$ remains constant at the lambda point at all pressures, while $\Delta_\lambda S$ decreases with increasing pressure if $k > 0$ (Fig. 2). Note that if T is greater than the transition temperature, T_λ^P, the integrations in (10) and (11) are performed between t_r and T_λ^P. The change in Gibbs free energy contributed by the increase in heat capacity associated with the lambda transition is calculated from

$$\Delta_\lambda G^{P,T} = \Delta_\lambda H^{P,T} - T \cdot \Delta_\lambda S^{P,T} \tag{12}$$

and $\Delta_\lambda G$ is added to the free energy function given by eqn (1).

It is important to note that, although the magnitude of Cp_λ is independent of P, the fact that the onset of the lambda transition does change with pressure (Figs. 1b, 1c) implies that $(\partial Cp/\partial P)_T \neq 0$, $(\partial S/\partial P)_T \neq 0$, and therefore that $(\partial V/\partial T)_P = -(\partial S/\partial P)_T$ is also non-zero. These relationships are most easily illustrated with reference to a particular lambda transition. Cp data for the α–β quartz transition are shown in Fig. 1b along with the computed Cp function using Eqns (4) and (8). The dotted curve in Fig. 1a shows volumes at 1 b for α-quartz computed from finite difference derivatives of $(\partial G_\lambda/\partial P)_T = V_\lambda$, with the assumption that the slope of this transition is constant ($k = (dT/dP) = 0.0237$ K/b). In order

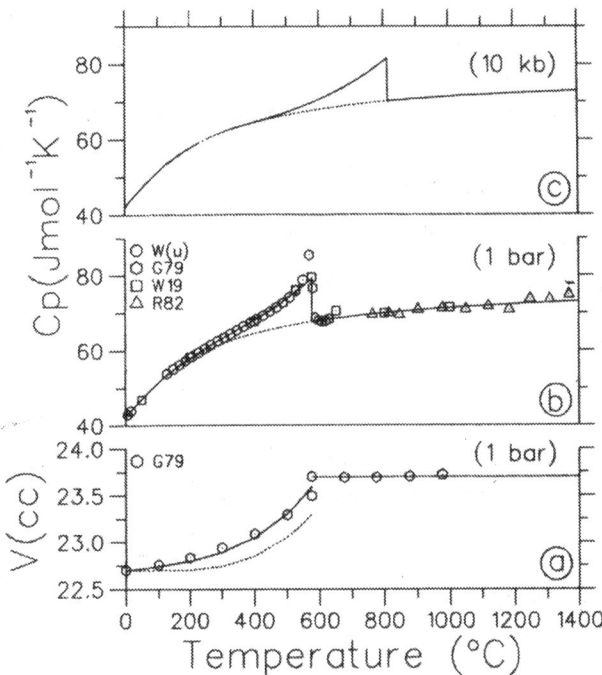

FIG. 1. Effect of the lambda transition on the heat capacity and volume of quartz. (a) Solid lines show the computed volume of quartz compared to Ghiorso et al.'s (1979) representation of experimental data. The dotted line shows that portion of the volume which results from the lambda transition (see text). (b) Heat capacity of quartz at 1 b compared to experimental data: W(u)—E. F. Westrum, Jr. (unpublished data); G79—Ghiorso et al. (1979); W19—White (1919); R82—Richet et al. (1982). The dotted line represents the 'lattice' heat capacity. (c) Heat capacity of quartz at 10 kb. Note that the magnitude of Cp_λ remains constant at elevated pressure, while the position of the onset of the transition moves to higher temperature.

to fit the measured volumes of α-quartz at 1 b, the *explicit* expansivity terms, v_3 and v_4 were fitted to the differences between the measured data and the volumes resulting from the lambda function. The same procedure was followed for other phases that undergo lambda transitions, and where there is information regarding the $P-T$ slope of the transition. Where this information is lacking, or where the data suggest a near-vertical slope (see the discussion of tridymite), it has been assumed that $(dT/dP) = 0$, so that $T_d = 0$, and therefore $V_\lambda = 0$.

The following points should be kept in mind in using the above equations and data presented below for calculation of the thermodynamic properties of polymorphic transitions. For those phases which undergo lambda transitions which are dependent on pressure, thermodynamic properties calculated for the low temperature polymorph with the above equations are valid up to T_λ^P. Thermodynamic properties at temperatures above the lambda point are computed using the standard state properties and 'lattice' Cp of the high temperature polymorph. The latter properties were derived from phase equilibrium constraints on the $P-T$ position of the transition. That such properties can be extracted is a necessary and crucial test of the consistency of the thermodynamic properties of both polymorphs, because most polymorph transitions involve very small entropy changes and thus small inconsistencies between the Gibbs free energy functions of both polymorphs result in large displacements or pronounced curvature of the computed transitions.

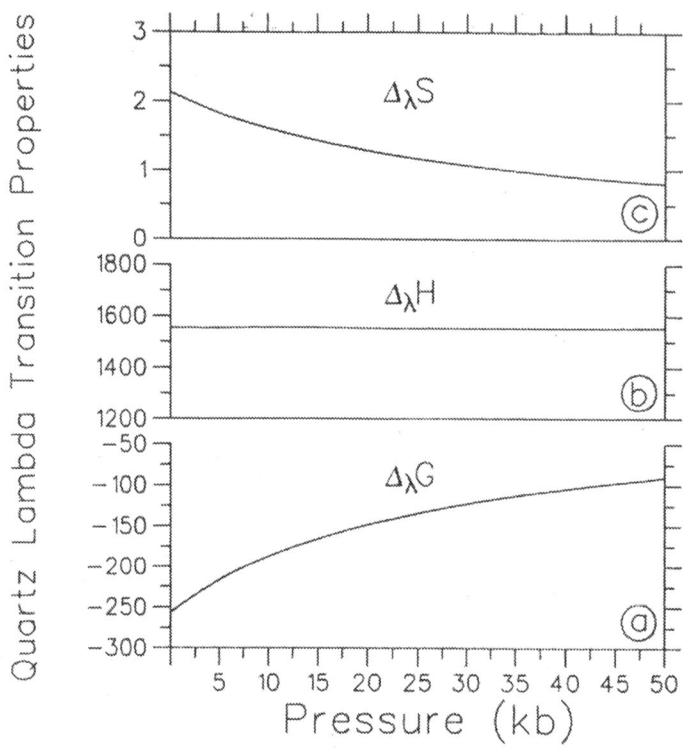

FIG. 2. Integral lambda transition properties (Jmol^{-1}) of quartz, calculated at the lambda points at each pressure.

For those phases that undergo lambda transitions which are assumed to be independent of P, standard state properties are provided only for the low temperature polymorph. The free energy function at temperatures above T_λ^P can be computed with

$$\Delta_a G^{P,T} = \Delta_a G^{P,T_\lambda} - (T - T_\lambda^P) \Delta_\lambda S^{P,T_\lambda} \tag{13}$$

where $\Delta_\lambda S^{P,T_\lambda}$ is given by eqn (11). Commonly, the best representation of Cp and heat content data for lambda transition phases is achieved if a first-order transition is assumed to be superimposed on the lambda transition. In such cases $\Delta_t H$ (over and above $\Delta_\lambda H$) is non-zero (499 Jmol^{-1} results from Berman & Brown's (1985) analysis of the α–β quartz transition), and the free energy function for the high temperature polymorph is computed from:

$$\Delta_a G^{P,T} = \Delta_a G^{P,T_\lambda} - (T - T_\lambda^P)(\Delta_t S + \Delta_\lambda S^{P,T_\lambda}). \tag{14}$$

Order–disorder

A number of important rock-forming minerals have disordered cation distributions, but the extent and temperature dependence of the disordering process are presently not well known for most phases. This problem has been treated here in two ways. Where the temperature dependence of this process is not constrained by available data, allowance has been made for a constant 'zero-point' contribution to the third law entropy. Where such data exist, or where representation of phase equilibrium data is improved by consideration of temperature dependent disordering, an approach has been adopted that is similar to that proposed by Helgeson *et al.* (1978). Here the temperature dependence of this process are accounted for by fitting the integrated form of an extended Cp function to the enthalpy differences between ordered and disordered phases. For $t < T < T_D$, where t is a reference temperature which marks the onset of disordering, and T_D is the temperature at which a phase is fully disordered,

$$Cp_{ds} = d_0 + d_1 T^{-0.5} + d_2 T^{-2} + d_3 T + d_4 T^2. \tag{15}$$

The enthalpy and entropy effects of disorder are computed by integrating (15) between t and T ($T < T_D$):

$$\Delta_{ds} H^T = d_0(T-t) + 2d_1(T^{0.5} - t^{0.5}) - d_2(T^{-1} - t^{-1})$$
$$+ d_3(T^2 - t^2)/2 + d_4(T^3 - t^3)/3 \tag{16}$$

$$\Delta_{ds} S^T = d_0(\ln T - \ln t)$$
$$- 2d_1(T^{-0.5} - t^{-0.5}) - d_2(T^{-2} - t^{-2})/2$$
$$+ d_3(T-t) + d_4(T^2 - t^2)/2 \tag{17}$$

while the volume and free energy changes are calculated by:

$$\Delta_{ds} V^T = \Delta_{ds} H^T / d_5 \tag{18}$$

$$\Delta_{ds} G^{P,T} = \Delta_{ds} H^T - T \Delta_{ds} S^T + \Delta_{ds} V^T (P - P_r). \tag{19}$$

where d_5 is a constant computed in such a way as to scale the disordering enthalpy to the volume of disorder. The apparent Gibbs free energy of the disordered phase is computed by adding the $\Delta_{ds} G^T$ values to the free energy function given by eqn (1). At temperatures above T_D, eqns (16) and (17) are integrated between t and T_D, and the free energy function is computed from

$$\Delta_a G^{P,T} = \Delta_a G^{P,T_D} - (T - T_D) \Delta_{ds} S^{T_D}. \tag{20}$$

An enlarged and updated internally consistent thermodynamic dataset with uncertainties and correlations: the system K_2O–Na_2O–CaO–MgO–MnO–FeO–Fe_2O_3–Al_2O_3–TiO_2–SiO_2–C–H_2–O_2

T. J. B. HOLLAND
Department of Earth Sciences, University of Cambridge, Downing Street, Cambridge CB2 3EQ, UK

R. POWELL
Department of Geology, University of Melbourne, Parkville, Victoria 3052, Australia

ABSTRACT We present, as a progress report, a revised and much enlarged version of the thermodynamic dataset given earlier (Holland & Powell, 1985). This new set includes data for 123 mineral and fluid end-members made consistent with over 200 P–T–X_{CO_2}–f_{O_2} phase equilibrium experiments. Several improvements and advances have been made, in addition to the increased coverage of mineral phases: the data are now presented in three groups ranked according to reliability; a large number of iron-bearing phases has been included through experimental and, in some cases, natural Fe:Mg partitioning data; H_2O and CO_2 contents of cordierites are accounted for with the solution model of Kurepin (1985); simple Landau theory is used to model lambda anomalies in heat capacity and the Al/Si order–disorder behaviour in some silicates, and Tschermak-substituted end-members have been derived for iron and magnesium end-members of chlorite, talc, muscovite, biotite, pyroxene and amphibole.

For the subset of data which overlap those of Berman (1988), it is encouraging to find both (1) very substantial agreement between the two sets of thermodynamic data and (2) that the two sets reproduce the phase equilibrium experimental brackets to a very similar degree of accuracy. The main differences in the two datasets involve size (123 as compared to 67 end-members), the methods used in data reduction (least squares as compared to linear programming), and the provision for estimation of uncertainties with this dataset. For calculations on mineral assemblages in rocks, we aim to maximize the information available from the dataset, by combining the equilibria from all the reactions which can be written between the end-members in the minerals. For phase diagram calculations, we calculate the compositions of complex solid solutions (together with P and T) involved in invariant, univariant and divariant assemblages. Moreover we strongly believe in attempting to assess the probable uncertainties in calculated equilibria and hence provide a framework for performing simple error propagation in all calculations in THERMOCALC, the computer program we offer for an effective use of the dataset and the calculation methods we advocate.

Key words: internally consistent thermodynamic data set; phase equilibrium calculations; thermodynamic data

List of symbols

$\Delta_f H_{1,298}$ = enthalpy of formation from the elements at 1 bar and 298 K. Units: kJ mol^{-1}
$S_{1,298}$ = entropy at 1 bar and 298 K. Units: kJ K^{-1} mol^{-1}.
$V_{1,298}$ = volume at 1 bar and 298 K. Units: kJ kbar^{-1} mol^{-1}.
C_p = heat capacity; $C_p = a + bT + cT^{-2} + dT^{-1/2}$. Units: kJ K^{-1} mol^{-1}.
α = coefficient of thermal expansion (usually grouped with V as αV). Units: K^{-1}.
β = coefficient of isothermal compressibility (usually grouped with V as βV). Units: kbar^{-1}.
$\Delta G°$ = Gibbs free energy change for a reaction among pure end-member phases at the pressure and temperature of interest. Units: kJ mol^{-1}.
R = the gas constant (0.0083143 kJ K^{-1} mol^{-1}).
T = absolute temperature. Units: K.
P = pressure. Units: kbar.
K = equilibrium constant for a balanced reaction. Units: dimensionless

INTRODUCTION

In previous publications (Powell & Holland, 1985; Holland & Powell, 1985; Powell & Holland, 1988: henceforth DS1, DS2 and DS3, respectively) we have presented an internally consistent thermodynamic dataset for end-members of rock-forming minerals, methods of application

Table 1. Formulae of end-members in the dataset, in alphabetical order. In the list, V denotes a site vacancy. The different sites in the mineral formulae are separated and, in some cases, parentheses are used to clarify the grouping of elements which share sites, with tetrahedral sites denoted usually by square brackets. For certain end-members (denoted by *) mixing-on-sites activities according to the above formulae were used in the data extraction process.

ab	equilibrium albite	$Na[AlSi_3]O_8$
abh	high albite	$Na[AlSi_3]O_8$
acm	acmite	$NaFe[Si_2]O_6$
ak	akermanite	$Ca_2[Si_2Mg]O_7$
alm	almandine	$Fe_3Al_2Si_3O_{12}$
ames	amesite	$Mg_4(Al_2)Si_2[Al_2]O_{10}(OH)_8$*
an	anorthite	$Ca[Al_2Si_2]O_8$
and	andalusite	Al_2SiO_5
andr	andradite	$Ca_3Fe_2Si_3O_{12}$
ann	annite	$KFe(Fe_2)Si_2[SiAl]O_{10}(OH)_2$*
anth	anthophyllite	$VMg_2Mg_3(Mg_2)Si_4[Si_4]O_{22}(OH)_2$
arag	aragonite	$CaCO_3$
bq	β-quartz	SiO_2
br	brucite	$Mg(OH)_2$
cats	Ca-Tschermak's pyroxene	$CaAl[SiAl]O_6$
cc	calcite	$CaCO_3$
cel	celadonite	$KV(MgAl)Si_2[Si_2]O_{10}(OH)_2$*
chr	chrysotile	$Mg_3Si_2O_5(OH)_4$
clin	clinochlore	$Mg_4(MgAl)Si_2[AlSi]O_{10}(OH)_8$*
coe	coesite	SiO_2
cor	corundum	Al_2O_3
crd	cordierite	$Mg_2[Al_4Si_5]O_{18}$
cumm	cummingtonite	$VMg_2Mg_3(Mg_2)Si_4[Si_4]O_{22}(OH)_2$
cz	clinozoisite	$Ca_2AlAl_2Si_3O_{12}(OH)$*
daph	daphnite	$Fe_4(FeAl)Si_2[AlSi]O_{10}(OH)_8$*
deer	deerite	$Fe_{12}Fe_6Si_{12}O_{40}(OH)_{10}$
di	diopside	$CaMg[Si_2]O_6$
di.o	Pbca-diopside	$CaMg[Si_2]O_6$
dia	diaspore	$AlOOH$
diam	diamond	C
dol	dolomite	$CaMg(CO_3)_2$
east	eastonite	$KMg(MgAl)Si_2[Al_2]O_{10}(OH)_2$*
ed*	edenite	$NaCa_2Mg_3(Mg_2)Si_4[Si_3Al]O_{22}(OH)_2$*
en	enstatite	$MgMg[Si_2]O_6$
en.c	C2/c-enstatite	$MgMg[Si_2]O_6$
ep	epidote	$Ca_2FeAl_2Si_3O_{12}(OH)$*
fa	fayalite	Fe_2SiO_4
fame	Fe-amesite	$Fe_4(Al_2)Si_2[Al_2]O_{10}(OH)_8$*
fath	Fe-anthophyllite	$VFe_2Fe_3(Fe_2)Si_4[Si_4]O_{22}(OH)_2$
fcar	Fe-carpholite	$FeAl_2Si_2O_6(OH)_4$
fcel	Fe-celadonite	$KV(FeAl)Si_2[Si_2]O_{10}(OH)_2$*
fcrd	Fe-cordierite	$Fe_2[Al_4Si_5]O_{18}$
fctd	Fe-chloritoid	$FeAl_2SiO_5(OH)_2$
fdol	Fe-dolomite	$CaFe(CO_3)_2$
fgl	Fe-glaucophane	$VNa_2Fe_3(Al_2)Si_4[Si_4]O_{22}(OH)_2$
fhb	Fe-hornblende	$VCa_2Fe_3(FeAl)Si_4[Si_3Al]O_{22}(OH)_2$*
fo	forsterite	Mg_2SiO_4
fs	ferrosilite	$FeFe[Si_2]O_6$
fst	Fe-staurolite	$Fe_4Al_{18}Si_{7.5}O_{48}H_4$
fta	Fe-talc	$Fe_2FeSi_2[Si_2]O_{10}(OH)_2$*
ftat	Fe-Tschermak's talc	$Fe_2AlSi_2[SiAl]O_{10}(OH)_2$*
ftr	Fe-tremolite	$VCa_2Fe_3(Fe_2)Si_4[Si_4]O_{22}(OH)_2$
geh	gehlenite	$Ca_2[Al_2Si]O_7$

Table 1. (Continued.)

gl	glaucophane	$VNa_2Mg_3(Al_2)Si_4[Si_4]O_{22}(OH)_2$
gph	graphite	C
gr	grossular	$Ca_3Al_2Si_3O_{12}$
grun	grunerite	$VFe_2Fe_3(Fe_2)Si_4[Si_4]O_{22}(OH)_2$
hb	hornblende	$VCa_2Mg_3(MgAl)Si_4[Si_3Al]O_{22}(OH)_2$*
hed	hedenbergite	$CaFe[Si_2]O_6$
hem	hematite	Fe_2O_3
herc	hercynite	$FeAl_2O_4$
ilm	ilmenite	$FeTiO_3$
iron	iron	Fe
jd	jadeite	$NaAl[Si_2]O_6$
kals	kalsilite	$K[AlSiO_4]$
ksp	equilibrium K-feldspar	$K[AlSi_3]O_8$
ky	kyanite	Al_2SiO_5
law	lawsonite	$CaAl_2Si_2O_7(OH)_2\cdot H_2O$
lc	leucite	$K[AlSi_2]O_6$
lime	lime	CaO
ma	margarite	$CaV(Al_2)Si_2[Al_2]O_{10}(OH)_2$
mag	magnesite	$MgCO_3$
mang	manganosite	MnO
mcar	Mg-carpholite	$MgAl_2Si_2O_6(OH)_4$
mctd	Mg-chloritoid	$MgAl_2SiO_5(OH)_2$
me	meionite	$Ca_4CO_3[Si_6Al_6]O_{24}$
merw	merwinite	$Ca_3MgSi_2O_8$
mgts	Mg-Tschermak's pyroxene	$MgAl[SiAl]O_6$
mont	monticellite	$CaMgSiO_4$
mrb	magnesioriebeckite	$VNa_2Mg_3(Al_2)Si_4[Si_4]O_{22}(OH)_2$
mst	Mg-staurolite	$Mg_4Al_{18}Si_{7.5}O_{48}H_4$
mt	magnetite	$FeFe_2O_4$
mu	muscovite	$KV(Al_2)Si_2[SiAl]O_{10}(OH)_2$*
naph	Na-phlogopite	$NaMg(Mg_2)Si_2[SiAl]O_{10}(OH)_2$
ne	nepheline	$Na[AlSi]O_4$
pa	paragonite	$NaV(Al_2)Si_2[SiAl]O_{10}(OH)_2$
parg	pargasite	$NaCa_2Mg_3(MgAl)Si_4[Si_2Al_2]O_{22}(OH)_2$*
per	periclase	MgO
phl	phlogopite	$KMg(Mg_2)Si_2[SiAl]O_{10}(OH)_2$*
pre	prehnite	$Ca_2AlSi_2[SiAl]O_{10}(OH)_2$
pswo	pseudo-wollastonite	$CaSiO_3$
pump	pumpellyite	$Ca_4Al_4(MgAl)Si_6O_{21}(OH)_7$
pxmn	pyroxmangite	$MnSiO_3$
py	pyrope	$Mg_3Al_2Si_3O_{12}$
pyhl	pyrophyllite	$Al_2Si_4O_{10}(OH)_2$
q	α-quartz	SiO_2
rhc	rhodochrosite	$MnCO_3$
rhod	rhodonite	$MnSiO_3$
rnk	rankinite	$Ca_3Si_2O_7$
ru	rutile	TiO_2
san	sanidine	$K[AlSi_3]O_8$
sdph	siderophyllite	$KFe(FeAl)Si_2[Al_2]O_{10}(OH)_2$*
sid	siderite	$FeCO_3$
sill	sillimanite	Al_2SiO_5
sp	spinel	$MgAl_2O_4$
sph	sphene	$CaTiSiO_5$
spu	spurrite	$Ca_4Si_2O_8\cdot CaCO_3$
ta	talc	$Mg_2MgSi_2[Si_2]O_{10}(OH)_2$*
tats	Tschermak's talc	$Mg_2AlSi_2[SiAl]O_{10}(OH)_2$*
teph	tephroite	Mn_2SiO_4
tr	tremolite	$VCa_2Mg_3(Mg_2)Si_4[Si_4]O_{22}(OH)_2$
ty	tilleyite	$Ca_3Si_2O_7\cdot (CaCO_3)_2$
usp	ulvospinel	$Fe(FeTi)O_4$
vsv	vesuvianite	$Ca_{19}Mg(MgAl_7)Al_4Si_{18}O_{69}(OH)_9$
wo	wollastonite	$CaSiO_3$
zo	zoisite	$Ca_2AlAl_2Si_3O_{12}(OH)$
CH4		CH_4
CO		CO
CO2		CO_2
H2		H_2
H2O		H_2O
O2		O_2

of the dataset, and a computer program (THERMOCALC), which performs calculations on the conditions of equilibration of mineral assemblages in rocks and allows the construction of phase diagrams. As well as being internally consistent, having been generated from high temperature–pressure mineral equilibria experiments, that dataset was in agreement with virtually all the high-temperature calorimetric data available. Thus, with confidence, we could predict mineral equilibria for reactions which have not been determined experimentally; moreover, we were able to estimate the likely uncertainties associated with such predictions, an absolute prerequisite in any (thermodynamic) calculations.

The previously published dataset was presented as a core (DS2, p. 350) of more reliable thermodynamic data, and was intended as a preliminary summary at an early stage in the longer term project. Natural rock assemblages are capable of yielding more information on their conditions of formation if reliable thermodynamic data exist for a larger set of mineral end-members, because it is then possible to define more independent conditions of chemical equilibrium in the mineral assemblage; thus the statistical advantage of the increased information available may be used in assessing *both* the conditions of formation *and* the degree of equilibrium attained in rocks (DS3, p. 190). Moreover, with a larger set of mineral end-members, for phase diagram calculations, we can now calculate the compositions of complex solid solutions (and P,T) involved in-, uni- and divariant equilibria, rather than being restricted to 'end-member' phase diagrams.

In our earlier dataset we presented data for 43 end-members (DS2 appendix A). In this work, which is essentially a progress report, we present new thermodynamic data for a further 80 end-members of common rock-forming minerals, bringing the total number to 123, a near-threefold increase over the original size. The mineral formulae and abbreviations used for the new set are given in Table 1.

THE ENLARGED AND UPDATED DATASET

We have built upon the principles and methods outlined in DS1 and 2; the details will be discussed only where they differ from those presented there, and only a brief resumé of the assumptions made will be given here. For a balanced chemical reaction the equilibrium relation is

$$0 = \Delta G° + RT \ln K.$$

For each end-member mineral phase the free energy contribution to $\Delta G°$ is

$$\Delta_f H_{1,298} - TS_{1,298} + \int_{298}^{T} C_p \, dT - T \int_{298}^{T} \frac{C_p}{T} dT + [V_{1,298} + \alpha V(T - 298)]P - \frac{\beta V}{2} P^2.$$

For fluid species end-members the volume–pressure integrals are replaced by the term $RT \ln f$, which may be evaluated from the equations given in Table 2.

Table 2. H_2O and CO_2 fugacity polynomials expressed as $RT \ln f$, in units of kJ mol^{-1}; T in Kelvin and P in kilobars.

$RT \ln f = a + bT + cT^2$

$a = a_1 + a_2 P + a_3 P^2 + a_4 P^{-1} + a_5 P^{-2}$
$b = b_1 + b_2 P + b_3 P^{-1} + b_4 P^{-2} + b_5 P^{-1/2} + b_6 P^{-3}$
$c = c_1 + c_2 P + c_3 P^{-2} + c_4 P^{-1/2} + c_5 P^{-1} + c_6 P^{-3}$

	CO_2		H_2O
a_1	−10.094	a_1	−40.338
a_2	2.6993	a_2	1.6474
a_3	−0.016983	a_3	−0.0062115
a_4	3.2804	a_4	2.0068
a_5	−0.24010	a_5	0.0562929
b_1	0.098597	b_1	0.117372
b_2	0.000865307	b_2	0
b_3	0.0161904	b_3	0
b_4	−0.0021663	b_4	−0.00046710
b_5	−0.046448	b_5	0
b_6	0.000143064	b_6	0
c_1	−3.6780 × 10^{-6}	c_1	−7.3681 × 10^{-6}
c_2	−1.189 × 10^{-7}	c_2	1.10295 × 10^{-7}
c_3	4.2467 × 10^{-7}	c_3	−9.8774 × 10^{-7}
c_4	0	c_4	−2.4819 × 10^{-5}
c_5	0	c_5	8.2948 × 10^{-6}
c_6	−3.9520 × 10^{-8}	c_6	8.33667 × 10^{-8}

Where mixed volatiles are involved (mixtures of H_2O and CO_2) we assume a non-ideal mixing relation expressed as a subregular model (DS1, p. 332) in order to calculate the activities of the fluid species at high pressure and temperature. As explained in DS1, we make the assumption that all the quantities in the above expressions are either well-known or can be estimated with relatively high precision, and that the term which we need to determine from phase equilibrium experiments is the enthalpy of formation of each mineral end-member.

For each reaction we have determined enthalpy of reaction brackets corresponding to the experimental P–T (X_{CO_2}, f_{O_2}) brackets. These individual enthalpy of reaction brackets are then processed to give an overall enthalpy of reaction bracket for each reaction (DS1, p. 333–334). The central value and with the width of these overall enthalpy of reaction brackets are used to set up a least squares problem. The goal is to determine a unique set of mineral enthalpies which reproduce, within error, the set of enthalpy brackets discussed above, and hence satisfy the original P–T (X_{CO_2}, f_{O_2}) brackets.

Changes to data extraction methods

As outlined above and in DS1, experimental brackets, such as temperature brackets at a series of pressures, are converted to enthalpy of reaction brackets, assuming that the remaining necessary thermodynamic data are relatively well-known. These enthalpy brackets are converted to one enthalpy bracket, representing this reaction, for inclusion in the least squares analysis. The less robust method of conversion advocated in DS1 has been replaced by taking the median of both the low-enthalpy ends of the brackets and of the high-enthalpy ends of the brackets, the average of these medians giving the preferred value of the enthalpy. The estimated standard deviation on the preferred enthalpy of reaction is taken as a quarter of the difference between the high and low bracket ends.

Following the appearance of criticisms in the literature of the least squares (LS) approach (Berman, 1988; Berman & Brown, 1987), it is worth reiterating and defending the advantages and validity of the LS method in processing these cumulative enthalpy of reaction brackets (see also DS1, p. 388). One of the main differences between the linear programming (LP) method and our use of LS is that in the LP method, *each* experimental bracket is treated separately, with the fit of the bracket being unconstrained within the bracket; in our method a *cumulative* bracket is represented by its centre, being weighted in the least squares sense according to the bracket width. Thus, the criticism of the least squares method is that it might tend to force the fit towards the centre of the bracket. We feel that this criticism is not justified partly because we use a cumulative bracket and also because the probabilities associated with the experimental bracket are unlikely to be as simple as assumed by the linear programmers. Whereas they assume an equal probability of a result lying anywhere within an experimental bracket, there is a finite probability that the result lies outside the bracket, or indeed, that the bracket is too wide; this is because the limits of experimental brackets are subject to experimental error. Moreover, from kinetic considerations, the value is less likely to be very close to the limits of the experimental bracket.

The probability curve is thus more likely to be bell-shaped, with a flat top whose width relates to the width of the bracket in relation to the experimental uncertainties (Demarest & Haselton, 1981). Indeed, ignoring the fuzziness associated with the limits of experimental brackets is one reason why the linear programmers have had to widen more or less arbitrarily the experimental brackets in order to create a feasible region in their analyses. In our approach of combining the enthalpy of reaction brackets into a cumulative bracket, the probability curve for this overall enthalpy of reacton bracket is given by multiplying together the probability curves of the component enthalpy of reaction brackets. The resulting probability distribution, even where the individual component brackets may have flat-topped distributions, will tend to be bell-shaped with only a narrow flat top or, more probably, without a flat top. Moreover, thinking about the cumulative enthalpy bracket in another way, the boundaries of the bracket are uncertain, not only because of the spread of values used for the medians (see earlier), but also because each of these values is uncertain from the experimental uncertainties. Again, the probability curve tends to be bell-shaped. As long as this probability curve does not have a flat top which is wide compared with the bracket width, the use of least squares will be valid. If the width of the bracket is not much wider than the experimental uncertainty on its ends (i.e. $d/s < 2$ in the nomenclature of Demarest & Haselton, 1981; DS1, p. 334) use of the least squares method is certainly justified because the probability distribution is close to Gaussian. In the diagnostic d/s referred to above, d is the half-width of the overall enthalpy of reaction bracket and s is the uncertainty in the location of the bracket ends. Moreover the use of the bracket with width as a weighting in the least squares will tend to minimize the damage caused by a flat top on the probability curve when $d/s \approx 2$ or even $d/s > 2$: the larger the bracket width, the more down-weighted is the reaction, and the less the reaction can contribute to the least squares analysis anyway.

The use of regression diagnostics (Belsley, Kuh & Welsch, 1980) together with new diagnostics discussed below can signal the forcing of a fitted enthalpy to be central to a bracket; in fact, no reaction with large d/s is indicated in Table 8. The majority of the reactions in Table 5 have d/s values less than 2, so the least squares method does appear to work with few problems. We feel that the criticisms of LS are largely illusory because in our experience the method has been demonstrably successful in fitting almost all the available experimental data within their uncertainty limits; additionally, the calculated error bounds at the 2σ level for individual reactions are very similar to the experimental bracket widths. Both LS and LP are capable of providing reliable and similar analyses of the experimental data, each with its advantages and disadvantages. However, we believe that the ability of the least squares method to provide the uncertainties, and their mutual correlations, of the calculated thermodynamic data makes this the preferable method of analysis.

Several levels of reliability are now involved in the data used in the least squares analysis, in particular, for reactions involving phases whose entropies cannot be estimated reliably because of, for instance, Al-Si disorder. For these reactions the overall enthalpy of reaction bracket has been arbitrarily widened, by a factor of up to 3, thus down-weighting every such experimental constraint.

As an aid in the data extraction process, some regression diagnostics (for a general discussion of this topic see Belsey *et al.*, 1980), specifically designed for this analysis, have been used. The sensitivity of the estimated enthalpy of formation of an end-member, $\Delta_f H$, to the enthalpy of the reaction, ΔH_R, is examined in three ways:

(a) via the change in $\Delta_f H$ which is a consequence of doubling the width of a ΔH_R bracket, scaled to an 'expected' $\sigma(\Delta_f H)$ using 0.15 kJ/atom. Rather than give all the values, only the values above a cut-off are presented; a cut-off of 0.25 is used;

(b) via the change in the uncertainty on $\Delta_f H$, $\sigma(\Delta_f H)$, which is a consequence of doubling the width of a ΔH_R bracket, scaled to $\sigma(\Delta_f H)$ for the undoubled ΔH_R bracket. A cut-off of 0.05 is used for the absolute value of this diagnostic;

(c) via the change in $\Delta_f H$ on moving a ΔH_R bracket by one quarter of its width, scaled to an 'expected' $\sigma(\Delta_f H)$, using 0.15 kJ/atom, and using a cut-off of 0.25.

All of these diagnostics can be calculated from intermediate results and the least squares solution itself, and are presented in Table 8. The algebraic derivation of these diagnostics is available from RP on request. These diagnostics have been valuable in signalling potential (and real) problems in individual experimental studies. For example, large values for diagnostics (a) and (c) indicate that the enthalpies of these end-members are strongly influenced by the reaction, while a large positive value for diagnostic (b) indicates that the (small) size of the uncertainty on the enthalpies of these end-members is strongly influenced by this reaction. Large values for diagnostic (a) may indicate aberrance of this reaction or competing influences (for instance calorimetric constraints in the case of diam = gph; or the influence of other experimental data on the calculated position of en + 2mag = 2fo + 2CO_2). Large values for diagnostics (b) and (c) indicate different aspects of the importance of these reactions for the enthalpies of the corresponding end-members, and do not indicate aberrance.

Changes to the existing data

A few changes have been made which slightly alter the results presented in DS2. The fugacity polynomials for CO_2 and H_2O have been improved so that they both cover an extended range (0.1–40 kbar, 300–1200 °C for H_2O and 0.1–k40 kbar, 300–1400 °C for CO_2). The new equation and polynomial terms are given in Table 2. Apart from the extended range of validity, the equations yield essentially identical results to those discussed in detail and displayed graphically in DS1.

The entropy of clinochlore has been measured calorimetrically (Henderson, Essene, Anovitz, Westrum, Hemingway & Bowman, 1983) and is much lower than the value used in DS2. Although the value used there was consistent with the mineral equilibrium studies used in that least squares analysis, that value is inconsistent with the slopes of a large number of reactions involving cordierite, as pointed out by Berman (1988) in his recent review. In contrast, the new value is consistent with these reactions. The entropy and heat capacity of grossular have also been changed; we now use, following the arguments of Berman (1988), the data measured for natural garnet in preference to those for synthetic garnet, although the differences are minor. The entropy of meionite has been reduced to 720 J K^{-1} to make it consistent with the Na/Ca exchange equilibria determined experimentally by Goldsmith & Newton (1977).

Other subtle differences between the dataset in DS2 and that presented here are induced by the change in the method of calculation of the enthalpy of reaction brackets, outlined above, and by the least squares analysis of the new experimental and calorimetric constraints.

Anchors and calorimetric data

In DS2, it was regarded as crucial to incorporate the minimum of calorimetric data in the least squares analysis, so that as much as possible of these data, particularly from high-T oxide-melt calorimetry, might provide an independent test of the dataset. This minumum of calorimetric data involved just one end-member for each chemical component of the dataset system; these were the so-called *anchors* of the dataset. With the demonstration of the consistency of the phase equilibrium data with most of the calorimetric measurements (Holland & Powell, 1985;

Table 3. Direct calorimetry constraints used in the fitting process. The data were taken from the compilation of Robie, Hemingway & Fisher 1979, with the exception of anorthite from Carpenter, McConnell & Navrotsky 1985. Units: kJ mol^{-1}.

	$\Delta H_f(1,298)$	σ
ksp	−3967.70	3.30
san	−3959.56	1.68
an	−4231.90	2.50
jd	−3029.94	2.10
fa	−1479.40	2.40
br	−924.50	1.00
dia	−1000.60	3.00
pyhl	−5643.76	4.40
cc	−1207.40	1.70
arag	−1207.40	1.40
mag	−1113.30	1.30
dol	−2324.50	1.70
mt	−1115.73	2.10
lime	−635.09	0.88
ru	−944.75	0.63
per	−601.49	0.30
mang	−385.22	0.46
cor	−1675.70	0.65
hem	−824.60	1.30
q	−910.70	0.50
iron	0.00	0.00
gph	0.00	0.00
diam	1.90	0.20
O_2	0.00	0.00
H_2	0.00	0.00

Berman, 1988), the anchor approach is no longer needed and most of the data from acid calorimetry and high-T oxide-melt calorimetry are now included as separate calorimetric constraints (Tables 3 and 4).

New data added to the dataset

We have added a further 80 end-members to the earlier core dataset, bringing the total to 123. Most of these could have been added using the methods of DS1 and 2; however, a few have required additions or modifications to the processing. Where entropies have not been measured, the new algorithm of Holland (1989) has been used to estimate third-law values. As in DS1, we have added the required minimum configurational increment to the third-law entropy in order to achieve agreement with experimental phase equilibria.

The dataset extraction now involves experimentally determined solid solution equilibria between minerals, particularly Tschermak (MgSi$Al_{-1}Al_{-1}$) and ferromagnesian (MgFe$_{-1}$) substitutions. In fact, we have found it necessary to include the Tschermak substitutions in talc, chlorite, muscovite, biotite, pyroxenes and calcic amphiboles in order to satisfy the experimental phase relations. In all such calculations ideal mixing on sites was assumed for mineral solid solutions according to the formulae in Table 1, with the exception of aluminous pyroxenes for which ideal charge balance was assumed. These aspects of the dataset are discussed at greater length in the next section.

Some iron-bearing minerals (iron, hem, mt) have lambda heat capacity anomalies at temperatures above 298 K due to magnetic disordering. In order to model the high-temperature thermodynamics, we have taken simple Landau theory for tricritical behaviour to describe the heat

Table 4. Oxide and reaction calorimetry. Units: kJ mol^{-1}. (a) Relative to oxides in borate melt solution calorimetry. Refs: Holm & Kleppa, 1966 (sill, and, ky); Charlu et al., 1975 (py, fo, crd); Brousse et al., 1984 (en, fo, ak, merw, mont); Charlu et al., 1978 (wo, pswo, di, gr); Charlu et al., 1981 (geh). (b) Relative to elements. Heats of formation from the elements from Robie et al., 1979. (c) Reaction 'calorimetry': the enthalpies of the first four reactions were taken to be zero with uncertainties of $\sigma = \pm 1$ kJ per mol FeMg$_{-1}$ exchange, thus assuming near linear dependence of end-members ftr, sdph, fame, ftat on the other end-members in the relevant reactions. Enthalpies for en.c = en and di.o = di are from Holland et al., 1979 and Lindsley, 1981.

(a) Relative to oxides.

Mineral	T(°C)	ΔH_{cal}	σ	ΔH_{298}
sill	697	−2.4	0.6	−0.8
py	697	−79.9	4.6	−79.4
fo	800	−59.5	1.9	−59.7
en	800	−67.8	3.5	−71.5
mont	800	−104.8	1.7	−105.3
ak	800	−178.2	1.6	−176.2
merw	800	−213.8	2.0	−211.4
crd	697	−68.1	3.0	−62.2
di	697	−146.4	1.7	−147.4
wo	697	−89.9	1.5	−89.8
pswo	697	−83.3	1.3	−82.4
geh	697	−128.2	1.3	−124.0
ky	701	−6.2	1.2	−7.9
and	701	−2.8	1.0	−3.9

(b) Relative to elements.

Reaction	T(°C)	ΔH_{cal}	σ
C + O$_2$ = CO$_2$	25	−393.5	0.1
H$_2$ + ½O$_2$ = H$_2$O	25	−241.8	0.0
C + 2H$_2$ = CH$_4$	25	−74.8	0.3
C + ½O$_2$ = CO	25	−110.5	0.2

(c) Reaction 'calorimetry'.

Reaction	T(°C)	ΔH_{298}	σ
4 ftr + 5 hb = 4 tr + 5 fhb	25	0.0	5.0
3 sdph + phl = 3 east + ann	25	0.0	3.0
5 fame + 4 clin = 5 ames + 4 daph	25	0.0	5.0
3 ftat + 2 ta = 3 tats + 2 fta	25	0.0	3.0
en.c = en	25	−6.1	2.0
di.o = di	25	−7.4	2.0

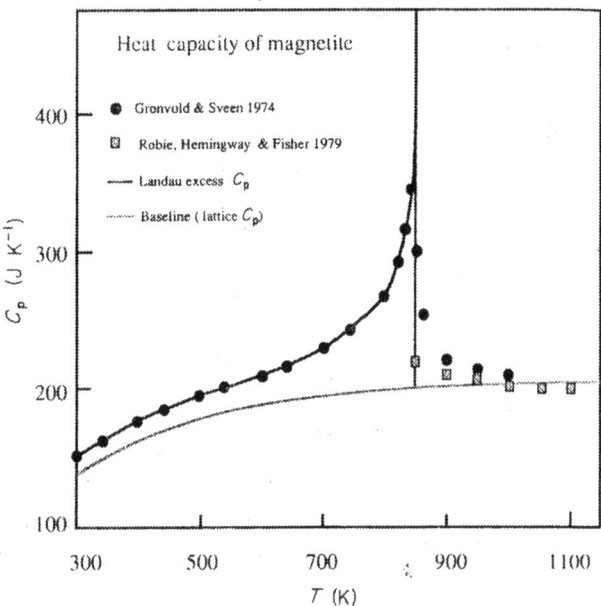

Fig. 1. Heat capacity of magnetite showing the lambda anomaly at 855 K and the Landau model used to represent it.

where T_c is the critical temperature of the phase transition, S_{max} is the maximum entropy due to the phase transition, and S^{ex} and H^{ex} are the entropy and enthalpy of disorder at any temperature T; Q is the macroscopic order parameter which is related to the free energy (of ordering) in conventional Landau terms by the expression

$$\Delta G^{ord} = -\frac{a}{2}(T_c - T)Q^2 + \tfrac{1}{6}cQ^6; \quad T_c = \frac{c}{a}; \quad S_{max} = \frac{a}{2}.$$

The Landau terms, S_{max} and T_c, are given in Table 7. The same Landau formalism can be used to describe Al–Si order–disorder, as discussed by Carpenter (1988); here we use it to describe the equilibrium behaviour of albite (ab), potassium feldspar (ksp), nepheline (ne) and leucite (lc). For ne and lc the Landau parameters were retrieved from measured heat capacities, whereas those for ab and ksp were taken from Carpenter's estimates of T_c, coupled with S_{max}, taken to be about three-quarters of the maximum configurational entropy. For albite, this simplification approximates the behaviour shown by Salje, Kuscholke, Wruck & Kroll (1985), and is quite adequate for most phase equilibrium calculations.

Fe/Mg partitioning experiments among pairs of silicate minerals have been used to extract enthalpies for a number of Fe-silicate and carbonate end-members (see Appendix A for details). For certain important mineral end-members no phase equilibrium experiments are available, and we have resorted to using Fe/Mg or Al/Fe^{3+} partitioning found in nature, together with reasonable estimates for temperature and pressure of formation, to retrieve enthalpies of formation. While this has proved a useful preliminary procedure, the data derived in this way should not be used to estimate rock temperatures through exchange thermometry, because of the obvious circular reasoning. However, these data may be quite robust in

capacity lambda anomalies. While this is only approximate, and an empirical device (it fails to explain the small excess heat capacity just above the critical temperature), the Landau model is both successful and practical in accounting for such heat capacity anomalies. A typical example, for magnetite, is shown in Fig. 1.

Although Landau theory is usually used to describe departure from the high-temperature, disordered state, we will use the petrological convention and describe the excess properties relative to the low-temperature, ordered state:

for $T < T_c$ $C_p^{ex} = \dfrac{TS_{max}}{2\sqrt{T_c}}(T_c - T)^{-1/2}$

$S^{ex} = S_{max}(1 - Q^2)$

$H^{ex} = 2S_{max}T_c\left(\dfrac{Q^6}{6} - \dfrac{Q^2}{2} + \dfrac{1}{3}\right)$,

where $Q = \left(1 - \dfrac{T}{T_c}\right)^{1/4}$

for $T > T_c$, $S_{ex} = S_{max}$, $H_{ex} = \tfrac{2}{3}S_{max}T_c$, $Q = 0$, calculations of net transfer reactions and for average pressure calculations on rocks, as described in DS3.

An example of the potential power of this line of reasoning is the extraction of thermodynamic data for siderite and ferrodolomite. The experimental data consist of phase equilibrium experiments on the reactions sid + hem = mt + CO_2 and sid = mt + gph + CO_2; the low-pressure experiments of French (1971) and Weidner (1972) are mutually inconsistent as well as not being in agreement with the measured entropy and volume data (Appendix A), whereas Weidner's high-pressure brackets do agree with the calculated $P-T$ slopes. In an attempt to reconcile these data with the experimental partitioning of Fe and Mg between siderite and ankerite (again two mutually inconsistent studies, Goldsmith, Witters & Northrup, 1962 and Rosenberg 1967; for a useful discussion see Anovitz & Essene 1987), we used natural Fe:Mg partitioning data for pyroxene/ankerite pairs (Klein 1978) and siderite/ankerite pairs (Anovitz & Essene 1987) to help discriminate among the available competing sets of experimental data. The result is that the natural data are in excellent agreement with the experimental data of Rosenberg (1987) and the high-pressure experiments (>4 kbar) of Weidner (1972) but are inconsistent with the experimental results of Goldsmith, Witters & Northrup (1962) and French (1971). Furthermore, the Fe:Mg partitioning data for natural dolomite/diopside pairs (Rice 1977) and dolomite/biotite pairs (Klein 1978) are in very good agreement with these results and the P,T location of our siderite breakdown reactions is almost identical to that deduced by Miyano & Klein (1986). Thus it is possible to use natural mineral compositions together with good estimates of equilibration temperatures to help constrain thermodynamic data and to discriminate among inconsistencies in experimental data.

The new dataset contains data for anhydrous Mg- and Fe-cordierite, derived from experimentally determined phase equilibria and the simple model for hydration and carbonation in cordierite described by Newton & Wood (1979) and extended by Kurepin (1985). Thus the reduction of the activity of anhydrous cordierite end-members in hydrous experimentally produced cordierites could be calculated. In the absence of data, we have assumed that Fe-cordierite behaves identically to Mg-cordierite in its uptake of volatiles, and have taken the equations directly from Kurepin (1985).

In this dataset we provide two sets of clinozoisite data cz and czl. Data for czl is essentially that of Holland & Powell (1985) and is based on the experiments of Jenkins, Newton & Goldsmith (1983). The data for cz is based on the experiments of Holland (1984) where kinetic extrapolation at 15 kbar leads to the conclusion that the cz/zo transformation occurs at around 500 °C (as opposed to <200 °C in Jenkins *et al.*, 1983). At present this problem is unresolved and we suggest that both alternatives be used in calculations on natural rocks, with a view to comparing and evaluating the effects of the two different assumptions. We have found good agreement in barometry with *cz* (as opposed to *czl*) with our limited data, but this may not be everyone's experience.

CHAPTER 12: FLUID DRIVEN METAMORPHISM

Ferry, J.M. (1983) Regional metamorphism of the Vassalboro Formation, south-central Maine, U.S.A.: a case study of the role of fluid in metamorphic petrogenesis.
Journal of the Geological Society of London, **140**, 551–576.

Metamorphism of rocks has classically been attributed to the addition (or loss) of heat and changes in pressure. During the first half of the 20th century, many petrologists became convinced by field evidence that seemed to indicate widespread metasomatic alteration of rocks (gain or loss of non-volatile elements) and, ultimately, "granitization". Landmark high-PT hydrothermal experiments in the late 1950s and early 1960s blunted the granitization movement by showing that partial melts of granitic composition could form from quartzo-feldspathic metasediments at temperatures as low as 650°C at mid- to lower-crustal pressures. At the same time, it became clear that the scale of possible mass transfer in metamorphic rocks by solid diffusion (= intragranular or lattice diffusion) and intergranular diffusion was limited to distances on the order of mm, and cm to dm, respectively (Thompson 1959). The pneumatolytic and hydrothermal alteration associated with silicic intrusions, skarns and ore deposits was obviously a special case of rock alteration by a local fluid source, a process not necessarily relevant to regional metamorphism. An idea that gained ground in the 1960s (e.g. Orville, 1962; Helgeson, 1967) was the possibility of alteration on a regional scale driven by exchange of H^+, K^+, Na^+, Ca^{2+} and Mg^{2+} with an infiltrating hydrothermal fluid. However, it was unclear whether these ideas applied widely to 'ordinary' metamorphism, the kind that turns metasediments into schists on a regional scale, yields isograds and metamorphic zones, and satisfies the requirements of the metamorphic facies (Chapter 3). Aside from the loss of H_2O and CO_2, the conventional view had been (and still is) that metamorphism is by-and-large isochemical.

Field evidence for the regional transport of chemical species in metamorphic rocks by fluids is subtle. Careful petrological and geochemical detective work is required to determine its existence and nature: whole-rock chemical analyses, 'before and after' comparisons, possible changes close to lithological boundaries, presence of relics, and so on. In any given case, insight into a possible role for chemically reactive fluid in determining the minerals present in our rocks, and their compositions, can usually be derived by writing balanced equations for chemical equilibrium among the minerals, mineral components, and species suspected of being present in the pore fluid. This step can help identify which externally imposed changes in chemical potential or P and T might have induced mineral reaction. A classic example is the study by Beach and Fyfe (1972), who described metamorphic alteration (oxidation and K-metasomatism) along shear zones at Scourie resulting from the focused flow of a hydrothermal fluid. But as recently as 1987, the jury was still out on the scale of fluid movement and element transport in the deep crust (Bridgwater, 1987).

For much of the 20th century, the pore fluid in ordinary metamorphism was viewed as playing a somewhat passive role. Metamorphic H_2O and CO_2 fluids evolved from metasediments were believed to escape upwards under the influence of buoyancy, not so much pervasively as via widely spaced channelways (Fyfe *et al.*, 1978), in what has become known as the single-pass model for fluid flow (Walther and Orville, 1982). Retrograde metamorphism had long been recognized as a limited and somewhat selective process of hydration or carbonation, driven by a fluid whose access is probably aided by deformation and decompression.

But could it be that these released (or introduced) H_2O-CO_2 fluids do not remain passive, but that circumstances might exist where instead they become chemically reactive? These questions were answered affirmatively in a challenging stream of papers starting in the late 1970s by John Ferry, who used changes in the modal amounts of minerals to measure the progress of reactions in samples of carbonate-bearing metasediment in the context of mixed volatile phase relations modelled by the isobaric T–X_{CO_2} diagram. Ferry recognized that such rocks are excellent monitors of fluid-rock interaction. We highlight here one of his many papers that demonstrated the magnitude and regional scale of infiltration of calcareous metasediments in prograde metamorphism by an external H_2O-rich fluid. Ferry's conclusions based on the field evidence for fluid-driven metamorphism have broad potential implications for our understanding of the evolution of metamorphic terranes. Metamorphic fluids are extremely important to issues such as heat and mass transfer in the crust, styles and mechanisms of rock deformation, seismology, physical properties of rocks, and the rates of reactions (Chapter 13). To what extent is the metamorphic mineral assemblage a product of the infiltration of fluid as opposed to simply changes in T and P? If fluid is implicated, what is its source, the nature of its pathways, and its ultimate destination?

Hugh Greenwood's treatment of metamorphic equilibria in mixed volatile systems (Chapter 6) provided criteria for distinguishing between situations where fluid is passively evolved along a progressive P-T-X_{CO_2} path of internal buffering with minimal reaction of low-variance assemblages, and those where readily detected reaction is driven by the introduction of an external fluid (Greenwood, 1975). Ferry made powerful use of this foundation.

Relatively early on, measured changes in oxygen, hydrogen and carbon isotope ratios of metamorphic rocks were interpreted as indicative of substantial exchange with an infiltrating fluid, one of meteoric origin in the vicinity of epizonal intrusions, or of inferred magmatic origin in deeper metamorphic complexes (Garlick and Epstein, 1967; Taylor and Forester, 1971; Taylor, 1974; Shieh and Schwarz, 1974; Rye *et al.*, 1976). Petrological studies by

Hewitt (1973), Moore and Kerrick (1976), and Rice (1977) also documented cases of significant fluid-driven reaction progress in metacarbonate rocks.

John Ferry began petrological studies of metacarbonate rocks in south-central Maine in the mid-1970s. Using a simple mathematical relation connecting the measured progress of dehydration-decarbonation reactions of known stoichiometry to the number of moles of fluid added to the rock, Ferry (1978) provided the first indication of the possible involvement of infiltrating fluid in driving ordinary metamorphism. In metamorphic petrology, petrographic modal analysis had fallen into disuse because mineral abundances were not required for the phase rule or for phase-equilibrium calculations; but now it acquired great importance as a quantitative measure of reaction progress. Geochemists had already recognized the importance of reaction progress measurements in the study of mass transfer involving aqueous solutions (e.g. Helgeson et al., 1970; Brimhall, 1979). Volumetric fluid/rock ratios determined by Ferry from reaction-progress in metacarbonate rocks in Maine were found to be on the order of 2 to 3, and integrated amounts of fluid were 1 to 2 orders of magnitude larger than the rock porosity (Ferry 1980, 1983a, 1983b). An even larger estimate of water/rock ratio was shown in a classic paper by Rumble et al. (1982) to have driven the wollastonite-forming reaction at an outcrop in Beaver Brook, New Hampshire. The maintenance of fluid-rock equilibrium at low, buffered values of X_{CO_2} in the pore fluid, despite the progress of decarbonation reactions, is the key observation pointing to the introduction of H_2O-rich fluid and simultaneous removal of evolved CO_2.

Ferry's 1983 paper reproduced here is a well illustrated, very readable account of the magnitude of fluid/rock interaction during ~3.5 kbar regional metamorphism of the Vassalboro Formation in south-central Maine, and its dependence on metamorphic grade. His estimated pore-fluid compositions (Fig. 4) indicate that, despite all reactions having evolved fluid with $X_{CO_2} \geqslant 0.5$ (Fig. 5), the metacarbonate rocks were in equilibrium with an H_2O-rich fluid at all grades. Fluid/rock ratios derived in this study from the measurement of reaction-progress are conservative estimates of integrated fluid flow because the infiltrating fluid was assumed to be pure H_2O, non-reactive fluid was not monitored, and possible reactive flow that took place below the biotite isograd was not included. It is important to note that the processes of buffering and infiltration associated with mixed-volatile fluids are in no way mutually incompatible; rather they proceed simultaneously. When, in prograde metamorphism, the buffer capacity of a reaction assemblage is eventually exhausted, another buffer reaction going to a higher temperature takes over (Fig. 5). Ferry showed that most of the infiltrating fluid in the Vassalboro formation passed through the rock at the biotite and amphibole grades, rather than at the higher Zo and Di grades (Figs 6 and 7). As a result, the heat added at low grades was consumed in driving metamorphic reactions rather than raising temperatures (Fig. 9). Ferry also found that, above the biotite zone, a substantial whole-rock loss of Na and K occurred in the metacarbonate rocks via hydrolysis reactions driven by the infiltrating fluid. This kind of regional metasomatism is something that is not necessarily obvious, but it does have large-scale geochemical implications; it is in fact very reasonable to expect that infiltration of aqueous fluid along a thermal gradient might induce whole-rock changes in some of the volatile constituents (see below). Ferry concluded that the infiltration of carbonate rocks during metamorphism is probably a general phenomenon, affecting rocks perhaps as deep as 30 km in the crust. His observations raised the issue of possible heat transfer during metamorphism by advecting fluid; conventional treatments of the heat budget of metamorphism assumed heat transfer exclusively by conduction (see Chapter 9).

Average shales contain carbonate minerals, but middle- and high-grade pelitic schists commonly do not; thus most metapelites do not register the passage of H_2O-rich fluid in terms of reaction progress. Ferry (1984) found that the biotite isograd in the pelitic Waterville Formation, also in south-central Maine, marked the limit of a decarbonation front related to a pervasive introduction of aqueous fluid similar to that which infiltrated the metacarbonate rocks.

Wood and Graham (1986) pointed out that estimates by Ferry of fluid/rock ratio of 1 to 2 in the Waterville pelites could be too high if the metamorphic temperatures on which the estimates are based are too low. Ferry (1986) defended his thermometry (see Chapter 10) and pointed out that his general conclusions were supported by oxygen isotope measurements of metamorphic fluid flow at low grades in metapelitic rocks in terranes worldwide.

In contrast to earlier results, Ferry (1987, 1988) detected systematic differences in the amounts of reactive fluid flow in different varieties of carbonate-bearing metasediment at the same outcrop, providing evidence for the lithological control of rock permeability. He found that reactive fluid flow was channellized parallel to lithological layering at low grades (Chl, Bt and Grt zones), and more pervasive at high grades (Sil zone). The issue of pervasive vs. channellized flow is important because the chemical and thermal effects of fluid-rock interaction will be greater in the former than in the latter, so the petrologist needs to know if the fluid driving metamorphism is moving predominantly parallel to layering or across it. Ferry's was not the only study of metacarbonate rocks to show that the flow of infiltrating aqueous fluid was channellized (e.g. Tracy et al., 1983; Nabelek et al., 1984; Bebout and Carlson, 1986). Because devolatilization reactions enhance porosity (Fyfe et al., 1978; Rumble and Spear, 1983), it is thought that reactive layers are likely to become the loci of high fluid flow (Ferry, 1986). As time went on, the structural attitudes of high-permeability planes (layering, faults, and joints) were found, not surprisingly, to exert an important control on flow directions.

Rumble (1989) felt confident that the existence of high fluid fluxes in metamorphism was well supported by studies of isotopic shifts and the solubilities of minerals, as well as by those of reaction progress. He noted that fluid flow was probably transient and caused by discrete events of short duration related to permeability enhancement, grain-scale dilatancy, hydrofracturing and deformation.

Fluid/rock ratios have been criticized because they are inadequate measures of the amount of fluid involved. They refer to static, zero-dimensional 'box' models that have no relevance to the dynamic process of flow in any direction, and unrealistically assume an idealized infiltrating fluid (e.g. pure H_2O) that is out of equilibrium at the reaction site (Baumgartner and Ferry, 1991). Time-integrated fluid flux (volume of fluid/area of rock cross-section) became the preferred measure of fluid flow because it can be linked to an integrated form of the one-dimensional mass conservation equation (Lichtner, 1985; Bickle and McKenzie, 1987) that incorporates the amount of reaction progress under conditions of local equilibrium for the duration of flow. In this continuum model, the nature of the mineral reaction involved (oxidation vs. reduction, hydration vs. dehydration) and the spatial orientation of the regional temperature (or pressure) gradient give a qualitative indication of the apparent direction of fluid flow (Baumgartner and Ferry, 1991; Ferry and Dipple, 1991; Ferry, 1994b). This one-dimensional model has classically been applied to the precipitation of quartz as veins (Fyfe et al., 1978; Walther and Orville, 1982; Yardley, 1986) and to alkali metasomatism (Orville, 1962; Hoffman, 1972). Pervasive fluid flow by the local equilibrium model is required to be in an up-temperature direction in decarbonation reactions because of the sign of the $\partial X_{CO_2}/\partial T$ gradients of the reactions involved; it led to a gradual increase in X_{CO_2} (Ferry, 1994a). The reverse will hold in fracture and fault zones in which there has been oxidation, hydration, and silicification. A useful general discussion of the physics of fluid flow through the crust was given by Spear (1993), and Ferry and Gerdes (1998) discussed important distinctions between the local equilibrium model of fluid flow which applies best to regional thermal gradients, and disequilibrium models which operate over shorter distances and tend to produce fronts of alteration.

The continuum model can also be used to account for regional gradients in stable isotope ratios of metamorphic rocks as a result of fluid flow along a regional temperature gradient under conditions of local fluid-rock equilibrium (Dipple and Ferry, 1992). In this model, isotope exchange is a consequence of the temperature dependence of stable-isotope fractionation factors (mineral vs. fluid), such that up-temperature fluid flow can be expected to create $\delta^{18}O$ and $\delta^{13}C$ depletions of several per mil. Infiltration by a chemically exotic non-equilibrium fluid is not required to induce isotopic exchange of this kind, although in some cases it does, as for example, in the Notch Peak and Alta aureoles, Utah (Nabelek and Nabotka, 1993; Bowman et al., 1994).

Investigation of a 2000 km^2 area of ~7 kbar Barrovian metamorphism of the Waits River formation in eastern Vermont by the mineralogical reaction-progress method (Ferry, 1992) and the analysis of O and C isotopes in calcite and ankerite (Stern et al., 1992) revealed both similar calculated time-integrated fluid fluxes (10^2–7×10^3 m^3 m^{-2}) and mutually consistent patterns of fluid flow in relation to metamorphic isograds and the locations of synmetamorphic granite bodies. This agreement is compelling evidence that the same infiltration process accompanying metamorphism was responsible for the observed reaction-progress and the isotopic shifts. Fluid flow was inferred to be subhorizontal, mostly layer-parallel, from the low-temperature flanks of major antiforms to the hot axial regions invaded by plutons (similar to conclusions by Yardley et al., 1991 for the Connemara schist), and then subvertically upwards out of the metamorphic terrane, down-T along fractures now filled by quartz, The infiltrating fluid was inferred to be an H_2O-rich mixture with CO_2 that was derived by prograde devolatilization reactions in pelites and metacarbonate rocks, augmented at high (diopside) grade by an acidic H_2O-fluid evolved from crystallizing granitic magma (Léger and Ferry, 1993). Regional metamorphism was driven by a combination of heat flow and pervasive infiltration by reactive H_2O-rich fluid. The average Darcy flux for this fluid in northern Vermont was about 0.1 mm/year. For a portion of the terrane in southeastern Vermont, Wing and Ferry (2002) used a three-dimensional inversion of the reaction progress and isotopic data to show that metamorphic fluid flow was locally complex, but predominantly upward and parallel to regional lithological layering.

The simplified, advective-continuum model employed by Ferry calls for up-T transport parallel to lithological layering, driving regional prograde reactions in metacarbonate rocks that liberate CO_2. Because of the very large variations in permeability from layer to layer, it is generally accepted that the main component of fluid flow in metamorphic rocks is layer-parallel. However, observations such as those by Hewitt (1973) of decarbonation reactions at the margins of calcareous layers in Connecticut, and by Graham et al. (1983) of carbonation reactions at the margins of greenschist-facies metabasites in the Scottish Dalradian, raise the possibility that reaction progress driven by local, cross-lithology fluid communication by some combination of advection, diffusion and dispersion of fluid may not be trivial. A term for these contributions to reaction progress can be included in the continuum equation. According to Ague and Rye (1999), cross-lithology diffusion can drive prograde CO_2 loss in metacarbonate rocks with time-integrated fluid fluxes considerably smaller than indicated by models based on one-dimensional fluid advection.

The magnitude of the component of fluid flux normal to lithological contacts is expressed by a displacement of a front of elemental, isotopic, or mineralogical change (Bickle and McKenzie, 1987). Some studies have found the shift to be barely detectable (Rumble and Spear, 1983; Ganor et al., 1989; Todd and Evans, 1993), whereas others have revealed shifts measured in metres (Bickle and Baker, 1990a; Skelton et al., 1997; Vyhnal and Chamberlain, 1996; Bickle et al., 1997), and occasionally much greater. According to Kohn and Valley (1994), oxygen isotope profiles in garnet and hornblende crystals in amphibolite-facies rock at the Townsend Dam, Vermont, are inconsistent with major pervasive fluid flow across strike. Bickle et al. (1997) concluded that cross-layer fluid flux was insufficient to have caused the observed reaction progress of Ms + Ank + Qtz → Bt + An + Cal in the Waterville limestone, Maine. Cross-layer homogenization of stable isotopes was found to be limited to between 1.5 and 6 m. Cartwright et al. (1995) presented a strong case for metamorphic fluid flow (time-integrated flux of 10^5 m^3/m^2) up a 50 km long temperature gradient in the Mt. Lofty Ranges, South Australia, based on oxygen-isotope depletions in metapelites and carbonate rocks, and on the progress of reactions in the carbonates. However, steep cross-lithology gradients in $\delta^{18}O$ and X_{CO_2} survived.

In folded terrane of greenschist to epidote amphibolite facies metamorphism in the SW Highlands of Scotland, Skelton et al. (1995) were able to integrate individual one-dimensional (1-D) estimates of net fluid flux based on reaction-front advection theory (Bickle and Baker, 1990b) into a 3-D picture of net fluid flux. They found that fluid was involved in an upward, single-pass movement, was channelled through the major phyllite unit, and focused into the axial zones of regional antiforms. The 1-D measurements were based on mafic sills such as the one described in greater detail in Skelton et al. (1997), who showed that the margin of a sill was penetrated 5 m on the inferred upstream side by infiltrating CO_2-bearing fluid derived from more permeable calc-phyllite at the contact, with diffusion/dispersion extending the reaction over another 1–3 m. Evans and Bickle (1999) used a two-dimensional model to examine in detail two Ky-zone samples from east-central Vermont studied earlier by Ferry. They questioned Ferry's estimates of fluid flux and the necessity for near-horizontal fluid flow over distances on the order of 10 to 100 km. They found that allowance for possible cross-layer flow from adjacent dehydrating pelites could greatly reduce the estimates of fluid flux based on Ferry's model. Ague (2000, 2002 and 2003b) has made a strong case for reactive transport of volatiles across pelite-carbonate contacts in rocks in the Wepawaug schist, Connecticut. This is an issue of great current interest and uncertainty.

Considering all the possible sources for infiltrating fluid in upper and middle crustal environments – surficial, magmatic and metamorphic – it is not surprising that Ferry has consistently found it to be hydrous, and therefore capable of driving reactions in metasediments containing carbonate minerals. There are only a few examples of the converse, where the infiltrating fluid drives a carbonate-forming reaction; some come from meta-peridotitic rocks, whose mineralogy generally equilibrates with a low-CO_2 hydrous fluid. A clear example is the formation of the silica-carbonate rock known as listwanite (magnesite + serpentine/talc/quartz) from serpentinite in areas of gold, pyrite, and mercury mineralization in the upper crust (e.g. Robinson et al., 2004; Hansen et al., 2005). Some of the reactions in ophicalcite in the contact aureole of the Bergell intrusion, Switzerland, were carbonate-forming (Ferry 1995). The formation of sagvandite (magnesite + enstatite/talc) from peridotite in the middle or lower crust (Evans and Trommsdorff, 1974; Ferry et al., 2005) requires infiltration of a CO_2-bearing fluid. The margins of metabasite sills in the SW Highlands underwent carbonation and hydration by reaction with fluid from the surrounding phyllite (Skelton et al., 1995).

Influx of CO_2-rich fluid into deep crustal granulites has been invoked to explain the presence of CO_2-rich fluid inclusions (Touret, 1971) and relatively low H_2O activities inferred from phase equilibria (Newton, 1992). However, convincing field examples of reaction progress in granulites driven by deep crustal CO_2 infiltration have been local rather than regional in scale (e.g. Harley and Santosh, 1995), and the overall importance of the process in granulite belts has been questioned (Valley et al., 1990). Vapour-absence caused by partial melting (Thompson, 1983; Waters, 1988), or the introduction of a hypersaline fluid (Newton et al., 1998), are alternative explanations currently more in favour. Of course, it is always possible that some dehydration reactions, not only in granulite terranes but also at lower grades, have been driven by the influx of H_2O-CO_2 fluid without the growth of a carbonate mineral. On the subject of melting, the up-T flow of aqueous fluid could have interesting consequences where temperatures reach 650–700°C. For example, Symmes and Ferry (1995) concluded that leucocratic veins in the innermost contact zone around the Onawa pluton, Maine, formed by "wet" melting of pelitic hornfels aided by this fluid.

John Ferry must be given credit for forcefully drawing our attention to the role played by infiltrating fluid in driving prograde metamorphism. There is still much to be learned, however, about where and when fluid is predictably involved driving metamorphism. Low permeabilities suggest that a single-pass fluid-flow model applies to middle and lower crust where deformation is ductile (Wood and Walther, 1986; Rumble, 1994; Ferry, 1994a; Ague, 2003a), whereas convective circulation remains a possibility in the upper crust where deformation results in brittle fractures (Etheridge et al., 1983; Wickham and Taylor, 1985). Patterns of fluid flow are clearly very dependent on the stratigraphical and structural disposition of lithologies with contrasting permeabilities. The hydrological driving forces are buoyancy and added fluid pressure from crystallizing magma and metamorphic reactions. Prograde metamorphic reactions supply much, if not all, of the fluid. There seems to be no universal flow direction during crustal metamorphism (Ferry and Gerdes, 1998).

Future research should include case studies in terranes with diverse geological parameters (presence/absence of plutonic activity, different ages, lithological mixes, crustal depths (Chapters 14 and 15), and P-T-time paths), so that we can build toward some sort of predictive model of fluid circulation during metamorphism in the crust (e.g. Symmes and Ferry, 1991; Powell et al., 2005). These studies should ideally integrate measurement of mineralogical, elemental and isotopic changes. To what extent does cross-lithology transfer of fluid detract from estimates of regional fluid fluxes based only on the regional T-gradient model? Will this factor alter conclusions about the sense of flow (up-T vs. down-T) inferred from the regional model?

References

Ague, J.J. (2000) Release of CO_2 from carbonate rocks during regional metamorphism of lithologically heterogeneous crust. *Geology*, **28**, 1123–1126.

Ague, J.J. (2002) Gradients in fluid composition across metacarbonate layers of the Wepawaug Schists, Connecticut, USA. *Contributions to Mineralogy and Petrology*, **143**, 38–55.

Ague, J.J. (2003a) Fluid infiltration and transport of major, minor, and trace elements during regional metamorphism of carbonate rocks, Wepawaug Schist, Connecticut, USA. *American Journal of Science*, **303**, 753–816.

Ague, J.J. (2003b) *Fluid flow in the deep crust*. Pp. 195–228 in: Treatise on Geochemistry, Elsevier Ltd., Amsterdam.

Ague, J.J. and Rye, D.M. (1999) Simple models of CO_2 release from metacarbonates with implications fore interpretation of directions and magnitudes of fluid flow in the deep crust. *Journal of Petrology*, **40**, 1443–1462.

Baumgartner, L.P. and Ferry, J.M. (1991) A model for coupled fluid-flow and mixed volatile reactions with applications to regional metamorphism. *Contributions to Mineralogy and Petrology*, **106**, 273–285.

Beach, A. and Fyfe, W.S. (1972) Fluid transport and shear zones at Scourie, Southerland: Evidence of overthrusting. *Contributions to Mineralogy and Petrology*, **36**, 175–180.

Bebout, G.E. and Carlson, W.D. (1986) Fluid evolution and transport during metamorphism: Evidence from the Llano uplift, Texas. *Contributions to Mineralogy and Petrology*, **892**, 518–529.

Bickle, M.J. and Baker, J. (1990a) Advective-diffusive transport of isotopic fronts: an example from Naxos, Greece. *Earth and Planetary Science Letters*, **97**, 78–93.

Bickle, M.J. and Baker, J. (1990b) Migration of reaction and isotopic fronts in infiltration zones; assessments of fluid flux in metamorphic terranes. *Earth and Planetary Science Letters*, **98**, 1–13.

Bickle, M.J. and McKenzie, D. (1987) The transport of heat and matter by fluid during metamorphism. *Contributions to Mineralogy and Petrology*, **95**, 384–392.

Bickle, M.J., Chapman, H.J., Ferry, J.M., Rumble, D. and Fallick, A.E. (1997) Fluid flow and diffusion in the Waterville limestone, south-central Maine: constraints from strontium, oxygen and carbon isotope profiles. *Journal of Petrology*, **38**, 1489–1512.

Bowman, J.R., Willet, S.D. and Cook, S.J. (1994) Oxygen isotopic transport and exchange during fluid flow: one-dimensional models and applications. *American Journal of Science*, **294**, 1–55.

Bridgwater, D. (1987) *Fluid Movements – Element Transport and the Composition of the Deep Crust*. nATO ASI Series C: Mathematical and Physical Sciences, **281**, 416 pp.

Brimhall, G.H. (1979) Lithologic determination of mass transfer mechanisms of multiple-stage porphyry copper mineralization at Butte, Montana: Vein formation by hypogene leaching and enrichment of potassium-silicate protore. *Economic Geology*, **74**, 556–589.

Cartwright, I., Vry, J. and Sandiford, M. (1995) Changes in stable isotope ratios of metapelites and marbles during regional metamorphism, Mount Lofty Ranges, South Australia: implications for crustal scale fluid flow. *Contributions to Mineralogy and Petrology*, **120**, 292–310.

Dipple, G.M and Ferry, J.M. (1992) Fluid flow and stable isotopic alteration in rocks at elevated temperatures with applications to metamorphism. *Geochimica et Cosmochimica Acta*, **56**, 3539–3550.

Etheridge, M.A., Wall, V.J. and Vernon, R.H. (1983) The role of the fluid phase during regional metamorphism and deformation. *Journal of Metamorphic Geology*, **1**, 205–226.

Evans, K.A. and Bickle, M.J. (1999) Determination of time-integrated metamorphic fluid fluxes from the reaction progress of multivariant assemblages. *Contributions to Mineralogy and Petrology*, **134**, 277–293.

Evans, B.W. and Trommsdorff, V. (1974) Stability of enstatite + talc, and CO_2-metasomatism of metaperidotite, Val d'Efra, Lepontine Alps. *American Journal of Science*, **274**, 274–296.

Ferry, J.M. (1978) Fluid interaction between granite and sediment during metamorphism, south-central Maine. *American Journal of Science*, **278**, 1035–1056.

Ferry, J.M. (1980) A case study of the amount and distribution of heat and fluid during metamorphism. *Contributions to Mineralogy and Petrology*, **71**, 373–385.

Ferry, J.M. (1983a) Regional metamorphism of the Vassalboro Formation, south-central Maine, U.S.A.: a case study of the role of fluid in metamorphic petrogenesis. *Journal of the Geological Society*, **140**, 551–576.

Ferry, J.M. (1983b) On the control of temperature, fluid composition, and reaction progress during metamorphism. *American Journal of Science*, **283-A**, 201–232.

Ferry, J.M. (1984) A biotite isograd in south-central Maine, U.S.A.: Mineral reactions, fluid transfer, and heat transfer. *Journal of Petrology*, **25**, 871–893.

Ferry, J.M. (1986) Infiltration of aqueous fluids and high fluid:rock ratios during greenschist facies metamorphism: A reply. *Journal of Petrology*, **27**, 695–714.

Ferry, J.M. (1987) Metamorphic hydrology at 13-km depth and 400–550°C. *American Mineralogist*, **72**, 39–58.

Ferry, J.M. (1988) Contrasting mechanisms of fluid flow through adjacent stratigraphic units during regional metamorphism, south-central Maine, U.S.A. *Contributions to Mineralogy and Petrology*, **98**, 1–12.

Ferry, J.M. (1992) Regional metamorphism of the Waits River Formation, eastern Vermont: Delineation of a new type of giant hydrothermal system. *Journal of Petrology*, **33**, 45–94.

Ferry, J.M. (1994a) A historical review of metamorphic fluid flow. *Journal of Geophysical Research*, **99**, B8, 15487–15498.

Ferry, J.M. (1994b) Overview of the petrologic evidence for fluid flow during regional metamorphism in northern New England. *American Journal of Science*, **294**, 905–988.

Ferry, J.M. (1995) Fluid flow during contact metamorphism of ophicarbonate rocks in the Bergell aureole, Val Malenco, Italian Alps. *Journal of Petrology*, **36**, 1039–1053.

Ferry, J.M. and Dipple, G.M. (1991) Fluid flow, mineral reactions, and metamorphism. *Geology*, **19**, 211–214.

Ferry, J.M. and Gerdes, M. (1998) Chemically reactive fluid flow during metamorphism. *Annual Reviews of Earth and Planetary Sciences*, **26**, 255–287.

Ferry, J.M., Rumble, D.III, Wing, B.A. and Penniston-Dorland, S.C. (2005) A new interpretation of centimeter-scale variations in the progress of infiltration-driven metamorphic reactions: Case study of carbonated metaperidotite, Val d'Efra, Central Alps, Switzerland. *Journal of Petrology*, **46**, 1725–1746.

Fyfe, W.S., Price, N.J. and Thompson, A.B. (1978) *Fluids in the Earth's Crust*. Elsevier, New York, 383 pp.

Ganor, J., Matthews, A. and Paldor, N. (1989) Constraints on effective diffusivity during oxygen isotope exchange at a marble-schist contact, Sifnos (Cyclades), Greece. *Earth and Planetary Sciences Letters*, **94**, 208–216.

Garlick, G.D. and Epstein, S. (1967) Oxygen isotope ratios in coexisting minerals of regionally metamorphosed rocks, *Geochimica et Cosmochimica Acta*, **31**, 181–214.

Graham, C.M., Greig, K.M., Sheppard, S.M.F. and Turi, B. (1983) Genesis and mobility of the H_2O-CO_2 fluid phase in greenschist metamorphism: a petrological and stable isotope study in the Scottish Dalradian. *Journal of the Geological Society of London*, **140**, 577–599.

Greenwood, H.J. (1975) Buffering of pore fluids by metamorphic reactions. *American Journal of Science*, **275**, 573–593.

Hansen, L.D., Dipple, G.M., Gordon, T.M. and Kellett, D.A. (2005) Carbonated serpentinite (listwanite) at Atlin, British Columbia: a geological analogue to carbon dioxide sequestration. *The Canadian Mineralogist*, **43**, 225–240.

Harley, S.L. and Santosh, M. (1995) Wollastonite at Nuliyam, Kerala, Southern India: a reappraisal of CO_2-infiltration and charnockite formation at a classic locality. *Contributions to Mineralogy and Petrology*, **120**, 83–94.

Helgeson, H.C. (1967) Solution chemistry and metamorphism. Pp. 362–404 in: *Researches in Geochemistry*, **2** (P.H. Abelson, editor). John Wiley, New York.

Helgeson, H.C., Brown, T.H., Nigrini, A. and Jones, T.A. (1970) Calculation of mass transfer in geochemical processes involving aqueous solutions. *Geochimica et Cosmochimica Acta*, **34**, 569–592.

Hewitt, D.A. (1973) The metamorphism of micaceous limestones from south-central Connecticut. *American Journal of Science*, **273-A**, 444–469.

Hoffmann, A.W. (1972) Chromatographic theory of infiltration metasomatism and its application to feldspars. *American Journal of Science*, **272**, 69–90.

Kohn, M.J. and Valley, J.W. (1994) Oxygen isotope constraints on metamorphic fluid flow, Townsend Dam, Vermont, USA. *Geochimica et Cosmochimica Acta*, **58**, 5551–5566.

Léger, A. and Ferry, J.M. (1993) Fluid infiltration and regional metamorphism of the Waits River Formation, north-east Vermont, USA. *Journal of Metamorphic Geology*, **11**, 3–29.

Lichtner, P.C. (1985) Continuum model for simultaneous chemical reactions and mass transport in hydrothermal systems. *Geochimica et Cosmochimica Acta*, **49**, 779–800.

Moore, J.N. and Kerrick, D.M. (1976) Equilibria in siliceous dolomites of the Alta aureole, Utah. *American Journal of Science*, **276**, 502–524.

Nabelek, P.I. and Nabotka, T.C. (1993) Implications of geochemical fronts in the Notch Peak contact-metamorphic aureole, Utah. *Earth and Planetary Sciences Letters*, **119**, 539–559.

Nabelek, P.I., Labotka, T.C, O'Neil, J.R. and Papike, J.J. (1984) Contrasting fluid/rock interaction between the Notch Peak granitic intrusion and argillites and limestones in western Utah: Evidence from stable isotopes and phase assemblages. *Contributions to Mineralogy and Petrology*, **86**, 25–34.

Newton, R.C. (1992) Charnockite alteration: Evidence for CO_2 infiltration in granulite facies metamorphism. *Journal of Metamorphic Geology*, **10**, 383–400.

Newton, R.C., Aranovich, L.Y., Hansen, J.C. and Vandenheuvel, B.A. (1998) Hypersaline fluids in Precambrian deep-crustal metamorphism. *Precambrian Research*, **91**, 41–63.

Orville, P.H. (1962) Alkali metasomatism and the feldspars. *Norsk Geologisk Tidsskrift*, **42**, 283–316.

Powell, R., Guiraud, M. and White, R.W. (2005) Truth and beauty in metamorphic phase equilibria: conjugate variables and phase diagrams. *The Canadian Mineralogist*, **43**, 21–34.

Rice, J.M. (1977) Progressive metamorphism of impure dolomitic limestone in the Marysville aureole. *American Journal of Science*, **277**, 1–24.

Robinson, P.T., Malpas, J., Zhou, M., Ash, C., Yang, J. and Bai, W. (2004) Geochemistry and origin of listwanites in the Sartohay and Luobusa ophiolites. *Serpentine and Serpentinites: Mineralogy, Petrology, Geochemistry, Ecology, Geophysics, and Tectonics*. International Book Series, **8**, Geological Society of America, pp. 475–502.

Rumble, D. (1994) Water circulation in metamorphism. *Journal of Geophysical Research*, **99**, B8, 15499–15502.

Rumble, D. and Spear, F.S. (1983) Oxygen isotope equilibration and permeability enhancement during regional metamorphism. *Journal of the Geological Society of London*, **140**, 619–628.

Rumble, D., Ferry, J.M., Hoering, T.C. and Boucot, A.J. (1982) Fluid flow during metamorphism at the Beaver Brook fossil locality, New Hampshire. *American Journal of Science*, **282**, 886–919.

Rumble, D. III (1989) Evidence for fluid flow during regional metamorphism. *European Journal of Mineralogy*, **1**, 731–737.

Rye, R.O., Schuiling, R.D., Rye, D.M. and Jansen, J.B.H. (1976) Carbon, hydrogen, and oxygen isotope studies of the regional metamorphic complex at Naxos, Greece. *Geochimica et Cosmochimica Acta*, **40**, 1031–1049.

Shieh, Y.N. and Schwarz, H.P. (1974) Oxygen isotope studies of granite and migmatite, Grenville Province of Ontario, Canada. *Geochimica et Cosmochimica Acta*, **38**, 21–45.

Skelton, A.D.L., Graham, C.M. and Bickle, M.J. (1995) Lithological and structural controls on regional 3-D fluid flow patterns during greenschist facies metamorphism of the Dalradian of the SW Scottish Highlands. *Journal of Petrology*, **36**, 563–586.

Skelton, A.D.L., Bickle, M.J. and Graham, C.M. (1997) Fluid flux and reaction rate from advective-diffusive carbonation of mafic sill margins in the Dalradian, southwest Scottish Highlands. *Earth and Planetary Science Letters*, **146**, 527–539.

Spear, F.S. (1993) *Metamorphic Phase Equilibria and Pressure-Temperature-Time Paths*. Monograph, Mineralogical Society of America, Washington, D.C., 799 pp.

Stern, L.A., Chamberlain, C.P., Barnett, D.E. and Ferry, J.M. (1992) Stable isotope evidence for regional-scale fluid migration in a Barrovian metamorphic terrane, Vermont, U.S.A. *Contributions to Mineralogy and Petrology*, **112**, 475–489.

Symmes, G.H. and Ferry, J.M. (1991) Evidence from observed mineral assemblages for filtration of pelitic schists by aqueous fluids during metamorphism. *Contributions to Mineralogy and Petrology*, **108**, 419–438.

Symmes, G.H. and Ferry, J.M. (1995) Metamorphism, fluid flow and partial melting in pelitic rocks from the Onawa contact aureole, Central Maine. *Journal of Petrology*, **36**, 587–612.

Taylor, H.P. (1974) The application of oxygen and hydrogen isotope studies to problems of hydrothermal alteration and ore deposition. *Economic Geology*, **69**, 843–883.

Taylor, H.P. and Forester, R.W. (1971) Low O^{18} igneous rocks from the intrusive complexes of Skye, Mull, and Ardnamurchan, western Scotland. *Journal of Petrology*, **12**, 465–497.

Thompson, A.B. (1983) Fluid absent metamorphism. *Journal of the Geological Society of London*, **140**, 533–546.

Thompson, J.B. (1959) Local equilibrium in metasomatic processes. Pp. 427–457 in: *Researches in Geochemistry* (P.H. Abelson, editor). John Wiley, New York.

Todd, C.S. and Evans, B.W. (1993) Limited fluid-rock interaction at marble-gneiss contacts during Cretaceous granulite-facies metamorphism, Seward Peninsula, Alaska. *Contributions to Mineralogy Petrology*, **114**, 27–41.

Touret, J.L.R. (1971) Le faciès granulite en Norvège méridionale. I: Les inclusions fluides. *Lithos*, **4**, 423–436.

Tracy, R.J., Rye, D.M., Hewitt, D.A. and Schiffries, C.M. (1983) Petrologic and stable isotope studies of fluid–rock interaction, south-central Connecticut: I. The role of infiltration in producing reaction assemblages in impure marbles. *American Journal of Science*, **283A**, 589–616.

Valley, J.W., Bohlen, S.R., Essene, E.J. and Lamb, W. (1990) Metamorphism in the Adirondacks: II. The role of fluids. *Journal of Petrology*, **31**, 555–596.

Vynhal, C.R. and Chamberlain, C.P. (1996) Preservation of early isotopic signatures during prograde metamorphism, Eastern Vermont. *American Journal of Science*, **296**, 394–419.

Walther, J.V. and Orville, P.M. (1982) Volatile production and transport in regional metamorphism. *Contributions to Mineralogy and Petrology*, **79**, 252–257.

Waters, D.J. (1988) Partial melting and the formation of granulite facies assemblages in Namaqualand, South Africa. *Journal of Metamorphic Geology*, **6**, 387–404.

Wickham, S.M. and Taylor, H.P. (1985) Stable isotope evidence for large-scale seawater infiltration in a regional metamorphic terrane: The Trois Seigneurs Massif, Pyrenees, France. *Contributions to Mineralogy and Petrology*, **91**, 122–137.

Wing, B.A. and Ferry, J.M. (2002) Three-dimensional geometry of metamorphic fluid flow during Barrovian regional metamorphism from an inversion of combined petrologic and stable isotope data. *Geology*, **30**, 639–642.

Wood, B.J. and Graham, C.M. (1986) Infiltration of aqueous fluid and high fluid:rock ratios during greenschist facies metamorphism: A discussion. *Journal of Petrology*, **27**, 751–761.

Wood, B.J. and Walther, J.V. (1986) Fluid flow during metamorphism and its implications for fluid-rock ratios. Pp. 89–108 in: *Fluid-Rock Interactions During Metamorphism* (J.V. Walther and B.J. Wood, editors). Advances in Physical Geochemistry, **5**, Springer Verlag, Berlin.

Yardley, B.W.D. (1986) Fluid migration and veining in the Connemara schists, Ireland. Pp. 109–131 in: *Fluid-Rock Interactions During Metamorphism* (J.V. Walther and B.J. Wood, editors). Advances in Physical Geochemistry, **5**, Springer Verlag, Berlin.

Yardley, B.W.D., Bottrell, S.H. and Cliff, R.A. (1991) Evidence for a regional-scale fluid loss event during mid-crustal metamorphism. *Nature*, **349**, 151–154.

Regional metamorphism of the Vassalboro Formation, south-central Maine, USA: a case study of the role of fluid in metamorphic petrogenesis

J. M. Ferry

SUMMARY: Progressively metamorphosed carbonate rocks of the Vassalboro Formation serve as a natural laboratory for the investigation of fluid/rock interactions during metamorphism. An integrated study of their whole-rock chemistry, mineralogy, mineral chemistry and mineral abundances has led to the following results. (a) Prograde mineral reactions involved hydrolysis as well as dehydration and decarbonation. The hydrolysis reactions have resulted in an almost quantitative extraction of K and Na from high-grade metacarbonates. (b) Prograde mineral reactions buffered the X_{CO_2} of coexisting metamorphic fluid during almost all of the rocks' metamorphic history. (c) The carbonate rocks were infiltrated by large volumes of H_2O-rich fluid while the buffering reactions progressed during the metamorphic event. Some high-grade metacarbonates were infiltrated by *at least* 3 rock volumes of fluid. Combined buffering and infiltration appears to be a general metamorphic phenomenon in carbonate-bearing rocks to depths of ~30 km in the crust. (d) There is an excellent positive correlation between the heat budget of individual metacarbonate rock samples and calculated fluid/rock ratios. Convective heat transfer by metamorphic fluids is therefore likely to be an important item in the heat budget of metamorphic terrains.

Consideration of these results leads to a model of regional metamorphism which involves infiltration of rock by large volumes of acid, H_2O-rich fluid (large-scale acid infiltration metasomatism?). In this model, the infiltrating fluid exerts enormous control over the mineralogical evolution of the carbonate rocks during the metamorphic event: it drives mineral reactions, it changes whole-rock chemistry and it supplies heat.

Studies of metamorphic mineral equilibria, of fluid inclusions in metamorphic minerals and of modern geothermal fields all indicate that a fluid phase is generally present during metamorphism. The fluid is probably composed of a mixture of components derived from a variety of possible sources: devolatilization reactions within the rock under consideration; devolatilization reactions in nearby rocks or in underlying rocks deep in the crust; magmatic fluid expelled from crystallizing synmetamorphic plutons; pristine fluid from the mantle; and fluid from the hydrosphere (meteoric water/seawater) which has penetrated deep into the crust. Because of buoyancy forces, hot metamorphic fluid will rapidly rise through the crust; consequently, the infiltration of rock during metamorphism by externally-derived fluid must be an integral part of the history of many metamorphic rocks.

Infiltration of rock by fluids during metamorphism exerts an important control over the prograde development of minerals in a variety of ways. Firstly, the infiltrating fluid influences the kinds of minerals which develop in a metamorphic rock because the stabilities of hydrates, carbonates, and sulphides are dependent on the composition of coexisting fluid. Some studies, in fact, have demonstrated that the prograde development of a sequence of metamorphic rocks can be exclusively controlled by fluid composition (Allen 1978; Grambling 1979). Secondly, infiltrating fluid may alter the chemical composition of rocks through metamorphic fluid/rock cation exchange reactions. The resulting change in bulk rock chemistry may significantly affect which prograde metamorphic minerals develop. Thirdly, infiltrating fluid may add heat to or extract heat from rocks during metamorphism. By controlling a rock's metamorphic heat budget, the fluid controls which prograde mineral reactions occur and, ultimately, the rock's mineralogical evolution. The ways in which infiltrating fluids affect the prograde development of minerals in metamorphic rocks are explored in detail by examining the metamorphic evolution of a single stratigraphic unit in south-central Maine, USA: the Silurian Vassalboro Formation. Of the various lithologies in the Vassalboro Formation, only the carbonate rocks will be considered. The metamorphic history of the unit is not an unusual one and may be taken as representative of the general case of regionally metamorphosed rocks. Details of the metamorphism of the metacarbonate rocks from the Vassalboro Formation are thus used to develop general concepts about the role of fluids in metamorphic petrogenesis.

Geological setting

Samples of the Vassalboro Formation were collected from the area illustrated in Fig. 1. The formation is composed of interbedded argillaceous

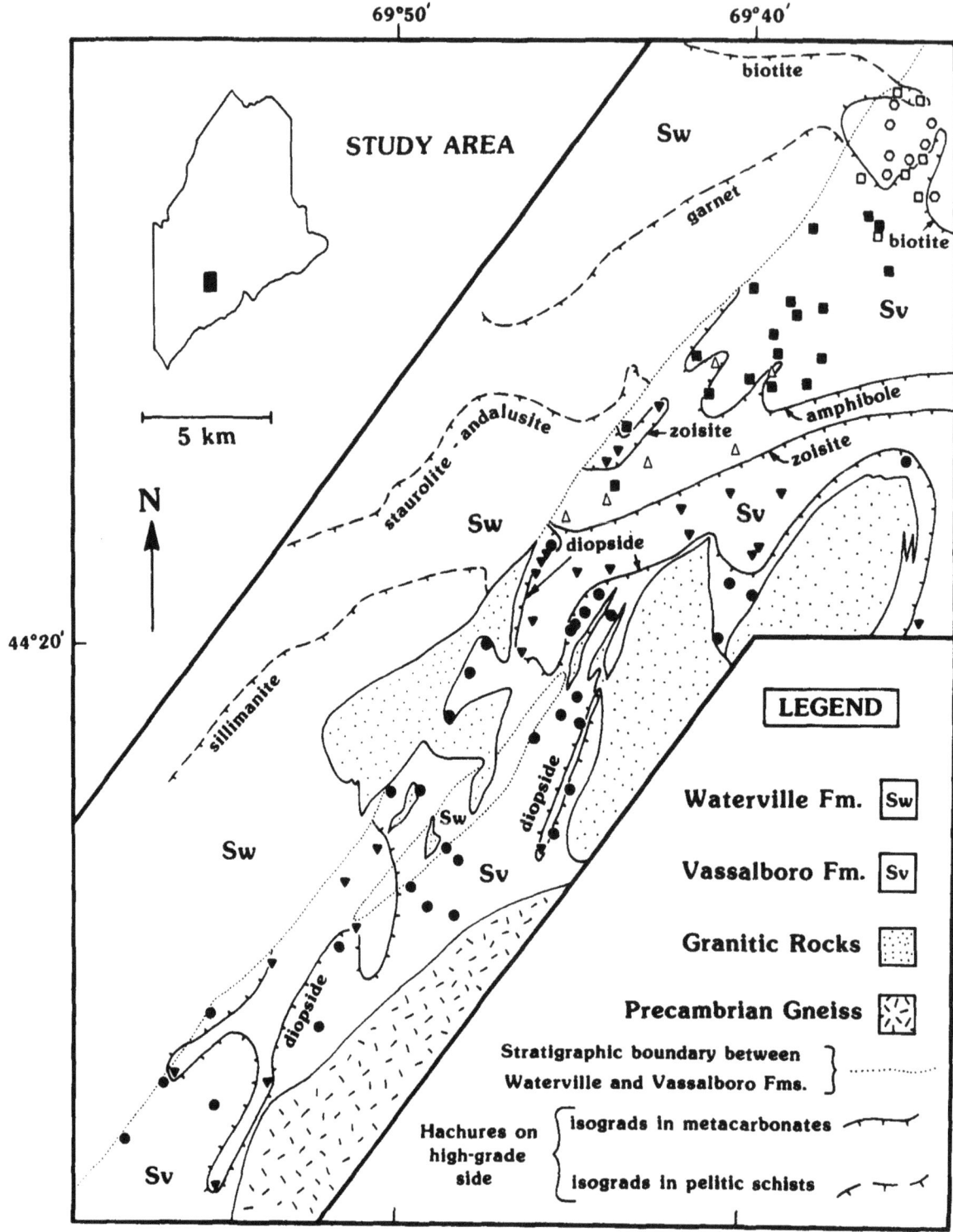

FIG. 1. Geological sketch map of area studied. Symbols are locations of outcrops sampled: open hexagons, ankerite zone; open squares, lower biotite zone; filled squares, upper biotite zone; open triangles, amphibole zone; filled triangles, zoisite zone; circles, diopside zone. Metamorphic zones in the Vassalboro Formation are separated by isograds based on mineral assemblages in the carbonate rocks (solid curves with hachures). Data from Osberg (1968, 1979, pers. comm.); Ferry (1980a).

sandstone, argillaceous carbonate rock and minor shale and their metamorphic equivalents. Compositional layering is on a scale of 2–15 cm. The metasediments are folded into tight, isoclinally refolded recumbent folds (Osberg 1979). Axes of the isoclinal folds trend NE–SW, and almost everywhere in outcrop lithological layering is near vertical with a NE–SW strike. Porphyroblasts cut across schistosity associated with the isoclinal folds and indicate that metamorphism followed almost all deformation. The metasediments are intruded by synmetamorphic quartz monzonite and granodiorite stocks. The grade of metamorphism, with reference to pelitic schists in the adjacent Silurian Waterville Formation, varies from chlorite zone conditions in the northernmost portion of Fig. 1 to sillimanite zone conditions in the S. Geochronological work on both the granitic and metamorphic rocks indicates that the metamorphism and emplacement of the granites occurred during the Devonian Acadian Orogeny (Dallmeyer & van Breemen 1981). The Acadian Orogeny was a Caledonian event during which most of the metamorphic rocks in northern New England developed (Thompson & Norton 1968). Evidence for 3 Acadian metamorphic episodes has been reported for rocks W of the exposure of the Vassalboro Formation in Fig. 1 (Novak & Holdaway 1981). One of the 3 episodes has only been reported well W of the area in Fig. 1. If there were 2 episodes of Acadian metamorphism of the Vassalboro Formation, they were closely spaced in time and P–T conditions (Holdaway et al. 1982). Metamorphism of the Vassalboro Formation will therefore be treated here as if it were the result of a single episode. The stratigraphy, structural geology and metamorphic history of the Vassalboro Formation have been discussed in more detail by Osberg (1968) as well as in references previously cited.

Methods of investigation

Approximately 200 samples of metacarbonate rock were collected from the Vassalboro Formation, of which 115 of the freshest were chosen for further chemical and petrographic study. Locations where the 115 samples were collected are shown in Fig. 1.

Thin sections were prepared from rock specimens that correspond in outcrop to individual layers of carbonate rock. At several outcrops more than one layer was sampled. Most samples appear homogeneous in hand specimen. Mineral assemblages were identified from petrographic observation of thin sections. Compositions of minerals were obtained by electron microprobe analysis, either at Harvard University (methods described by Ferry 1976a) or at the Geophysical Laboratory (methods described by Ferry 1980a). The same natural and synthetic minerals were used as standards for all analyses.

Modal amounts of all minerals in most specimens were determined by counting 500–1000 points in thin section. Modes for biotite, plagioclase, amphibole, zoisite, microcline, and diopside were later determined in selected samples by counting 1000–2000 points. Volume amounts of minerals were converted to molar amounts (per reference volume of rock) using molar volume data (Robie et al. 1967; Hewitt & Wones 1975).

Chemical analyses of rock samples for major element metal oxides were performed using inductively coupled argon plasma methods by Technical Service Laboratories (TSL), Mississauga, Ontario, Canada (A. Debnam and W. Grondin, analysts). Whole-rock samples were also analysed by TSL for total carbonate carbon using a LECO carbon analyser. Samples for chemical analysis, approximately 5 cm^3 in volume, were cut from hand specimens as close as possible to where chips were removed to prepare thin sections. A more detailed discussion of analytical procedures is presented by Ferry (1983).

Mineralogy and mineral chemistry

The Vassalboro Formation (Fig. 1) has been divided into 5 zones based on the mineralogy of the metamorphosed carbonate rocks. The mineralogy of the metacarbonates is summarized here by zone. Boundaries between zones are isograds which are based upon metamorphic mineral reactions (Ferry 1976b).

Ankerite zone

The lowest grade rocks studied (open hexagons, Fig. 1) contain the assemblage ankerite + quartz + albite + muscovite ± calcite ± chlorite. Accessory minerals include various combinations of pyrite, pyrrhotite, apatite, tourmaline, ilmenite, and graphitic material. Rocks with this mineralogy constitute the ankerite zone of the Vassalboro Formation which occurs in the extreme NE portion of Fig. 1. Average compositions of muscovite, ankerite, calcite, feldspar, chlorite, and ilmenite in rocks of the ankerite zone are listed in Table 1. Mineral compositions in Table 1 symbolically represent Fe-Mg-Mn solid solution by the term (Fe, Mg, Mn). This representation of mineral composition simplifies mineral formulae but, of course, obscures Fe-Mg-Mn partitioning among minerals and obscures systematic changes in the concentration of Fe-, Mg-, and Mn-components as a function of metamorphic grade.

Biotite zone

Immediately S of the ankerite zone is an area in Fig. 1 in which biotite appears in metamorphosed carbonate rocks. The biotite isograd represents a line mapped between occurrences of metacarbonate rocks with and without biotite (Ferry 1976b). Rocks which contain biotite but not amphibole are represented by squares in Fig. 1 and their occurrences constitute the biotite zone. Rocks in the biotite zone have been subdivided into 2 groups. The lower biotite zone (open squares, Fig. 1) consists of rocks with the assemblage biotite

TABLE 1. *Average composition of minerals in metamorphosed carbonate rocks from the Vassalboro Formation*

Mineral	Zone	Composition
Ankerite	Average: Ankerite and Biotite Zones	$Ca(Fe, Mg, Mn)(CO_3)_2$
Calcite	1. Average: Ankerite and Biotite Zones	$Ca_{0.93}(Fe, Mg, Mn)_{0.07}CO_3$
	2. Average: Biotite and Amphibole Zones	$Ca_{0.96}(Fe, Mg, Mn)_{0.04}CO_3$
	3. Average: Amphibole, Zoisite, and Diopside Zones	$Ca_{0.97}(Fe, Mg, Mn)_{0.03}CO_3$
Biotite	1. Average: Biotite Zone	$(K_{0.88}Na_{0.02})\{(Fe, Mg, Mn)_{2.33}Ti_{0.12}Al_{0.40}\}(Al_{1.22}Si_{2.78})O_{10}(OH)_2$
	2. Average: Biotite, Amphibole, and Zoisite Zones	$(K_{0.89}Na_{0.02})\{(Fe, Mg, Mn)_{2.36}Ti_{0.10}Al_{0.38}\}(Al_{1.22}Si_{2.78})O_{10}(OH)_2$
Amphibole	1. Average: Amphibole and Zoisite Zones	$Na_{0.19}Ca_{1.93}(Fe, Mg, Mn)_{4.60}Ti_{0.05}Al^{VI}_{0.43}Al^{IV}_{0.82}Si_{7.18}O_{22}(OH)_2$
	2. Average: Zoisite Zone	$Na_{0.15}Ca_{1.93}(Fe, Mg, Mn)_{4.67}Ti_{0.03}Al^{VI}_{0.38}Al^{IV}_{0.68}Si_{7.32}O_{22}(OH)_2$
	3. Average: Zoisite and Diopside Zones	$Na_{0.16}Ca_{1.94}(Fe, Mg, Mn)_{4.58}Ti_{0.03}Al^{VI}_{0.44}Al^{IV}_{0.75}Si_{7.25}O_{22}(OH)_2$
Diopside	Average: Diopside Zone	$Ca_{0.97}Na_{0.01}(Fe, Mg, Mn)_{1.01}Al^{VI}_{0.01}Al^{IV}_{0.01}Si_{1.99}O_6$
Chlorite	Average: Ankerite, Biotite, and Amphibole Zones	$(Fe, Mg, Mn)_{4.50}Al_{2.70}Si_{2.71}O_{10}(OH)_8$
Muscovite	Average: Ankerite and Biotite Zones	$(K_{0.90}Na_{0.05})\{(Fe, Mg, Mn)_{0.23}Ti_{0.03}Al_{1.78}\}(Al_{0.85}Si_{3.15})O_{10}(OH)_2$
Microcline	Average: Zoisite and Diopside Zones	$Or_{0.94}Ab_{0.06}$
Plagioclase	1. Average: Ankerite Zone	An_{01}
	2. Average: Biotite Zone	An_{29}
	3. Average: Amphibole Zone	An_{70}
	4. Average: Zoisite Zone	An_{74}
	5. Average: Diopside Zone	An_{79}
Zoisite	Average: Zoisite and Diopside Zones	$Ca_{2.00}(Al_{2.68}Fe_{0.32})Si_{3.00}O_{12}(OH)$
Scapolite	Average: Zoisite and Diopside Zones	$Ca_{3.01}Na_{0.96}K_{0.03}Al_{5.01}Si_{6.99}O_{24}CO_3$
Sphene	Average: Biotite, Amphibole, Zoisite, and Diopside Zones	$CaTiSiO_5$
Ilmenite	All Zones	$(Fe, Mg, Mn)TiO_3$
Quartz	All Zones	SiO_2

+ ankerite + quartz + albite + muscovite + calcite + chlorite. Biotite occurs in very small amounts (<3 modal per cent). The upper biotite zone (filled squares, Fig. 1) consists of rocks with the assemblage biotite + quartz + oligoclase/labradorite + calcite + chlorite ± muscovite. Rocks of the upper biotite zone contain abundant biotite (16–30 modal per cent) but no ankerite. Both subgroups contain, as accessory minerals, various combinations of pyrite, pyrrhotite, apatite, tourmaline, ilmenite, sphene and graphitic matter. Average compositions of biotite, ankerite, calcite, muscovite, chlorite, plagioclase, ilmenite, and sphene for rocks of the biotite zone are listed in Table 1.

Amphibole zone

Immediately S of the biotite zone in Fig. 1 is an area in which calcic amphibole appears in metamorphosed carbonate rocks. The amphibole isograd represents a line mapped between occurrences of metacarbonate rock with and without amphibole (Ferry 1976b). Rocks which contain amphibole but not zoisite are represented by open triangles in Fig. 1 and their occurrences constitute the amphibole zone, which consists of rocks with the assemblage calcic amphibole + quartz + andesine/anorthite + calcite + biotite ± chlorite. Accessory minerals include combinations of pyrrhotite, apatite tournaline, ilmenite, sphene, and graphic material. Average compositions of amphibole, plagioclase, calcite, biotite, and chlorite are listed in Table 1. The amphibole zone contains 2 anomalous outcrops (filled squares S of the amphibole isograd in Fig. 1) which contain rocks with mineralogy characteristic of the upper biotite zone. The preservation of these biotite zone assemblages in the amphibole zone may either be due to anomalies in bulk rock composition or due to relative isolation of the outcrops from the regional flow of metamorphic fluid. To avoid excessive complexity, tiny closed amphibole isograds have not been drawn around the locations of the 2 anomalous outcrops.

Zoisite zone

Immediately S of the amphibole zone in Fig. 1 is an area in which zoisite appears in metamorphosed carbonate rocks. The zoisite isograd represents a line mapped between occurrences of metacarbonate rock with and without zoisite (Ferry 1976b, 1978, 1980b). Rocks which contain zoisite but not diopside are represented by filled triangles in Fig. 1 and their occurrences constitute the zoisite zone, which consists of rocks which contain the assemblage zoisite + calcic amphibole + quartz + andesine/anorthite + calcite ± biotite ± microcline. Accessory minerals include combinations of muscovite (2 samples only), scapolite, garnet, pyrrhotite, apatite, tourmaline, sphene and graphite. Average compositions of zoisite, amphibole, plagioclase, biotite, microcline and scapolite are listed in Table 1 for rocks of the zoisite zone. S of the granitic stocks there is a complicated distribution of zoisite-bearing rocks with and without diopsode. N of the stocks the distribution of zoisite-bearing rocks is somewhat more regular but there nevertheless exists an isolated 'island' of metacarbonate rocks with zoisite in the amphibole zone.

Diopside zone

Immediately S of the zoisite zone in Fig. 1 is an area in which diopside appears in metamorphosed carbonate rocks. The diopside isograd represents a line mapped between occurrences of metacarbonate rocks with and without diopside (Ferry 1976b, 1978, 1980a). Rocks which contain diopside are represented by filled circles in Fig. 1 and their occurrences constitute the diopside zone, which consists of rocks containing the mineral assemblage diopside + zoisite + calcic amphibole + calcite + quartz + andesine/anorthite ± biotite ± microcline. Accessory minerals include combinations of scapolite, garnet, pyrrhotite, apatite, tourmaline, sphene, and graphite. Table 1 lists average compositions of diopside, zoisite, amphibole, calcite, plagioclase, microcline, and scapolite.

Whole-rock chemistry

115 samples of metamorphosed carbonate rock were chemically analysed for major element metal oxides and for carbonate CO_2. The purpose of the analyses was to test for systematic changes in bulk rock composition with changes in metamorphic grade. Results are presented in Fig. 2. Analyses are grouped by zones. Analyses for rocks from the lower biotite zone were grouped with analyses for rocks of the ankerite zone for the following reasons: (a) the rocks contain identical mineralogy except for 0.1–2.7 modal % biotite in rocks of the lower biotite zone; (b) the compositions of minerals in the 2 groups of rocks are indistinguishable. In order to compare the composition of metamorphic rocks from different zones, some basis for comparison must be established. Previous studies (Carmichael 1969; Thompson 1975) have suggested that relatively little mass transfer of aluminium may occur during metamorphism. An aluminium-based reference frame is therefore adopted in Fig. 2. The ratios of the atomic concentrations of major elements to that of aluminium were calculated for each analysed sample, and these ratios were averaged for each zone. Average values are the filled circles in Fig. 2. There is no detectable change in whole-rock Fe/Al, Mg/Al, Ti/Al, Si/Al or Ca/Al with increasing grade of metamorphism. There is, however, a significant decrease in both Na/Al and K/Al with increasing metamorphic grade. A similar trend has been reported for progressively metamorphosed Moinian calc-silicate rocks (Tanner & Miller 1980). In the Vassalboro Formation average diopside zone rocks contain only 31% as much Na and only 18% as much K as do average rocks from the lowest grades. The most depleted rocks in the diopside zone contain 14% as much Na and 7% as much K as do average low-grade rocks. Because all carbonate carbon was originally in the rocks as Ca-Fe-Mg-Mn carbonates (ankerite ± calcite), the carbonate carbon contents of rocks are compared in Fig. 2 using the variable $C/(Ca + Fe + Mg + Mn)$ rather than the variable C/Al. The average value of $C/(Ca + Fe + Mg + Mn)$ is less than one in the ankerite zone because in these rocks some Fe, Mg, and Mn is contained in muscovite and chlorite as well as in carbonate. Metacarbonate rocks in the diopside zone, on average, contain 30% as much carbonate carbon as do rocks in the ankerite zone.

The difference in average ratios between the different zones in Fig. 2 was statistically evaluated using Student's t test. Average K/Al, Na/Al, and $C/(Ca + Fe + Mg + Mn)$ are less in the zoisite and diopside zones than average values in the ankerite zone at a >99.9% confidence level. In contrast, average Ti/Al, Fe/Al, Mg/Al, Si/Al, and Ca/Al in the diopside and zoisite zones are not different from average values in the ankerite zone at more than a 95% confidence level. Changes in Na/Al, K/Al and $C/(Ca + Fe + Mg + Mn)$ with increasing grade are therefore statistically significant at a very high level of confidence.

In principle, the changes in Na/Al, K/Al, and $C/(Ca + Fe + Mg + Mn)$ with increasing grade could

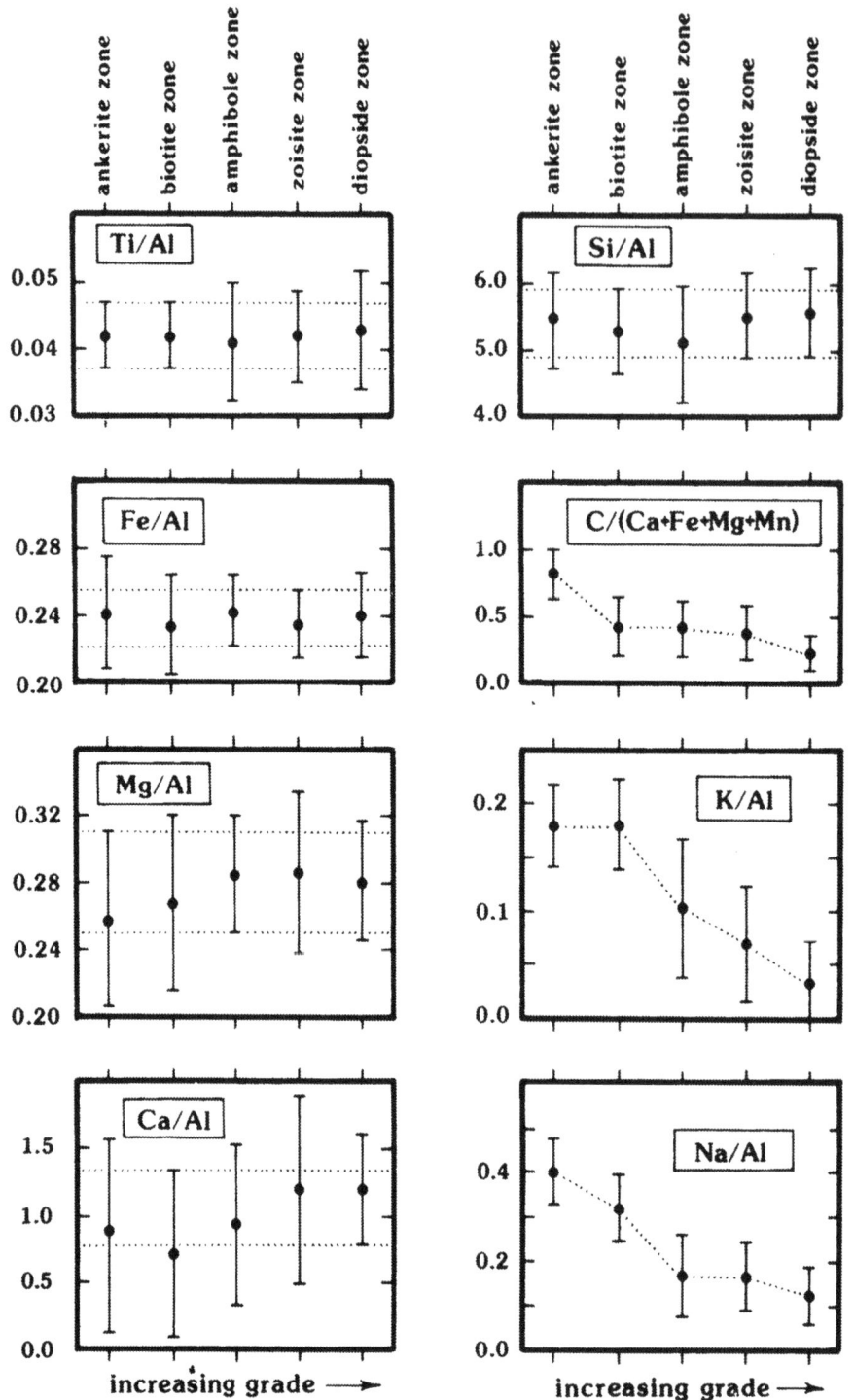

FIG. 2. Average atomic ratio of i/Al and of C/(Ca + Mg + Fe + Mn) for metacarbonate rocks in each metamorphic zone where i = K, Na, Si, Ca, Fe, Mg, Ti. Length of error bar represents 2 standard deviations.

be due either to sedimentary or metamorphic processes. Arguments have been presented (Ferry 1983) which favour a metamorphic origin for the observed whole-rock chemical changes. Metamorphic processes systematically depleted the carbonate rocks in Na, K, and C during the metamorphic event.

Prograde mineral reactions

Prograde changes in mineralogy and in whole-rock chemistry of the carbonate rocks can be directly related to a set of prograde mineral reactions. Mineralogical and chemical changes which occur in a rock during metamorphism, however, may be represented by a number of different but equivalent sets of mineral reactions (Thompson 1982; Thompson et al. 1982). The set which is used is to some degree arbitrary and subject only to these 2 constraints: the reactions must be linearly independent and they must be sufficient in number to account for all prograde changes in mineralogy which are of concern. One such set is presented here for metamorphosed carbonate rocks from the Vassalboro Formation. It is emphasized that the reactions are only one set of a number of possible, alternative ways in which the prograde chemical and mineral changes could be represented. The reactions which follow were derived from a variety of information about the metacarbonates: (a) appearance and disappearance of minerals with increasing grade; (b) changes in whole-rock chemistry with increasing grade (Fig. 2); (c) changes in mineral composition with increasing grade (Table 1); and (d) quantitative determination of changes in molar proportions of minerals with increasing grade. Because the derivation of the prograde mineral reactions is described in detail elsewhere (Ferry 1983), only a brief summary of the results is presented here. The mineral reactions which occurred in each of the zones are considered separately. Only those reactions needed to describe the principal changes in carbonate rock chemistry and mineralogy are discussed. Reactions which affect the abundance of accessory minerals or which only result in small changes in mineral composition have been neglected. A summary of all reactions appears in Fig. 3. The mineral reactions reported here are different from those in an earlier study of the metamorphism of the Vassalboro Formation (Ferry 1976a) because the earlier study erroneously assumed that metamorphism was isochemical.

Reactions in the biotite zone

The difference in mineralogy between rocks of the ankerite and biotite zones may be largely attributed to a mineral reaction which produced biotite and $CaAl_2Si_2O_8$ components

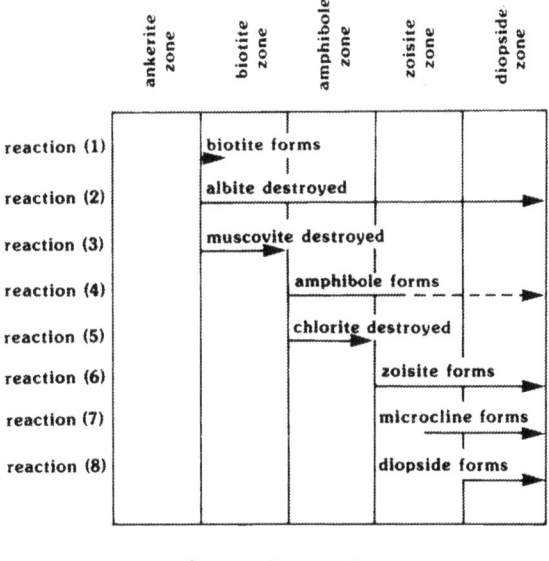

FIG. 3. Distribution of mineral reactions in metamorphosed carbonate rocks within the different metamorphic zones.

of plagioclase at the expense of ankerite and muscovite:

$$0.98 \text{ muscovite} + 2.23 \text{ ankerite} + 0.65 \text{ quartz}$$
$$+ 0.09 \text{ ilmenite} + 0.03 \text{ HCl} + 0.01 \text{ H}_2\text{O}$$
$$= 1.00 \text{ biotite (1)} + 1.89 \text{ calcite (1)}$$
$$+ 0.48 \text{ CaAl}_2\text{Si}_2\text{O}_8 + 0.03 \text{ NaCl} + 2.57 \text{ CO}_2 \quad (1)$$

The reaction utilizes average mineral compositions listed in Table 1; numbers in parentheses refer to a particular formula when more than one for that mineral appears in Table 1. The reaction (and all others which follow) are slightly imbalanced for some elements because reaction coefficients have been rounded to the nearest 0.01. Because cations are almost completely associated in fluids at the P–T conditions of metamorphism (Helgeson 1969; Helgeson et al. 1981), reaction (1) (and all others) are written with associated complexes rather than with charged species. The complexes have been somewhat arbitrarily written as chlorides. Reaction (1) (and all others) conserve Al, Si, Ti, Fe, Mg, Mn, and Ca in mineral phases, but not necessarily K, Na, CO_2, or H_2O (Fig. 2). In rocks of the lower biotite zone, reaction (1) has barely occurred: the rocks contain little biotite and abundant ankerite. In rocks of the upper biotite zone, reaction (1) has gone to completion: the rocks contain abundant biotite but no ankerite.

Reaction (1) did not release sufficient NaCl to account for the observed decrease in whole-rock Na/Al with increasing metamorphic grade in the biotite zone (Fig. 2). Furthermore, even though reaction (1) released $CaAl_2Si_2O_8$, many metacarbonate rocks in the biotite zone contain too calcic a plagioclase (e.g. An_{40}) to be explained solely by progress of reaction (1). This and other data (Ferry 1983) led to the conclusion that incongruent destruction of plagioclase occurred during prograde metamorphism according to:

$$2\,NaAlSi_3O_8 + CaCO_3 + 2\,HCl$$
$$= CaAl_2Si_2O_8 + 4\text{ quartz} + 2\,NaCl + CO_2 + H_2O \quad (2)$$

The reaction evidently occurred not only in the biotite zone but at higher metamorphic grade as well. Reaction (2) was the principal mechanism by which Na was liberated from the carbonate rocks during metamorphism.

Muscovite disappears in all but 2 samples of metacarbonate rock in the upper biotite zone. The inferred mineral reaction is:

$$1.00\text{ muscovite} + 1.28\text{ calcite (2)} + 0.84\,HCl$$
$$= 0.12\text{ biotite (1)} + 0.02\text{ sphene} + 0.36\text{ quartz}$$
$$+ 1.22\,CaAl_2Si_2O_8 + 0.79\,KCl + 0.05\,NaCl$$
$$+ 1.30\,H_2O + 1.28\,CO_2 \quad (3)$$

Because rocks in the upper biotite zone contain only a few modal per cent muscovite, reaction (3) must have played a relatively minor role in the mineralogical evolution of the metacarbonate rocks.

Reactions in the amphibole zone

The principal Fe-Mg mineral in rocks from the biotite zone is biotite. The difference in mineralogy between rocks of the biotite and amphibole zones may be largely attributed to a mineral reaction which produced calcic amphibole and $CaAl_2Si_2O_8$ component of plagioclase at the expense of biotite. No new K-rich mineral formed with amphibole; data in Fig. 2 indicate that K released from destruction of biotite was lost from the metacarbonate system. A reaction which is consistent with all these constraints is:

$$1.90\text{ biotite (2)} + 3.16\text{ calcite (2)} + 3.53\text{ quartz}$$
$$+ 0.15\,NaAlSi_3O_8 + 1.69\,HCl$$
$$= 1.00\text{ amphibole (1)} + 0.14\text{ sphene} + 0.97\,CaAl_2Si_2O_8$$
$$+ 1.69\,KCl + 3.16\,CO_2 + 1.74\,H_2O \quad (4)$$

Reaction (4) is spread out over a range of metamorphic grades. Reactants and products are found in all rocks represented as open triangles in Fig. 1 and in many of the rocks represented by filled triangles. Reaction (4) was the principal mechanism by which K was liberated from the carbonate rocks during metamorphism.

Chlorite disappears from metacarbonate rocks in the amphibole zone. The inferred mineral reaction is:

$$1.00\text{ chlorite} + 2.94\text{ calcite (2)} + 5.73\text{ quartz}$$
$$+ 0.05\text{ ilmenite} + 0.19\,NaAlSi_3O_8$$
$$= 1.02\text{ amphibole (1)} + 0.86\,CaAl_2Si_2O_8 + 2.94\,CO_2$$
$$+ 2.98\,H_2O \quad (5)$$

Rocks in the amphibole zone contain at most a few modal per cent chlorite; reaction (5) must have played a relatively minor role in the mineralogical evolution of the carbonate rocks during metamorphism.

Destruction of the albite component of plagioclase and production of the anorthite component of plagioclase according to reaction (2) occurred in the amphibole zone as well as in the biotite zone.

Reactions in the zoisite zone

The difference in mineralogy between rocks of the amphibole and zoisite zones may in part be attributed to a mineral reaction which produced zoisite at the expense of the $CaAl_2Si_2O_8$ component of plagioclase:

$$0.07\text{ Fe-component of amphibole (2)}$$
$$+ 0.59\text{ Fe-component of calcite (3)}$$
$$+ 1.31\,CaAl_2Si_2O_8 + 0.01\,HCl + 0.59\,H_2O$$
$$= 1.00\text{ zoisite} + 0.002\text{ sphene} + 0.10\text{ quartz}$$
$$+ 0.01\,NaCl + 0.59\,CO_2 + 0.16\,H_2 \quad (6)$$

The participation of amphibole is required to account for the small amount of iron in zoisite. Amphibole was chosen for the source of iron because it is the only Fe-Mg mineral which occurs in sufficient quantities in all rocks from the amphibole and zoisite zones. Reaction (6) is quenched in all samples of metacarbonate rock collected in the zoisite zone and at higher grades. Reactants and products are found in all samples from locations represented by filled triangles and circles in Fig. 1.

A second reaction accounts for the formation of microcline in rocks from the high-grade portions of the zoisite zone. The only K-bearing mineral from which microcline could have formed is biotite; the inferred microcline-forming reaction is:

$$1.06\text{ biotite (2)} + 1.28\text{ calcite (3)} + 3.94\text{ quartz}$$
$$+ 0.13\,NaAlSi_3O_8 = 1.00\text{ microcline} + 0.55\text{ amphibole (3)}$$
$$+ 0.09\text{ sphene} + 0.08\,CaAl_2Si_2O_8 + 0.50\,H_2O + 1.28\,CO_2$$
$$\quad (7)$$

Scapolite occurs in one rock and garnet occurs in another collected from the zoisite zone. Because of the rarity of the 2 minerals, the reactions by which they formed are not considered. In the zoisite zone biotite was also converted to calcic amphibole according to reaction (4) and albite was converted to the anorthite component of plagioclase according to reaction (2) as at lower metamorphic grade.

Reactions in the diopside zone

The difference in mineralogy between rocks of the zoisite and diopside zones may in part be attributed to a mineral reaction which produced diopside at the expense of calcic amphibole:

$$0.22\text{ amphibole (3)} + 0.70\text{ calcite (3)} + 0.67\text{ quartz}$$
$$+ 0.03\,HCl = 1.00\text{ diopside} + 0.01\text{ sphene}$$
$$+ 0.12\,CaAl_2SiO_8 + 0.03\,NaCl + 0.23\,H_2O + 0.70\,CO_2$$
$$\quad (8)$$

Reaction (8) is spread out over the entire diopside zone. Reactants and products are found in each specimen from locations represented by a filled circle in Fig. 1.

In the diopside zone zoisite also formed according to reaction (6); microcline formed according to reaction (7); biotite was consumed according to reaction (4); and the albite component of plagioclase was converted to anorthite according to reaction (2). Garnet occurs in one rock and scapolite in another 2 rocks collected from the diopside zone.

Because the minerals occur in such a small number of rocks, the reactions which produced the minerals have been ignored.

Pressure, temperature, and fluid composition

Pressure and temperature

Pressure and temperature during metamorphism of the Waterville Formation in the area of Fig. 1 has been previously characterized (Ferry 1980b). Pressure near the sillimanite isograd was 3500 ± 200 bar (based on 4 independent geobarometers). Mean temperatures (based on the biotite-garnet geothermometer) for zones of the Waterville Formation defined by isograds in pelitic schists are: garnet zone, 460°C; staurolite-andalusite zone, 490°C; sillimanite zone, 550°C. Pressure and temperatures in the Vassalboro Formation were undoubtedly very similar. Calcite-dolomite geothermometry supplements these temperatures at lower grades (Ferry 1976a, 1979). Calcite-dolomite pairs in 2 outcrops from the middle of the garnet zone of the Waterville Formation both record a temperature near 460°C. 5 samples of metacarbonate rock from the lower biotite zone of the Vassalboro Formation record an average temperature of 408°C.

Numerous estimates of temperature have been calculated from mineral equilibria in metamorphosed carbonate rocks collected in the zoisite and diopside zones of the Vassalboro Formation (Ferry 1976a, 1978, 1980a). All these temperatures have been recalculated using new thermochemical data for the anorthite-calcite-zoisite-fluid equilibrium extracted from recent experiments by Allen & Fawcett (1982). New values of $\Delta \bar{H}^0$ and $\Delta \bar{S}^0$ for the reaction

$$2\ Ca_2Al_3Si_3O_{12}(OH) + CO_2 = 3\ CaAl_2Si_2O_8 + CaCO_3 + H_2O, \quad (9)$$

derived from the experimental results of Allen & Fawcett and used in the calculations, are 13 346 cal and 24.653 cal/degree, respectively (calculated values refer to a standard state of one bar pressure and are assumed to be independent of temperature within the range of temperatures at which the reaction occurs). Recalculated temperatures also reflect a new activity-composition model for metamorphic plagioclase solid solutions (Grove et al. 1983). They postulated a miscibility gap in the plagioclase solid solution series between $\sim An_{40}$ and $\sim An_{90}$ at temperatures between $\sim 400°C$ and $\sim 575°C$. (Average values of plagioclase composition in Table 1 of An_{70}, An_{74}, and An_{79} do not represent *actual* measured feldspar compositions; they represent the average of numerous feldspar compositions from both sides of the miscibility gap. They may represent average plagioclase composition in a rock without referring to the composition of individual grains.) Consequently the following activity-composition model was used: $a_{CaAl_2Si_2O_8,plagioclase} = X_{CaAl_2Si_2O_8,plagioclase}$ for $X \geq 0.9$; $a_{CaAl_2Si_2O_8,plagioclase} = 0.9$ for $0.4 \leq X \leq 0.9$. In all other respects procedures for calculating temperature were the same as in earlier reports (Ferry 1976a, 1978, 1980a). For 4 samples of metacarbonate rock from the zoisite zone of the Vassalboro Formation, calculated temperature was in the range 475–520°C with an average of $\sim 510°C$. For 43 samples from the diopside zone of the Vassalboro Formation, calculated temperature was in the range 490–545°C with an average of $\sim 520°C$.

Temperatures calculated from mineral equilibria both in the pelitic schists of the Waterville Formation and in the metacarbonate rocks suggest the following temperatures were attained during metamorphism of the Vassalboro Formation (Fig. 4): ankerite zone <400°C; lower biotite zone: $\sim 400°C$; upper biotite zone: ~ 400–$440°C$; amphibole zone ~ 400–$500°C$; zoisite zone: ~ 500–$520°C$; diopside zone $> \sim 520°C$. These values are approximate both because of inherent uncertainties in the geothermometers themselves and because there almost certainly existed some overlap in temperature between adjacent zones (especially the zoisite and diopside zones).

Fluid composition

Previous calculations suggest that metacarbonate rocks in the Vassalboro Formation were in equilibrium during metamorphism with approximately binary CO_2-H_2O fluids (Ferry 1976a 1979). 2 mineral equilibria, illustrated in the T-X_{CO_2} diagram of Fig. 4, indicate that the fluids were H_2O-rich at all grades. Because carbonate rocks in the biotite zone and at higher grades contain sphene, equilibrium metamorphic fluids must have been characterized by an X_{CO_2} less than that of the rutile + calcite + quartz + sphene + fluid equilibrium curve. Because rocks in the biotite and amphibole zone contain calcic plagioclase + calcite without zoisite, equilibrium metamorphic fluids must have been characterized by an X_{CO_2} greater than that for reaction (9), corrected for solid solution. Corrections for solid solution, expressed in terms of lnK_s for reaction (9) (Ferry 1976a), are in the range −2.22 to +0.71. The 2 values of lnK_s result in 2 calculated curves for the plagioclase-calcite-zoisite-fluid equilibrium in Fig. 4. In the biotite and amphibole zones, fluid composition must have been in the shaded area between left-hand plagioclase-calcite-zoisite curve and the rutile-calcite-quartz-sphene curve. Metamorphic fluids in the lower biotite zone had composition $X_{CO_2} = 0.03$–0.05; fluids in the upper biotite zone, $X_{CO_2} = 0.02$–0.10; fluids in the amphibole zone, $X_{CO_2} = 0.01$–0.29. Fluid composition could be calculated directly for rocks of the zoisite and diopside zones from the plagioclase-calcite-zoisite fluid equilib-

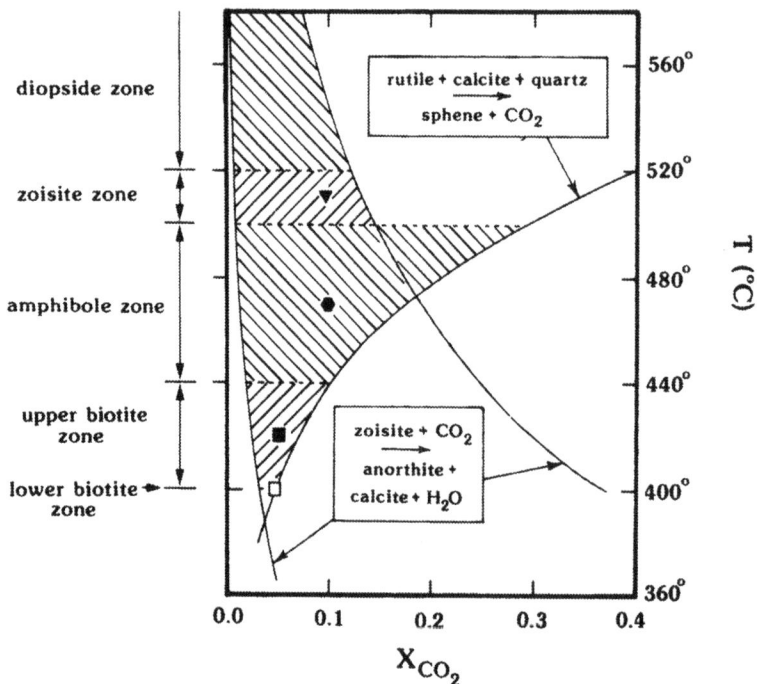

FIG. 4. Inferred temperatures and fluid compositions during metamorphism of the carbonate rocks (shaded area). Sphene-calcite-rutile-quartz curve calculated from data of Jacobs & Kerrick (1981). Zoisite-calcite-plagioclase curves are calculated from data of Allen & Fawcett (1982) and reflect the range in observed zoisite, calcite, and plagioclase compositions in the metacarbonate rocks (left-hand curve: $\ln K_s = -2.217$; right-hand curve: $\ln K_s = +0.713$).

rium at inferred or calculated temperatures of metamorphism. Results are: zoisite zone, $X_{CO_2} = 0.09-0.12$; diopside zone, $X_{CO_2} = 0.06-0.13$. Estimated fluid compositions, summarized in Fig. 4, suggest that carbonate rocks of the Vassalboro Formation were in equilibrium with H_2O-rich fluids at all grades during the metamorphic event.

Fluid/rock interaction during metamorphism

Buffering of fluid composition by mineral reactions

Reactions (1), (3), (4), (6), (7) and (8) are schematically represented on an isobaric T-X_{CO_2} diagram (Fig. 5). The reactions may rigorously be represented as curves only if there is no change in mineral composition during reaction and if the concentration of each dissolved electrolyte in the equilibrium fluid is constant. During metamorphism there were some changes not only in mineral composition but also in the concentration of electrolytes in fluid during progress of the reactions. Fig. 5, however, may be considered appropriate in a qualitative fashion, and some general deductions about fluid/rock interaction during metamorphism may be made from it. The reactions account for the major mineralogical changes that occurred in the carbonate rocks during metamorphism. The curve for each reaction is positioned in order of increasing temperature in accordance with field observations.

Reactants and products for reactions (1), (3), (4), (6), (7) and (8) are spread out in the field over a range of grades and hence were in equilibrium over a range of temperatures. If progressive metamorphism of the carbonate rocks, in terms of T and X_{CO_2}, is represented by a path through Fig. 5, then the path, at least partially, must follow the reaction curves in Fig. 5. The dotted curve is such a path and its position is consistent with (a) the prograde sequence in which minerals appear and disappear in the carbonate rocks, (b) the distribution of mineral reactants and products in the field, and (c) the observation that metamorphic fluids were H_2O-rich at all grades. Along the portions of the dotted path which follow the reaction curves, X_{CO_2} of fluid in equilibrium with reacting minerals is buffered by the mineral reaction (Greenwood 1975). It is evident from Fig. 5 that for almost all of the metamorphic history of the carbonate rocks, mineral reactions internally controlled the X_{CO_2} of coexisting fluid. This conclusion is supported by a study of

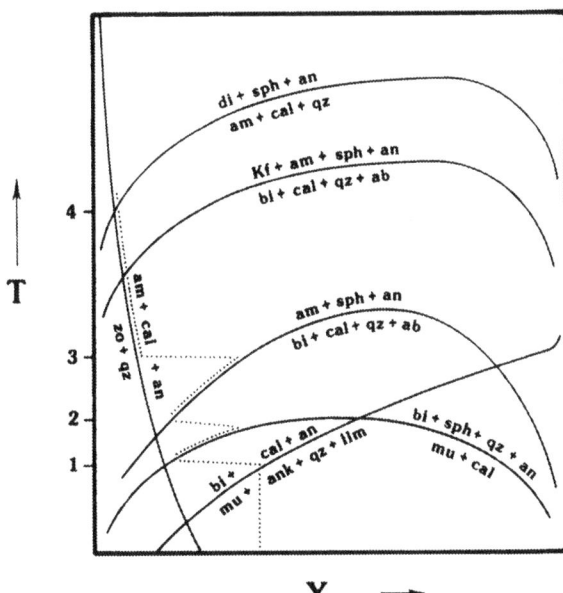

FIG. 5. Qualitative T-X_{CO_2} diagram illustrating mineral equilibria which correspond to reactions (1), (3), (4) and (6)–(8) in the text. Dotted path explained in text. The many other curves and metastable extensions required by Schreinemakers' rules have been omitted for simplicity. Abbreviations: mu = muscovite; ank = ankerite; qz = quartz; ilm = ilmenite; bi = biotite; cal = calcite; an = anorthite; sph = sphene; ab = albite; am = amphibole; zo = zoisite; Kf = microcline; di = diopside. Temperature scale: lower biotite zone, $T \sim T_1$; upper biotite zone, $T \sim T_1$–T_2; amphibole zone, $T \sim T_2$–T_3; zoisite zone, $T \sim T_3$–T_4; diopside zone, $T > T_4$.

metacarbonate rocks in a portion of the Waterville Formation in Fig. 1 in which it was demonstrated that differences in X_{CO_2} existed during metamorphism between adjacent beds within the same outcrop (Ferry 1979). The differences in X_{CO_2} within the outcrop further argue that fluid composition was controlled internally by mineral reactions in the carbonate rocks during the metamorphic event.

Infiltration of carbonate rock by H_2O-rich fluids

Metamorphism of the carbonate rocks presents an interesting paradox. It is evident from Fig. 5 that mineral reactions controlled fluid composition during metamorphism. The mineral reactions (1)–(8) all produced CO_2-rich fluid ($X_{CO_2} \geq 0.5$). On the other hand, fluids with which the rocks were in equilibrium were H_2O-rich ($X_{CO_2} \leq 0.3$). This apparent paradox has been observed before (Ferry 1980a; Rumble et al. 1982) and has been resolved through consideration of infiltration of the carbonate rocks by H_2O-rich fluids. Although the mineral reactions buffered X_{CO_2} during metamorphism, the H_2O-rich fluid, occupying a rock's pore space at any one instant and in equilibrium with the rock, was actually a mixture. Part of the mixture was a CO_2-rich component produced internally by mineral reaction; the other part of the mixture was an H_2O-rich component introduced into the carbonate rock from some (unknown) external source. The 2 components mixed in the pore space of the carbonate rocks during metamorphism and minerals in the rock equilibrated with the mixture. In this fashion mineral reactions in the carbonates could internally produce CO_2-rich fluids but the rock could record equilibrium with an H_2O-rich fluid. The mechanism by which H_2O-rich fluid was introduced into the carbonate rocks during metamorphism is considered to be infiltration. The subhorizontal portions of the dotted T-X_{CO_2} path in Fig. 5 represent the effect of infiltration on fluid composition in the carbonate rocks at points in their metamorphic evolution when they contained no mineral assemblages which buffered fluid composition. Because some equilibrium fluids themselves are very H_2O-rich ($X_{CO_2} \leq 0.052$ in the lower biotite zone; $X_{CO_2} = 0.06$–0.13 in the diopside zone), the infiltrating fluid must have been nearly pure H_2O (or an aqueous brine). The source of the infiltrating H_2O fluid is uncertain; it may represent the product of dehydration reactions in adjacent beds, fluid expelled by crystallizing granitic magmas (now the granitic stocks in Fig. 1), metamorphic fluids from deeper in the crust, fluid from the hydrosphere, or some combination of these 4 (Ferry 1976b, 1978).

The minimum *amount* of fluid which infiltrated any particular metacarbonate rock sample may be estimated if the following information is known (Ferry 1980a): (a) the mineral reactions which proceeded during infiltration; (b) the measured progress of each reaction; (c) the composition of fluid with which the rock was in equilibrium during infiltration; and (d) the composition of the infiltrating fluid. It was argued above that the infiltrating fluid was very H_2O-rich. Calculations have been simplified by assuming that the infiltrating fluid was pure H_2O (a conservative action which leads to minimum calculated amounts of fluid). If n_{H_2O} moles H_2O infiltrates a carbonate rock, buffer reaction i will proceed until:

$$X^{eq}_{CO_2,i} = \frac{\nu_{CO_2,i}\xi_i}{(\nu_{CO_2,i} + \nu_{H_2O,i})\xi_i + n_{H_2O}} \quad (10)$$

where $X^{eq}_{CO_2,i}$ is the composition of fluid in equilibrium with the rock during progress, ξ_i, of reaction i; $\nu_{CO_2,i}$ and $\nu_{H_2O,i}$ are the stoichiometric coefficients for CO_2 and H_2O, respectively, in the mineral-fluid reaction. Equation (10) can be rewritten substituting a certain volume of H_2O, V_{H_2O}, for a certain number of moles H_2O, and rearranged:

$$V_{H_2O} = \bar{V}_{H_2O,i} \left\{ \frac{\nu_{CO_2,i}\xi_i - X^{eq}_{CO_2,i}(\nu_{CO_2,i} + \nu_{H_2O,i})\xi_i}{X^{eq}_{CO_2,i}} \right\}$$

$$(11)$$

where $\bar{V}_{H_2O,i}$ is the molar volume of H_2O at the physical conditions of the reaction. If more than one mineral reaction proceeds during infiltration, then

$$V^T_{H_2O} = \sum_i \bar{V}_{H_2O,i} \left\{ \frac{\nu_{CO_2,i}\xi_i - X^{eq}_{CO_2,i}(\nu_{CO_2,i} + \nu_{H_2O,i})\xi_i}{X^{eq}_{CO_2,i}} \right\} \quad (12)$$

where $V^T_{H_2O}$ is the total volume of H_2O fluid that infiltrated the rock sample during progress of all mineral reactions.

Equation (12) was used to estimate the minimum amount of H_2O which infiltrated the Vassalboro Formation during metamorphism in the lower biotite zone and at higher grades. Only reactions (1), (2), (4), (6), (7) and (8) were considered; other mineral reactions progressed to such small extents that they could safely be neglected. Progress variables for each reaction were defined as follows:

$\xi_6 \equiv$ moles zoisite/(1000 cm³ metamorphic rock);   (13)

$\xi_7 \equiv$ moles microcline/(1000 cm³ metamorphic rock);   (14)

$\xi_8 \equiv$ moles diopside/(1000 cm³ metamorphic rock);   (15)

$\xi_4 \equiv$ moles amphibole/(1000 cm³ metamorphic rock)
$- 0.55\xi_7 + 0.22\xi_8 + 0.07\xi_6$;   (16)

$\xi_1 \equiv$ moles biotite/(1000 cm³ metamorphic rock)
$+ 1.06\xi_7 + 0.418\xi_8 + 0.13\xi_6 + 1.90\xi_4$;   (17)

$\xi_2 \equiv$ moles anorthite/(1000 cm³ metamorphic rock)
$- 0.589\xi_7 - 0.534\xi_8 + 1.178\xi_6 - 1.882\xi_4 - 0.48\xi_1$.   (18)

Expression (16) involves ξ_6, ξ_7, and ξ_8 both because not all amphibole in a high-grade rock may have been produced by reaction (4) and because not all amphibole produced by reaction (4) may remain in the rock. More specifically, some amphibole may have been produced by reaction (7). For every increment of progress of reaction (7), 0.55 moles amphibole are produced. An amount of amphibole equal to $0.55\xi_7$ therefore must be subtracted from the total amount of amphibole in the rock to obtain ξ_4. Furthermore, some amphibole produced by reaction (4) in a rock may have been consumed by reactions (6) and/or (8). For every increment of progress of reactions (6) and (8) 0.07 and 0.22 moles amphibole are consumed, respectively. An amount of amphibole equal to $0.07\xi_6$ and $0.22\xi_8$ therefore must be added to the total amount of amphibole in the rock to obtain ξ_4. Combining these corrections for reactions (6), (7) and (8) results in equation (16) which relates ξ_4 to the total amount of amphibole in a rock and measured values of ξ_6, ξ_7, and ξ_8. Similarly equations (17) and (18) contain terms which involve ξ_1, ξ_4, ξ_6, ξ_7, and ξ_8 because biotite and $CaAl_2Si_2O_8$ are involved in a number of the prograde mineral reactions. Molar amounts of minerals were calculated from measured modes and molar volume data for minerals. The ν_{CO_2} and ν_{H_2O} in equation (12) are taken directly from reactions (1), (2), (4), (6), (7) and (8). Table 2 summarizes the inferred $T - X^{eq}_{CO_2,i}$ conditions at which each reaction (1), (2), (4), (6), (7) and (8) occurred. These inferred conditions are plotted as the various symbols in Fig. 4.

Volumes of H_2O that infiltrated each of 89 samples of carbonate rock from the Vassalboro Formation

TABLE 2. T-$X^{eq}_{CO_2}$ conditions during progress of mineral reactions in metamorphosed carbonate rocks of the Vassalboro Formation

Reaction Number	$T(°C)$	$X^{eq}_{CO_2}$
(1)	400°	0.05
(4)	470°	0.10
(6)	510°	calculated for each sample
(7)	calculated for each sample	calculated for each sample
(8)	calculated for each sample	calculated for each sample
(2)*	400°	0.05
(2)†	420°	0.10
(2)‡	470°	0.10
(2)§	480°	0.10

* lower biotite zone; † upper biotite zone; ‡ amphibole zone; § zoisite and diopside zones.

were calculated from equation (12). Results were divided by 1000 to express them in terms of volumetric fluid/rock ratios. At a number of outcrops more than one lithologic layer of carbonate rock was sampled. Within any particular outcrop, calculated fluid/rock ratios varied by no more than $\sim \pm 0.3$, while over the entire terrain, fluid/rock ratios varied between 0.01 and 3.08. Consequently any sample collected from an outcrop may be considered to approximately record the amount of H_2O that infiltrated the entire outcrop (see also discussion in Ferry 1980a). Fluid/rock ratios for each outcrop were therefore plotted on a map of the area and contoured (Fig. 6). The biotite isograd in Fig. 6 constitutes a contour of zero fluid/rock ratio because calculations only consider fluid which infiltrated carbonate rock under metamorphic conditions of the biotite zone and at higher grades. Results portray the general amount and distribution of H_2O fluid that infiltrated the Vassalboro Formation during metamorphism. Fluid/rock ratios in Fig. 6 are minimum estimates for 3 reasons: (a) fluid could have flowed through rocks during metamorphism without chemically reacting with them; (b) the composition of infiltrating fluid was chosen as pure H_2O; larger fluid/rock ratios would be calculated if infiltrating fluid was characterized by $X_{CO_2} > 0$; (c) the equilibrium fluid in the lower biotite zone was taken as the most CO_2-rich permitted by the mineral equilibria; larger fluid/rock ratios would be calculated if $X^{eq}_{CO_2,i} < 0.05$ were chosen.

Equation (12) can not only be used to calculate fluid/rock ratios integrated over the entire metamorphic event, but it can also be used to look at fluid/rock ratios *during any portion of the metamorphic event*. For example, if one were interested in fluid/rock ratios at the peak of high-grade metamorphism, equation (12) would be applied, summing

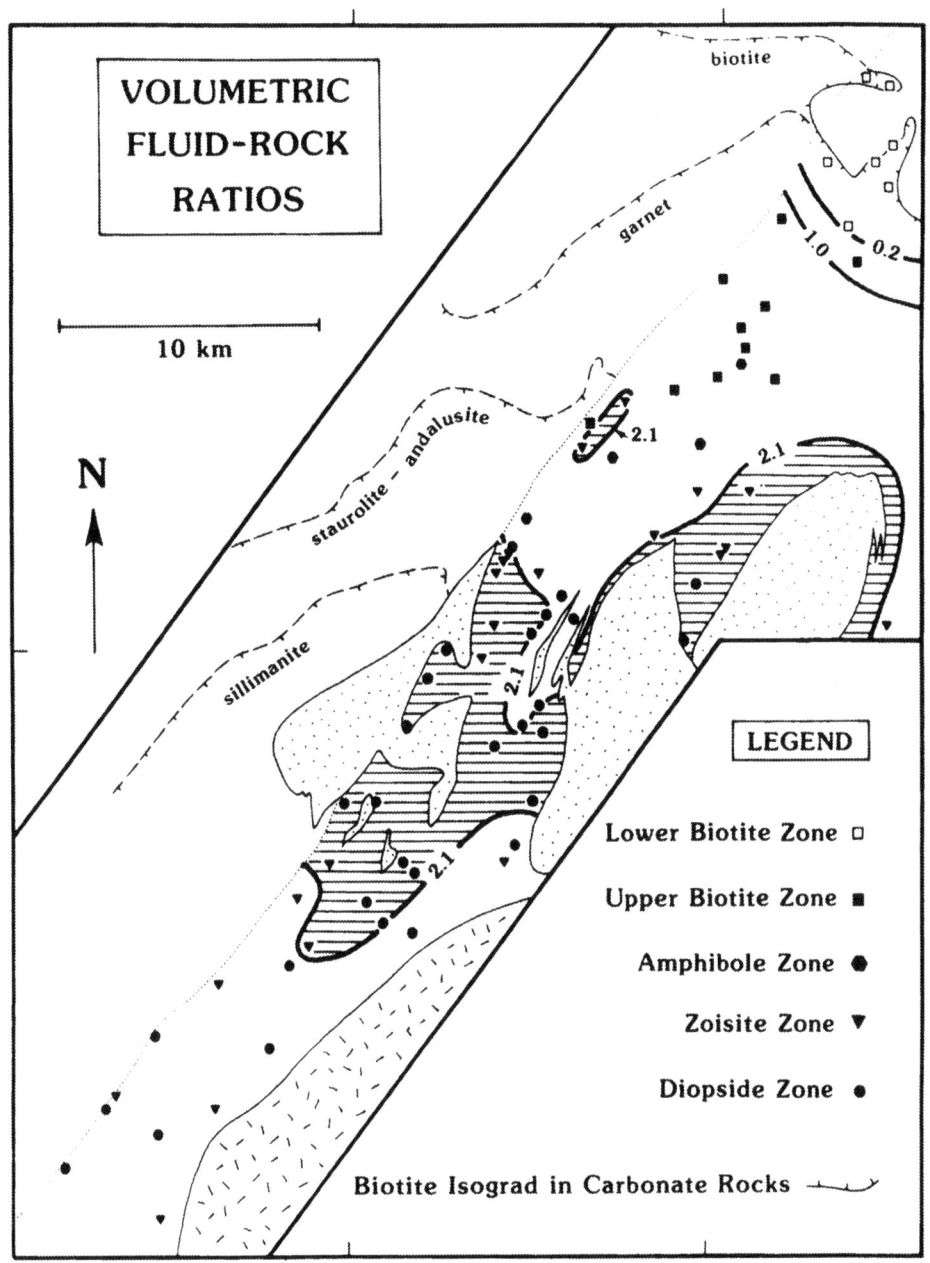

FIG. 6. Regional pattern of volumetric fluid/rock ratios during metamorphism. Symbols are the data set used to determine the contour lines. Fluid/rock ratios are integrated over the entire metamorphic event.

over the effects of reactions (6), (7) and (8) and ignoring the effects of reactions (1)–(4) which occurred early in the metamorphic evolution of the high-grade rocks. Such an exercise was carried out for carbonate rocks from the zoisite and diopside zones of the Vassalboro Formation in Fig. 1. Results were plotted on a map of the area and contoured (Fig. 7).

The zoisite isograd in Fig. 7 constitutes a contour of zero fluid/rock ratio because calculations only consider fluid which infiltrated carbonate rock under metamorphic conditions of the zoisite zone and at higher grades. Fig. 7 provides more detailed information about fluid/rock interaction at the peak of metamorphism in the zoisite and diopside zones than does

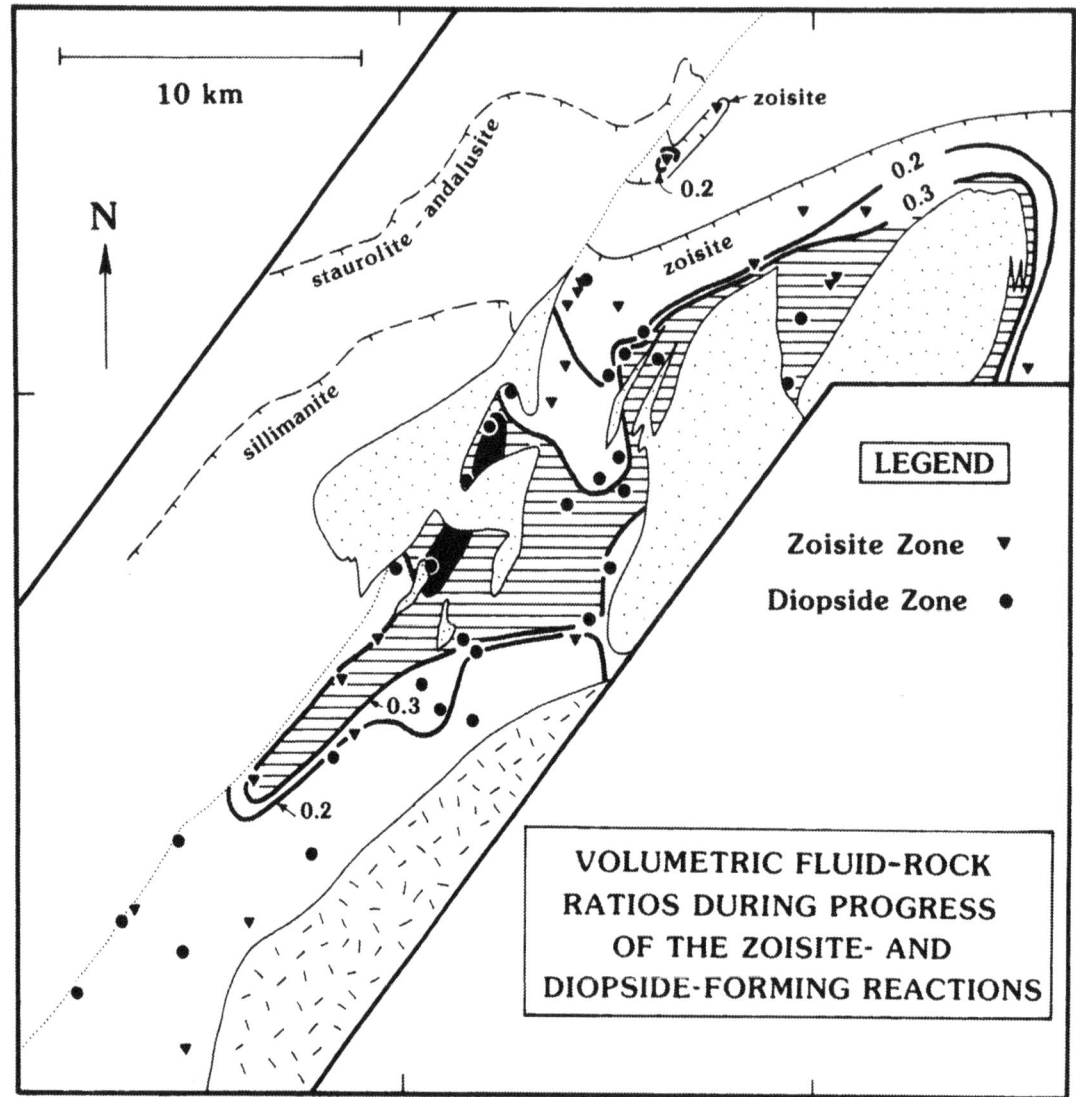

FIG. 7. Regional pattern of volumetric fluid/rock ratios during the portion of the metamorphic event in which mineral reactions (6)–(8) produced zoisite, diopside, and microcline. Symbols are the data set used to determine the contour lines. Black area refers to a region in which fluid/rock ratios >0.5.

Fig. 6. Comparison of Figs 6 and 7 illuminates differences in behaviour of fluids between early and late times during the metamorphic event. For example, it is clear from Figs 6 and 7 that most of the fluid that infiltrates a rock passes through the rock under low-grade conditions during the early stages of the metamorphic episode. Fig. 7 is similar to a figure published earlier (Ferry 1980a). The principal difference is that new thermochemical data for the zoisite-calcite-anorthite-fluid equilibrium has resulted in smaller values of calculated $X^{eq}_{CO_2,i}$ and hence of larger values of calculated fluid/rock.

Heat/rock interaction during metamorphism

Metamorphism is, in part, a consequence of the interaction of rocks with heat. The metamorphic heat budget of a particular rock sample may be calculated if the following information is known: (a) the peak temperature of metamorphism; (b) the mineral reactions which occurred in the specimen during metamorphism; and (c) the extent to which each reaction progressed. The quantitative formulation of the total heat budget, Q_T, of a rock then is (Ferry 1980a):

$$Q_T = c\Delta T + Q_{rxn}, \quad (19)$$

where c is the volumetric heat capacity of the rock, ΔT is the change in temperature that occurred during metamorphism, and Q_{rxn} is heat consumed or liberated as chemical work driving mineral reactions.

$$Q_{rxn} = \sum_i (\Delta \bar{H}_i^{P,T})\xi_i, \quad (20)$$

where $\Delta \bar{H}_i^{P,T}$ is the enthalpy of reaction i at the pressure–temperature conditions of reaction, and ξ_i is the extent to which reaction i progressed during metamorphism.

Amounts of heat that interacted with each of 89 samples of carbonate rock from the Vassalboro Formation were calculated from equation (19). Only reactions (1), (2), (4), (6), (7) and (8) were considered; other mineral reactions progressed to such a small degree that they contributed negligibly to the total heat budget. The ξ were determined by equations (13)–(18). Values of $\Delta \bar{H}_i^{P,T}$ were estimated from data for model reactions listed in Table 3 and from molar volume data for CO_2 and H_2O (Shmonov & Schmulovich 1974; Burnham et al. 1969). The value of c was taken as 0.77 cal/cm³-degree for all samples (Ferry 1980a). The change in temperature during metamorphism, ΔT, was taken with reference to the biotite isograd (T ~ 400°C); values of ΔT therefore were calculated as T-400 where T(°C) is the inferred maximum temperature attained by a particular rock during the metamorphic event (Table 2).

Calculated results for each sample were divided by 1000 and expressed in terms of cal/cm³ metamorphic rock. The sign of all calculated values of Q_T was positive, which signifies that heat was added to all rocks during the metamorphic episode. The range in calculated Q_T for all samples was between 1 and 432 cal/cm³; the range in calculated Q_T for samples collected at the same outcrop is ~±50 cal/cm³. Consequently any sample from a particular outcrop may be considered to approximately record the amount of heat that interacted with the outcrop as a whole. Heat/rock ratios for each outcrop were therefore plotted on a map of the area and contoured (Fig. 8). The biotite isograd in Fig. 8 constitutes a contour of zero heat/rock ratio because calculations only considered heat which interacted with carbonate rock at conditions of the biotite zone and at higher grades.

Discussion

Distribution of heat and fluid during metamorphism

Figs 6 and 8 illustrate the distribution of fluid and heat within the carbonate rocks of the Vassalboro Formation integrated over the entire metamorphic event. The amount of heat and fluid which interacted with rock increased rapidly S of the biotite isograd within the biotite zone. Amounts of heat and fluid continued to increase southward of the biotite zone with increasing grade but at a rate less than the rate of increase in the biotite zone. It is interesting to note from Figs 6 and 8 that approximately half the fluid/rock and heat/rock interaction which high-grade rocks experienced occurred under biotite-zone conditions. Heat and fluid transfer appear to have been much more intense at lower grades than at higher grades.

The shaded areas in Figs 6–8 refer to regions in the diopside and zoisite zone which interacted with substantial amounts of heat and fluid. The average fluid/rock ratio over an area of ~100 km² is >2 by volume. The area of high fluid/rock and high heat/rock ratios are spatially closely related to the granitic stocks. The syn-metamorphic granitic stocks

TABLE 3. *Model reactions used in the estimation of* $\Delta \bar{H}_i^{P,T}$ *for mineral reactions in metamorphosed carbonate rocks of the Vassalboro Formation*

Reaction Number	Model Reaction	Source of $\Delta \bar{H}_i^{P,T}$ for Model Reaction
(1)	$3 CaMg(CO_3)_2 + KAl_3Si_3O_{10}(OH)_2 + 2 SiO_2$ $= KMg_3AlSi_3O_{10}(OH)_2 + CaAl_2Si_2O_8 + 2 CaCO_3 + 4 CO_2$	Extracted from experimental data of Hewitt (1973b); Puhan (1978)
(2)	$2 NaAlSi_3O_8 + CaCO_3 + 2 H^+ = CaAl_2Si_2O_8 + 4 SiO_2 + CO_2$ $+ H_2O + 2 Na^+$	Helgeson et al. (1978)
(4)	$1.667 KMg_3AlSi_3O_{10}(OH)_2 + 2.833 CaCO_3 + 4.667 SiO_2 + 1.667 H^+$ $= Ca_2Mg_5Si_8O_{22}(OH)_2 + 0.833 CaAl_2Si_2O_8 + 1.500 H_2O$ $+ 2.833 CO_3 + 1.667 K^+$	Helgeson et al. (1978)
(6)	$1.5 CaAl_2Si_2O_8 + 0.5 CaCO_3 + 0.5 H_2O = Ca_2Al_3Si_3O_{12}(OH) + 0.5 CO_2$	Extracted from experimental data of Allen & Fawcett (1982)
(7)	$KMg_3AlSi_3O_{10}(OH)_2 + 1.2 CaCO_3 + 4.8 SiO_2$ $= KAlSi_3O_8 + 0.6 Ca_2Mg_5Si_8O_{22}(OH)_2 + 1.2 CO_2 + 0.4 H_2O$	Ferry (1976a)
(8)	$0.2 Ca_2Mg_5Si_8O_{22}(OH)_2 + 0.4 SiO_2 + 0.6 CaCO_3$ $= CaMgSi_2O_6 + 0.6 CO_2 + 0.2 H_2O$	Ferry (1976a)

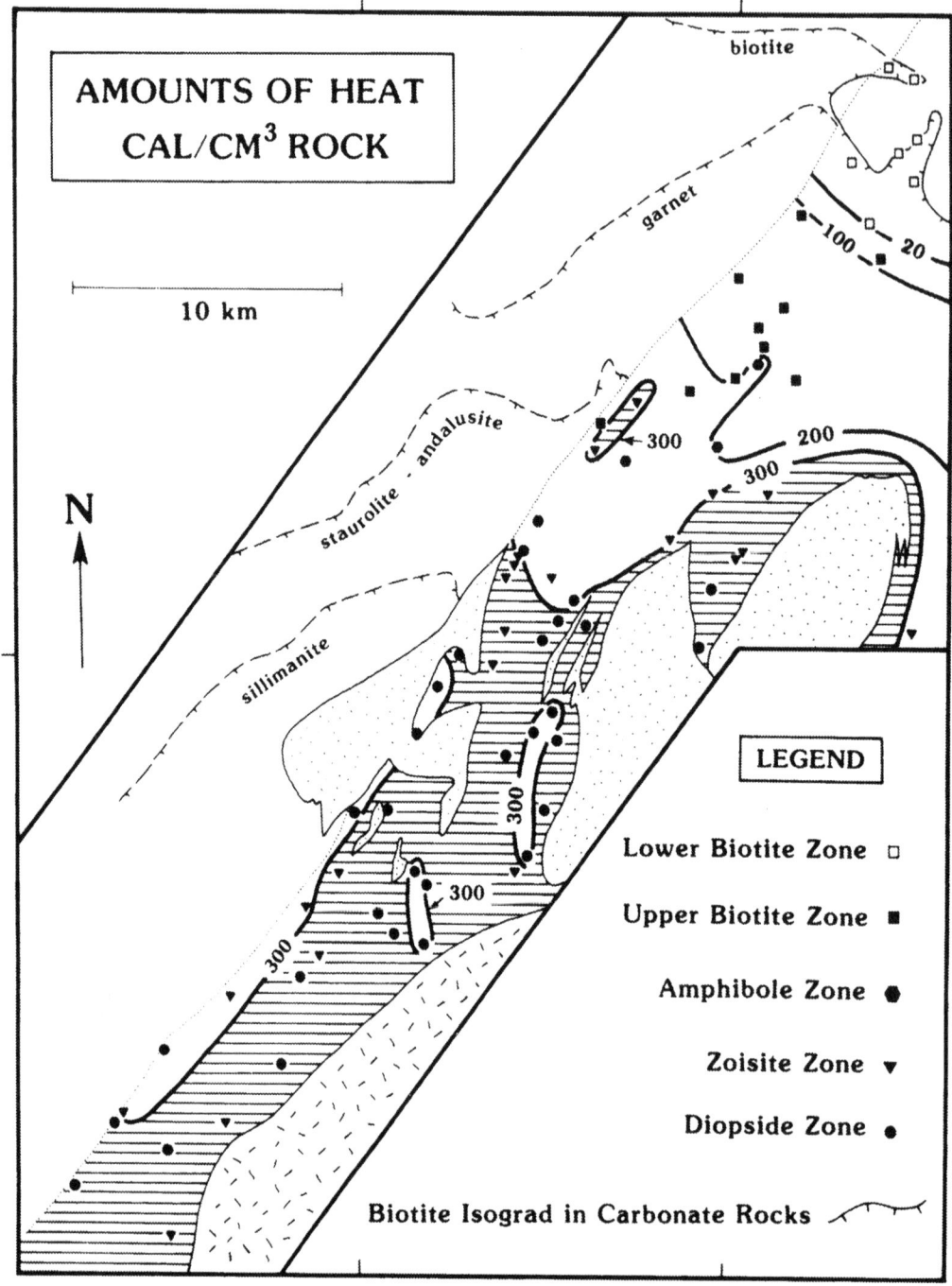

FIG. 8. Regional distribution of the amount of heat involved in metamorphism. Symbols are the data set used to determine the contour lines. Amounts of heat were determined from calculations which integrate over the entire metamorphic event.

appear to have been foci for high levels of heat and fluid transport during metamorphism. The contours in Figs 6–8 are believed to delineate a metamorphic hydrothermal cell analogous to those developed around epizonal plutons (Taylor 1977; Norton & Knight 1977). If the area in Fig. 1 is not atypical, syn-metamorphic plutons may drive convective heat and mass transfer deep in the crust with an intensity that matches or exceeds that which occurs and is well-documented at shallower depths.

Figs 7 and 8 illustrate that there is some fabric to the distribution of heat and fluid in the high-grade portion of the Vassalboro Formation. In Fig. 7 there is an elongated area of high fluid/rock ratio which strikes NE–SW (a metamorphic channelway of high fluid flow?). A similar fabric is seen in Fig. 8 in which elongated areas of low heat/rock ratio also strike roughly NE–SW. The strike of the elongated features in Figs 7 and 8 parallels a regional schistosity and orientation of vertical beds that are a result of isoclinal folding in the area. The isoclinal folds formed well before the peak of metamorphism, and hence the structures existed when the heat and fluid transfer occurred during metamorphism. It would not be unreasonable to expect that heat and fluid transfer would occur more readily in directions parallel to bedding and schistosity than in directions at high angles to them. The parallel orientation of pre-metamorphic foliation and bedding and of elongated contours which relate to metamorphic heat and fluid transfer therefore suggests that the structural geology may have exerted significant control over mass and energy transport during metamorphism.

Equations (19) and (20) permit investigation of the partitioning of heat during metamorphism between chemical work (Q_{rxn}) and changes in rock temperature ($c\Delta T$). Fig. 9 illustrates the relation between Q_{rxn} and Q_T for the metacarbonate rocks of the Vassalboro Formation. The boxes represent data for samples in each of the metamorphic zones: the centre of each box is the mean for all samples from that zone and the length of the sides of the box represents 2 standard deviations of the calculated quantities. Straight lines indicate where data points would lie if various percentages of the total metamorphic heat budget (Q_T) were used to drive mineral reactions. Actual data demonstrate that a large proportion of metamorphic heat is dissipated as chemical work driving mineral reactions. At low grades almost all heat added to rock is consumed as chemical work. With increasing grade, a progressively smaller fraction of heat is used to drive mineral reactions and a larger proportion causes changes in the temperature of the rock to occur.

Fluid/rock interaction involving buffering and infiltration

There is compelling evidence to expect that under many circumstances devolatilization reactions in rocks control or buffer the composition of coexisting metamorphic fluid (Greenwood 1975). Greenwood's theoretical arguments have been thoroughly substantiated by studies of a variety of rock types (e.g. Rice 1977a, b; Rumble 1978; Ferry 1981). In fact, a recent review documents that the buffering of fluid composition by metamorphic mineral reactions is an almost universal phenomenon (Rice & Ferry 1982). Although the phenomena of buffering and infiltration are often considered somehow incompatible, they occurred simultaneously during metamorphism of the carbonate rocks of the Vassalboro Formation. It is appropriate, though, to question whether combined infiltration and buffering is a common metamorphic phenomenon or not. Many studies of metamorphosed carbonate rock report buffering of metamorphic fluid composition by mineral reactions. If these reactions occurred without infiltration, almost all mineral reaction should have occurred either at isobaric invariant points or at temperature maxima along isobaric univariant T-X_{CO_2} reaction curves (Greenwood 1975; Rice 1977a, b). At the invariant points and temperature maxima, mineral reaction must produce a fluid whose composition matches that of fluid in equilibrium with minerals at the invariant point or temperature maximum. Consequently, for those studies which report buffering and in which infiltration did not occur, reported mineral reactions should be capable of producing a fluid whose composition matches that of fluid with which the rocks were in equilibrium as the reactions proceeded. Conversely, if reported mineral reactions are incapable of producing a fluid whose composition matches that of the equilibrium fluid, infiltration must have accompanied mineral reaction. For example, mineral reactions produced CO_2-rich fluid ($X_{CO_2} \geq 0.5$) during metamorphism of the Vassalboro Formation, yet mineral equilibria record equilibrium with H_2O-rich fluid ($X_{CO_2} \leq 0.3$); infiltration of the rocks by H_2O-rich fluid must have occurred while the reactions proceeded.

Table 4 lists 15 studies of the metamorphism of carbonate rocks in which authors reported both prograde mineral reactions and the compositions of fluid in equilibrium with the rocks when the reactions occurred. In all but 3 cases at least some equilibrium fluids were H_2O-rich ($X_{CO_2} \leq 0.2$) while in all cases the mineral reactions produced CO_2-rich fluids ($X_{CO_2} > 0.5$). In most cases, therefore, the mineral reactions were incapable of producing an equilibrium fluid and metamorphism must have involved infiltration. Furthermore, because the equilibrium fluid was more H_2O-rich than the fluid produced internally by the buffer reactions, the infiltrating fluids must have been H_2O-rich. The metamorphic events listed in Table 4 range from Tertiary to Precambrian in age. Depth of metamorphism ranges from ~4 to ~30 km. I conclude that infiltration of carbonate rock by H_2O

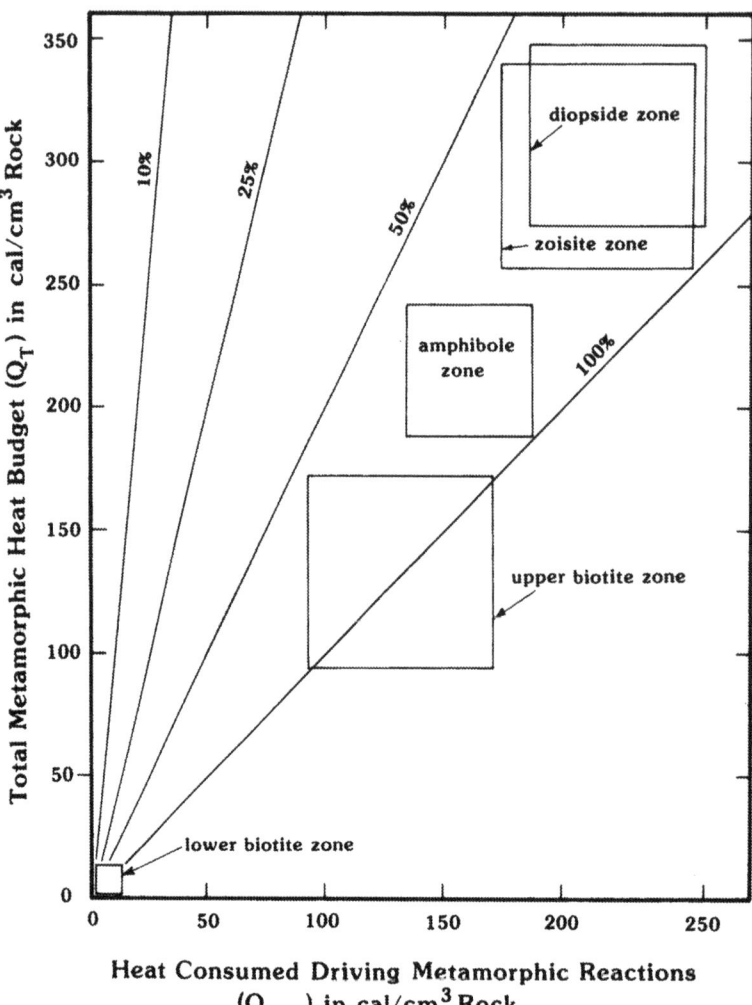

Fig. 9. Calculated relationship between the total heat budget of metamorphism (Q_T) and the amount of heat used to drive mineral reactions (Q_{rxn}), both in cal/cm³ metamorphic rock. Boxes represent data for each metamorphic zone in Fig. 1; straight lines explained in text.

fluid during metamorphism is a general phenomenon in space and time at least as long ago as 1 billion years and as deep as 30 km in the crust.

Infiltration as a control on whole-rock chemistry

With increasing grade, carbonate rocks in the Vassalboro Formation (a) were infiltrated by increasingly large volumes of fluid, and (b) were depleted in their whole-rock alkali contents. The relation between calculated fluid/rock ratios associated with the infiltration and the measured alkali contents of the carbonates is illustrated in Fig. 10. Boxes represent samples from each zone with centres denoting mean values of plotted variables and sides denoting 2 standard deviations of the variables. The negative correlation of fluid/rock ratio with whole-rock atomic (K + Na)/Al suggests that infiltration was the mechanism by which the carbonate rocks changed bulk composition during metamorphism. The alkalis, K and Na, were probably removed from metacarbonates by through-flowing, H_2O-rich metamorphic fluids.

The occurrence of Cl-free scapolite in carbonate rocks from the zoisite and diopside zones (Table 1) appears to be a contradiction to the model of alkali transport by aqueous chloride solutions during metamorphism. Scapolite, however, is a very uncommon mineral; it occurs in only 3 of the 70 samples of carbonate rock studied from the zoisite and diopside

TABLE 4. *Relation of the composition of equilibrium metamorphic fluid* ($X^{eq}_{CO_2}$) *to the composition of fluid produced internally by mineral reactions* ($X^{rxn}_{CO_2}$) *in metamorphosed carbonate rocks*

Locality	Age of Metamorphism	Depth of Metamorphism	$X^{rxn}_{CO_2}$	$X^{eq}_{CO_2}$	Reference
Mount Royal, Quebec	Cretaceous	~2 km	≥0.75	>0.75	Williams-Jones (1981)
Elkorn, Montana	Tertiary	~4 km	>0.75	0.03–0.8	Bowman & Essene (1982)
Helena, Montana	Cretaceous	~4 km	≥0.75	≥0.75	Rice (1977a, b)
Alta, Utah	Tertiary	4–7 km	>0.5	0.2–0.7	Moore & Kerrick (1976)
Salmo, British Columbia	Cretaceous	4–9 km	1.0	~0.1	Greenwood (1967)
Sierra Nevada, California	Mesozoic	~7 km	1.0	<0.1	Kerrick et al. (1973)
Stanhope area, Quebec	Devonian	6–10 km	>0.75	0.05–0.15	Erdmer (1981)
Waterville area, Maine	Devonian	13 km	≥0.5	0.01–0.3	This study
Beaver Brook, New Hampshire	Devonian	13 km	>0.83	0.09–0.14	Rumble et al. (1982)
Whetstone Lake, Ontario	Precambrian	15–19 km	>0.75	0.06–0.25	Carmichael (1970); Skippen & Carmichael (1977)
Tudor, Ontario	Precambrian	~19 km	>0.75	<0.5	Allen (1978)
Lepontine Alps	Tertiary	~20 km	>0.50	>0.50	Trommsdorf (1972)
South-central Connecticut	Devonian	~22 km	>0.75	0.1–0.4	Hewitt (1973a)
Mica Creek, British Columbia	Mesozoic	20–28 km	>0.75	0.23–0.27	Ghent et al. (1979)
Adirondack Mountains, New York	Precambrian	22–28 km	>0.63	0.1–?	Valley & Essene (1980)

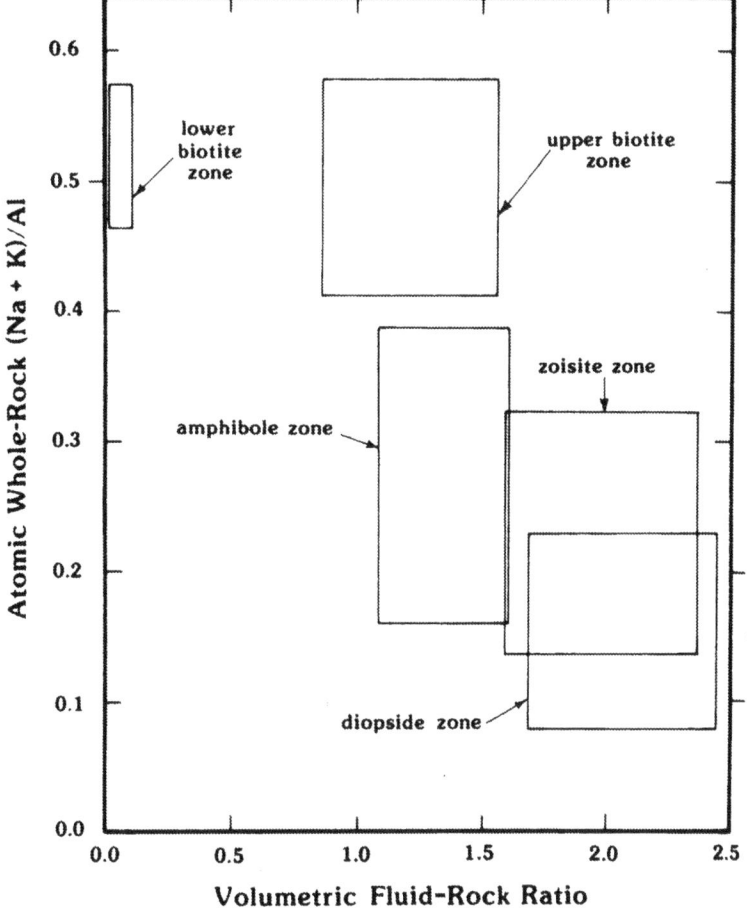

FIG. 10. Relationship between the calculated volumetric fluid/rock ratio and the measured atomic whole-rock ratio (K + Na)/Al for metamorphosed carbonate rocks. Boxes represent data for each metamorphic zone in Fig. 1.

zones. Furthermore, scapolite occurs in rocks which have K- and Na- contents close to those of average rocks in the ankerite zone. The composition of the scapolite and its occurrence in K-rich and Na-rich high-grade rocks indicate that its host rocks evolved, for some unknown reason, as systems closed to interaction with externally-derived aqueous chloride solutions. The scapolite-bearing rocks, in fact, emphasize that alkali depletion was not a thoroughly pervasive process in the high-grade portion of the terrane. A number of outcrops appear to have escaped the effects of infiltration metasomatism.

Fig. 10 demonstrates that fluids flowing through rocks under biotite zone conditions had very little effect on rock chemistry. Fluids which flowed through rocks at conditions of the amphibole zone and at higher grades were more effective in leaching K and Na from the carbonate rocks. The difference is not simply a question of fluid volume, because almost as much fluid flowed through rocks during biotite zone conditions as did at higher grade conditions. Infiltrating fluids at high grades must have been much more reactive chemically than fluids which infiltrated rock at low grades.

If the depletion of K and Na in the carbonate rocks during metamorphism was the result of infiltration by acid, aqueous fluids, the measured changes in whole-rock composition provide an independent estimate on the amount of fluid with which the rocks interacted. The observed depletion of rocks in K provides a more stringent constraint than corresponding data for Na because natural fluids generally contain greater concentrations of dissolved Na than of K. Average metacarbonate rocks in the zoisite and diopside zones of Fig. 1 lost 0.5–1.0 moles K/1000 cm^3 rock during the metamorphic event. The amount of fluid required to transport this much K as a function of fluid composition is illustrated in Fig. 11. Fluid/rock ratios can therefore be determined from the curves in Fig. 11, but they represent minimum estimates: the estimates assume that the infiltrating fluids contained no K before fluid/rock interaction and the amount specified by the abscissa in Fig. 11 after fluid/rock interaction. It is unlikely that natural processes would have this perfect efficiency. Nevertheless, Fig. 11 indicates that if metamorphic fluids contained not unreasonable concentrations of K, similar to those of Salton Sea geothermal fluids, fluid/rock ratios required by whole-rock chemical data roughly match those independent estimates in Fig. 6 based on measured progress of the prograde dehydration/decarbonation reactions. Thus inferences about fluid/rock interaction based on consideration of whole-rock chemistry are consistent with inferences based on consideration of mineral equilibria and mineral abundances.

Reactions (2) and (4), the reaction mechanisms which liberated Na and K from minerals in the carbonate rocks, were driven by the infiltration of rock

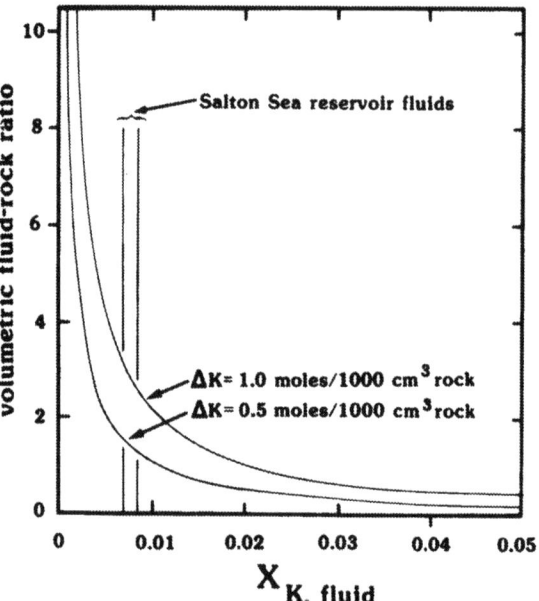

FIG. 11. Volume of fluid (per reference rock volume) required to transport 0.5–1.0 moles K as a function of the potassium concentration in the fluid. Vertical lines refer to compositions of fluids from the Salton Sea geothermal system (Helgeson 1968).

by aqueous fluids which were not in equilibrium with the carbonates. Infiltration of carbonate rocks evidently exerted a profound control on bulk metamorphic rock composition. Infiltration has obviously occurred in many metamorphic terrains (Table 4). It is not unreasonable to suspect that infiltration has influenced rock composition in at least some of these other areas as well. Infiltration may have had an important, but poorly characterized, effect on metamorphic rock composition in many metamorphic terranes and during many metamorphic events.

Heier (1965, 1973) has demonstrated that rocks of the lower crust are depleted in incompatible elements relative to rocks of the upper crust. This difference in chemistry has been often attributed to leaching of the lower crust of incompatible elements by magmas which then are emplaced in the upper crust. The results of this study are consistent with an alternative mechanism by which the incompatible elements could become depleted in the lower crust through infiltration (Tarney & Windley 1977; Weaver & Tarney 1981). Infiltration has almost quantitatively extracted K and Na from the carbonate rocks of the Vassalboro Formation. Infiltration at some stage during metamorphism of rocks which now constitute the lower crust may have leached significant quantities of the incompatible elements from them. The incompatible elements, dissolved in aqueous fluid, would be transported rapidly to the upper crust (or hydrosphere) where they would become fixed in minerals or seawater. Infiltration during metamorphism, therefore, may have played a significant role in the geochemical evolution and stratification of the Earth's crust.

Hydrolysis reactions during metamorphism

It has long been known that the high temperature stability limit of hydrous alkali-bearing minerals is lower for decomposition by hydrolysis reactions than for isochemical decomposition by dehydration. This led Eugster (1970) to propose that petrologists should consider alkali-rich hydrous minerals, such as muscovite, breaking down in metamorphic rocks first by hydrolysis and then, at higher grades, by isochemical dehydration. Eugster's prediction is confirmed in a remarkable fashion by the behaviour of biotite in the metacarbonate rocks from the Vassalboro Formation. Biotite begins to decompose in the amphibole zone by hydrolysis reaction (4). Only at higher grades and at higher temperatures in the zoisite zone does biotite decompose isochemically by dehydration reaction (7). A remarkable aspect of metamorphism of the Vassalboro Formation is the important role that hydrolysis reactions played in the mineralogical evolution of the rocks. Most of the amphibole and much of the $CaAl_2Si_2O_8$ in the rocks were produced by hydrolysis reactions (4) and (2).

The region in Fig. 1 contains a large area of carbonate rock that has been depleted during metamorphism of K and Na. The Na and K were released by hydrolysis reactions (2) and (4). Infiltration of rocks by acid, aqueous metamorphic fluid was the process which caused the mineral reactions and consequent element depletion. In this regard the metamorphic event in the area of Fig. 1 may be considered to have involved acid metasomatism. Normally acid metasomatism is associated with hydrothermal activity on a relatively small scale (Meyer & Hemley 1967). Acid metasomatism during metamorphism of the Vassalboro Formation, however, has affected hundreds of square kilometres of rock. Perhaps it is appropriate to conceive of some instances of regional metamorphism as large-scale acid metasomatism.

Infiltration as a driving force behind mineral reactions during metamorphism

A necessary driving force behind hydrolysis reactions (2) and (4) was infiltration of the carbonate rocks by acid fluid. The infiltrating fluids were H_2O-rich (Fig. 4) as well as acid. The infiltration of carbonate rocks by aqueous fluid was probably an important driving force behind all the mineral reactions (1)-(8). Consider a rock containing amphibole + calcite + quartz + diopside + plagioclase + sphene. At a particular pressure and temperature the mineral assemblage buffers the X_{CO_2} of coexisting CO_2-H_2O fluid to one of 2 specific values (uppermost curve, Fig. 5). If the rock is infiltrated by a quantity of H_2O-rich fluid, reaction (8) will proceed until the fluid in contact with the rock is restored to the left-hand equilibrium composition (Fig. 5). At constant temperature, reaction (8) thus will be driven by interaction of rock with H_2O-fluid. Similar arguments can be made for each of the other reactions (1)-(8). Calculations have already demonstrated that at all grades above the biotite isograd, carbonate rocks were infiltrated by large amounts of H_2O fluid. The infiltration of carbonate rock by aqueous fluid was thus an important phenomenon which *caused* the mineral reactions to occur (increasing temperature was, of course, an additional cause). Reaction (6) is probably characterized by $\Delta \bar{H} < 0$ (Ferry 1976a; Allen & Fawcett 1982). If so, the reaction cannot be driven by increasing temperature (it would be inhibited by increasing temperature). For at least one reaction, (6), infiltration appears to have been the exclusive force driving mineral reaction. Because infiltration is a significant phenomenon in many terranes (Table 4), infiltration may be an important general control on the progress of mineral reaction during metamorphism.

Fluid transfer and heat transfer

Fig. 12 illustrates the relationship between the calculated fluid/rock ratio and total heat budget for metamorphism of carbonate rocks in the Vassalboro Formation. Each box represents samples from a different zone: centres of boxes denote the mean of all samples in the zone; sides of boxes denote 2 standard deviations of the calculated quantities. There is a strong positive correlation between the total metamorphic heat budget of the carbonate rocks and the amount of fluid with which they have interacted. The relationship is illustrated in another way by the similarity of the contours in Figs 6 and 8. The correlation between rocks which record high fluid/rock ratios and rocks which record large heat budgets may be explained by the role that circulating fluids can play in heat transfer during metamorphic and hydrothermal events (Norton & Knight 1977; Schuiling & Kreulen 1979). Flowing fluids are probably an important transport agent of metamorphic heat, and heat transfer during metamorphism probably occurs to a significant degree not only by conduction but also by convection.

Conclusions

The interaction of rocks with externally-derived fluids is an important general phenomenon, as illustrated by a study of the petrogenesis of metacarbonate rocks

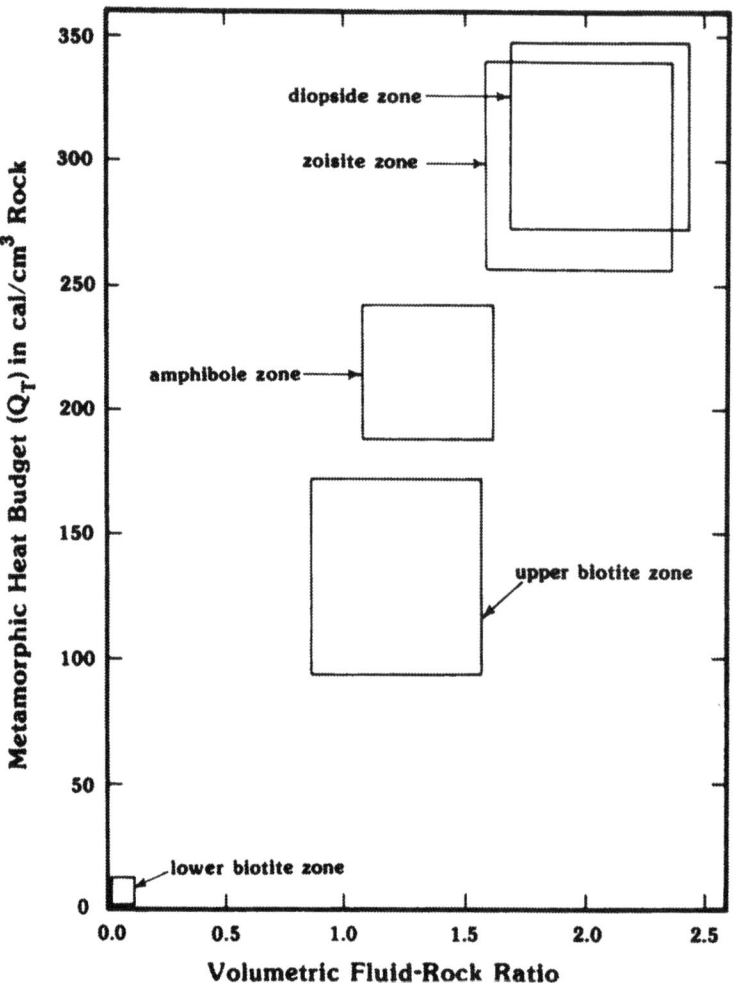

FIG. 12. Calculated relationship between the total heat budget of metamorphism (Q_T, cal/cm^3 rock) and the volumetric fluid/rock ratio. Boxes represent data for each metamorphic zone.

from the Vassalboro Formation. Infiltration during metamorphism may act as a significant control of certain whole-rock compositional parameters. Infiltration may serve as a driving force behind many mineral reactions. As a driving force, the infiltrating fluid has a profound influence on the mineralogical evolution of metamorphic rocks both in terms of which minerals develop and in terms of the amounts in which minerals form. Because circulating fluids transport heat, convective heat transfer by infiltrating fluid is probably an important item in the heat budget of many metamorphic terranes.

ACKNOWLEDGMENTS. Research was supported by a Cottrell Grant from Research Corporation and NSF Grant EAR 80-20567 (Division of Earth Sciences). Electron microprobe analyses were obtained at the Geophysical Laboratory with the kind permission of H. S. Yoder, Jr., Director. The manuscript was improved by helpful reviews from C. M. Graham and J. M. Rice.

References

ALLEN, J. M. 1978. Calc-silicate equilibria in a remetamorphosed aureole: grade is related to gas phase composition. *EOS*, **59**, 407.

—— & FAWCETT, J. J. 1982. Zoisite-anorthite-calcite stability relations in H_2O-CO_2 fluids at 5000 bars: An experimental and SEM study. *J. Petrol.* **23**, 215–39.

BOWMAN, J. R. & ESSENE, E. J. 1982. P-T-X(CO_2) conditions of contact metamorphism in the Black Butte aureole, Elkorn, Montana. *Am. J. Sci.* **282**, 311-41.

BURNHAM, C. W., HOLLOWAY, J. R. & DAVIS, N. F. 1969. Thermodynamic properties of water to 1,000°C and pressures from 2000 to 10,000 bars. *Spec. Pap. geol. Soc. Am.* **132**.

CARMICHAEL, D. M. 1969. On the mechanism of prograde metamorphic reactions in quartz-bearing pelitic rocks. *Contrib. Mineral. Petrol.* **20**, 244-67.

—— 1970. Intersecting isograds in the Whetstone Lake area, Ontario. *J. Petrol.* **11**, 147-81.

DALLMEYER, R. D. & VAN BREEMAN, O. 1981. Rb-Sr whole-rock and $^{40}K/^{39}Ar$ mineral ages of the Togus and Hallowell quartz monzonite and Three Mile Pond grandiorite plutons, south-central Maine: Their bearing on Post-Acadian cooling history. *Contrib. Mineral. Petrol.* **78**, 61-73.

ERDMER, P. 1981. Metamorphism at the northwest contact of the Stanhope pluton, Quebec Appalachians: Mineral equilibria in interbedded pelite and calc-schist. *Contrib. Mineral. Petrol.* **76**, 109-15.

EUGSTER, H. P. 1970. Thermal and ionic equilibria among muscovite, K-feldspar, and aluminosilicate. *Fortsch. Mineral.* **47**, 106-23.

FERRY, J. M. 1976a. P, T, f_{CO_2}, and f_{H_2O} during metamorphism of calcareous sediments in the Waterville-Vassalboro area, south-central Maine. *Contrib. Mineral. Petrol.* **57**, 119-43.

—— 1976b. Metamorphism of calcareous sediments in the Waterville-Vassalboro area, south-central Maine: Mineral reactions and graphical analysis. *Am. J. Sci.* **276**, 841-82.

—— 1978. Fluid interaction between granite and sediment during metamorphism, south-central Maine. *Am. J. Sci.* **278**, 1025-56.

—— 1979. A map of chemical potential differences within an outcrop. *Am. Mineral.* **64**, 966-85.

—— 1980a. A case study of the amount and distribution of heat and fluid during metamorphism. *Contrib. Mineral. Petrol.* **71**, 373-85.

—— 1980b. A comparative study of geothermometers and geobarometers in pelitic schists from south-central Maine. *Am. Mineral.* **65**, 720-32.

—— 1981. Petrology of graphitic sulfide-rich schists from south-central Maine: An example of desulfidation during prograde regional metamorphism. *Am. Mineral.* **66**, 908-30.

—— 1983. Mineral reactions and element migration during metamorphism of calcareous sediments from the Vassalboro Formation, south-central Maine. *Am. Mineral.* **68**, 334-54.

GHENT, E. D., ROBBINS, D. B. & STOUT, M. Z. 1979. Geothermometry, geobarometry and fluid compositions of metamorphosed calc-silicates and pelites, Mica Creek, British Columbia. *Am. Mineral.* **64**, 874-85.

GRAMBLING, J. A. 1979. Isothermal prograde metamorphism and fluid diffusion, Truchas Peaks region, northern New Mexico. *Abs. Prog. geol. Soc. America*, **11**, 435.

GREENWOOD, H. J. 1967. Wollastonite: Stability in H_2O-CO_2 mixtures and occurrence in a contact metamorphic aureole near Salmo, British Columbia, Canada. *Am. Mineral.* **52**, 1669-80.

—— 1975. Buffering of pore fluids by metamorphic reactions. *Am. J. Sci.* **275**, 573-93.

GROVE, T. L., FERRY, J. M. & SPEAR, F. S. 1983. Phase transitions and decomposition relations in calcic plagioclase. *Am. Mineral.* **68**, 41-59.

HEIER, K. S. 1965. Metamorphism and chemical differentiation in the crust. *Forh. Geol. Foren. Stockholm*, **87**, 249-56.

—— 1973. Geochemistry of granulite facies rocks and problems of their origin. *Philos. Trans. R. Soc. London*, **273A**, 429-42.

HELGESON, H. C. 1968. Geologic and thermodynamic characteristics of the Salton Sea geothermal system. *Am. J. Sci.* **266**, 729-804.

—— 1969. Thermodynamics of hydrothermal systems at elevated temperatures and pressures. *Am. J. Sci.* **266**, 729-804.

——, DELANY, J. M., NESBITT, H. W. & BIRD, D. K. 1978. Summary and critique of the thermodynamic properties of rock-forming minerals. *Am. J. Sci.* **278A**, 1-229.

——, KIRKHAM, D. H. & FLOWERS, G. C. 1981. Theoretical prediction of the thermodynamic behavior of aqueous electrolytes at high pressures and temperatures: IV. Calculation of activity coefficients, osmotic coefficients, and apparent molal and standard and relative partial molal properties to 600°C and 5 kb. *Am. J. Sci.* **281**, 1249-1516.

HEWITT, D. A. 1973a. The metamorphism of micaceous limestones from south-central Connecticut. *Am. J. Sci.* **273A**, 444-69.

—— 1973b. Stability of the assemblage muscovite-calcite-quartz. *Am. Mineral.* **58**, 785-91.

—— & WONES, D. R. 1975. Physical properties of some synthetic Fe-Mg-Al trioctahedral biotites. *Am. Mineral.* **60**, 854-62.

HOLDAWAY, M. J., GUIDOTTI, C. V., NOVAK, J. M. & HENRY, W. E. 1982. Polymetamorphism in medium- to high-grade pelitic metamorphic rocks, west-central Maine. *Bull. geol. Soc. Am.* **93**, 572-84.

JACOBS, G. K. & KERRICK, D. M. 1981. Devolatilization equilibria in H_2O-CO_2 and H_2O-CO_2-NaCl fluids: An experimental and thermodynamic evaluation at elevated pressures and temperatures. *Am. Mineral.* **66**, 1135-53.

KERRICK, D. M., CRAWFORD, K. E., & RANDAZZO, A. F. 1973. Metamorphism of calcareous rocks in three roof pendants in the Sierra Nevada, California. *J. Petrol.* **14**, 303-25.

MEYER, C. & HEMLEY, J. J. 1967. Wall rock alteration. In: BARNES, H. L. (ed.) *Geochemistry of Hydrothermal Ore Deposits*. Holt, Rinehart and Winston, New York, 166-235.

MOORE, J. N. & KERRICK, D. M. 1976. Equilibria in siliceous dolomites of the Alta aureole, Utah. *Am. J. Sci.* **276**, 502-24.

NORTON, D. A. & KNIGHT, J. 1977. Transport phenomena in hydrothermal systems: Cooling plutons. *Am. J. Sci.* **277**, 937-81.

NOVAK, J. M. & HOLDAWAY, M. J. 1981. Metamorphic petrology, mineral equilibria, and polymetamorphism in the Augusta Quadrangle, south-central Maine. *Am. Mineral.* **66**, 51-69.

OSBERG, P. H. 1968. Stratigraphy, structural geology, and metamorphism of the Waterville-Vassalboro area, Maine. *Bull. Maine Geol. Surv.* **20**, 64 pp.

—— 1979. Geologic relationships in south-central Maine. In:

OSBERG, P. H. & SKEHAN, J. W., (eds) *The Caledonides in the U.S.A.* Weston Observatory of Boston College, Weston, Massachusetts, 37–62.

PUHAN, D. 1978. Experimental study of the reaction: dolomite + K-feldspar + H_2O = phlogopite + calcite + CO_2 at the total gas pressures of 4000 and 6000 bars. *Neues Jahrb. Mineral. Monatshefte*, 110–27.

RICE, J. M. 1977a. Contact metamorphism of impure dolomitic limestone in the Boulder aureole, Montana. *Contrib. Mineral. Petrol.* **59**, 237–59.

—— 1977b. Progressive metamorphism of impure dolomitic limestone in the Marysville aureole, Montana. *Am. J. Sci.* **277**, 1–14.

—— & FERRY, J. M. 1982. Buffering, infiltration, and the control of intensive variables during metamorphism. *In*: FERRY, J. M. (ed.) *Characterization of Metamorphism through Mineral Equilibria*. Mineralogical Society of America, Washington, D.C., 263–326.

ROBIE, R. A., BETHKE, P. M. & BEARDSLEY, K. M. 1967. Selected X-ray crystallographic data, molar volumes, and densities of minerals and related substances. *Bull. U.S. geol. Surv.* **1248**, 87 pp.

RUMBLE, D. 1978. Mineralogy, petrology, and oxygen isotopic geochemistry of the Clough Formation, Black Mountain, New Hampshire, U.S.A. *J. Petrol.* **19**, 317–40.

—— FERRY, J. M., HOERING, T. C. & BOUCOT, A. J. 1982. Fluid flow during metamorphism at the Beaver Brook fossil locality, New Hampshire. *Am. J. Sci.* **282**, 886–919.

SHMONOV, V. M. & SCHMULOVICH, V. P. 1974. Molar volumes and equation of state of CO_2 at temperatures from 100 to 1,000°C and pressures from 2000 to 10,000 bars. *Dokl. Akad. Nauk SSSR*, **217**, 206–9.

SHUILING, R. D. & KREULEN, R. 1979. Are thermal domes heated by CO_2-rich fluids from the mantle? *Earth planet. Sci. Lett.* **43**, 298–302.

SKIPPEN, G. B. & CARMICHAEL, D. M. 1977. Mixed-volatile equilibria. *In*: GREENWOOD, H. J. (ed.) *Short Course in Application of Thermodynamics to Petrology and Ore Deposits*. Mineralogical Association of Canada, Vancouver, 109–125.

TANNER, P. W. G. & MILLER, R. G. 1980. Geochemical evidence for loss of Na and K from Moinian calc-silicate pods during prograde metamorphism. *Geol. Mag.* **117**, 267–75.

TARNEY, J. & WINDLEY, B. F. 1977. Chemistry, thermal gradients and evolution of the lower continental crust. *J. geol. Soc. London*, **134**, 153–72.

TAYLOR, H. P. JR. 1977. Water/rock interactions and the origin of H_2O in granitic batholiths. *J. geol. Soc. London*, **133**, 509–558.

THOMPSON, A. B. 1975. Calc-silicate diffusion zones between marble and pelitic schist. *J. Petrol.* **16**, 314–46.

THOMPSON, J. B. JR. 1982. Reaction space: An algebraic and geometric approach 33–52 *In*: FERRY, J. M. ed.) *Characterization of Metamorphism through Mineral Equilibria*. Mineralogical Society of America Washington, D.C., 33–52.

—— & NORTON, S. A. 1968. Paleozoic regional metamorphism in New England and adjacent areas. *In*: ZEN, E-an, WHITE, W. S., HADLEY, J. B. & THOMPSON, J. B. Jr. (eds.) *Studies of Appalachian Geology: Northern and Maritime*. Interscience Publishers, New York, 203–18.

——, LAIRD, J. & THOMPSON, A. B. 1982. Reactions in amphibolite, greenschist and blueschist. *J. Petrol.* **23**, 1–27.

TROMMSDORF, V. 1972. Change in T-X during metamorphism of siliceous dolomite rocks of the central Alps. *Schweiz. mineral. petrograph. Mitt.* **52**, 567–71.

VALLEY, J. W. & ESSENE, E. J. 1980. Calc-silicate reactions in Adirondack marbles: the role of fluids and solid solution: Summary. *Bull. geol. Soc. Am.* **91**, 114–17.

WEAVER, B. L. & TARNEY, J. 1981. Lewisian gneiss geochemistry and Archaean crustal development models. *Earth planet. Sci. Lett.* **55**, 171–80.

WILLIAMS-JONES, A. E. 1981. Thermal metamorphism of siliceous limestones in the aureole of Mount Royal, Quebec. *Am. J. Sci.* **281**, 673–96.

Received 8 April 1982; revised typescript received 29 November 1982.

J. M. FERRY, Department of Geology, Arizona State University, Tempe, Arizona 85287, U.S.A.

Discussion

DR C. M. GRAHAM commented that on the basis of the observed extraction of K and Na from carbonate lithologies of the Vassalboro Formation, Dr Ferry has speculated that infiltration may provide an alternative model to partial melting to account for the depletion of these and other incompatible elements in the lower crust. Carbonate rocks comprise only a small proportion of most metamorphic terrains, and deductions based on data obtained from them may give a biased impression of the general importance of infiltration as a process capable of producing incompatible element depletion on a *regional scale*. The question, therefore, is whether there is any evidence from non-carbonate lithologies in the area studied by Dr Ferry, particularly from pelitic lithologies, either for the operation of infiltration mechanism or for accompanying incompatible element depletions?

DR FERRY responded: I have studied two other lithologies in the Waterville Formation (the stratigraphic unit adjacent to the Vassalboro Formation) for evidence of incompatible element depletions. Sulphide-rich schists in the Waterville Formation show evidence for infiltration by 1–5 rock volumes C–O–H fluid during metamorphism. The high-grade sulphide-rich schists are depleted in S and Na but not in K relative to their low-grade equivalents (Ferry 1981). Normal pelitic schists from the high-grade portion of the Waterville Formation are depleted in neither Na nor K relative to normal low-grade phyllites (Ferry 1982). The normal pelitic schists, however, show no evidence for infiltration (at least none has been detected yet). There is no question that infiltration and element depletion has yet to be demonstrated as a general process for all or even many rock types during metamorphism. Data on the carbonate rocks from the Vassalboro Formation, however, suggest that it is an issue worth further investigation.

Dr B. Harte asked the author to explain exactly how he had erected the mineral reaction equations. He noted that the author had presented certain textural evidence, but that the author had not indicated a complete chain of local textural reactions (similar to those introduced by D. M. Carmichael 1969) from which a net reaction had been determined. To what extent did the author use the differences in zonal mineral assemblages to deduce the reactions given? Also, why does the author prefer to write reactions using NaCl rather than Na^+? For example, is it because of some tangible evidence in the rocks or because the author thinks that it is generally likely to be more realistic?

Dr Ferry responded: first let me answer the question concerning speciation of the fluid. Mineral reactions were written with NaCl, KCl, and HCl because numerous fluid inclusion studies of metamorphic rocks report trapped fluids which contain significant dissolved Cl. Metamorphic fluids therefore are often aqueous chloride solutions. Laboratory experiment and thermodynamic calculations show that species in an aqueous chloride fluid at the elevated temperatures and pressures of metamorphism are associated chloride complexes. The Na and K therefore probably existed in solution during metamorphism as associated NaCl and KCl species.

The question concerning formulation of the prograde mineral reactions is a complicated one worthy of an hour's talk itself. I formulated the reactions following a strategy recently developed by J. B. Thompson (Thompson 1982; Thompson et al. 1982). I will illustrate the strategy by considering mineral reactions in the biotite zone of the Vassalboro Formation. Rocks from the biotite zone contain combinations of these minerals: ankerite, calcite, plagioclase, quartz, chlorite, biotite, muscovite, ilmenite, and sphene (ignoring trace amounts of apatite, sulfide, tourmaline, and graphitic material). We wish to know how many reactions among these minerals is needed to represent prograde mineral changes in the biotite zone.

We begin with an unconventional choice of components to describe the composition of each of the nine minerals. One component is the mineral formula in Table 1 of the paper with 'Mg' replacing '(Fe, Mg, Mn)'. These are called *additive* components. Let $NaAlSi_3O_8$ be the additive component for plagioclase. The other components are all *exchange* components, and, for these minerals, include $FeMg_{-1}$ (ankerite, muscovite, biotite, ilmenite, calcite, chlorite); $MnMg_{-1}$ (ankerite, muscovite, biotite, ilmenite, calcite, chlorite); KNa_{-1} (muscovite, biotite, fluid); $Al_2Mg_{-1}Si_{-1}$ (muscovite, biotite, chlorite); $CaAlNa_{-1}Si_{-1}$ (plagioclase); $NaAl\square_{-1}Si_{-1}$ (muscovite, biotite); $CaMg_{-1}$ (calcite, ankerite); $TiAl_2Mg_{-1}Si_{-2}$ (muscovite, biotite); and $\square Al_2Mg_{-3}$ (muscovite, biotite). In addition we consider the fluid as an $HCl-NaCl-H_2O-CO_2-KNa_{-1}$ solution. Considering all components in all phases (9 minerals + fluid), there are a total of 40, C_p, *phase components* in the system. We can also choose a set of system components; the standard oxides (+HCl) serve this purpose: K_2O, Na_2O, CaO, Al_2O_3, SiO_2, FeO, MgO, MnO, TiO_2, H_2O, CO_2, HCl. There are 12, C_s, *system components*. All mass transfer among minerals and fluid may be described by $C_p - C_s = 40 - 12 = 28$ linearly independent reaction relationships among the 40 phase components.

These 28 reactions may be factored into n_{ex} *exchange reactions* and n_{nt} *net-transfer reactions*. Exchange reactions refer to reactions such as $FeMg_{-1}$ (biotite) = $FeMg_{-1}$ (chlorite) and describe the exchange of atoms between minerals or between mineral and fluid. Exchange reactions may change the compositions of phases but not their molar amounts. There are 18 linearly independent exchange reactions among minerals and fluid in the biotite zone. Net-transfer reactions refer to reactions which involve the additive components and therefore are those reactions which change the *amounts* of minerals and fluid during metamorphism. There are $28 - 18 = 10$ linearly independent net-transfer reactions among minerals and fluid in the biotite zone. Which reactions are chosen is completely arbitrary. The only specification is that they be 10 in number and linearly independent.

5 of the 10 net-transfer reactions involve writing a separate reaction relationship among each of these 5 exchange components and the additive components (plus KNa_{-1}): $Al_2Mg_{-1}Si_{-1}$; $NaAl\square_{-1}Si_{-1}$; $CaMg_{-1}$; $TiAl_2Mg_{-1}Si_{-2}$; and $\square Al_2Mg_{-3}$. The concentrations of the 5 exchange components do not vary significantly in minerals from the carbonate rocks of the biotite zone. This in turn means that the 5 reactions involving these exchange components did not proceed to any appreciable degree during metamorphism and may be neglected.

5 additional linearly-independent net-transfer reactions may be written among the additive components and the exchange component $CaAlNa_{-1}Si_{-1}$ (= $CaAl_2Si_2O_8 - NaAlSi_3O_8$). The 5 that I have chosen, which seem to correspond to textural relations

observed in thin section, are:

0.98 muscovite + 2.23 ankerite + 0.65 quartz
  + 0.09 ilmenite + 0.03 HCl + 0.01 H_2O

= 1.00 biotite + 1.89 calcite + 0.48 $CaAl_2Si_2O_8$
  + 0.03 NaCl + 2.57 CO_2;   (1)

2 $NaAlSi_3O_8$ + calcite + 2 HCl = $CaAl_2Si_2O_8$
  + 4 quartz + CO_2 + H_2O + 2 NaCl;   (2)

1.00 muscovite + 1.28 calcite + 0.84 HCl

= 0.12 biotite + 0.02 sphene + 0.36 quartz
  + 1.22 $CaAl_2Si_2O_8$ + 0.79 KCl + 0.05 NaCl
  + 1.30 H_2O + 1.28 CO_2;   (3)

2.77 muscovite + 11.04 ankerite + 0.26 ilmenite
  + 1.86 quartz + 0.08 HCl + 4.20 H_2O

= 1.00 chlorite + 2.83 biotite + 11.87 calcite
  + 0.08 NaCl + 10.21 CO_2; and   (4)

ilmenite + 2 calcite + quartz = sphene
  + ankerite.   (5)

I have cheated slightly in reactions (2) and (5) by representing calcite as $CaCO_3$. This could be corrected by simply adding an appropriate amount of $MgCa_{-1}$ (= $-CaMg_{-1}$) to the 2 reactions.

Reactions (1)–(3) correspond to reactions (1)–(3) in my paper. Rocks on the biotite zone contain on average 2 modal per cent chlorite and <1% sphene. Reactions (4) and (5) therefore contribute negligibly to the mineralogical changes which occurred in the rocks during metamorphism. In addition, rocks in the upper biotite zone contain on average only ~3% muscovite. Like reactions (4) and (5), reaction (3) can be considered much less important than reactions (1) and (2) in accounting for the overall mineralogical changes in the biotite zone. Thus of all the 10 net-transfer reactions needed, in principle, to describe changes in amounts of minerals and fluid during metamorphism, only reactions (1) and (2) are needed to adequately account for the *major* changes in mineralogy which occurred in the biotite zone.

The overall mineralogical and chemical changes which a rock in the biotite zone experienced during metamorphism can be represented by the reaction $\Sigma_i (\Sigma_j v_j(j)\xi)_i = 0$ summed over the $i = 28$ exchange and net-transfer reactions. The v_j is the stoichiometric coefficient of species j in the ith reaction (positive for products and negative for reactants) and ξ_i is the progress of the ith reaction (measured, for example, by modal analysis). The overall changes in *amounts* of minerals and fluid which a rock experienced during metamorphism may be represented by the reaction $\Sigma_{i'} (\Sigma_j v_j(j)\xi)_{i'} = 0$ summed over the $i' = 10$ net-transfer reactions. Total amounts of mineral or fluid species produced during metamorphism may also be monitored by the net-transfer reactions. The total molar amount of CO_2 produced during metamorphism in the biotite zone was $\Sigma_{i'} v_{CO_2,i'}\xi_{i'}$ which corresponds to the numerator of equation (10) in the paper (equation 10 refers to the effects of one net-transfer reaction). The total molar amount of $CO_2 + H_2O$ produced during metamorphism in the biotite zone was $\Sigma_{i'} ((v_{CO_2,i'} + v_{H_2O,i'})\xi_{i'})$ which corresponds to the first term in the denominator of equation (10) (in equation 10, $i' = 1$). The net-transfer reactions, in addition, are useful in reconstructing the heat budget of metamorphic rocks. The value of ΔH for exchange reactions is small relative to $\Delta \bar{H}$ for the net-transfer reactions. The amount of heat involved in driving mineral reactions in the biotite zone, therefore, may be closely approximated by $\Sigma_{i'} \Delta\bar{H}_{i'}^{P,T}\xi_{i'}$ where $\Delta\bar{H}_{i'}^{P,T}$ is the enthalpy of reaction for net-transfer reaction i'. The summation corresponds to the right-hand side of equation (20).

Of the 10 net-transfer reactions which, in principle, may be of interest for rocks in the biotite zone of the Vassalboro Formation only reactions (1) and (2) above are of practical concern. Measured progress of the other eight reactions is sufficiently small that they may be ignored. Thus calculation of fluid/rock ratios and heat/rock ratios for rocks in the biotite zone involved application of equations (12) and (20) summed over reactions (1) and (2). Similar (lengthy) considerations have deduced the major reactions of interest in the amphibole, zoisite, and diopside zones (reactions 4–8, in the paper). These, in turn, were used in conjunction with equations (12) and (20) to calculate fluid/ratios and heat/rock ratios at higher grades.

CHAPTER 13: KINETICS: EXPERIMENT AND THEORY

Wood, B.J. and Walther, J.V. (1983) Rates of hydrothermal reactions. *Science*, **222**, 413–415.

It has been known since the early days of petrography that crystalline rocks commonly reach Earth's surface in mountain belts in a pristine state. Their textures and mineralogy bear witness to conditions of elevated temperature and deep burial, preserved in a manner that Rosenbusch (1877) and Barrow (1893), for example, could recognize as evidence of a fossil thermal structure. The rates of mineralogical change during the exhumation of crystalline mountain belts can thus be zero, or nearly so. These rates are in vast contrast to the speed at which the properties of sediments succumb to heat and pressure when taken to great depths during orogenesis. Most rocks undoubtedly recrystallize at rates between these two extremes. It is not surprising, therefore, that the kinetics of metamorphism continue to be a subject of theoretical, experimental and field investigation in many quarters. Isograds have been successfully exploited as indicators of metamorphic grade for more than a century (Chapter 2), but do we really know if they record the actual P and T of the reaction that describes them? Or are they only a crude approximation? What use is mineral thermometry and barometry if reaction rates vary significantly among the equilibria being measured in a sample? Are there chemical systems or subsystems that equilibrate readily and others that do not? These are issues of fundamental importance to valid interpretations of metamorphic rocks.

An isogradic reaction proceeds via a sequence of steps: dissolution of reactants; transport of components to growth sites; nucleation if necessary; and growth of product minerals. The first and last of these steps take place on mineral surfaces. Surface or interface kinetics are rate-determining for net-transfer reactions among pure minerals that proceed in a porous hydrothermal milieu in the presence of reactants and products: transport of components is fast, and no nucleation is required. In dry, fluid-undersaturated, or low-porosity crystalline rocks, prograde or retrograde, intergranular transport or other mechanisms are more likely to be rate determining. This paper by Wood and Walther presented a quantitative expression based on laboratory experiments for the rate constant of silicate mineral reactions taking place in a fluid-rich medium at elevated temperatures and pressures. It provided a robust and universal equation for the temperature dependence of reaction rate in hydrothermal systems. Controversial to this day, however, is the extent to which interface rates apply to prograde metamorphic systems in nature, where fluid is periodically released onto grain boundaries but porosity is small.

In the 1960s and 1970s, students of W.S. Fyfe followed his advice to determine the equilibrium P-T conditions of experimentally sluggish reactions by measuring weight changes in a single crystal of a reactant or product phase in an otherwise powdered reaction assemblage over a temperature range spanning the equilibrium T. They did this without realizing that they were providing a database usable later by Wood and Walther (W&W) in an analysis of hydrothermal reaction kinetics.

A zero-order rate equation (W&W, p. 473) is valid for a reaction among phases that have unit concentration or activity. Wood and Walther found that an interface-controlled rate equation in inverse T could be fit to more than 40 high-P-T single-crystal + powder laboratory experiments with an uncertainty in rate constant of better than one-half log unit. The minerals in the experiments were end-member phases with tecto-, neso- and phyllo-silicate structures in systems involving SiO_2, Al_2O_3, CaO and K_2O. Allowing for uncertainties, the results of many of the experiments were consistent with rates of reaction (not the rate constant) that are a linear function of reaction affinity ($-\Delta G_r$) within a few tens of °C of the equilibrium T at the P of the experiment, where $|\Delta G_r| < RT$ (Lasaga, 1981b, p. 146). Forward and reverse reaction rates in many cases were similar (Fig. 1a), but in other cases different (Fig. 1b; Kerrick, 1972; Lasaga, 1981a); there is no reason that they should be the same. There appeared to be no obvious dependence of rate constant on pressure. On the assumption that surface reaction at activated sites on the single crystal was rate controlling, they extracted rate constants and normalized them to gram atoms of oxygen per cm^2 s^{-1}. The rate constants of high-T experiments were found to be continuous with those for dissolution experiments at low T, a reflection of the similarity in the processes involved.

In a short end-paragraph, Wood and Walther applied their hydrothermal rate constant equation to a typical prograde metamorphic dehydration reaction. They concluded that an overstep of the reaction boundary of 1°C to produce 2 mm crystals would have reaction times of 330 y at 500°C and 70 y at 700°C – brief on any sort of geological time-scale.

Response time was correspondingly swift. Rubie and Thompson (1985, pp. 63–64), citing the very much slower rates for grain-boundary diffusion in rocks compared to diffusion in bulk water, concluded that "kinetic data from such (hydrothermal) experiments are not likely to be directly applicable to rocks undergoing metamorphism". Ridley and Thompson (1986, p. 168) believed that Wood and Walther's rate data might be biased on the fast side by the inclusion of only successful experiments, the exclusion of solid solution phases with partially ordered structures, and the presence of strain in the crushed mineral grains.

Most net-transfer reactions among metamorphic minerals, except those taking place at low temperatures, are in fact found to proceed at least to a minimally detectable degree on the time-scale of laboratory experiments. The single-crystal rate studies were done expressly on those reactions known to be particularly sluggish, and less amenable to a clear interpretation based on microscopic examination or X-ray powder diffraction. The mineral thermodynamic databases discussed in Chapter 11 utilize some 200 detectably reversed high P-T experimental boundaries (including a few by the more sensitive single-crystal method) in compositionally diverse systems. It is hard to believe, therefore, that incorporation of additional single-crystal hydrothermal experiments would greatly change the equation given by Wood and Walther. It would be reassuring perhaps to have available similar weight-change experiments on systems containing Fe, Mg and other components. However, in the last 20 y, other kinds of experimental reaction-rate studies have produced results in general agreement with those of Wood and Walther (see Baxter 2003).

The rate equation (Wood and Walther, 1983; Walther and Wood, 1984, p. 248) is linear close to the equilibrium T_{eq}:

$$dm/dt = -k_r \Delta G_r / RT_{eq} \quad (1)$$

where dm/dt is the rate of mass transfer in gram atoms of oxygen cm^{-2} s^{-1}, k_r the rate constant, and ΔG_r the free energy driving force of the reaction. For a thermally driven reaction with reaction entropy ΔS_r, equation 1 can be written as:

$$dm/dt = k_r[\Delta S_r(T-T_{eq})]/RT_{eq} \quad (2)$$

A more useful form of equation 2 expresses $T-T_{eq}$ as the rate of increase of $T \times t$ (the time since crossing the equilibrium boundary), and dm/dt in terms of the rate of change of the radius dr/dt of a typical crystal such as quartz (Walther and Wood, 1986, p. 204):

$$dr/dt = k_r \Delta S_r \times \text{(rate of } T \text{ increase)} \times t/0.9RT_{eq} \quad (3)$$

Integration of equation 3 gives (Walther and Wood, 1986, p. 305):

$$t = \sqrt{\frac{(r-r^0) \times 0.9RT_{eq}}{0.5 k_r \Delta S_r \times \text{(rate of increase in } T)}} \quad (4)$$

where r^0 is the initial radius and r the final. Reaction overstep in °C is given by $t \times$ rate of increase in T.

Assuming that a typical regional metamorphic event has a rate of temperature increase of 10°C/my, Walther and Wood (1986) using equation 4 found that growing and dissolving crystals on the order of 0.1 cm radius in a devolatilization reaction with $\Delta S_r = 20$ cal K^{-1} (= 83.6 JK^{-1}) per mol of fluid would require 13,000 y at 500°C, and that the T overstep would be 0.13°C. At 700°C, the corresponding numbers are 6200 y and 0.06°C. Their conclusion was that the rate of surface reaction keeps pace with the rate of temperature increase in a regional metamorphic event. In contact metamorphism, the rate of increase of T is faster, so that reaction time is shorter and oversteps are greater (Walther and Wood, 1984). For example, at a rate of increase of T of 5000°C/my, the same reaction at 500°C is completed in 600 y, with an overstep of 3°C, and at 700°C, 277 y and 1.4°C. For a solid-solid reaction with smaller reaction entropy, such as Ky → Sil in regional metamorphism at 600°C ($\Delta S_r = 10$ J K^{-1} per mol), the reaction time is 24,600 y and overstep 0.25°C. The transition And → Sil ($\Delta S_r = 1.4$ J K^{-1} per mol) in contact metamorphism at 500°C at 10000°C/y would take place in 3260 y with an overstep of 33°C. These rates of reaction will apply to metamorphic reactions to the extent that interface kinetics are rate limiting and that Wood and Walther's linear rate equation is valid for them. As correctly pointed out by Ridley and Thompson (1986, p. 186), these calculated rates are questionable because porosity and permeability in crystalline schists are very much lower than in laboratory hydrothermal systems. The high fluid/rock ratios in the experiments are not necessarily a good model for prograde metamorphism in nature.

Heterogeneous nucleation was not considered a significant rate limiting factor (Walther and Wood, 1984), on the grounds that an infinite number of different sites is available in a metamorphic mineral assemblage, and, when sluggish completion of reaction is found in nature, reactant minerals remain as relics among the product (index) minerals which seem by contrast to form with consistent regularity.

Estimates of reaction oversteps of <1°C should prompt us to re-examine the assumption of a 10°C/my rate of T rise for regional metamorphism. For such a small overstep, a mapped isograd is the place where conditions reached a maximum, the 'high tide' mark of the thermal event, where the rate of increase of T is close to zero. If we adopt a rate of 1°C/my instead of 10°C/my, the reaction time increases and overstep decreases by a factor of $\sqrt{10}$. A similar, more modest, rate of T rise at isograds in contact metamorphism would yield numbers comparable to those of regional metamorphism.

During the progress of devolatilization reactions, fluid will exist, at least transiently, in the pores of crystalline rocks. Fluid inclusions in metamorphic rocks (Crawford, 1981; Roedder, 1984; Touret, 2001) are testament to the presence of fluid at various times in the course of a metamorphic cycle. Furthermore, there has developed a general consensus that prograde metamorphism of a package of metasediments is accompanied by a flux of aqueous fluid integrated over time on the order of 10^3 m^3/m^2 (Ferry, 1984; Hanson, 1997; Ague, 2003; Chapter 12). Fluid may exist as a thin coating along grain boundaries, even when it is not otherwise present as a pore phase with the properties of fluid. Diffusion rates in this intergranular medium will fall somewhere between those appropriate for an open fluid ($\sim 10^{-8}$–10^{-9} m^{2}s^{-1}) and those for grain-boundary diffusion ($\sim 10^{-20}$ m^{2}s^{-1}). Careful petrographic study of porphyroblasts grown during prograde metamorphism, notably garnet, has shown that nearest-neighbour separations are commonly greater than in a random arrangement, and that crystal size is correlated with the degree of isolation of a crystal from its neighbours (Carlson, 1989, 1991; Carlson and Johnson, 1991; Carlson et al., 1995; Spear and Daniel, 2001). These observations indicate textural control by intergranular diffusion. They are consistent with Carmichael's (1969) interpretation of the textures of prograde metapelites in terms of control by the sluggish rate of intergranular diffusion of Al. However, textural observations in other studies (e.g. Holness et al., 1991; Kretz, 1993) have suggested control by surface-reaction rates.

The equations of Wood and Walther can be used to estimate approximate reaction times and T oversteps in regional metamorphism when effective rates of reaction (k_r) are lower than hydrothermal laboratory, surface-controlled rates (k_h), because of: (1) reduced intergranular transport of elements in grain-boundary pore-fluid with much decreased interconnectedness and porosity; (2) nucleation difficulties; (3) several other reasons, as discussed in Baxter and DePaolo (2002, pp. 505–511). It may be that no great decrease in porosity from that in the hydrothermal experiments is required for intergranular diffusion (or nucleation) to become rate limiting (Walther and Wood, 1984, Figs 6 and 7; Matthews and Goldsmith, 1984; Brearley and Rubie, 1990; Hacker et al., 1992). For every 2 orders of magnitude decrease in k_r below k_h, the time for reaction and its T overstep are increased by one order of magnitude. Thus, when $k_r = k_h \times 10^{-4}$, the dehydration reaction time and overstep at 500°C for a T-rise rate of 10°C/my become 1,343,000 yr and 13.4°C. At 700°C, the corresponding numbers are 620,000 yr and 6.2°C. These degrees of overstepping would still be hard to detect by means of textural interpretation and thermobarometry (Chapter 10). In other words, the petrographic evidence for consistent isograd behaviour does not rule out reaction rates in regional metamorphism as much as four orders of magnitude slower than those of interface control; in these cases there may be a non-random distribution of reactant minerals in a rock. In contact metamorphism, for $k_r = k_h \times 10^{-1}$, the And → Sil reaction could be overstepped by more than 100°C, and for $k_r = k_h \times 10^{-4}$ the dehydration of Ms + Qtz similarly delayed. These amounts of reaction overstep can easily lead to progress of metastable reactions. This is especially true for rapidly heated xenoliths, as found by Grapes (1986), for example. Rubie and Brearley (1987) found that experimental treatment of a muscovite-quartz rock produced mullite and a "sub-solidus" granitic melt. Similar phenomena may well be encountered in the inner zones of contact aureoles (Rubie 1998).

In very carefully conducted high-T weight-change experiments at 0.5 to 5 kbar, Schramke et al. (1987) found that the rate of the important metamorphic reaction: Ms + Qtz = And + Ksp + H_2O varied exponentially with the T overstep (and reaction free energy). Lasaga (1986) expressed a general rate law as: $R = kA\Delta G^n/\varphi$, where A is the specific area of the reaction-limiting mineral (cm^2/cm^3) and φ the porosity. Taking the data of Schramke et al. (1987), with exponent $n = 2.7$, he showed that overstep of this reaction in contact metamorphism would be larger than that calculated by Walther and Wood (1994) using their linear equation. He used a numerical model for conductive heat flow and reaction to calculate the Ksp + And isograd position, "T overstep", and field width of the reaction assemblage, in aureoles marginal to 1200°C intrusions, 0.2 to 2 km in size (Lasaga 1986, Table III). For a 1 km intrusion, and a reaction producing 5% And, he found an overstep of ~20°C for the Ms-out isograd and ~5°C for the And-in isograd (Fig. 9), and a reaction time of 6000 y. These numbers changed little when arbitrarily selected nucleation oversteps of were added. The rate of T rise at the position of the isograd declined from >300,000°C/my to ~35,000°C/my when T_{eq} was reached, and then to zero shortly thereafter. Clearly, an average rate of T rise is not a very practical basis for calculating possible reaction rates in short-lived contact metamorphic events. The co-existence of reactant and product phases of the reaction represents kinetic disequilibrium not divariant equilibrium. Lasaga's graphs of reaction progress vs. distance from the igneous contact (Fig. 9) tend to be sigmoidal, because A for a product mineral is zero at the start, and A for a reactant mineral is zero at the end of the reaction.

Kerrick et al. (1991, p. 591) concluded that there was enough evidence to indicate that heterogeneous metamorphic reaction kinetics cannot be adequately quantified by a single rate equation, as suggested by Wood and Walther. This view recognized the influence of experimental variables not considered by Wood and Walther, and the apparent prevalence of non-linear rate relations. Exponential rate laws result in a slower reaction response for small values of the driving force ΔG_r and faster reaction for large values of ΔG_r. The newer experiments reviewed by Kerrick et al. (1991) suggested that the presence of surface crystal defects was enough to render the reaction rate non-linear soon after ΔG_r became finite. At the very least, it seems important for the purpose of application to nature to apply whenever possible the appropriate form of the rate equation, whether linear or non-linear. Experiments indicate that each can apply in different cases (Lasaga and Rye, 1993).

It is then legitimate to ask: What is the relevance to metamorphic petrology of the Wood and Walther (1983) equation for surface-controlled reactions? Laboratory rates are surely appropriate to consider for metamorphism developed in convective hydrothermal systems at ocean spreading centres and on the continents (e.g. Bird et al., 1984; Steefel and Lasaga, 1994). Opinions differ as to whether the regional metamorphism of metasediments in orogenic belts is appropriately labelled hydrothermal (Ferry, 1992). Lasaga and Rye (1993) assumed that interface-controlled rates apply to the metamorphism of mixed volatile systems, but that the transport of CO_2 and H_2O in infiltrating fluid could have the effect of slowing reaction progress and introducing petrologically significant oversteps.

Studies of porphyroblast growth (e.g. garnet, staurolite, biotite) show that mineral growth in a porphyroblast's immediate (mm to cm scale) surroundings may be diminished because reaction affinity (driving force) is lowered by the sluggish intergranular transport of relatively insoluble Al. According to Carlson (2002), transport rates are greatest for Fe and Mg (to which I would add Na and K), and least for elements of 3^+, 4^+ and 5^+ valency. Thus, metamorphic reactions that require intergranular transport only of relatively soluble Mg, Fe, Na, and K may proceed at rates approaching hydrothermal. Many isogradic reactions in regionally metamorphosed carbonate sediments, ironstones, mafic and ultramafic rocks fall into this category. Complex-system rocks may show different sub-system states of partial equilibrium (Loomis, 1983), depending on the presence or absence of slow diffusing species exchanged in the independent reactions. The overall rate of reaction is likely to depend on the second lowest diffusing species when intergranular kinetics are rate determining (Walther, 1990).

Whereas intergranular diffusion rates in nature probably vary over orders of magnitude (Brady, 1983; Lasaga, 1989; Lasaga and Rye, 1993), from totally dry systems (Chapter 15; also Rubie, 1986, 1990) to systems verging on fluid saturated (Walther, 1990), interface reaction rates are represented by a quantifiable relation that appears to show minimal dependence on the system composition and mineral structures; it has a well defined dependence on temperature, and provides an upper bound for heterogeneous reactions in fluid-rock systems; it has been used as a reference line for comparison with rates measured in nature by isotopic and geological means (Vance and O'Nions, 1990; Skelton et al., 1997; Baxter and DePaolo, 2002; Waters and Lovegrove, 2002). Such comparisons help in enquiries into the kinetics of metamorphism that may be determined by different processes. Skelton et al. (1997) showed that the rates of reaction resulting from fluid infiltration into metabasites in the SW Highlands were between 1 and 6 orders of magnitude lower than hydrothermal rates. Baxter and DePaolo (2002) estimated a rate of reaction for the growth of garnet in amphibolite to be five to six orders of magnitude lower than laboratory-derived rates. Waters and Lovegrove (2002) found the rate of porphyroblast growth in hornfels from the Bushveld complex to be about two orders of magnitude smaller. It is important that field tests of reaction rate be done on well constrained situations and clearly distinguish among the reaction processes involved: *in situ* devolatilization, infiltration inducing net vapour loss (see Chapter 12), and infiltration consuming vapour (Skelton et al., 1997), etc. In a general review of field-determined rates, Baxter (2003, Fig. 2) found reaction rates in regional metamorphism to be four to six orders of magnitude slower than laboratory rates, but for contact metamorphism the disagreement was much smaller. Thus, the metastability and obvious reaction overstepping in contact metamorphism is a result of high heating rates rather than reaction rates greatly slowed by intergranular diffusion.

Consider three recent assessments: "There is abundant empirical evidence that chemical equilibrium is often closely approached at the centimeter scale at the peak of metamorphism (the state of local equilibrium in geochemistry and petrology)" (Ferry and Gerdes, 1998, p. 258); "Concrete examples of chemical disequilibrium for prograde metamorphic devolatilization reactions and fluid flow are still relatively rare" (Ague, 2003, p. 208). These statements basically reiterate classical views on metamorphism going back to Goldschmidt and Eskola (Chapters 1 and 3). A somewhat different view was expressed by Carlson (2002, p. 185): "Mounting evidence suggests that partial disequilibrium – meaning disequilibrium for some elements, but not for others – may be a common but rarely detected phenomenon during metamorphic mineral growth, even in ordinary prograde reactions that progress to completion." These apparently conflicting views reflect a difference in emphasis: on the kinetics of devolatilization reactions on the one hand; on the other, on the more formidable problems associated with the elimination of all inherited chemical potential gradients among components within and among the grains of a rock. The first two views do not necessarily imply hydrothermal reaction rates for prograde regional metamorphism, but would be consistent with rates within perhaps four orders of magnitude of them, as well as textural evidence for intergranular or other controls.

If we extend our purview to retrograde metamorphism, fluid-absent metamorphism, and metamorphism of dry protoliths, the evidence for disequilibrium or local equilibrium is overwhelming. Reaction products tend to be localized at spatially restricted reaction sites (Fisher, 1977; Foster, 1981; Joesten, 1991), producing reaction zones, segregations, nodules, coronas and symplectites. The kinetic study of these natural situations can nonetheless be rewarding (e.g. Carlson and Johnson, 1991; Ashworth, 1993; Ashworth and Sheplev, 1997; Ashworth and Chambers, 2000). With the aid of intracrystalline diffusion rates in garnet, Carlson (2002) was able to conclude that that the diffusivity of Al in a fluid undersaturated system is 2.6 log units lower than in a fluid-saturated system.

The widespread application to metamorphic rocks of the petrogenetic grid (Chapter 4), the techniques of mineral thermometry and barometry (Chapter 8), and calculations of fluid and chemical flux, presupposes that there are enough instances in nature where mineral reactions have proceeded without an excessively large overstep in T or P. The fact that the results of our phase-equilibrium investigations have the internal consistency that they do suggests at least local equilibrium (it has become common practice to make this assumption), but this consistency does not rule out modest P-T shifts resulting from a measure of kinetic control as well. The reaction-rate experiments and accompanying theory have shown that short-lived thermal events, such as those of contact metamorphism and those at thrust contacts in rapidly deforming orogens, probably result in potentially detectable overstepping and metastability. This is exactly what we are finding in the field.

One point of agreement in all of this is the critical need for more kinetic data on metamorphic systems, especially from the field.

References

Ague, J.J. (2003) Fluid flow in the deep crust. Pp. 195–228 in: *Treatise on Geochemistry. Volume 3, The Crust* (R.L. Rudnick, editor). Elsevier, Amsterdam.

Ashworth, J.R. (1993) Fluid-absent diffusion kinetics of Al inferred from retrograde metamorphic coronas. *American Mineralogist*, **78**, 331–337.

Ashworth, J.R. and Chambers, A.D. (2000) Symplectic reaction in olivine and controls of intergrowth spacing in symplectites. *Journal of Petrology*, **41**, 285–304.

Ashworth, J.R. and Sheplev, V.S. (1997) Diffusion modelling of metamorphic layered coronas with stability criterion and consideration of affinity. *Geochimica et Cosmochimica Acta*, **61**, 3671–3689.

Barrow, G. (1893) On an intrusion of muscovite-biotite gneiss in the south-eastern Highlands of Scotland, and its accompanying metamorphism. *Geological Society of London Quarterly Journal*, **49**, 330–358.

Baxter, E.F. (2003) Natural constraints on metamorphic reaction rates. Pp. 182–202 in: *Geochronology: Linking the Isotopic Record with Petrology and Textures* (D. Vance, W. Müller and I.M.Villa, editors). Special Publication **220**, Geological Society, London.

Baxter, E.F. and DePaolo, D.J. (2002) Field measurement of high temperature bulk reaction rates II: Interpretation of results from a field site near Simplon Pass, Switzerland. *American Journal of Science*, **302**, 465–516.

Bird, D.K, Schiffman, P., Elders, W.A., Williams, A.E. and McDowell, S.D. (1984) Calc-silicate mineralization in active geothermal systems. *Economic Geology*, **79**, 671–695.

Brady, J.B. (1983) Intergranular diffusion in metamorphic rocks. In: Greenwood H.J. (ed) Studies in metamorphism and metasomatism, *American Journal of Science (Orville volume)*, **283A**, 181–200.

Brearley, A.J. and Rubie, D.C. (1990) Effects of H_2O on the disequilibrium breakdown of muscovite + quartz. *Journal of Petrology*, **31**, 925–956.

Carlson, W.D. (1989) The significance of intergranular diffusion to the mechanisms and kinetics of porphyroblast crystallization. *Contributions to Mineralogy and Petrology*, **103**, 1–24.

Carlson, W.D. (1991) Competitive diffusion-controlled growth of porphyroblasts. *Mineralogical Magazine*, **55**, 317–330.

Carlson, W.D. (2002) Scales of disequilibrium and rates during metamorphism. *American Mineralogist*, **87**, 185–204.

Carlson, W.D. and Johnson, C.D. (1991) Coronal reaction textures in garnet amphibolites of the Llano Uplift. *American Mineralogist*, **76**, 757–772.

Carlson, W.D., Denison, C. and Ketcham, R.A. (1995) Controls on the nucleation and growth of porphyroblasts: Kinetics from natural textures and numerical models. *Geological Journal*, **30**, 207–225.

Carmichael, D.M. (1969) On the mechanism of prograde metamorphic reactions in quartz-bearing rocks. *Contributions to Mineralogy and Petrology*, **20**, 244–267.

Crawford, M.L. (1981) Fluid inclusions in metamorphic rocks – low and medium grade. Pp. 157–181 in: *Fluid inclusions: Applications to Petrology* (L.S. Hollister and M.L. Crawford, editors). Short Course Notes, Mineralogical Association of Canada.

Ferry, J.M. (1984) A biotite isograd in south-central Maine, U.S.A.: Mineral reactions, fluid transfer, and heat transfer. *Journal of Petrology*, **25**, 871–893.

Ferry, J.M. (1992) Regional metamorphism of the Waits River Formation, eastern Vermont: Delineation of a new type of giant hydrothermal system. *Journal of Petrology*, **33**, 45–94.

Ferry, J.M. and Gerdes, M. (1998) Chemically reactive fluid flow during metamorphism. *Annual Reviews of Earth and Planetary Sciences*, **26**, 255–287.

Fisher, G.W. (1977) Non-equilibrium thermodynamics in metamorphism. Pp. 381–403 in: *Thermodynamics in Geology* (D.G. Fraser, editor). D. Reidel Publishing Company, Dordrecht, The Netherlands.

Foster, C.T. Jr. (1981) A thermodynamic model of mineral segregations in the lower sillimanite zone near Rangely, Maine. *American Mineralogist*, **69**, 848–857.

Grapes, R.H. (1986) Melting and thermal reconstitution of pelitic xenoliths, Wehr Volcano, East Eifel, Germany. *Journal of Petrology*, **27**, 343–396.

Hacker, B.R., Kirby, S.H. and Bohlen, S.R. (1992) Time and metamorphic petrology: calcite to aragonite experiments. *Science*, **258**, 1109–1112.

Hanson, R.B. (1997) Hydrodynamics of regional metamorphism due to continental collision. *Economic Geology*, **92**, 880–891.

Holness, M.B., Bickle, M.J. and Graham, C.M. (1991) On the kinetics of textural equilibration in forsterite marbles. *Contributions to Mineralogy and Petrology*, **108**, 356–367.

Joesten, R. (1991) Grain boundary diffusion kinetics in silicate and oxide minerals. *Advances in Physical Geochemistry*, **8**, 345–395.

Kerrick, D.M. (1972) Experimental determination of muscovite + quartz stability with $P_{H_2O} < P_{TOTAL}$. *American Journal of Science*, **272**, 946–958

Kerrick, D.M., Lasaga, A.C. and Raeburn, S.P. (1991) Kinetics of heterogeneous reactions. Pp. 583–671 in: *Contact Metamorphism* (D.M. Kerrick, editor). Reviews in Mineralogy, **26**, Mineralogical Society of America, Washington, D.C.

Kretz, R. (1993) A garnet population in Yellowknife schists, Canada. *Journal of Metamorphic Geology*, **11**, 101–120.

Lasaga, A.C. (1981a) Rate laws of chemical reactions. Pp. 1–68 in: *Kinetics of Geochemical Processes* (A.C. Lasaga and R.J. Kirkpatrick, editors). Reviews in Mineralogy, **8**, Mineralogical Society of America, Washington, D.C.

Lasaga, A.C. (1981b) Transition state theory. Pp. 135–170 in: *Kinetics of Geochemical Processes* (A.C. Lasaga and R.J. Kirkpatrick, editors). Reviews in Mineralogy, **8**, Mineralogical Society of America, Washington, D.C.

Lasaga, A.C. (1986) Metamorphic reaction rate laws and development of isograds. *Mineralogical Magazine*, **50**, 359–373.

Lasaga, A.C. (1989) Fluid flow and chemical reaction kinetics in metamorphic systems: a new simple model. *Earth and Planetary Science Letters*, **94**, 417–424.

Lasaga, A.C. and Rye, D.M. (1993) Fluid flow and chemical reaction kinetics in metamorphic systems. *American Journal of Science*, **293**, 361–404.

Loomis, T.P. (1983) Compositional zoning of crystals: a record of growth and reaction history. Pp. 1–60 in: *Kinetics and Equilibrium in Mineral Reactions* (S.K. Saxena, editor). Advances in Physical Geochemistry, **3**, Springer-Verlag, New York.

Matthews, A. and Goldsmith, J.R. (1984) The influence of metastability of reaction kinetics involving zoisite formation from anorthite at elevated pressures and temperatures. *American Mineralogist*, **69**, 848–857.

Ridley, J. and Thompson, A.B. (1986) The role of mineral kinetics in the development of metamorphic microtextures. Pp. 154–193 in: *Fluid-Rock Interactions during Metamorphism* (J.V. Walther and B.J. Wood, editors). Advances in Physical Geochemistry, **5**, Springer-Verlag, New York.

Roedder, E. (1984) *Fluid Inclusions*. Reviews in Mineralogy, **12**, Mineralogical Society of America, Washington, D.C., 644 pp.

Rosenbusch, H. (1877) Die Steiger Schiefer und ihre Kontaktzone an der Graniten von Barr-Andlau und Hohwald. *Abhandlungen zür geologischse Specialkarte von Elsass-Lothringen*, **1**, 79–393.

Rubie, D.C. (1986) The catalysis of mineral reactions by water and restrictions on the presence of aqueous fluid during metamorphism. *Mineralogical Magazine*, **50**, 399–415.

Rubie, D.C. (1990) Role of kinetics in the formation and preservation of eclogites. Pp. 180–203 in: *Eclogite Facies Rocks* (D.A. Carswell, editor). Blackie & Son, Glasgow, UK.

Rubie, D.C. (1998) Disequilibrium during metamorphism: the role of nucleation kinetics. Pp. 199–214 in: *What Drives Metamorphism and Metamorphic Reactions?* (P.J. Treloar and P.J. O'Brien, editors). Special Publications, **138**, Geological Society, London.

Rubie, D.C. and Brearley, A.J. (1987) Metastable melting during the breakdown of muscovite + quartz at 1 kbar. *Bulletin de Minéralogie*, **110**, 533–549.

Rubie, D.C. and Thompson, A.B. (1985) Kinetics of metamorphic reactions at elevated temperatures and pressures: An appraisal of available experimental data. Pp. 27–79 in: *Metamorphic Reactions* (A.B. Thompson and D.C. Rubie, editors). Advances in Physical Geochemistry, **4**, 291 pp., Springer-Verlag, New York.

Schramke, J.A., Kerrick, D.M. and Lasaga, A.C. (1987) The reaction muscovite + quartz = andalusite + K-feldspar + water. Part 1. Growth kinetics and mechanism. *American Journal of Science*, **287**, 517–559.

Skelton, A.D.L., Bickle, M.J. and Graham, C.M. (1997) Fluid flux and reaction rate from advective–diffusive carbonation of mafic sill margins in the Dalradian, southwest Scottish Highlands. *Earth and Planetary Science Letters*, **146**, 527–539.

Spear, F.S. and Daniel, C. (2001) Diffusion control of garnet growth, Hapswell Neck, Maine. *Journal of Metamorphic Geology*, **19**, 179–195.

Steefel, C.I. and Lasaga, A.C. (1994) A coupled model for transport of multiple chemical species and kinetic precipitation/dissolution reactions with applications to reactive flow in single phase hydrothermal systems. *American Journal of Science*, **294**, 529–592.

Touret, J.L.R. (2001) Fluids in metamorphic rocks. *Lithos*, **55**, 1–27.

Vance, D. and O'Nions, R.K. (1990) Isotopic chronometry of zoned garnets: growth kinetics and metamorphic histories. *Earth and Planetary Science Letters*, **97**, 227–240.

Walther, J.V. (1990) Fluid-rock reaction rates during metamorphism at mid-crustal conditions. *Journal of Geology*, **102**, 559–570.

Walther, J.V. and Wood, B.J. (1984) Rate and mechanism in prograde metamorphism. *Contributions to Mineralogy and Petrology*, **88**, 246–259.

Walther, J.V. and Wood, B.J. (1986) Mineral-fluid reaction rates. Pp. 194–211 in: *Fluid-Rock Interactions during Metamorphism* (J.V. Walther and B.J. Wood, editors). Advances in Physical Geochemistry,

5, Springer-Verlag, New York.
Waters, D.J. and Lovegrove, D.P. (2002) Assessing the extent of disequilibrium and overstepping of prograde metamorphic reactions in metapelites from the Bushveld Complex aureole, South Africa. *Journal of Metamorphic Geology*, **20**, 135–149.

Rates of Hydrothermal Reactions

Abstract. *The rates of reactions of silicates and aqueous fluids follow zero-order kinetics controlled by the reacting surface area with the rate constant given by the equation: log k ≏ −2900/T − 6.85, where T is temperature and where k has the unit's gram-atoms of oxygen per square centimeter per second. This expression appears to hold for all silicates and for reactions involving dissolution, fluid production, or solid-solid transformations in the presence of a fluid of moderate to high pH.*

The rate relations of hydrothermal reactions are of considerable interest in geologic and materials sciences. A full understanding of these processes would enable us to predict weathering phenomena, reactions in geothermal and metamorphic systems, and the aqueous dissolution behavior of, for example, nuclear waste–bearing ceramics. In the course of reviewing published data (*1–19*) on reaction rates, we have discovered a general Arrhenius relation between reaction rate and temperature that appears to hold over temperatures from 25° to 710°C. It also holds for a large number of different silicate and related mineral species.

In the most detailed studies of mineral reaction rates (*1–7*), investigators have examined low-temperature dissolution phenomena of the type associated with weathering at the earth's surface. Such reactions may be represented as follows:

mineral + H_2O → aqueous solution
    + secondary solids   (1)

Although many experimenters have found complex dissolution behavior, it has recently been shown (*4, 20, 21*) that apparent nonlinear relations are artifacts of sample preparation. Once fine-grained material is removed, dissolution follows a zero-order rate relation (*3–7*) controlled by the reacting surface area:

$$\frac{dm}{dt} = k_{dis} \quad (2)$$

The rate of removal of mass from the dissolving mineral (dm/dt, in moles per square centimeter per second) is therefore a constant at fixed temperature, pressure, and activities of solution species. The rate constant, k_{dis}, is, however, a function of temperature (*1, 7, 21*) and is pH-dependent in acidic solutions. Dissolution rates increase with increasing temperature (*1*) and with decreasing pH. At moderate to high pH [> 5 for feldspar at 25°C (*4, 20*)], however, the dissolution rate becomes pH-independent. The reason for the change in behavior reflects the nature of the rate-determining step. At high activities of H_3O^+ the hydrolysis process involves H_3O^+ in the rate-determining step, whereas at very low activities of this species undissociated solvent (H_2O) is the agent of hydrolysis (*20*). The extent of the pH-dependent region varies from phase to phase (*4, 20, 21*), presumably because of different interfacial properties and extents of H_3O^+ adsorption.

Most dissolution experiments have been performed in the temperature range from 0° to 90°C, with a few studies (*1, 2, 7*) extending up to 300°C. The higher temperature experiments were predominantly directed toward establishing equilibrium rather than toward estimating kinetic properties of reactions. Many of these "equilibrium" studies may be used to estimate reaction rates, provided certain assumptions are made about the nature of the rate-controlling process.

Let us consider a generalized high-temperature reaction involving solid and fluid phases:

phase A + phase B =
    phase C + phase D   (3)

Such reactions are generally studied in order to establish the temperature and pressure conditions of equilibrium between phases A, B, C, and D, one of which is typically a fluid. One of the most precise and successful methods of determining equilibrium in such cases (*8–19, 22*) is to use a fine-grained matrix of all but one of the product and reactant solids and a single crystal of the remaining phase. The experimental method requires only a few percent of fluid so that the production of fluid species is minimized. One then determines the direction of reaction by weighing the single crystal before and after the experiment to determine whether it is in the stable or unstable assemblage (Fig. 1). This method yields equilibrium at the point of zero weight change and, near equilibrium, an approximately linear relation between weight change and temperature (Fig. 1) (*23*). Linear behavior close to equilibrium is predicted by transition state theory (*20, 23–25*) which, under these circumstances, yields an equation of the form

$$\frac{dm}{dt} = \frac{k_r [\Delta S_r(T - T_{eq})]}{RT_{eq}} \quad (4)$$

where k_r is the rate constant for the forward reaction at the equilibrium temperature T_{eq} (in kelvins), ΔS_r is the overall entropy change of the reaction, and R is the gas constant. Using Eq. 4, we have extracted k_r values for a large number of single-crystal studies (*8–19*) performed at temperatures from 300° to 710°C (Fig. 2). Values of ΔS_r were computed from data of Helgeson *et al.* (*26*). Values of ΔS_r for reactions in H_2O–CO_2 were corrected for nonideal fluid behavior. The linear approximation breaks down far from equilibrium because of the temperature dependence of k_r (*23*) (Fig. 1). Nevertheless, the linear portion generally spans a wide enough temperature range to permit estimation of k_r within a factor of 2 (Fig. 2).

In order to compare the widely differing studies, we found it necessary to make some assumptions about the rate-determining steps that controlled the rates of weight change. Because the surface area of the single crystal is relatively small (~ 0.5 cm^2) relative to the other fine-grained phases and because the presence of a fluid provides an abundant medium for diffusion, we assumed that the rate of surface reaction at activated sites on the single crystal is rate-determining. This applies for either growth or dissolution of the single crystal. Thus, given approximate crystal masses, geometries, and hence surface areas by the various researchers, we cast Eq. 4 to yield (dm/dt) in terms of moles per square centimeter per second. In order to compare studies with different minerals, we normalized each rate constant to gram-atoms of oxygen per square centimeter per second (Fig. 2), that is, we multiplied each molar rate by the number of oxygens per mole. Although the grain size of the powdered material has some effect on the rate (9), we assumed that this was of second order since it appears to be less than a factor of 3 (9).

The assumption of surface reaction control is vindicated by the consistency between k_r values derived from different studies (Fig. 2). These values show an Arrhenius temperature dependence with relatively little scatter and no consistent difference between different mineral species. This result implies that the k_r of surface reaction for a wide range of silicates (and for corundum) is similar at any fixed temperature.

A final test of the hypothesis of surface-reaction control of the hydrothermal reactions is provided by the low-temperature dissolution data. If, as suggested here, surface detachment (or attachment) is rate-controlling, then from transition state theory (20, 23) k_{dis} in Eq. 2 (corrected to gram-atoms of oxygen) and k_r in Eq. 4 should, under the same conditions, be equivalent. Figure 2 shows available dissolution data for conditions of near-neutral pH where the rates are not substantially pH-dependent (4, 20, 27). The available dissolution data for this particular pH regime are consistent with an extrapolation of the high-temperature "reaction rate" data, confirming our hypothesis of surface-reaction control.

In this set of studies (Fig. 2), the experimental temperature of most of the scattered and inconsistent data is 25°C, a temperature at which the crystals must be fine-grained (< 120 mesh) in order for dissolution to proceed at readily measurable rates. Grinding crystals to this size introduces substantially increased surface defects and very fine particles (4), which dissolve more rapidly than the bulk equilibrium surface. Etching to remove these high-surface-energy phenomena (4) probably also results in enhanced dissolution rates because this process produces etch pits and a thin leached layer (27). Thus, most of the 25°C data refer to abnormally high exposures of defects or leached-layer phenomena, whereas the higher temperature, single-crystal experiments and most of the high-temperature dissolution experiments involve cleaved (unground) crystals (7) with minimized concentrations of surface defects. Despite these uncertainties, the data in Fig. 2 exhibit remarkable consistency throughout the temperature range of interest.

In contrast to the case at 25°C, fluid composition appears to have little effect in the high-temperature regime. A number of the single-crystal experiments (18) (Fig. 2) were performed in H_2O–CO_2 fluids with up to 0.95 mole fraction of CO_2. The rates for these reactions are indistinguishable from those in pure H_2O. This finding suggests either that

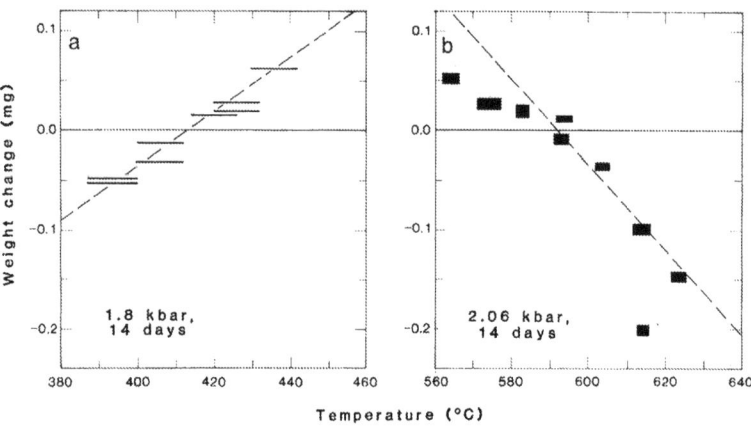

Fig. 1. (a) Weight changes of single crystals of quartz for the reaction pyrophyllite = andalusite + 3 quartz + H_2O (11). (b) Weight changes of single crystals of clinozoisite for the reaction 4 clinozoisite + quartz = grossularite + 5 anorthite + 2 H_2O (10).

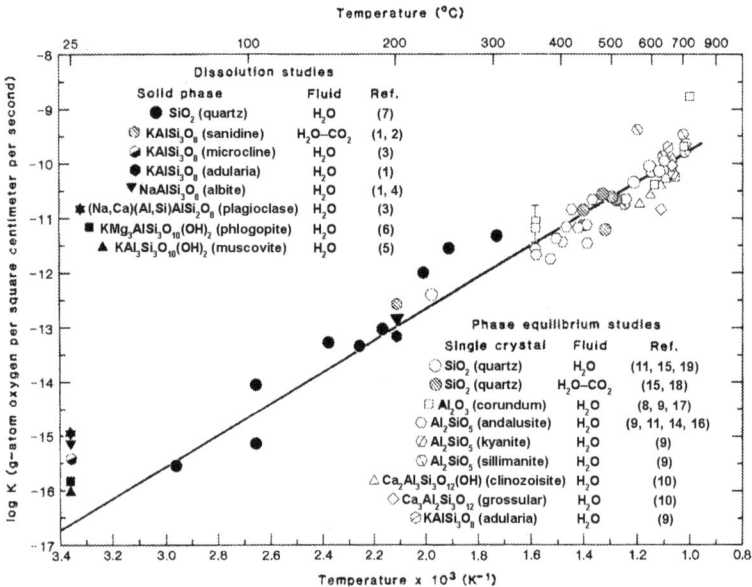

Fig. 2. Arrhenius plot of the reciprocal of absolute temperature (1/T) versus the logarithm (to the base 10) of the rate constant, determined from both dissolution and phase-equilibrium studies.

the rate-controlling step does not involve solvation or that H_2O partitions so strongly into the mineral interface that its concentration at the interface is not greatly affected by 95 percent dilution in the bulk of the fluid. In either case, the result greatly extends the applicability of the observed rate relation. A similar conclusion applies to the effect of total pressure; we find no systematic effect of total pressure on rate constants at pressures from 1 to 7000 bar. Thus, it appears that temperature dominates over fluid composition and pressure effects in the single-crystal studies.

The solid line in Fig. 2 describes a rate constant given by the equation

$$\log k = -2900/T - 6.85 \quad (5)$$

where T is temperature and where k has the units gram-atoms of oxygen per square centimeter per second. The possible applications of this expression to geologic phenomena are many, but let us consider as an example a prograde metamorphic dehydration reaction with $\Delta S_r = 80$ J K^{-1}. At 500°C overstep of this reaction boundary by 1°C with 2-mm crystals would cause the reaction to go to completion in 330 years, given the presence of all phases involved. At 700°C the reaction would go to completion in 70 years. These rapid rates support the classic observation (28) that prograde mineral assemblages remain very close to thermodynamic equilibrium when a fluid phase is present.

BERNARD J. WOOD
JOHN V. WALTHER

Department of Geological Sciences,
Northwestern University,
Evanston, Illinois 60201

References and Notes

1. M. Lagache, *Bull. Soc. Fr. Mineral. Cristallogr.* **88**, 223 (1965).
2. ———, *Geochim. Cosmochim. Acta* **40**, 157 (1976).
3. E. Busenburg and C. V. Clemency, *ibid.*, p. 41.
4. G. R. Holden, Jr., and R. A. Berner, *ibid.* **43**, 1161 (1979).
5. F.-C. Lin and C. V. Clemency, *ibid.* **45**, 571 (1981).
6. ———, *Clays Clay Miner.* **29**, 101 (1981).
7. J. D. Rimstidt and H. L. Barnes, *Geochim. Cosmochim. Acta* **44**, 1683 (1980).
8. W. S. Fyfe and M. A. Hollander, *Am. J. Sci.* **262**, 709 (1964).
9. B. W. Evans, *ibid.* **263**, 647 (1965).
10. M. J. Holdaway, *ibid.* **264**, 643 (1966).
11. D. M. Kerrick, *ibid.* **266**, 204 (1968).
12. A. B. Thompson, *ibid.* **269**, 267 (1970).
13. ———, *ibid.* **271**, 72 (1971).
14. M. J. Holdaway, *ibid.*, p. 97.
15. D. M. Kerrick, *ibid.* **272**, 946 (1972).
16. H. H. Haas and M. J. Holdaway, *ibid.* **273**, 449 (1973).
17. N. D. Chatterjee and W. Johannes, *Contrib. Mineral. Petrol.* **48**, 89 (1974).
18. J. Slaughter, D. M. Kerrick, V. J. Wall, *Am. J. Sci.* **275**, 143 (1975).
19. A. B. Thompson, in *Progress in Experimental Petrology*, G. M. Biggar, Ed. (Natural Environment Research Council, London, 1976), p. 9.
20. P. Aagaard and H. C. Helgeson, *Am. J. Sci.* **282**, 237 (1982).
21. J. Schott, R. A. Berner, E. L. Sjoberg, *Geochim. Cosmochim. Acta* **45**, 2123 (1981).
22. W. S. Fyfe, *J. Geol.* **68**, 553 (1960).
23. A. C. Lasaga and G. W. Fisher, in *Kinetics of Geochemical Processes*, A. C. Lasaga and R. J. Kirkpatrick, Eds. (Mineralogical Society of America, Washington, D.C., 1981), p. 135.
24. H. Eyring, *J. Chem. Phys.* **3**, 107 (1935).
25. M. Boudart, *J. Phys. Chem.* **80**, 2869 (1976).
26. H. C. Helgeson, J. M. Delany, H. W. Nesbitt, D. K. Bird, *Am. J. Sci.* **278A**, 1 (1978).
27. R. Wollast, personal communication.
28. V. M. Goldschmidt, *Z. Anorg. Chem.* **71**, 313 (1911).
29. Supported by NSF grants EAR80-24146 and 82-12502.

17 January 1983; accepted 23 June 1983

CHAPTER 14: ULTRA-HIGH PRESSURE METAMORPHISM

Chopin, C. (1984) Coesite and pure pyrope in high-grade blueschists of the Western Alps: a first record and some consequences. *Contributions to Mineralogy and Petrology*, **86**, 107–118.

Until about the middle of the 20th century, petrologists had little concept of the magnitude of effective pressures in metamorphism. Nineteenth century ideas were of a 'metamorphic layer' in the upper part of the crust extending down about 10 km; this would translate into load pressures of no more 3 kbar (Touret and Nijland, 2002). As we saw in Chapter 1, Goldschmidt (1912, Fig. 1) contemplated metamorphic pressures for katazonal rocks as high as 15,000 atm (15 kbar), but he believed that these were load pressures raised far upwards by tectonic stresses. Nevertheless, many years later, Eskola (1939, p. 345) still made no attempt to attach either pressure or temperature values to his metamorphic facies. In his textbook, Ramberg's facies diagram (1952, Fig. 73) had a temperature scale, but the pressure axis was left without values. Barth's text (1952, Fig. 137) suggested a geostatic upper pressure limit for metamorphic rocks of ~6000 atm, although he allowed the possibility that some "tectonic" eclogites may have sustained higher values of effective pressure. Turner and Verhoogen (1960, p. 458) considered metamorphism to occur in the range 0 to 10,000 bars. Winkler (1965, Fig. 28) extended his facies diagram to 11 kbar. Mountain roots were known to exist, but there were no indications that metamorphic rocks in them ever made it back to the surface. The relative importance and magnitude of lithostatic, tectonic and volatile pressures were poorly understood. This had been a problem with our interpretation of blueschists (Chapter 7), but it was more serious for eclogites because it has a bearing on possible protoliths and the source regions of eclogites (deep crust vs. mantle).

For many good reasons, eclogites have held a special place in the minds of generations of petrologists and mineralogists. Godard (2001) addresses this point in his delightful review of the two-century history of eclogites and what they have meant for the earth sciences. Eclogites and glaucophane schists were known to represent higher-than-normal pressures because they are denser than gabbros and greenschists, respectively (Eskola, 1921, 1939; Miyashiro and Banno, 1958). The pressures implied by the presence of omphacite and garnet at less than mantle temperatures were not clear, however (Yoder and Tilley, 1962). Even when the gabbro-to-eclogite boundary was finally determined experimentally (Ringwood and Green, 1966; Green and Ringwood, 1967), a long and risky extrapolation was required to define the *P-T* stability field of eclogite at modest metamorphic temperatures.

Eclogites and garnet peridotites, long thought to be the sole compositional representatives of the eclogite facies, were described through the years as occurring as bands, lenses, and blocks in metaperidotites, high-grade gneisses, and glaucophane schists (Eskola, 1921). Eclogitic rocks were typically much overprinted by minerals representative of lower-pressure facies. Their exotic nature and high density encouraged the belief that most, if not all, eclogites were fragments of the eclogite shell of the upper mantle (e.g. Holmes, 1936), transported tectonically or magmatically into the crust (Eskola, 1939; Fyfe *et al.*, 1958; Yoder and Tilley, 1962; O'Hara and Mercy, 1963; Lappin, 1974). Thus, the range of possible pressures and temperatures in the crust appeared to have little relevance to the conditions under which eclogites form.

Long ago, some writers proposed that their eclogites were metamorphosed basalts and gabbros (e.g. Backlund, 1936), although they attributed an important role to concentrated shearing stresses. More recently, a number of workers in western Norway described the conversion of basalt, dolerite and gabbro to eclogite during regional metamorphism in what is now called the Western Gneiss region of Norway (Gjelsvik, 1952; Kolderup, 1960; Schmitt, 1964). According to O'Hara and Mercy (1963), a similar origin for eclogite was proposed by Chevenoy in 1957 and Udovkina in 1959. In 1959, Bearth wrote (p. 28, translated): "Studies extending across a large section of the ophiolites of the Western Alps have shown that a substantial part of the widely distributed eclogites and glaucophane schists originated from pillow basalts and associated breccias." His Fig. 3 was a field drawing of pillow basalt structure overprinted by eclogite in the Zermatt-Saas ophiolite. Excellent outcrops of eclogitized pillow basalt at Pfulwe can actually be viewed after a short hike up from Zermatt. But O'Hara and Mercy (1963, p. 301) evidently found all these descriptions unconvincing – Bearth (1970, p. 30) took them to task for this unnecessarily conservative view. Coleman and Lee (1963) labelled Franciscan type IV eclogite blocks as metabasalt. In support of this view, Coleman *et al.* (1965) emphasized the compositional differences between coexisting garnet and pyroxene in blueschist-related (group C) eclogites and those in eclogites clearly of mantle origin (group A), as well as the apparent textural equilibrium between glaucophane, garnet and omphacite in the California eclogites, and the presence of textural zoning and abundant mineral inclusions in the garnets. They thought in terms of tectonic overpressure. On the basis of their association with glaucophane-schist facies rocks and the similar composition of omphacite and garnet in blueschists and eclogites, Essene *et al.* (1965) agreed that the California eclogites were metamorphosed metabasites. Church (1964) reported that the sheared margins of metagabbro in Co. Donegal, Ireland, were converted to eclogite. Bearth (1965) described in detail the conversion of the spectacular Allalin olivine gabbro, Switzerland, into eclogite. Use of the electron microprobe showed that the garnets in some eclogites possessed the same kind of concentric compositional zoning as the garnets in metapelites (Evans, 1965; see Ghent, 1988, Fig. 2.1). Miller (1970) described kyanite-eclogite in the Ötztal Altkristallin, Austria, which had clearly formed from gabbro. Bryhni et al. (1970) argued the case for the crustal origin of the eclogites of western Norway. Descriptions of eclogites from many terranes were beginning to suggest a degree of petrological diversity that made it difficult to sustain the idea that they all were fragments of mantle eclogite (Coleman et al., 1965; Church, 1968; Turner, 1968). The deep-crustal, prograde metamorphic origin of some eclogites was thus finally accepted, but there was a reluctance to invoke pressures much greater than 10 kbar, because that seemed to be about the limit of lithostatic pressure reached in the crust. For this reason, the obviously crustal eclogites associated with glaucophane schists were interpreted to be the product of dry conditions of metamorphism (Fry and Fyfe, 1969; Carswell, 1990, Fig. 1.1); under wet conditions, such rocks would be composed of hydrated greenschist, blueschist, or amphibolite assemblages. Bearth (1959, 1965, 1970) correlated the sequence: eclogite → glaucophane schist → albite-epidote-barrroisite-amphibolite with increasing introduction of H_2O without serious change in pressure. (We now know, based on the thermodynamic database of minerals, that at pressures of 10–14 kbar, wet eclogite can occur at ~500°C: Evans, 1990; Will et al., 1998; Wei et al., 2003). Consult Bucher et al. (2005) for the latest information on the Zermatt-Saas eclogites.

As discussed in Chapter 7 on the subject of blueschists, the advent of plate tectonics in the late 1960s, and our appreciation of the subduction process, raised the bar on estimates of the magnitude of possible lithostatic pressures experienced by crustal rocks in the glaucophane-schist and eclogite facies.

But it was not until the mid-1970s that rocks of eclogite facies were found both to be components of structurally coherent terranes and to comprise lithologies other than metabasites. Brothers (1974) and Black (1977) were able to map regional isograds for garnet and omphacite in mafic and pelitic rocks in a 70×20 km terrane of progressive high P/T metamorphism in the Ouégoa district, northern New Caledonia. Their estimate of maximum pressure was what now seems a conservative 9 kbar, but this was later revised upwards to 15 kbar with the finding of jadeite in meta-acidic rocks (Brothers and Yokoyama, 1982). Granitic rocks metamorphosed under eclogite-facies conditions, containing jadeite, garnet, zoisite, phengite, quartz and K-feldspar, were described from the Sesia Zone of the Western Alps by Compagnoni and Maffeo (1973). The suggested pressure was 14.5 kbar, or lower than this if the temperature was <600°C. Chinner and Dixon (1973) suggested 10 to 15 kbar for the Allalin metagabbro, based on the presence of kyanite + talc.

The next important step was recognition of the high-pressure nature of the assemblage talc + phengite in pelites of the low-temperature part of the eclogite facies (Abraham and Schreyer, 1976; Schreyer, 1977; Schreyer, 1988). Massonne and Schreyer (1989) confirmed a minimum pressure of 11 kbar for this mineral pair, which takes the place of the chlorite + biotite typical of low-grade Barrovian pelites. Chopin (1981) found talc + phengite, and the even higher-pressure talc + garnet and talc + chloritoid pairs, to be widespread in pelitic schists in the Western Alps.

Evans and Trommsdorff (1978) described garnet peridotite formed by prograde Alpine metamorphism of serpentinite associated with eclogite and metarodingite at Cima di Gagnone in the Cima-Lunga nappe, Lepontine Alps. Their suggested pressure was >20 kbar for the metaperidotite (recently refined to 30 kbar, Nimis and Trommsdorff, 2001), and 25 kbar for the eclogite-metarodingite suite with MORB trace element chemistry (Evans et al., 1979). Heinrich (1986a,b) found a regional P-T gradient in prograde eclogites in the related Adula nappe, and showed that the enclosing continental basement rocks had experienced the same burial history. Similar regional field relationships and P-T gradients (up to 800°C and 18–20 kbar) were emerging from work on the prograde Caledonian eclogites in the Western Gneiss Region, Norway (Griffin et al., 1985).

A further advance for eclogite petrology was Holland's (1979a) experimental reversal of the high-pressure breakdown reaction of paragonite to jadeite + kyanite + H_2O. This clearly revealed the relatively high-pressure nature of the kyanite + omphacite paragenesis common in eclogites in the deep crust. Holland (1979b) argued convincingly for the presence of H_2O-rich fluid and a pressure of 19.5 kbar for eclogite of volcanic origin in the Tauern Window.

Despite this escalation in pressure estimates for crustal eclogites, the discovery by Chopin (1984) of coesite as inclusions in pyrope megacrysts in whiteschist layers of the Dora-Maira Massif in the Western Alps was sensational. With this revelation, we were talking about the transport of basement granites and associated sediments down to depths corresponding to at least 28 kbar. We were confronted with convincing evidence that continental crust could be buried to depths of 100 km and later returned to the surface by orogenic processes. It was a paper with far-reaching implications for subduction geodynamics, seismology, geochemical cycling, and arc magmatism. Some simple numerical modelling by Molnar and Gray (1979), based on the densities of crustal and mantle rocks, had earlier shown that it was certainly possible for continental lithosphere to be conducted down into subduction zones following oceanic lithosphere, reaching depths corresponding to the eclogite facies.

Also in 1984, Smith found coesite in eclogite in the Western Gneiss Region of Norway. Previously, coesite had been recognized only in impact structures and diamond-bearing xenoliths in kimberlite. Its stability relative to quartz (>28 kbar at 800°C) was known from experiments in several laboratories. Nearly 20 y earlier, Chesnokov and Popov (1965) had described (in English) radial expansion cracks in garnet and omphacite around quartz in eclogite from the Maksyutov complex, South Urals, Russia; their preferred explanation was the former presence of coesite. Their paper had no great impact because, for many people at the time, eclogites originated in the mantle. These authors recognized the same texture in an eclogite from the Münchberg gneiss mass, Bavaria; coesite has not yet been found there, although pressures in the coesite field (31 kbar, 630°C) have been claimed for associated calc-silicate rocks

(Klemd et al., 1994). The status of the Maksyutov eclogites remains controversial (Glodny et al., 2002), although the discovery of fine-grained diamond inclusions in garnet may have changed matters (Bostick et al., 2003).

Apart from its geodynamic implications, an impressive feature of Chopin's 1984 paper was the robustness of the evidence for ultra-high pressure conditions in what were clearly crustal rocks: *in situ* coesite in a non-stress environment jacketed inside garnet, plus the presence of additional ultra-high pressure indicators, namely, nearly pure pyrope + coesite/quartz, jadeite + kyanite, and high Si-phengite. My students would call this paper a 'slam dunk'. The stability fields of these assemblages were known thanks to the efforts of experimental petrologists working with the piston-cylinder apparatus, as reviewed, for example, by Schreyer (1985). Quartz in a distinctive polycrystalline, feathery or palisade texture, as in xenolithic coesite (Smyth, 1977), had partially replaced coesite during decompression of the Dora-Maira Massif, and radial racks were generated in the host garnet due to the 10% volume expansion. Chopin's conclusions met with instant general approval and acclaim.

This paper ushered in what has become a new, exciting and expanding branch of research in metamorphic petrology: ultra-high pressure metamorphism, or UHPM (defined by the presence of coesite in rocks of the eclogite facies). It is an area of interest that has encouraged of a variety of laboratory investigations − experimental, analytical and computational − as well as a large component of vital field research into the occurrence of relics of UHPM. Chopin's reinforcement of the significance of radiating cracks in garnet and inclusions of polycrystalline and palisade quartz greatly facilitated the search for UHPM in existing high-P terranes, where attention focused on the interiors of garnet and zircon. The search soon brought to light several UHPM terranes, some of which dwarf in size the coesite-unit of the Dora-Maira massif (10 × 15 km), for example, the Dabie-Sulu Belt, China (400 × 75 km), the Western Gneiss Region, Norway (350 × 70 km), and the Kokchetav Massif, Kazakhstan (120 × 10 km).

Chopin et al. (1991) described the coesite-bearing UHPM unit as part of a roughly 1 km thick nappe complex juxtaposed against other Alpine metamorphosed HP (but not UHPM) tectonics units along flat-lying faults. The coesite-bearing whiteschist, thought perhaps to have been an evaporite sediment, occurs as discontinuous layers and boudin trails within orthogneiss, above and below a varied series of metapelites, metabasite, and marble. Coesite was also found in the metapelites and eclogites, but not in the enclosing acidic gneisses (Chopin, 1987), although these experienced the same UHP metamorphism, as evidenced by the presence of relics of a former grossular-rich garnet + rutile assemblage (Chopin et al., 1991). The UHP metamorphism was originally thought to be Eo-alpine in age, but Tilton et al. (1991) found a U/Pb concordia age of 38 ± 1.4 Ma for zircon and 30−34 Ma ages for monazite and the new UHP mineral ellenbergerite, all coexisting with coesite inside pyrope in the whiteschist. Puzzlement surrounded the seemingly conflicting radiometric evidence for some time (Compagnoni et al., 1995). Schertl et al. (1991) evaluated the prograde and retrograde P-T paths for the Dora-Maira whiteschist and jadeite-quartzite, and suggested maximum metamorphic conditions of 37 kbar and 800°C, thus very close to the diamond stability field. They also reported the rare presence of talc + jadeite, the high-pressure equivalent of glaucophane. The decompression path was not adiabatic but involved a moderate degree of conductive cooling. Kienast et al. (1991) suggested maximum conditions in the range 27−35 kbar and 680−750°C. Chopin et al. (1992) proposed a minimum pressure of 28 kbar (at 725°C) based on the stability field of the ellenbergerite found in the whiteschist. Avigad (1992) recognized that the structure of the massif was shaped during decompression of the UHP rocks. He found ductile shear zones above the coesite unit, operating with a normal sense of motion. Based on oxygen-isotope thermometry and phase equilibria, Sharp et al. (1993) suggested maximum conditions of 34±2 kbar and 700−750°C, and postulated the former presence of a partial melt to account for H_2O activities in the range 0.4 to 0.75. Michard et al. (1995) concluded that the latest part of the exhumation process was probably related to gravity collapse, but for the major part of it they could not decide between extension and a forced-flow extrusional model involving imbricate slices. Gebauer et al. (1997) confirmed the age of *in situ* UHP metamorphic zircon in whiteschist by ion microprobe to be 35.4±1.0 Ma. Earlier magmatic oscillatory-zoned zircon in the same sample demonstrated that the whiteschist protolith was Hercynian granite, metasomatically altered along shear zones. A prograde path based on thermobarometry of eclogites in the coesite-bearing unit was shown approaching very close to the diamond field (Chopin and Schertl, 1999, Fig. 2; Nowlan et al., 2000). Rubatto and Hermann (2001) and Rubatto and Scambelluri (2003) confirmed the previous UHP whiteschist ages with one at 35.1 Ma for a peak metamorphic titanite in a metacarbonate rock. An initial exhumation rate of 3.4 cm/y was inferred when these ages were combined with those of titanite formed during decompression. This rate compares with plate velocities rather than those for erosion. Similar fast rates have recently been determined in some of the other UHPM terranes. Experiments in the system KCMASH by Hermann (2003) indicated that the peak whiteschist assemblage of garnet+phengite+kyanite+coesite required a pressure >45 kbar at 730°C, and that phengite barometry based on these experiments shows similar pressures for eclogites and metapelites associated with the whiteschist. Therefore, the entire UHP unit of the Dora-Maira Massif apparently experienced conditions in the diamond field. As far as I know, inclusions of diamond have not been found in these rocks, so this question remains undecided.

It was only a matter of time after Chopin's discovery, for metamorphic diamond to be discovered in crustal metamorphic rocks, namely as inclusions in garnet and zircon in garnet-biotite gneisses in the Kokchetav massif (Sobolev and Shatsky, 1990). This extended the pressures for crustal metamorphism up to 40 kbar, thus burial, metamorphism, and exhumation from depths of 140 km or more, an even greater challenge for geodynamics. Later, metamorphic microdiamonds were found in metamorphosed crustal rocks in other coesite-bearing terranes:

Dabie-Shan, China (Xu et al., 1992), the Western Gneiss Region, Norway (Dobrzhinetskaya et al., 1995), Sulawesi, Indonesia (Parkinson et al., 1999), the Erzgebirge, Germany (Nasdala and Massonne, 2000), Rhodope, Greece (Mposkos and Kostopoulos, 2001, later confirmed as diamond, C. Chopin pers. comm., 2004) and Qinling, China (Yang et al., 2002). C-O-H-N fluids are believed to play a role in the precipitation of some of the finer diamonds.

The new field of UHPM study has spawned numerous national and international conferences and symposia, workshops, books, special volumes, a mushrooming literature, and back-to-back review papers. Unquestionably, these have been pushing a scientific frontier, and this effort is not over yet. The latest and most comprehensive review is that published by the EMU in 2003 (Carswell and Compagnoni, editors).

As many as 22 UHPM terranes have now been recognized worldwide; they are lacking only on the continents of North America and Australia (Liou et al., 2002; Chopin, 2003). I have focused here on developments in our knowledge of the Dora-Maira occurrence, for that is where crustal coesite was first identified. Some of the other terranes have been intensely studied as well, but space limitations preclude any detailed review of them here. Most UHPM terranes are associated with type A continental-plate collision belts; their ages range from late Precambrian (620 Ma, Mali: Jahn et al., 2001) and Cambrian (540–530 Ma, Claoue-Long et al., 1991) to 35 Ma in the Dora-Maira. Older ages reported earlier for Mali (1045–820 Ma) resulted from inherited argon, a recurring problem in the geochronology of HP and UHP belts. It is possible then that the onset of deep crustal subduction and exhumation was delayed in geological time in comparison with the formation of blueschist belts (Chapter 7, Stern, 2005). While the geographical distribution of confirmed UHP rocks in any one terrane may be extensive, they continue to be found, like most eclogites, and the coesite-bearing rocks of Dora-Maira, rather sporadically as boudins or layers in host gneisses showing lower pressure facies mineralogy. However, armoured relics of coesite or coesite pseudomorphs in surrounding gneisses of the major UHP terranes (Western Gneiss Belt, Kokchetav Massif, Dabie-Sulu Mountains, Western Alps, Bohemian Massif) confirm that the UHP rocks are not simply isolated tectonic fragments but belong to nappe units composed of coherent lower crustal material (± portions of cover sequence) previously consolidated as part of an older continental fragment (Coleman and Wang, 1995).

In all these UHPM terranes, there is a predominance of granitic and granodioritic gneisses. Other lithologies are clearly of supracrustal continental origin: metapelites, marbles, quartzites. Metabasites and garnet peridotites may have been derived from crustal or mantle sources. Mineralogy and phase equilibria are indicative of a relatively low activity of H_2O, and maximum conditions – for standard coesite-bearing terranes, $T = 650–900°C$ and confining $P = 2.8–4.0$ Gpa – point to low geotherms of 7–8°C/km and a subduction-zone setting. Exhumed UHP units now present in the upper crust occur as thin subhorizontal slabs, commonly bounded by normal faults on the top and reverse faults on the bottom, and sandwiched between HP and lower-grade units. UHPM relics occur predominantly in the eclogites, which tend to be resistant to penetrative deformation and the introduction of fluid. Cool P-T-time paths are accompanied by high rates for both convergence and exhumation (Perchuk et al., 1998; Liou et al., 1998; Rubatto and Hermann, 2001), at least for the smaller UHP terranes. This suggests uplift tectonic processes related to buoyancy, slab break-off, and imbricate faulting. Failure of the dry granitic rocks to convert to UHPM mineralogy during subduction (Chapter 15) may be a contributory cause of rapid buoyant exhumation (Carswell and Compagnoni, 2003). Whereas post-collision and late-stage granite plutons are common in some terranes, coeval calc-alkali volcanic and plutonic rocks are not found.

The list of mineral indicators of UHPM has expanded in the last 20 y beyond the SiO_2 and C polymorphs, coesite and diamond (or pseudomorphs of quartz and graphite). Some of the unusual minerals found in the Dora-Maira Massif owe their presence to the combination of UHP conditions and high Mg/Fe bulk composition, and they are not necessarily going to be present elsewhere. These include the new minerals ellenbergerite (Chopin, 1986; Chopin et al., 1986) and bearthite (Chopin et al., 1993), as well as high-Mg staurolite and end-member chloritoid (Simon et al., 1997). Mineral solid solutions extended by very high pressure, reminiscent of xenoliths in kimberlites, have been found in some places: e.g. K-rich clinopyroxene (or exsolved K-feldspar or phengite, Zhang and Liou, 1998), supersilicic clinopyroxene (Katayama et al., 2000; Schmädicke and Müller, 2000; see Page et al., 2005, re quartz inclusions), majoritic garnet (or garnet with clinopyroxene lamellae, van Roermund et al., 2000, Ye et al., 2000), supersilicic titanite (Ogasawara et al., 2002), and OH-bearing topaz (Wunder et al., 1993; Zhang et al., 2002).

The unlikely preservation of coesite and diamond during the decompression cycle can largely be attributed to their inclusion in mechanically strong and stable host minerals acting like pressure vessels, preventing fluid entry and maintaining buffered pressure levels. It seems that very dry conditions helped, because coesite has also been found inside weak minerals and as a matrix phase, and there is phase-equilibrium evidence for very low activity of H_2O during decompression (Liou and Zhang, 1996; Simon and Chopin, 2001; Mosenfelder et al., 2005). Based on pressure-dependent shifts in laser micro-Raman spectra, Parkinson and Katayama (1999) found spectacular proof of present-day pressures of up to 2 Gpa in coesite inclusions in unfractured garnet and zircon from a number of UHPM terranes. This is powerful confirmatory evidence for the stable origin of coesite.

It may be that we will not find many more UHPM terranes; the search seems to have been thorough. There is still much to be done, however. A high priority should be further experimental calibration and thermodynamic (rather than empirical) optimization of the many equilibria (preferably H_2O conserved) in the KNCMFASH system among the minerals of UHP eclogites: Grt, Omp, Ky, Phe, Ep, Coe/Qtz in the pressure range 3 to 10 Gpa, a range that can be duplicated in the multi-anvil press. The latest and best solution models for the garnet components (Chapter 10) need to be incorporated into present thermobarometric calibrations (Ravna and Paquin, 2003). It seems unlikely to me that minerals with compositions extended by high pressure will make sufficiently sensitive barometers (e.g. K, Si, Ca-Eskola in Cpx, Na, Si in Grt). Whenever possible, the thermobarometry should be applied in the context of datable microstructures and *in situ* radiometric dating. Perhaps we shall see some breaks-through in the issue of elemental and isotopic recycling between deeply subducted crustal material and the mantle.

There is no question that the application of cutting-edge laboratory techniques has been responsible for much of the progress made in the last several years in our understanding of UHPM rocks. But we should not lose sight of the fact that it all started with careful and enlightened microscopic petrography. The next 'revolution' may not happen at all if curriculum innovators succeed in eliminating optical mineralogy from any place in the university curriculum.

References

Abraham and Schreyer, W. (1976) A talc-phengite assemblage in piemontite schist from Brezovica, Serbia, Yugoslavia. *Journal of Petrology*, **17**, 421−439.

Avigad, D. (1992) Exhumation of coesite-bearing rocks in the Dora-Maria massif (western Alps, Italy). *Geology*, **20**, 947−950.

Backlund, H.G. (1936) Zur genetischen Deutung der Eklogite. *Geologische Rundschau*, **28**, 47−61.

Barth, T.F.W. (1952) *Theoretical Petrology*. John Wiley and Sons, Inc., New York, 387pp.

Bearth, P. (1959) Über Eklogite, Glaukophanschiefer und metamorphen Pillowlaven. *Schweizerische Mineralogische und Petrographische Mitteilungen*, **39**, 267−286.

Bearth, P. (1965) Zur Entstehung der alpinotyper Eklogite. *Schweizerische Mineralogische und Petrographische Mitteilungen*, **45**, 179−188.

Bearth, P. (1970) Zur Eklogitbildung in den Westalpen. *Fortschritte der Mineralogie*, **47**, 27−33.

Black, P.M. (1977) Regional high-pressure metamorphism in New Caledonia: phase equilibria in the Ouegoa district. *Tectonophysics*, **43**, 89−107.

Bostick, B.C., Jones, R.E., Ernst, W.G., Chen, C., Leech, M.L. and Beane, R. (2003) Low-temperature microdiamond aggregates in the Maksyutov Metamorphic Complex, South Ural Mountains, Russia. *American Mineralogist*, **88**, 1709−1717.

Brothers, R.N. (1974) High-pressure schists in northernmost New Caledonia. *Contributions to Mineralogy and Petrology*, **46**, 109−127.

Brothers, R.N. and Yokoyama, K. (1982) Comparison of the high-pressure schist belts of New Caledonia and Sanbagawa, Japan. *Contributions to Mineralogy and Petrology*, **79**, 219−229.

Bryhni, I., Green, D.H., Heier, K.S. and Fyfe, W.S. (1970) On the occurrence of eclogite in Norway. *Contributions to Mineralogy and Petrology*, **26**, 12−19.

Bucher, K., Fazis, Y., de Capitani, C. and Grapes, R. (2005) Blueschists, eclogites, and decompression assemblages of the Zermatt-Saas ophiolite: High-pressure metamorphism of subducted Tethys lithosphere. *American Mineralogist*, **90**, 821−835.

Carswell, D.A. (1990) Eclogites and the eclogite facies: definitions and classification. Pp. 1−13 in: *Eclogite Facies Rocks* (D.A. Carswell, editor). Blackie, Glasgow, UK.

Carswell, D.A. and Compagnoni, R. (2003) Introduction with review of the definition, distribution and geotectonic significance of ultrahigh pressure metamorphism. Pp. 3−9 in: *Ultrahigh Pressure Metamorphism* (D.A. Carswell and R. Compagnoni, editors). EMU Notes in Mineralogy, **5**. European Mineralogical Union.

Chesnokov, B.V. and Popov, V.A. (1965) Increase in the volume of quartz grains in South Urals eclogites (English Translation). *Doklady of the Academy of Sciences USSR, Earth Sciences Section*, **162**, 176−178.

Chinner, G.A. and Dixon, J.E. (1973) Some high-pressure parageneses of the Allalin gabbro, Valais, Switzerland. *Journal of Petrology*, **14**, 185−202.

Chopin, C. (1981) Talc-phengite: a widespread assemblage in high-grade pelitic blueschists. *Journal of Petrology*, **22**, 628−650.

Chopin, C. (1984) Coesite and pure pyrope in high-grade blueschists of the Western Alps: a first record and some consequences. *Contributions to Mineralogy and Petrology*, **86**, 107−118.

Chopin, C. (1986) Phase relationships of ellenbergerite, new high-pressure Mg-Al-Ti-silicate in pyrope-coesite-quartzite from the Western Alps. *Geological Society of America Memoir*, **164**, 31−42.

Chopin, C. (1987) Very high-pressure metamorphism in the Western Alps: implications of subduction of continental crust. *Transactions of the Royal Society of London*, series A, **321**, 183−197.

Chopin, C. (2003) Ultrahigh-pressure metamorphism: tracing continental crust into the mantle. *Earth and Planetary Science Letters*, **212**, 1−14.

Chopin, C. and Schertl, H.-P. (1999) The UHP unit in the Dora Maira massif, Western Alps. *International Geology Review*, **41**, 765−780.

Chopin, C., Klaska, R., Medenbach, O. and Dron, D. (1986) Ellenbergerite, a new high-pressure Mg-Al-(Ti,Zr)-silicate with a novel structure based on face-sharing octahedra. *Contributions to Mineralogy and Petrology*, **92**, 316−321.

Chopin, C., Henry, C. and Michard, A. (1991) Geology and petrology of the coesite-bearing terrain, Dora Maira massif, Western Alps. *European Journal of Mineralogy*, **3**, 263−291.

Chopin, C., Schreyer, W. and Baller, T. (1992) Ellenbergerite stability, a reappraisal. *Terra Nova Abstracts Supplement* **4**, 10.

Chopin, C., Brunet, F., Gebert, W., Medenbach, O. and Tillmanns, E. (1993) Bearthite, $Ca_2Al[PO_4]_2(OH)$, a new mineral from high-pressure terranes of the western Alps. *Schweizerische Mineralogische und petrographische Mitteilungen*, **73**, 1−9.

Church, W.R. (1964) Metamorphic eclogites from Co. Donegal, Eire. *Indian Mineralogist, International Mineralogical Association Special* no. **22-23**.

Church, W.R. (1968) Eclogites, Pp. 755−798 in: *BASALT, The Poldervaart Treatise on Rocks of Basaltic Composition*, vol. 2. Interscience Publishers, John Wiley & Sons, New York.

Claoue-Long, J.C., Sobolev, N.V., Shatsky, V.S. and Sobolev, A.V. (1991) Zircon response to diamond-pressure metamorphism in the Kokchetav massif, USSR. *Geology*, **19**, 710−713.

Coleman, R.G. and Lee, D.E. (1963) Glaucophane-bearing metamorphic rock types of the Cazadero area, California. *Journal of Petrology*, **4**, 260−301.

Coleman, R.G. and Wang, X. (1995) Overview of the geology and tectonics of UHPM. Pp. 1−32 in: *Ultrahigh Pressure Metamorphism* (R.G. Coleman and X. Wang, editors). Cambridge University Press, Cambridge, UK.

Coleman, R.G., Lee, D.E., Beatty, L.B. and Brannock, W.W. (1965) Eclogites and eclogites: Their differences and similarities. *Geological Society of America Bulletin*, **76**, 483−508.

Compagnoni, R. and Maffeo, B. (1973) Jadeite-bearing metagranites *l.s.* and related rocks in the Mount Mucrone area (Sesia-Lanzo zone, western Italian Alps). *Schweizerische Mineralogische und Petrographische Mitteilungen*, **53**, 355−378.

Compagnoni, R., Hirajima, T., and Chopin, C. (1995) Ultra-high pressure metamorphic rocks in the Western Alps. Pp. 206−243 in: *Ultrahigh Pressure Metamorphism* (R.G. Coleman and X. Wang, editors). Cambridge University Press, Cambridge, UK.

Dobrzhinetskaya, L.F., Eide, E.A., Larsen, R.B., Sturt, B.A., Tronnes, R.G., Smith, D.C., Taylor, W.R. and Posukhova, T.V. (1995) Microdiamond in high-grade metamorphic rocks of the western gneiss region, Norway. *Geology*, **23**, 597−600.

Eskola, P. (1921) On the eclogites of Norway. *Videnskapsselskaps Skrifter I, Matematisk-Naturvidenskapelig Klasse*, **8**, 1−118.

Eskola, P. (1939) Die metamorphen Gesteine. Pp. 263−407 in: *Die Entstehung der Gesteine* (T.F.W. Barth, C.W. Correns, and P. Eskola, editors). Julius Springer, Berlin (reprinted in 1960 and 1970), 422 pp.

Essene, E.J., Fyfe, W.S. and Turner, F.J. (1965) Petrogenesis of Franciscan glaucophane schists and associated metamorphic rocks, California. *Contributions to Mineralogy and Petrology*, **11**, 695−704.

Evans, B.W. (1965) Microprobe study of zoning in eclogite garnets. *Geological Society of America Special Paper*, **87**, 54 (abstract).

Evans, B.W. (1990) Phase relations of epidote blueschists. *Lithos*, **25**, 3–23.
Evans, B.W. and Trommsdorff, V. (1978) Petrogenesis of garnet lherzolite, Cima di Gagnone, Lepontine Alps. *Earth and Planetary Science Letters*, **34**, 333–348.
Evans, B.W., Trommsdorff, V. and Richter, W. (1979) Petrology of an eclogite-metarodingite suite at Cima di Gagnone, Ticino, Switzerland. *American Mineralogist*, **64**, 15–31.
Fry, N. and Fyfe, W.S. (1969) Eclogites and water pressure. *Contributions to Mineralogy and Petrology*, **24**, 1–6.
Fyfe, W.S., Turner, F.J. and Verhoogen, J. (1958) Metamorphic reactions and metamorphic facies. *Geological Society of America Memoir*, **73**, 259 pp.
Gebauer, D., Schertl, H.-P., Brix, M. and Schreyer, W. (1997) 35 Ma old ultrahigh-pressure metamorphism and evidence for very rapid exhumation in the Dora Maira Massif, Western Alps. *Lithos*, **41**, 5–24.
Ghent, E.D. (1988) A review of zoning in eclogite garnets. Pp. 207–236 in: *Eclogites and Eclogite-Facies Rocks* (D.C. Smith, editor). Developments in Petrology, 12, Elsevier, Amsterdam.
Gjelsvik, T. (1952) Metamorphosed dolerites in the gneiss area of Sunnemore, Norway. *Norsk Geologisk Tidsskrift*, **30**, 33–134.
Glodny, J., Bingen, B., Austrheim, H., Molina, J.F. and Rusin, A. (2002) Precise eclogitization ages deduced from Rb/Sr mineral systematics: The Maksyutov complex, southern Urals, Russia. *Geochimica et Cosmochimica Acta*, **66**, 1221–1235.
Godard, G. (2001) Eclogites and their geodynamic interpretation: a history. *Journal of Geodynamics*, **32**, 165–203.
Goldschmidt, V.M. (1912) Die Gesetze der Gesteinsmetamorphose. *Norsk Videnskapsselskaps Skrifter I, Matematisk-Naturvidenskapelig Klasse*, **22**, 16 pp.
Green, D.H. and Ringwood, A.E. (1967) An experimental investigation of the gabbro-eclogite transformation and its petrological implications. *Geochimica et Cosmochimica Acta*, **31**, 767–833.
Griffin, W.L., Austrheim, H., Brastad, K., Bryhni, I., Krill, A.G., Krogh, E.J., Mørk, M.B., Kvale, H., and Tørudbakken, B. (1985) High-pressure metamorphism in the Scandinavian Caledonides. Pp. 783–801 in: *The Caledonide Orogen* (D.G. Gee and B.A. Sturt, editors). John Wiley, New York.
Heinrich, C.A. (1986a) Eclogite facies regional metamorphism of hydrous mafic rocks in the Central Alpine Adula Nappe. *Journal of Petrology*, **27**, 123–154.
Heinrich, C.A. (1986b) Kyanite eclogite to amphibolite facies evolution of hydrous mafic and pelitic rocks, Adula nappe, central Alps. *Contributions to Mineralogy and Petrology*, **81**, 30–38.
Hermann, J. (2003) Experimental evidence for diamond-facies metamorphism in the Dora-Maira massif. *Lithos*, **70**, 163–182.
Holland, T.J.B. (1979a) Experimental determination of the reaction paragonite = jadeite + kyanite + H$_2$O, and internally consistent thermodynamic data for part of the system Na$_2$O-Al$_2$O$_3$-SiO$_2$-H$_2$O, with applications to eclogite and blueschists. *Contributions to Mineralogy and Petrology*, **68**, 175–198.
Holland, T.J.B (1979b) High water activities in the generation of high pressure kyanite eclogites of the Tauern Window, Austria. *Journal of Geology*, **87**, 1–27.
Holmes, A. (1936) A contribution to the petrology of kimberlite and its inclusions. *Transactions of the Geological Society of South Africa*, **39**.
Jahn, B., Caby, R. and Monié, P. (2001) The oldest UHP eclogites of the world: age of UHP metamorphism, nature of protolith and tectonic implication. *Chemical Geology*, **178**, 143–158.
Katayama, I., Parkinson, C.D., Okamoto, K., Nakajima, Y. and Maruyama, S. (2000) Supersilicic clinopyroxene in ultrahigh-pressure metamorphic rocks from the Kokchetav massif, Kazakhstan. *American Mineralogist*, **85**, 1368–1374.
Kienast, J.R., Lombardo, B., Biino, G. and Pinardon, J.L. (1991) Petrology of very-high-pressure eclogitic rocks from the Brossasco-Isasca Complex, Dora-Maira Massif, Italian Western Alps. *Journal of Metamorphic Geology*, **9**, 19–34.
Klemd, R., Matthes, S. and Schüssler, U. (1994) Reaction textures and fluid behaviour in very high pressure calc-silicate rocks of the Münchberg gneiss complex, Bavaria, Germany. *Journal of Metamorphic Geology*, **12**, 735–745.
Kolderup, N.H. (1960) Origin of Norwegian eclogites in gneiss. *Norsk Geologisk Tidsskrift*, **40**, 73–76.
Lappin, M.A. (1974) Eclogites from the Sunndal-Grubse ultramafic mass, Almklovdalen, Norway, and the T-P history of the Almklovdalen masses. *Journal of Petrology*, **15**, 567–601.
Liou, J.G. Zhang, R.Y., Ernst, W.G. and Rumble, D. III and Maruyama, S. (1998) High-pressure minerals from deeply subducted metamorphic rocks. Pp. 33–96 in: *Ultrahigh Pressure Mineralogy* (R. Hemley and D. Mao, editors). Reviews in Mineralogy, 37, Mineralogical Society of America, Washington, D.C.
Liou, J.G and Zhang, R.Y. (1996) Occurrences of intergranular coesite in ultrahigh-P rocks from the Su-Lu region, eastern China: implications for lack of fluid during exhumation. *American Mineralogist*, **81**, 1217–1221.
Liu, F., Zhiqin, X.U., Liou, J.G., Katayama, I., Masago, H., Maruyama, S. and Yang, J. (2002) Ultrahigh pressure mineral inclusions in zircons from gneissic core samples of the Chinese continental scientific drilling site. *European Journal of Mineralogy*, **14**, 499–512.
Massonne, H.-J. and Schreyer, W. (1989) Stability field of the high-pressure assemblage talc + phengite and two new phengite barometers. *European Journal of Mineralogy*, **1**, 391–410. QE 351 E88
Michard, A., Henry, C. and Chopin, C. (1995) Structures in UHPM rocks: A case study from the Alps. Pp. 132–158 in: *Ultrahigh Pressure Metamorphism* (R.G. Coleman and X. Wang, editors). Cambridge University Press, Cambridge, UK.
Miller, C. (1970) Zur metamorphose der Eklogite und Metagabbros im Oetztaler Altkristallin. *Fortschritte der Mineralogie*, **47**, Beiheft 1, 43–44.
Miyashiro, A. and Banno, S. (1958) Nature of glaucophanitic metamorphism. *American Journal of Science*, **256**, 97–110.
Molnar, P. and Gray, D. (1979) Subduction of continental lithosphere: some constraints and uncertainties. *Geology*, **7**, 58–62.
Mosenfelder, J.L., Schertl, H.-P., Smyth, J.R. and Liou, J.G. (2005) Factors in the preservation of coesite: The importance of fluid infiltration. *American Mineralogist*, **90**, 779–789.
Mposkos, E.D. and Kostopoulos, D.K. (2001) Diamond, former coesite and supersilicic garnet in metasedimentary rocks from the Greek Rhodope: a new ultrahigh-pressure metamorphic province established. *Earth and Planetary Sciences*, **192**, 497–506.
Nasdala, L. and Massonne, H.-J. (2000) Microdiamonds from the Saxonian Erzgebirge, Germany; in situ micro-Raman characterization. *European Journal of Mineralogy*, **12**, 495–498.
Nimis, P. and Trommsdorff, V. (2001) Revised thermobarometry of Alpe Arami and other garnet peridotites of the Central Alps. *Journal of Petrology*, **42**, 103–115.
Nowlan, E.U., Schertl, H.P. and Schreyer, W. (2000) Garnet-omphacite-phengite thermobarometry of eclogites from the coesite-bearing unit of the southern Dora-Maira Massif, Western Alps. *Lithos*, **52**, 197–214.
Ogasawara, Y., Fugasawa, K. and Maruyama, S. (2002) Coesite exsolution from supersilicic titanite in UHP marble from the Kokchetav massif, northern Kazakhstan. *American Mineralogist*, **87**, 454–461.
O'Hara, M.J. and Mercy, E.L.P. (1963) Petrology and petrogenesis of some garnetiferous peridotites. *Transactions of the Royal Society of Edinburgh*, **65**, 251–314.
Page, F.Z., Essene, E.J. and Mukasa, S.B. (2005) Quartz exsolution in clinopyroxene is not proof of ultrahigh pressures: Evidence from eclogites from the Eastern Blue Ridge, Southern Appalachians. *American Mineralogist*, **90**, 1092–1099.
Parkinson, C.D. and Katayama, I. (1999) Present-day ultrahigh-pressure conditions of coesite inclusions in zircon and garnet: evidence from laser Raman spectroscopy. *Geology*, **27**, 979–982.
Perchuk, A.L., Yaspaskurt, V.O. and Podlesskii, S.K. (1998) Genesis and exhumation dynamics of eclogites in the Kokchetav Massif near Mount Sulu-Tyube, Kazakhstan. *Geochemistry International*, **36**, 877–885.
Ramberg, H. (1952) *The Origin of Metamorphic and Metasomatic Rocks*. The University of Chicago Press, Chicago, USA, 317 pp.
Ravna, E.J.K. and Paquin, J. (2003) Thermobarometric methodologies applicable to eclogites and garnet ultrabasites. Pp. 229–259 in: *Ultrahigh Pressure Metamorphism* (D.A. Carswell and R. Compagnoni, editors). EMU Notes in Mineralogy 5, European

Mineralogical Union.

Ringwood, A.E. and Green, D.H. (1966) An experimental investigation of the gabbro to eclogite transformation and some geophysical implications. *Tectonophysics*, **3**, 383−428.

Rubatto, D. and Hermann, J. (2001) Exhumation as fast as subduction. *Geology*, **29**, 3−6.

Schertl, H.-P., Schreyer, W. and Chopin, C. (1991) The pyrope-coesite rocks and their country rocks at Parigi, Dora-Maira Massif, Western Alps: detailed petrography, mineral chemistry and *PT*-path. *Contributions to Mineralogy and Petrology*, **108**, 1−21.

Rubatto, D. and Scambelluri, M. (2003) U-Pb dating of magmatic zircon and metamorphic baddeleyite in the Ligurian eclogites (Voltri Massif, Western Alps). *Contributions to Mineralogy and Petrology*, **146**, 341−355.

Schmitt, H. (1964) Metamorphic eclogites of the Eiksund area, Sunnemore, Norway. *American Geophysical Union Transactions*, **45**, 128.

Schreyer, W. (1977) Whiteschists: Their compositions and pressure-temperature regimes based on experimental, field, and petrographic evidence. *Tectonophysics*, **43**, 127−144.

Schreyer, W. (1985) Metamorphism of crustal rocks at mantle depths: High-pressure minerals and mineral assemblages in pelites. *Fortschritte der Mineralogie*, **63**, 227−261.

Schreyer, W. (1988) Experimental studies on metamorphism of crustal rocks under mantle pressure. *Mineralogical Magazine*, **52**, 1−26.

Sharp, Z.D., Essene, E.J. and Hunziger, J.C. (1993) Stable isotope geochemistry and phase equilibria of coesite-bearing whiteschist, Cora Maira massif, Western Alps. *Contributions to Mineralogy and Petrology*, **114**, 1−12.

Simon, G. and Chopin, C. (2001) Enstatite-sapphirine crack-related assemblages in ultrahigh-pressure pyrope megablasts, Dora-Maira massif, western Alps. *Contributions to Mineralogy and Petrology*, **140**, 637−652.

Simon, G., Chopin, C. and Schenk, V. (1997) Near endmember magnesiochloritoid in prograde zoned pyrope, Dora-Maira Massif, western Alps. *Lithos*, **41**, 37−57.

Smith, D.C. (1984) Coesite in clinopyroxene in the Caledonides and its implications for geodynamics. *Nature*, **310**, 641−644.

Smyth, J.R. (1977) Quartz pseudomorphs after coesite. *American Mineralogist*, **62**, 828−830.

Sobolev, N.V. and Shatsky, V.S. (1990) Diamond inclusions in garnet from metamorphic rocks: a new environment for diamond formation. *Nature*, **343**, 742−746.

Stern, R.J. (2005) Evidence from ophiolites, blueschists, and ultrahigh-pressure metamorphic terranes that the modern episode of subduction tectonics began in Neoproterozoic time. *Geology*, **33**, 557−560.

Tilton, G.R., Schreyer, W. and Schertl, H-P. (1991) Pb-Sr-Nd isotopic behavior of deeply subducted crustal rocks from the Dora Maira Massif, Western Alps, Italy − II: What is the age of the UHP metamorphism? *Contributions to Mineralogy and Petrology*, **108**, 22−37.

Touret, J.R.L. and Nijland, T.G. (2002) Metamorphism today: new science, old problems. Pp. 113−142 in: *The Earth Inside and Out: Some Major Contributions to Geology in the Twentieth Century* (D.R. Oldroyd, editor). Special Publications **192**, The Geological Society, London..

Turner, F.J. (1968) *Metamorphic Petrology*, 1st edition. McGraw-Hill, New York, 403 pp.

Turner, F.J. and Verhoogen, J. (1960) *Igneous and Metamorphic Petrology*. McGraw-Hill Book Company, Inc., New York, 694 pp.

Wei, C.J., Powell, R. and Zhang, L.F. (2003) Eclogites from the south Tianshan, N. China: petrological characteristics and calculated mineral equilibria in the Na_2O-CaO-FeO-MgO-Al_2O_3-SiO_2-H_2O system. *Journal of Metamorphic Geology*, **21**, 167−179.

Will, T.M. Okrusch, M., Schmädicke, E. and Chen, G. (1998) Phase relations in greenschist-blueschist-amphibolite-eclogite facies: Calculated mineral equilibria in the system Na_2O-CaO-FeO-MgO-Al_2O_3-SiO_2-H_2O (NCFMASH), with applications to the PT evolution of metamorphic rocks from Samos, Greece. *Contributions to Mineralogy and Petrology*, **132,** 85−102.

Winkler, H.G.F. (1965) *Petrogenesis of Metamorphic Rocks*. Springer-Verlag, New York Inc., 220 pp.

Wunder, B., Rubie, D.C., Ross, C.R., Medenbach, O., Seifert, F. and Schreyer, W. (1993) Synthesis, stability and properties of $Al_2SiO_4(OH)$; A fully hydrated analog of topaz. *American Mineralogist*, **78**, 285−297.

Xu, S., Okay, A.I., Ji, S., Sengor, A.M.C., Su, W., Liu, Y. and Jiang, L. (1992) Diamond from the Dabie Shan metamorphic rocks and its implications for tectonic setting. *Science*, **256**, 80−82.

Yang, J., Xu, Z., Pei, X., Shi, R. and Wu, C. (2002) Discovery of diamond in North Qinling: evidence for a giant UHPM belt across central China and recognition of Paleozoic and Mesozoic dual deep subduction between the North China and Yangtze Plates. *Acta Geologica Sinica*, **76**, 484−495 (with English abstract).

Yoder, H.S. and Tilley, C.E. (1962) Origin of basalt magmas: An experimental study of natural and synthetic rock systems. *Journal of Petrology*, **3**, 342−532.

Zhang, R.Y. and Liou, J.G. (1998) Ultrahigh-pressure metamorphism of the Sulu Terrane, eastern China: A perspective view. *Continental Dynamics*, **3**, 32−53.

Zhang, R.Y., Liou, J.G. and Shu, J.F. (2002) Hydroxyl-rich topaz in high-pressure and ultrahigh-pressure kyanite quartzites, with retrograde woodhouseite, from the Sulu terrane, eastern China. *American Mineralogist*, **87**, 445−453.

Coesite and pure pyrope in high-grade blueschists of the Western Alps: a first record and some consequences*

Christian Chopin
ER 224, Laboratoire de Géologie, Ecole Normale Supérieure, 46 rue d'Ulm, 75005 Paris, France

Abstract. A pyrope-quartzite originally described by Vialon (1966) from the Dora Maira massif was resampled and reinvestigated. Garnet (up to 25 cm in size), phengite, kyanite, talc and rutile are in textural equilibrium in an undeformed matrix of polygonal quartz. The garnet is a pyrope-almandine solid solution with 90 to 98 mol % Mg end-member. It contains inclusions of coesite which has partially inverted to quartz, resulting in a typical radial cracking of the host garnet around the inclusions. Several lines of evidence show that coesite crystallised under nearly static pressure conditions and that the whole matrix has once been coesite.

The formidable pressures of formation implied (≥ 28 kbar) are independently indicated by i) the coexistence of nearly pure pyrope with free silica and talc, ii) the coexistence of jadeite with kyanite, iii) the high Si content of phengite. Water activity must have been low. The stability of talc-phengite and the presence of rare glaucophane inclusions in pyrope point to low formation temperatures (about 700 °C) and to a probable Alpine age for the assemblage.

This is evidence that low temperature gradients, how essentially transient they are, may nevertheless persist to considerable depths. Moreover, the upper crustal (evaporite-related?) origin of the quartzite and its interbedding within a continental unit implies that continental crust may also be subducted to depths of 90 km or more. The return back to the surface is problematic; the retrograde assemblages observed show that it must be tectonic. If the rocks remain at depth, new perspectives open for the genesis of intermediate to acidic magmas. Eventually, the role of continental crust in geodynamics may have to be reconsidered.

Introduction

Coesite, one of the high-pressure polymorphs of silica, was first synthesized and described by Coes (1953). The first natural occurrence was reported by Chao et al. (1960) from Meteor Crater, Arizona, and many similar occurrences have then been reported from impact structures. The high-pressure stability field of coesite, as investigated by many workers (from Mac Donald 1956, to Massonne and Mirwald 1981, with references), precluded the occurrence of this mineral in crustal rocks except at impact sites, but suggested that coesite is a stable phase in the upper mantle provided free silica exists there. Indeed, Sobolev et al. (1976) and Smyth and Hatton (1977) reported the first occurrences of coesite from a natural, static pressure environment, from xenoliths in diamond-bearing kimberlites, a now almost common type of occurrence (e.g. Lazko et al. 1983). Likewise very pyrope-rich garnet had exclusively been found in mantle-derived rocks, in accordance with its high-pressure stability field (Boyd and England 1959; Schreyer 1968).

At variance with the former results, an occurrence of coesite inclusions in nearly pure pyrope garnets is reported in this paper from a metasedimentary rock in high-grade blueschist terrane of the Western Alps. It implies that both coesite and pyrope may also be metamorphic products by burial of upper crustal material (cf. Chopin 1983).

Geological setting and the pioneering work of P. Vialon

The pyrope-bearing samples have been found in the Dora Maira massif which is, with Monte Rosa and Gran Paradiso, one of the three internal crystalline massifs of the Western Alps (Fig. 1). The massif consists of a Paleozoic basement and a Mesozoic cover series both of which underwent Alpine high-pressure, low-temperature metamorphism. The basement itself is composed of several volcanosedimentary series. One of them is polymetamorphic, intruded by granite bodies and unconformably overlain by the other series which are presumably of Late Paleozoic age (Fig. 1). This outline is made after Vialon (1966) who achieved the most comprehensive geologic and petrographic study of the massif to date.

Vialon noted, within the polymetamorphic series, a remarkable quartzite layer – probably mentioned as micascisti splendenti by Stella (1895) and Franchi (1900) – containing large phengite flakes and greenish or pale pink nodules up to 25 cm in size. He gave a detailed mineralogical description and established that the nodules were huge, monocrystalline, nearly pure pyrope garnets at various stages of chloritisation. He gave a wet chemical analysis of the garnet which, despite the probable presence of kyanite impurities, show that the garnet was indeed the closest to the pyrope end-member ever found (V16, Table 1). Vialon was also able to map the quartzite layer, which is a few metres thick, over about 15 kilometres and established that it was spatially associated with a sill (?) of metagranite on one side of which it systematically occurs. The surrounding rocks are a series of fine-grained felsic (para-?) gneisses with some marble and metabasite lenses, the latter occasionally containing relic assemblages of glaucophane eclogite (Franchi 1900; Vialon 1966).

The quartzite layer was resampled at Parigi, near Martiniana Po, Italy, in what is given by Vialon as the most spectacular out-

---

* This paper is dedicated to Pierre Vialon who first discovered these fascinating rocks

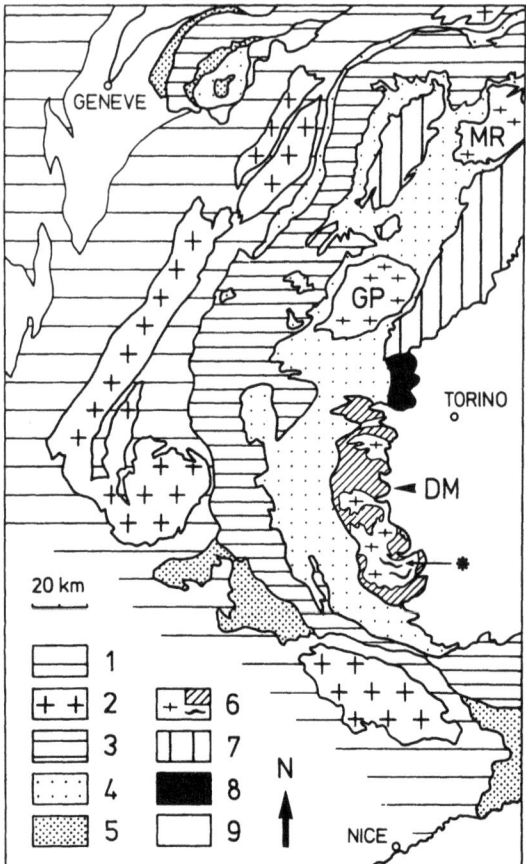

Fig. 1. Structural sketch map of the Western Alps. External zone: (1) folded post-Paleozoic cover; (2) External crystalline massifs (basement, Paleozoic and older). Penninic domain: (3) Briançonnais-St Bernard zone, basement and post-Paleozoic cover undifferenciated; (5) Upper Cretaceous Helminthoid flysch nappe; Piémont zone with (4) Schistes lustrés nappe (Mesozoic, mainly oceanic material) and (5) internal crystalline massifs, Monte Rosa (MR) and Gran Paradiso (GP) undifferenciated, in Dora Maira (DM) the cross-hatched area denotes the Late Paleozoic series, crosses the polymetamorphic, Early Paleozoic (?) basement with the pyrope quartzite layer (in black, the star showing the studied locality). Austroalpine units: (7) Sesia zone, Dent Blanche nappe and Monte Emilius klippe. (8) Lanzo ultramafic body. (9) Tertiary molassic basins

crop. Indeed the quartzite looks locally like an unsorted conglomerate with very pale pink rounded pebbles that could be mistaken for rose quartz were it not the higher density and occasional euhedral shapes. The size, the distribution and the alteration state of these "nodules" are quite variable, while the groundmass appears as a rather homogeneous quartzite containing up to one centimetre large white mica flakes which hardly show preferred orientation. The size of the garnets ranges from 0.2 to 25 cm. The garnets are evenly distributed in the groundmass regardless of their size but they may form impressive clusters that give the appearance of a conglomerate. The garnets are fresh, especially the larger ones, but they may also be completely transformed into a greenish phyllitic aggregate, these extreme states occurring nearly side by side. Rare, centimetre large kyanite crystals may also be found in the outcrop.

At the outcrop scale the minimum apparent thickness of the quartzite "layer" is about 30 metres. In fact, the quartzite occupies the core of a recumbent fold. This fold is a few hundred metres in magnitude, it has an E–W axis and an axial plane slightly dipping to the south. The very faint foliation of the quartzite is concordant with the axial plane foliation. These structures correspond to the first, main tectonic phase recognised by Vialon (1966) which created several-kilometre-large recumbent folds of northward vergence in this part of the massif. The only lithological variation consists of rare, decimetre-thick layers (or veins?) of a weathered bluish rock devoid of mica. They occur twice (due to folding?) and are also concordant with the axial plane foliation. The contact between quartzite and surrounding gneisses can be well observed along the fold crest. It is sharp, there is no transitional lithology and the quartzite is nearly pure just below the contact. The gneisses are mostly phengite-, biotite-, albite- and garnet-bearing, and form a thick monotonous series. Some variational lithologies are reported to contain pseudomorphs after or even relics of glaucophane (Franchi 1900; Vialon 1966).

Another outcrop was visited at Brossasco, Val Maira. Here, the granite sill mentioned by Vialon occurs in the immediate vicinity but the only contacts observed are between quartzite and fine-grained gneiss, just as in Parigi. The structural situation is somewhat different, probably along the limb of a great fold, and the garnets have been completely transformed into a phyllitic aggregate.

Petrography

The present report is concerned with a single representative sample (2–39) in which garnet and the whole primary assemblage escaped any subsequent retrogression. The garnets are 2 to 10 mm large and nearly colourless, thus leaving the rock a white colour. The quartz groundmass shows a perfect, undeformed polygonal texture. White mica occurs as large, commonly up to 5 mm flakes, and talc as smaller ones (up to 1 mm) mostly on the rim of, or interlayered with the white mica. Both may be slightly bent and define a very faint foliation. Abundant kyanite is medium-sized (up to 2 mm). Rutile occurs as rounded grains, zircon is accessory. A sodic pyroxene was found twice in one out of four thin sections of the same sample. It is partly replaced by a rim of brownish, extremely fine-grained material but has formerly been in contact with kyanite. Chlorite is conspicuously absent. If one excepts pyroxene, all the matrix minerals mentioned coexist in contact with each other, and there is little doubt that the assemblage quartz-garnet-white mica-talc-kyanite-rutile was stable as quartz acquired its well recrystallised texture (Fig. 2).

The garnets contain mainly inclusions of kyanite, quartz and rutile, more rarely of talc sometimes associated with kyanite. The inclusions do not show any preferred orientation nor do they define a growth texture in the garnet. As in the matrix, the kyanite and rutile crystals are short prisms with smooth edges. In contrast the quartz inclusions have irregular shapes and a texture sharply differing from that of the matrix (see Fig. 3). Quartz is finely polycrystalline; on the inclusion boundary the grains or subgrains are strained elongated bands forming a radial pattern. The inner part of the inclusions looks like crosscut by "veinlets" also made of quartz, resulting in a very characteristic feathery texture (Fig. 3a–d). In each thin section several of these inclusions still contain relics of a colourless mineral with distinctly higher relief (Fig. 3). This mineral is obviously replaced by quartz as shown by the preservation of various replacement stages in different inclusions. The radial pattern on the inclusion rim corresponds to a peripheral transformation of the more highly refringent mineral; the "veinlets" in the middle part of the inclusion suggest a selective replacement along cleavages and cracks (Fig. 3). Another remarkable feature is the existence in the garnet of radial cracks around the quartz inclusions, which do not exist

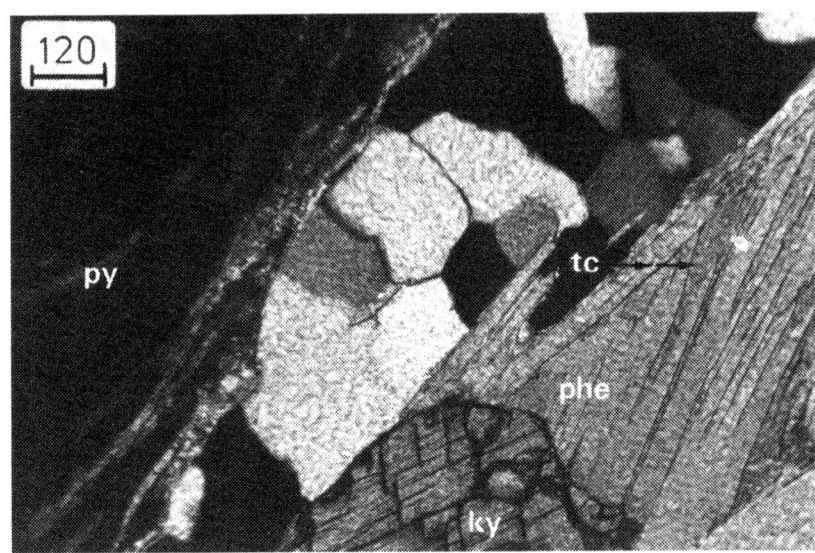

Fig. 2. The matrix assemblage quartz-pyrope-kyanite-phengite-talc. Note the polygonal texture of quartz. Crossed nicols, scale bar in μm

around the rutile and kyanite inclusions (Fig. 3). This unusual texture comparable with that sometimes observed around mineral inclusions undergoing metamictisation, suggests that the inclusions underwent a considerable volume change and overcame the mechanical strength of the garnet. This radial jointing, the high relief of the relic mineral and the nature of its transformation product (quartz!) suggested that it could be a high-pressure polymorph of silica, and this was confirmed by the electron microprobe giving pure SiO_2. The polymorph was further characterised as coesite by its Raman spectrum obtained in thin section using a MOLE microprobe, and by an electron diffraction study carried out by Ph. Gillet and M. Madon and on which will be reported elsewhere.

Beside the inversion of coesite to quartz and the partial breakdown of jadeitic pyroxene, the only other retrograde reaction has been observed between garnet and white mica. All the former contacts between garnet and mica have reacted to form a narrow rim of phlogopite, growing radially to garnet, and of kyanite (sometimes associated with minute talc flakes) on the white mica side (Fig. 4). The contacts between garnet and quartz matrix are sharp but, in the vicinity of mica flakes, both minerals may be separated by a very narrow rim of phlogopite on the garnet side and kyanite ($\pm$ talc) on the quartz side, probably reflecting potassium migration along grain boundaries.

For the sake of completeness it should be mentioned that, in other pyrope-quartzite samples, the relative sizes and amounts of talc and white mica are quite variable, as well as the breakdown products of garnet. In some cases the latter are only chlorite, in other cases zoisite, margarite and an omphacitic pyroxene are also present, suggesting that the Ca content of garnet may have been quite variable from one sample to the other.

The bluish weathered rock interlayered within the pyrope quartzite deserves a further article for itself, but some of its features will be used here in the petrologic discussion. It is an almandine-kyanite-jadeite quartzite, and iron-rich chloritoid is produced by the retrogression of garnet.

Mineral chemistry

The microprobe analyses of *coesite*, of *quartz* in the matrix and of quartz in inclusions are undistinguishable, the main impurity being Al_2O_3 which amounts to near 0.1 wt %. Na lies near the detection limit, the other elements analysed (K, Mg, Ca, Ti, Fe) rather below it. Analyses of the other minerals making up this unusual rock are listed in Table 1.

Garnet is indeed nearly pure pyrope with 90 to 98 mol % of the end member, the rest being essentially almandine. Chromium is typically absent, and sodium is close to the detection limit. The smaller garnets are the more Mg-rich, their Fe content decreases toward the rim and their Ca content continuously increases but remains very low (Table 1, two first garnet analyses). The range of compositional variations is greater in the larger garnets and the zonation pattern is a little more complex, with a slight iron enrichment in the rim (Table 1, next three analyses).

The composition of *talc* is close to the ideal one. The iron content represents at most one mole per cent of the iron end member. The noticeable alumina content of about 0.45 wt.% is not linked to a sodium incorporation as in other cases (Råheim and Green 1974; Abraham and Schreyer 1976; Schreyer et al. 1980). It is probably in octahedral position and would represent about one mole per cent of the pyrophyllite end member (compare Newton 1972).

Kyanite is pure Al_2SiO_5, free of chromium and of iron. This together with the garnet analyses suggests that the iron in the rock might be essentially divalent.

The *white mica* analyses show a perfectly dioctahedral mica with a very important extent of the substitution SiMg $\rightleftharpoons$ AlAl. It is thus a phengite, with 3.55 Si per 12 oxygen formula unit. This value remains constant throughout the thin section and also within a crystal, no matter whether phengite occurs in the matrix or in contact with garnet. The same holds for the K $\rightleftharpoons$ Na substitution, with a nearly constant value between 2.5 and 3 mol % of paragonite in phengite. Although always very small, the iron content is somewhat higher in phengite coexisting with garnet than in phengite in the matrix.

The rare *pyroxene* grains analysed have a jadeite content of over 70 mol %, they are thus close to the jadeite limb of the omphacite-jadeite solvus as drawn, for example, by Carpenter (1980). Noteworthy is the slight but systematic excess of Al over Na in the structural formula (Table 1). It is not clear as yet whether it reflects a peraluminous

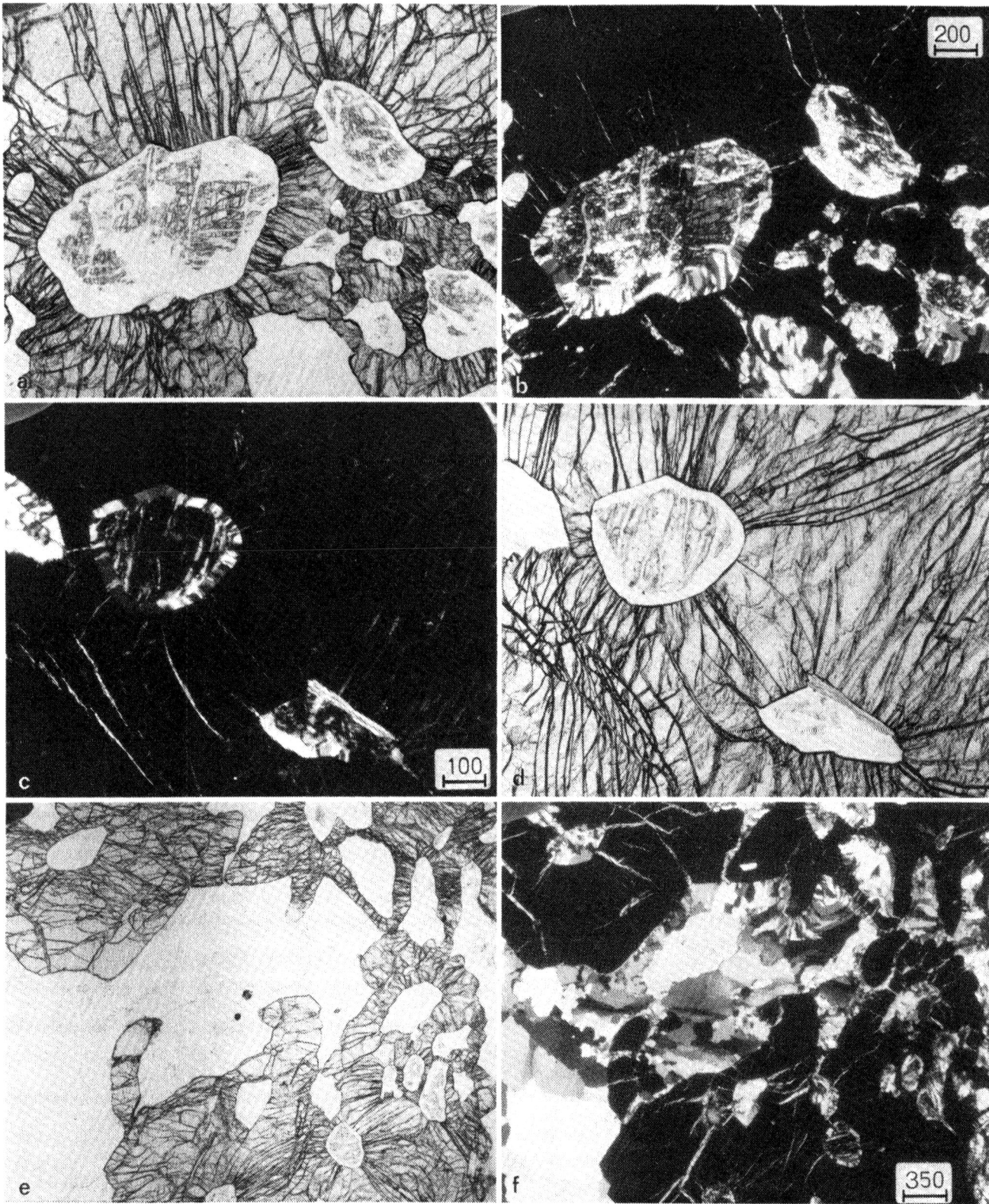

Fig. 3a–f. Microphotographs from sample 2–39. Scale bars in microns. **a–b, c–d** Relics of coesite (higher relief) partly inverted to quartz (lower relief), as inclusions in pyrope. Note the irregular shape of the inclusions, the absence of preferred orientation and of growth texture (also *E–F*), and the radial fractures of garnet around the inclusions. The replacement texture of coesite by quartz in garnet is characteristic with a radial growth of polycrystalline quartz around the inclusion rim (**c–d**) and a development along cleavages and cracks of coesite (**a–b**). Compare Smyth (1977). **e–f** The same texture of quartz pseudomorph after coesite exists in a part of matric embayed in a poikilitic pyrope crystal (centre of the picture) but nowhere else in the matrix (lower left corner)

Fig. 4. Characteristic reaction rim between pyrope and phengite (in the matrix). Aluminous phlogopite grows on the garnet side, kyanite bundles (sometimes with talc) on the phengite side. Plane polarised light, scale bar in μm. This may illustrate the water-conserving reaction (8) discussed in the text but also, because of the occasional presence of talc, the quartz-absent hydration reaction pyrope + phengite + $H_2O \rightarrow$ phlogopite + kyanite + talc

Table 1. Mineral analyses

V 16 is a wet chemical analysis of pyrope concentrate (+kyanite?) given by Vialon (1966). The rest is microprobe analyses from sample 2-39. The first two garnet analyses are from a small garnet (about 2 mm), the three next ones from a large garnet (1 cm) containing coesite inclusions. The phengite analysed coexists with talc and garnet. Phlogopite is retrograde. Camebax electron microprobe (15 kV, 10 nA); standards used are forsterite (Mg), anorthite (Al, Ca), Fe_2O_3, $MnTiO_3$, K-feldspar (K), jadeite (Na). Forsterite was used for Si in most analyses, but quartz was used instead for the two first garnet analyses. The discrepancy in Si content between the two sets of garnet analyses does not allow any statement on a hydropyrope component (Ackermann et al. 1983)

	V 16	Garnet					Jadeite	Phengite	Talc	Phlogopite
		rim	core	rim		core				
SiO_2	41.90	45.21	45.13	44.46	44.85	43.84	58.26	53.81	63.54	40.13
TiO_2	tr.	0.00	0.02	0.01	0.00	0.01	0.07	0.25	0.00	0.02
Al_2O_3	29.15	25.55	25.39	25.35	25.52	25.15	18.87	24.12	0.46	20.61
FeO	1.87	0.91	1.35	1.89	1.58	4.20	0.17	0.19	0.35	0.52
MnO	tr.	0.02	0.03	0.00	0.03	0.09	0.02	0.00	0.00	0.02
MgO	26.05	28.83	28.98	29.26	29.83	27.10	5.04	5.66	30.55	23.10
CaO	0.30	0.28	0.09	0.26	0.11	0.23	5.81	0.06	0.02	0.27
Na_2O	0.45						11.03	0.19	0.02	0.43
K_2O	0.10						0.00	11.33	0.03	10.22
	100.35	100.80	100.99	101.23	101.92	100.62	99.27	95.61	94.97	95.32
Si		3.020	3.015	2.977	2.978	2.982	2.002	3.550	4.022	2.779
Ti			0.001	0.001		0.001	0.002	0.012		0.001
Al		2.012	2.000	2.001	1.997	2.017	0.765	1.876	0.034	1.683
Fe		0.051	0.075	0.106	0.088	0.239	0.005	0.010	0.019	0.030
Mn		0.001	0.002		0.002	0.005	0.001			0.001
Mg		2.870	2.885	2.920	2.951	2.748	0.258	0.557	2.882	2.384
Ca		0.020	0.006	0.019	0.008	0.017	0.214	0.004	0.001	0.020
Na							0.735	0.024	0.002	0.058
K								0.954	0.002	0.903
Σ cat.		7.974	7.984	8.022	8.024	8.009	3.981	6.988	6.963	7.859
Pyr		97.6	97.2	95.9	96.8	91.3				
Alm		1.7	2.5	3.5	2.9	7.9				
Gro		0.7	0.2	0.6	0.3	0.6				
Spe		–	0.1	–	–	0.2				

character of the pyroxene, as in the kimberlite xenolith described by Smyth and Hatton (1977), or 2 to 5 mol % of spodumene component in solid solution.

The secondary *phlogopite* is more aluminous than the ideal end member. The excess of octahedral Al is compensated by a deficiency of octahedral filling.

Petrologic discussion

The significance of coesite

Given the negligeable solid solutions in coesite and quartz and the independence of the phase transition relative to water activity and fluid composition, at least in the realm of metamorphic conditions, the occurrence of coesite should be readily interpreted in terms of geobarometry. Under static conditions it would imply unusually high metamorphic pressure values, between 25 and 30 kbar, depending on temperature. However, it is crucial to know whether coesite crystallised under static conditions or not. It has been experimentally shown that coesite can grow from highly strained quartz as much as 10 kbar below the transition determined under static conditions (Hobbs 1968; Naka et al. 1972; Green 1974). It might then be questioned whether the beautiful, non deformed, hardly oriented texture of the sample does not result from a complete recrystallisation subsequent to the coesite breakdown, the coesite growth having taken place under high-stress conditions. However, it seems unconceivable that under such conditions 15 cm large garnets could grow developing euhedral faces and including monocrystalline coesite, talc and kyanite grains devoid of preferred orientation. Besides, the deviatoric stress in the experiments mentioned exceeded 10 kbar, i.e. was at least one order of magnitude larger than the usual estimates for geological processes. As stated by Green, "it probably is not possible to produce a dislocation density anywhere near that proposed here by natural means other than meteorite impact". Therefore, I favour the idea that coesite grew in this rock under nearly static pressure conditions.

The presence of coesite inclusions in garnet suggests that the whole groundmass now made of quartz has formerly also been coesite. The idea is supported by the fact that the very peculiar texture of quartz pseudomorph after coesite (Fig. 3 and Smyth 1977) has been observed, although in an advanced recrystallisation stage, in a part of the groundmass which had been nearly isolated within a cluster of small garnets. It has also been observed in quartz inclusions in kyanite and in a part of matrix embayed in garnet. In the latter case, even the radial growth pattern along the garnet rim is preserved (Fig. 3e-f), which I consider to be a definite evidence for a former coesite matrix. In fact, the same result may be obtained from mechanical considerations. On the basis of the much greater thermal expansion of quartz as compared to garnet, one might think that, if *quartz* inclusions are trapped in garnet on a *prograde* path, a considerable pressure will develop in the inclusions (and not in the matrix) upon further temperature increase, which may lead to coesite growth in the inclusions, and only in the inclusions. However, calculations performed using an elastic model (Gillet et al. 1984; compare Rosenfeld and Chase 1961) show that along metamorphic geotherms the pressure in the quartz inclusion reaches the transition value only if the external pressure does, the effect of the large thermal expansion of quartz being cancelled by its large compressibility. Therefore, if coesite is now present in garnet, it must have been present in the matrix.

Conversely, let us now consider a *coesite* inclusion in garnet (in a coesite matrix) on a *retrograde* metamorphic path starting within the coesite stability field. Upon pressure release, the coesite *matrix* will invert to quartz very quickly since it may be in contact with a fluid phase and since temperature is still rather high, two factors which are known to enhance the inversion kinetics (e.g. Boyd and England 1960). On the contrary, the inversion of coesite to quartz, which proceeds with a large volume increase, will be counteracted *in the garnet* by the low compressibility and the mechanical strength of the host mineral. Once initiated, for example by large overstepping of the phase transition, the inversion results in the development of a high pressure in the inclusion (which may explain the strained state of quartz) and eventually in the disruption of the garnet, i.e. radial jointing around the inclusion. This gives way to external fluids and makes possible the expansion of the inclusion; coesite inversion should therefore go to completion very quickly. The fact that relics are still preserved indicates that the fracturing of garnet around the inclusions must have occurred at very low temperatures and is thus quite a late process.

So, it is believed that this rock has endured unusually high lithostatic pressures which were then released at relatively low temperature. The preservation of coesite in garnet and the contrast in quartz texture between matrix and inclusions in garnet result from the fact that garnet acted during decompression as a pressure vessel on the inclusions and isolated them from any penetration of external fluids.

Phase relations of pyrope

The most Mg-rich garnets known so far were from xenoliths in kimberlite. They rarely exceeded 80 mol % pyrope and, if more Mg-rich, contained considerable amounts of chromium as knorringite component (e.g. Meyer and Boyd 1972). The chromium-free garnet analysed here must be of quite different origin and represents by far the closest approximation of the pyrope end-member.

Its coexistence with free silica is an important feature. It gives support to the experimental work of Hensen and Essene (1971) after which the pyrope-quartz pair has a stability field in the $MgO-Al_2O_3-SiO_2-H_2O$ system, a result which has been confirmed, refined and extended to the pyrope-coesite pair by Massonne (1983). As seen in the inset of Fig. 5, several alternative assemblages are possible, among which talc-Mg-chloritoid, talc-kyanite or enstatite-kyanite. The lower stability limit of pyrope + free silica can be approximately located using Massonne's data (1983 and pers. comm.) on the curve of the reaction

$$\text{pyrope} + \text{quartz} = \text{enstatite} + \text{kyanite}. \qquad (1)$$

Its intersection with the upper stability limit of talc defines an invariant point (Fig. 6) from which emanates the curve of the reaction

$$\text{talc} + \text{kyanite} = \text{pyrope} + \text{quartz} + H_2O \qquad (2)$$

which must have a steep negative slope. This invariant point lies close to the quartz-coesite transition and, considering the experimental uncertainties and the still poorly known influence of Al-incorporation on talc stability, it is in fact not clear whether it lies in the quartz or in the coesite field, i.e. whether it is stable or metastable. Whereas the existence of a talc-pyrope-coesite stability field is certain, that of a talc-pyrope-quartz stability field is thus still questionable and requires additional experimental work. At any rate, if such a stability field exists, it represents a very restricted P−T domain near 28 kbar and 800–830° C.

Considered in the system $MgO-FeO-Al_2O_3-SiO_2-H_2O$, reaction (2) or its equivalent with coesite instead of quartz is divariant and is relevant to our assemblage. Indeed, natural pyrope being more Fe-rich than talc, the Mg-enrichment of garnet proceeds through the reaction

$$Mg-Fe-\text{talc} + \text{kyanite} = Mg-Fe-\text{garnet} + SiO_2 + H_2O. \qquad (2')$$

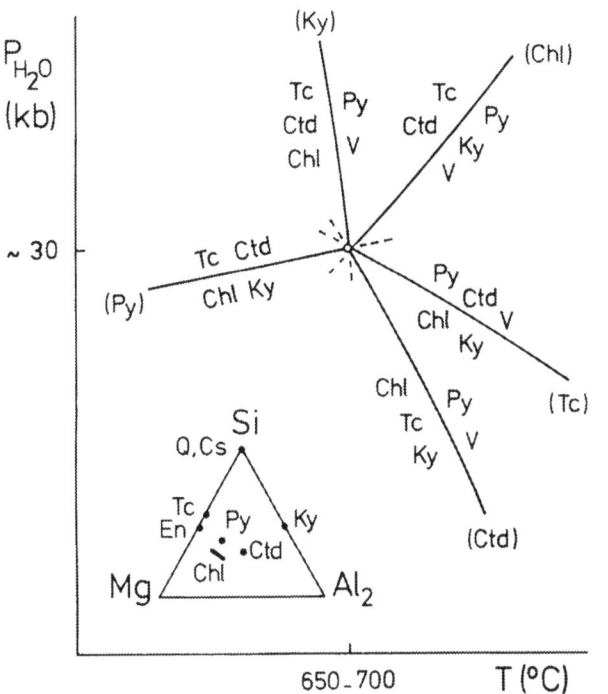

Fig. 5. The lower temperature stability limit of pyrope. The quartz-absent invariant point can be approximately located using the experimental data of Schreyer (1968) on the chloritoid-absent reaction and of Chopin and Schreyer (1983) on the pyrope-absent one. As inset the $MgO - Al_2O_3 - SiO_2 - H_2O$ system is projected on the water-free plane. Abbreviations used throughout this paper: *Ab* albite; *Alm* almandine; *Chl* chlorite; *Ctd* chloritoid; *Cs* coesite; *En* enstatite; *Jd* jadeite; *Ky* kyanite; *Par* paragonite; *Phe* phengite; *Phl* phlogopite; *Py* pyrope; *Q* quartz; *V* hydrous fluid

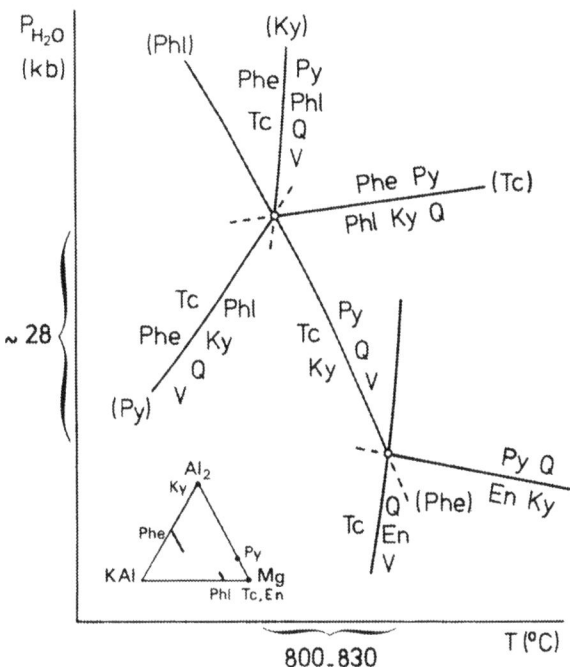

Fig. 6. High-pressure phase relations in the system $K_2O - MgO - Al_2O_3 - SiO_2 - H_2O$: a tentative outline. The invariant points have been located using the experimental data of Kitahara et al. (1966) for the talc breakdown, of Massonne (1983) for the reaction $En + Ky = Py + Q$, and of Massonne and Schreyer (pers. comm. 1983) for the reaction $Tc + Phe = Phl + Ky + Q + V$. See text for discussion. The main feature is the existence of a stability field for the assemblage free silica-pyrope-talc ($\pm$ phengite) at very high pressure and on a rather narrow temperature interval (compare Fig. 8). The inset projection is made through SiO_2 and H_2O. Abbreviations: see Fig. 5

The trace of this reaction in the P−T plane depends both on the actual composition of the solid phases and on water activity. Iron incorporation in the system and lowered water activity will both shift the curve toward lower temperatures as compared to the Mg-end-member reaction curve drawn in Fig. 6. Since in our assemblage garnet and talc are close to the end-member composition, the position of the relevant curve is essentially dependent on water activity. However, calculations performed using the P−V−T data of H_2O from Halbach and Chatterjee (1982) show that even at very low water activity the curve cannot lie much below 700° C.

The phase relations observed in this sample may be compared with those observed in other high-pressure assemblages from Mg-rich metapelites of the Gran Paradiso and Monte Rosa massifs. There, the characteristic feature is the stable coexistence of talc with chloritoid, the chloritoid containing up to 74 mol % of the Mg end-member (Chopin 1979, 1981; Chopin and Schreyer 1983; Chopin and Monié 1984). Depending on metamorphic grade, chlorite may be stable in the presence of quartz in the critical assemblage talc-chloritoid-chlorite-quartz, or not, leading to the talc-chloritoid-kyanite assemblage (Fig. 7, GP and MR, respectively). In either case the coexistence of garnet with kyanite is forbidden in Mn- and Ca-free systems by the stability of Fe−Mg-chloritoid + quartz and by the talc-chloritoid tie-line. The phase relations observed in the Dora Maira sample (Fig. 7, DM) imply the instability of the talc-chloritoid join and of very Mg-rich chloritoid in the presence of quartz. In their experimental study of the pure magnesium system Chopin and Schreyer (1983) had already shown that Mg-chloritoid is not stable in the presence of quartz due to the greater stability of talc + kyanite or of Mg-carpholite, but they had not considered the pyrope-kyanite pair, which is also alternative to talc + chloritoid. Figure 5 shows the phase relations around the quartz-absent invariant point which is determined by the intersection of the lower stability limit of pyrope and of the lower pressure stability limit of talc + chloritoid. Respectively, the relevant reactions are

$$\text{chlorite} + \text{kyanite} + \text{talc} = \text{pyrope} + H_2O \qquad (3)$$

and

$$\text{chlorite} + \text{kyanite} = \text{chloritoid} + \text{talc} \qquad (4)$$

the latter being water-conserving (Chopin and Schreyer 1983). The pyrope-kyanite assemblage appears as a lower-pressure but essentially higher-temperature equivalent of the talc-chloritoid pair. Again, iron incorporation in the solids or low water activity will extend the field of pyrope + kyanite relative to the others.

Two other talc-kyanite-garnet-quartz occurrences have been reported, one from Kazakhstan (Udovkina et al. 1978, 1980) where, noteworthy, garnet contains inclusions of Mg-rich chloritoid (35 mol % Mg end-member), the other from Tasmania (Råheim and Green 1974). They may be referred to as "intermediate pressure whiteschists" (Chopin and

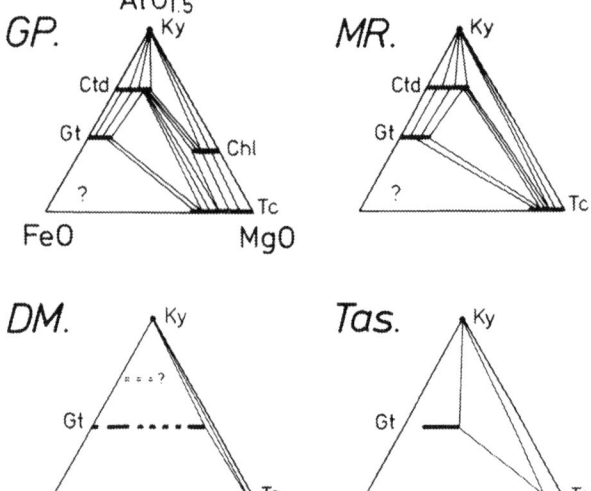

Fig. 7. Phase relations in some high-pressure, Mg-rich metapelites. *GP* Gran Paradiso massif, after Chopin (1979, 1981); *MR* Monte Rosa massif, overlying ophiolites of the Zermatt unit, or Hohe Tauern (Eastern Alps), after Chinner and Dixon (1963) and Miller (1977), respectively; *DM* Dora Maira massif, this paper; *Tas* Tasmania, after Råheim and Green (1974). The talc-chloritoid stability is a characteristic feature in GP and MR whereas the garnet-kyanite-talc coexistence suggests the instability of any chloritoid (+ quartz) in DM and Tas. The projection is made through SiO_2, H_2O and an eventual phengite component. Abbreviations as in Fig. 5

Schreyer 1983), the order of magnitude of the pressure estimates being 10 kbar. The Mg/(Mg+Fe) ratio of garnet, buffered according to reaction (2'), reaches about 0.4 in the former case and 0.5 in the latter. Given the steep negative slope of reaction (2), the much higher pyrope content of the Dora Maira garnet implies considerably higher pressures of formation, or higher temperatures, or lower water activity.

Phase relations with phengite

Considering now the K-bearing system, the coexistence of phengite with talc and pyrope and the absence of phlogopite in the primary assemblage are remarkable features. The talc-phengite pair has been shown to be a high-pressure assemblage (Abraham and Schreyer 1976; Schreyer and Baller 1977) and to be widespread in the Gran Paradiso area where it coexists with Mg-rich chloritoid (Chopin 1981). Since then it has also been found in the Monte Rosa where it coexists with magnesiochloritoid and kyanite (Chopin and Monié 1984). As shown by Fig. 6, the talc-phengite-pyrope-quartz assemblage is not only a very high-pressure assemblage but also a relatively low-temperature one, all the more because low water activity would favour the appearance of phlogopite according to the reaction

talc + phengite = phlogopite + pyrope + quartz + H_2O. (5)

The extent of phengitic substitution is also quite meaningful. In the assemblage phengite-talc-kyanite-(pyrope)-quartz, it is buffered to relatively low values as compared to kyanite-free, phlogopite- or K-feldspar-bearing assemblages (see inset of Fig. 6). However the value reached by the substitution in this rock is among the highest recorded ($Si_{3.55}$), implying that the maximum possible value reached under the same conditions (in the assemblage phengite-phlogopite-K-feldspar-quartz) was exceptionnally high. According to Velde (1967) and Massonne (1981), this points again to unusually high pressure conditions and/or low temperatures. Noteworthy, water activity has no influence on the extent of phengitic substitution in the assemblage phengite-pyrope-kyanite-SiO_2 since the relevant buffering reaction

pyrope + muscovite + SiO_2 = phengite + kyanite (6)

is water-conserving. In fact talc also participates to the natural assemblage and the high Si content of phengite may be also partly due to low water activity, through the reaction

talc + muscovite = phengite + kyanite + SiO_2 + H_2O. (7)

Furthermore, if one assumes the less hydrated products to be on the high-temperature side, volume considerations show that the isopleths relevant to the latter reaction have a negative slope, i.e. phengite substitution in the assemblage talc-kyanite-quartz increases with pressure but, in contrast with phengite in the classical buffering assemblage K-feldspar-phlogopite-quartz, also increases with temperature. Thus the high Si content of phengite in this sample may result from three non exclusive factors: high pressure, relatively high temperature, or low water activity. For comparison, in the assemblage phengite-talc-kyanite-Mg-chloritoid-chlorite-quartz from the Monte Rosa (Chopin and Monié 1984; Chopin and Schreyer 1983) which is equivalent as far as the buffering of phengite substitution is concerned, phengites are between $Si_{3.35}$ and $Si_{3.42}$ and pressure has been independently estimated at 16 to 18 kbar for temperature and water activity varying between 500 and 560° C, 0.6 and 1, respectively.

The retrograde formation of phlogopite + kyanite from pyrope + phengite (Fig. 4) deserves particular attention. It corresponds to a univariant reaction in the pure magnesium system (Fig. 6), namely

phengite + pyrope = phlogopite + kyanite + quartz. (8)

This reaction is water-conserving regardless of the extent of phengite substitution and it proceeds with a large positive volume change of 4 to 5%. It is therefore strongly dependent on pressure and insensitive to variations of water activity (see Fig. 6). Thus, the two main retrograde reactions observed, the inversion of coesite to quartz and the breakdown of pyrope + phengite, although they may not have proceeded at the same time, are both indicative of a pressure drop and are independent of the presence and composition of a fluid phase.

In this respect two other reaction curves have to be considered, that of reaction (2) and that of the reaction

talc + phengite = phlogopite + kyanite + quartz + H_2O (9)

which has been experimentally determined by Massonne and Schreyer (pers. commun.). During isothermal decompression both curves are crosscut, the former in the hydration direction, the latter in the dehydration direction (Fig. 6 and 8). In the sample studied, the lack of evidence for reaction (2) proceeding in the retrograde direction implies a considerable overstepping of the reaction which may result both from a low water activity in the fluid (if any) and from the unfavourable kinetics of the hydration reaction. Conversely, a large overstepping of reaction (9) during decompression is very unlikely because dehydration is favoured both kinetically and by low water activity. Therefore, the stability of talc-phengite throughout the evolution of the rock implies that the curve of reaction (9) has not been crosscut and thus puts severe constraints on any decompressional path. The constraints are all the more severe as water activity is lower; for example calculations show that the curve of reaction (9) is shifted toward lower temperatures by about 100° for a water activity of 0.2.

Significance of sodic pyroxene

Although an accessory mineral, the jadeite-rich pyroxene is an important petrologic indicator because of its coexistence with kyanite and white mica. The reference reaction for the components is

paragonite = jadeite + kyanite + H_2O (10).

It is very pressure-dependent and proceeds at 24–25 kbar water pressure for temperatures between 550 and 700° C, as experimentally determined by Holland (1979b). Its location in the P–T plane is dependent on water activity and on additional components in pyroxene and white mica, the effects of the latter counteracting each other. A calculation using the actual mineral compositions of the sample is made difficult by the very low paragonite content of mica (<3%) and the strong non-ideality of solid solutions on the jadeite-diopside and paragonite-muscovite joins. Assuming a jadeite activity of 0.7 in pyroxene, a paragonite activity of 0.1 in phengite and a water activity of 0.2, the calculation yields a pressure value of 26 kbar, which is nearly independent of the assumed temperature. If jadeite or water activities are in fact higher, or paragonite activity lower, this pressure estimate is only a minimum value. Moreover, the thin beds (?) of bluish rock occurring within the pyrope-quartzite layer also contain the equilibrium assemblage jadeite-kyanite-phengite (+quartz+almandine) with similar paragonite contents in phengite and 90 mol % end member in jadeite. Whatever temperature or water activity might have been, this assemblage implies formation pressures much in excess of any estimate ever made in the realm of regional metamorphism; it confirms the exceptional character of the sample studied.

Resulting evidence

The petrologic investigation and qualitative comparisons with other high-pressure assemblages yield a series of independent and consistent evidences concerning the petrogenetic conditions of this rock. High rock pressures are indicated by the occurrence of the assemblage free silica-pyrope-talc and of coesite, which cannot have crystallised under extreme shear-stress conditions, by the jadeite content of pyroxene coexisting with kyanite and white mica, and, accessorily, by the extent of phengite substitution. Even in the absence of coesite, minimum pressure estimates of 25 kbar are hardly avoidable. A relatively low temperature is implied by the stability of talc and the absence of primary phlogopite which yield an absolute upper limit of about 800° C for the pressures considered (compare Fig. 6 and 8). This upper limit depends on water activity, it is shifted down to about 700° C for a water activity reduced

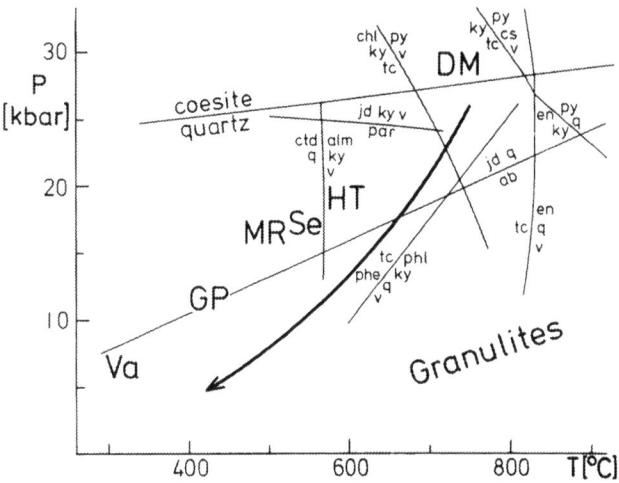

Fig. 8. The pyrope-coesite-talc assemblage as product of high-pressure, low-temperature metamorphism. P–T estimates from the highest-grade high-pressure assemblages from several areas of the Alpine chain: *Va* Vanoise (Goffé 1982; Goffé and Velde 1984); *GP* Gran Paradiso (Chopin and Maluski 1980, for T, P from phengite composition using Massonne's 1981 data); *MR* Monte Rosa (Chopin and Monié 1984); *Se* Sesia zone (Compagnoni 1977; Desmons and O'Neil 1978); *HT* Hohe Tauern (Holland 1979a; compare Miller 1977); *DM* Dora Maira, this study. For comparison with an other extreme of metamorphism, a granulite field is indicated. The originality of Alpine-type high-pressure metamorphism is obvious. The arrow drawn for the decompression path only represents an upper temperature limit for it (see text). The upper pressure stability limits of paragonite and albite are from Holland (1979b, 1980), the breakdown of chloritoid + quartz is derived from Ganguly (1969). See text for the other curves. All the curves drawn for $P_{H_2O} = P$. Abbreviations as in Fig. 5

to 0.2. The divariant assemblage pyrope-coesite-talc-kyanite also implies temperatures between about 700 and 800° C, depending on water activity and on the exact location of the relevant reaction curve (2). Actually, for temperatures close to 800° C and high water activity, melting would occur in K- or Na-bearing systems. The absence so far of evidence for melting in the sample studied and in the country-rocks supports the assumption of low water activity and thus the lower temperature estimates.

Therefore, the pyrope-coesite-talc-kyanite-phengite assemblage is thought to have formed under conditions of low water activity, at pressures in excess of 28 kbar and temperatures probably close to 700° C. This represents a very low thermal gradient and, together with the existence of rare glaucophane inclusions in huge pyrope crystals, confirms the definitely Alpine character of this rock. Obviously, blueschist facies metamorphism may reach extremely high grades and persist to considerable depths (Fig. 8).

The decompression path is well constrained on the high-temperature side by the stability of the talc-phengite pair, all the more as water activity is lower. Clearly, the uplift must proceed under decreasing temperature conditions and must therefore be of tectonic origin, the upward thrusting upon cooler units preventing a temperature rise. An additional evidence for rather low-temperature, high-pressure conditions prevailing during uplift is given by the retrograde breakdown of garnet in the almandine-kyanite-jadeite quartzite, which produces chloritoid and no staurolite, suggesting pressures above the staurolite-quartz stability field (compare Fig. 8 and Rao and Johannes 1979). Since the uplift path remains poorly constrained on the low-temperature side, the possibility even arises that most of the decompression took place under P–T conditions corresponding to the blueschist facies, and thus that the blueschist assemblages observed in the terranes now juxtaposed in and on the Dora Maira massif were acquired in part on a prograde path, in part on a retrograde path.

Additional petrographic, experimental and field studies are clearly needed in order to define more accurately the evolution of the mineral assemblage described here. An investigation of the compositional variations and of the inclusions in the huge garnets, dating of phengite by the $^{40}Ar-^{39}Ar$ method, and an experimental determination of the equilibrium curves of the reactions (2) and (3) are presently undertaken.

Conclusions

Pressure values in excess of 28 kbar typically represent upper mantle conditions. Therefore, the problem arises of the origin of the pyrope-quartzite and of the enclosing series which do not have a mantle chemistry.

The protolith

The bulk composition of the sample studied is characterised by an extremely high Mg/Fe ratio, by a high silica content and by the virtual absence of Ca and Na, the latter point not holding true throughout the layer (see Petrography). This composition could be approximated by a mixture of quartz, Mg-chlorite and illite, and it reminds one of the highly magnesian, metaevaporitic whiteschists described and reviewed by Schreyer (1977). In the more restricted Alpine frame, very low Na and Ca contents and high Mg/(Mg+Fe) ratios (between 0.6 and 0.85) are characteristic for the layer of talc-chloritoid schist which occurs both in the Gran Paradiso (Chopin 1981) and the Monte Rosa massifs (Chopin and Monié 1984) as the first unit of a sedimentary and volcanic sequence of Paleozoic age transgressive on a granite and gneiss basement (see Bearth 1952). The highly magnesian character of this schist, its occasional contents of magnesite, dolomite and of the Ce–Al-phosphate florencite (Chopin 1979 and unpubl. data) suggest, together with its geological situation, an epicontinental, shallow-depth, possibly evaporitic origin. The correlation of the Dora Maira quartzite with this layer, and thus an evaporitic origin for it, is one of the most interesting working hypothesis as yet. The presence of metabasite and marble lenses in the series enclosing the quartzite is a common feature with the series overlying the magnesian layer in the Gran Paradiso and Monte Rosa massifs. However, neither carbonates nor florencite have been found in the Dora Maira quartzite and their is no obvious cover/basement distinction possible within the surrounding gneiss series. Besides, the Gran Paradiso and Monte Rosa series concerned are not affected by granite intrusions that they overlie discordantly. On the contrary, the Dora Maira series considered is intruded by granite bodies (and is much thicker). Vialon was even struck by the close spatial association of one of the granite sills and the quartzite layer, and he suggested that the latter might be genetically related to the granite intrusion as a sort of metasomatic marginal alteration. In fact, in Parigi as well as in Brossasco, the granite-quartzite association is not as close and as systematic as suggested by Vialon (1966). Furthermore, careful observation of the contacts between quartzite and surrounding gneiss supports the idea that the quartzite existed as such prior to granite intrusion in what would become the gneiss series. In any event these considerations make the crustal derivation of this material obvious.

The sedimentary origin of the series and the fact that the series acted as a basement during the Late Paleozoic sedimentation even imply that it probably remained at rather high levels of the continental crust up to Alpine times (Late Cretaceous?). Therefore, the generation of the high-pressure assemblage studied during the Alpine metamorphism implies a burial of upper crustal continental units of at least 90 km.

Plate tectonic consequences

Considering the contrasting densities of average continental and oceanic material, it was commonly accepted that oceanic material could be subducted to considerable depths under continental or oceanic plates while continental material always remained nearer to the earth's surface. It has been then realised that the overthrusting of oceanic crust on continental crust (obduction) was possible and able to generate at the edges of continental plates low-temperature, high-pressure mineral assemblages implying depths of 15 to 40 km. Besides, a crustal thickening resulting from continental collision, as that detected below Tibet, may lead to a Moho as deep as 60 to 70 km but is unlikely to produce a low-temperature metamorphism there. The fact that continental crust may be buried – necessarily rapidly in order to preserve a low temperature gradient – at such considerable depth as 90 km implies a revision of the current ideas on the relationships between oceanic and continental crust. A subduction process is not necessarily blocked by the collision of continental masses and continental lithosphere may be subducted as well.

In fact, a similar result seems to arise from geophysical work. According to Hamilton (1979), the Australian continental plate exists at depth below Timor and extends about 200 km north of the Timor through. In that case, however, it remains at shallow depth due to the gentle dip of the subducting plate (see also Silver et al. 1983). Besides, unpublished data presented by D.S. Cowan at the 1983 Penrose Conference illustrated the collision and partial subduction of a "microcontinental terrane" called Yakutat block, in the Gulf of Alaska. This terrane would extend about 300 km westward beneath the Alaskan continental margin.

Nevertheless, contrasting densities may result in a contrasted behaviour with regard to uplift: continental crust may return back to the surface more easily, sometimes perhaps more rapidly in order to preserve some exceptional high-pressure assemblages like that described here. In the case where the deeply subducted continental unit is not rapidly thrust upward, thermal reequilibration results in advanced melting. Therefore, the evidence presented here for subducted continental material opens new possibilities for the genesis of acidic to intermediate magmas. In particular, the occurrence of an acidic igneous activity during early Alpine times must be envisaged.

Other occurrences

Even if several geologic processes exist which enable crustal rocks to be buried at depths sufficient for coesite to crystallise, their return back to the surface and the preservation of coesite imply a rapid and considerable uplift, therefore uncomfortable tectonic problems, and very peculiar circumstances such as inclusion in deformation-resistant minerals. The simultaneous fulfilment of these conditions is rather unlikely and, accordingly, the occurrence of preserved metamorphic coesite should remain exceptional. An indication or even an evidence for its presence or former presence may be the existence of radial jointing around quartz inclusions in other minerals. Thus, after having examined thin section of the Dora Maira quartzite, D.C. Smith (pers. comm.) found these features in a Norwegian eclogite, which led him to the discovery of relic coesite in small quartz inclusions in pyroxene. In fact, as early as 1965, Chesnokov and Popov had drawn attention to the existence of radial cracks around quartz inclusions in eclogites of crustal derivation and had proposed as the most likely explanation that these inclusions formerly were coesite, now inverted to quartz. The discovery reported here and that of Smith give considerable support to their hypothesis. According to their observations coesite might have been present in eclogites from southern Ural and from the Münchberger Gneissmasse, Bavaria, among other regions. A careful search for coesite in these regions is a very desirable step for our understanding of the origin of high-pressure assemblages. Several of the geobarometers used for high-pressure assemblages often yield minimum pressure estimates due to the absence of a buffered assemblage; the range of pressure endured by crustal rocks of continental derivation might be in fact much wider than presently thought. Some of our present concepts about earth tectonics might have to be modified by these coesite discoveries and other possible ones to come.

Acknowledgements. H.-J. Massonne, Bochum, provided many unpublished experimental results which promoted the interpretation of the mineral assemblage. Sincere thanks are also due to W. Schreyer for critical comments on an earlier version of the manuscript and for the many facilities offered in his institute, and to D.S. Cowan, T.J.B. Holland, P.W. Mirwald, R.C. Newton and D.C. Smith for stimulating discussions. An anonymous reviewer, O. Medenbach who contributed most of the pictures, E. Grew and A. Mottana who have drawn my attention to Russian literature are also gratefully acknowledged.

References

Abraham K, Schreyer W (1975) A talc-phengite assemblage in piemontite schist from Brezovica, Serbia, Yugoslavia. J Petrol 17:421–439

Ackermann L, Cemič L, Langer K (1983) Hydrogarnet substitution in pyrope: a possible location for "water" in the mantle. Earth Planet Sci Lett 62:208–214

Bearth P (1952) Geologie und Petrographie des Monte Rosa. Beitr Geol Karte Schweiz NF 132:p 130

Boyd FR, England JL (1959) Pyrope. Carnegie Inst Washington Yearb 58:83–87

Boyd RF, England JL (1960) The quartz-coesite transition. J Geophys Res 65:749–756

Carpenter MA (1980) Mechanisms of exsolution in sodic pyroxenes. Contrib Mineral Petrol 71:289–300

Chao ETC, Shoemaker EM, Madsen BM (1960) First natural occurrence of coesite. Science 132:220–222

Chesnokov BV, Popov VA (1965) Increase in the volume of quartz grains in South Urals eclogite. Dokl Akad Nauk SSSR 162:176–178

Chinner GA, Dixon J (1973) Some high-pressure parageneses of the Allalin gabbro, Valais, Switzerland. J Petrol 14:185–202

Chopin C (1979) De la Vanoise au massif du Grand Paradis. Thèse 3ème cycle Univ P et M Curie Paris:p 145

Chopin C (1981) Talc-phengite: a widespread assemblage in high-grade pelitic blueschists of the Western Alps. J Petrol 22:628–650

Chopin C (1983) High-pressure facies series in pelitic rocks: a review. Terra Cognita 3:183 (abstr)

Chopin C, Maluski H (1980) $^{40}Ar-^{39}Ar$ dating of high-pressure metamorphic micas from the Gran Paradiso area (Western Alps): evidence against the blocking temperature concept. Contrib Mineral Petrol 74:109–122

Chopin C, Monié P (1984) A unique Mg-chloritoid-bearing, high-pressure assemblage from the Monte Rosa, Western Alps: a petrologic and $^{40}Ar-^{39}Ar$ radiometric study. (Submitted to Contrib Mineral Petrol)

Chopin C, Schreyer W (1983) Magnesiocarpholite and magnesiochloritoid: two index minerals of pelitic blueschists and their preliminary phase relations in the model system $MgO-Al_2O_3-SiO_2-H_2O$. Am J Sci Orville vol:72–96

Coes L (1953) A new dense crystalline silica. Science 118:131–152

Compagnoni R (1977) The Sesia-Lanzo zone: high-pressure, low-temperature metamorphism in the Austroalpine continental margin. Rend Soc Ital Mineral Petrol 33:325–374

Desmons J, O'Neil JR (1978) Oxygen and hydrogen isotope compositions of eclogites and associated rocks from the eastern Sesia zone (Western Alps, Italy). Contrib Mineral Petrol 67:79–85

Franchi S (1900) Sopra alcuni giacimenti di roccie giadeitiche nelle Alpi occidentali e nell'Appennino ligure. Boll R Comit Geol Ital (IV) 1:119–158

Ganguly J (1969) Chloritoid stability and related parageneses: theory, experiments and applications. Am J Sci 267:910–944

Gillet P, Ingrin J, Chopin C (1984) Coesite in subducted continental crust: (P, T) history deduced from an elastic model. (Submitted to Earth Planet Sci Lett)

Goffé B (1982) Définition du faciès à Fe-Mg-carpholite-chloritoïde, marqueur du métamorphisme de haute pression-basse température. Thèse d'état Univ P et M Curie Paris

Goffé B, Velde B (1984) Contrasted metamorphic evolutions in thrusted cover units of the Briançonnais zone (French Alps): a model for the conservation of HP–LT metamorphic mineral assemblages. Earth Planet Sci Lett (in press)

Green HW (1974) Metastable growth of coesite in highly strained quartz. J Geophys Res 77:2478–2482

Halbach H, Chatterjee ND (1982) An empirical Redlich-Kwong type equation of state for water to 1000° C and 200 kbar. Contrib Mineral Petrol 79:337–345

Hamilton W (1979) Tectonics of the Indonesian region. US Geol Survey Prof Paper 1078

Hensen BJ, Essene EJ (1971) Stability of pyrope-quartz in the system $MgO-Al_2O_3-SiO_2$. Contrib Mineral Petrol 30:72–83

Hobbs BE (1968) Recrystallization of single crystals of quartz. Tectonophysics 6:353–401

Holland TJB (1979a) High water activities in the generation of high-pressure kyanite eclogites of the Tauern Window, Austria. J Geol 87:1–27

Holland TJB (1979b) Experimental determination of the reaction paragonite = jadeite + kyanite + water, and internally consistent thermodynamic data for part of the system $Na_2O-Al_2O_3-SiO_2-H_2O$, with applications to eclogites and blueschists. Contrib Mineral Petrol 68:293–301

Holland TJB (1980) The reaction albite = jadeite + quartz determined experimentally in the range 600–1200° C. Am Mineral 65:129–134

Kitahara S, Takenouchi S, Kennedy GC (1966) Phase relations in the system $MgO-SiO_2-H_2O$ at high temperatures and pressures. Am J Sci 264:223–233

Lazko YY, Koptil VI, Serenko VP, Tsepin AI (1983) Fassaite

clinopyroxenes from diamond-bearing kyanite eclogite xenoliths. Dokl Earth Sci Section 258:138–142
MacDonald GJF (1956) Quartz-coesite stability relations at high temperatures and pressures. Am J Sci 254:713–721
Massonne HJ (1981) Phengite: eine experimentelle Untersuchung ihres Druck-Temperatur-Verhaltens im System $K_2O-MgO-Al_2O_3-SiO_2-H_2O$. Dissertation Ruhr-Universität Bochum
Massonne HJ (1983) Experiments on melting to 50 kbar in the system $MgO-Al_2O_3-SiO_2-H_2O$ (MASH) with excess SiO_2 and H_2O (abstract). EOS Trans AGU 64:875
Massonne HJ, Mirwald PW, Schreyer W (1981) Experimentelle Überprüfung der Reaktionskurve Chlorit + Quarz = Talk + Disthen im System $MgO-Al_2O_3-SiO_2-H_2O$. Fortschr Mineral 59:122–123
Meyer HOA, Boyd FR (1972) Composition and origin of crystalline inclusions in diamonds. Geochim Cosmochim Acta 36:1255–1273
Miller C (1977) Chemismus und phasenpetrologische Untersuchung der Gesteine aus der Eklogitzone des Tauernfensters, Österreich. Tschermaks Mineral Petrogr Mitt 24:221–277
Mirwald PW, Massonne HJ (1980) The low-high quartz and quartz-coesite transition to 40 kbar between 600° C and 1600° C and some reconnaissance data on the effect of $NaAlO_2$ component on the low quartz-coesite transition. J Geophys Res 85:6983–6990
Naka S, Ito S, Inagaki M (1972) Effect of shear on the quartz-coesite transition. J Am Ceramic Soc 55:323–324
Newton RC (1972) An experimental determination of the high-pressure stability limits of magnesian cordierite under wet and dry conditions. J Geol 80:398–420
Råheim A, Green DH (1974) Talc-garnet-kyanite-quartz schist from an eclogite-bearing terrane, Western Tasmania. Contrib Mineral Petrol 43:223–231
Rao BB, Johannes W (1979) Further data on the stability of staurolite + quartz and related assemblages. Neues Jahrb Mineral Monatsh 10:437–447
Rosenfeld JL, Chase AB (1961) Pressure and temperature of crystallisation from elastic effects around solid inclusions in minerals. Am J Sci 259:519–541
Schreyer W (1968) A reconnaissance study of the system $MgO-Al_2O_3-SiO_2-H_2O$ at pressures between 10 and 25 kbar. Carnegie Inst Wash Yearb 66:380–392
Schreyer W (1977) Whiteschists: their compositions and pressure-temperature regimes based on experimental, field, and petrographic evidence. Tectonophysics 43:127–144
Schreyer W, Abraham K (1976) Three stage metamorphic history of a whiteschist from Sar e Sang, Afganistan, as part of a former evaporite deposit. Contrib Mineral Petrol 59:111–130
Schreyer W, Abraham K, Kulke H (1980) Natural sodium phlogopite coexisting with potassium phlogopite and sodian aluminian talc in a metamorphic evaporite sequence from Derrag, Tell Atlas, Algeria. Contrib Mineral Petrol 74:223–233
Schreyer W, Baller T (1977) Talc-muscovite: synthesis of a new high-pressure phyllosilicate assemblage. Neues Jahrb Mineral Monatsh 9:421–425
Silver EA, Reed D, McCaffrey R (1983) Back arc thrusting in the Eastern Sunda arc, Indonesia: a consequence of arc-continent collision. J Geophys Res 88:7429–7448
Smyth JR (1977) Quartz pseudomorphs after coesite. Am Mineral 62:828–830
Smyth JR, Hatton CJ (1977) A coesite-sanidine grospydite from the Roberts Victor kimberlite. Earth Planet Sci Lett 34:284–290
Sobolev NV et al. (1976) Coesite, garnet and omphacite inclusions in Yakut diamonds – first finding of coesite paragenesis. Dokl Akad Nauk SSSR 230:1442–1444 (in Russian) Abstr in Mineral Abstr 78–818
Stella A (1895) Sul rilevamento geologico eseguito nel 1894 in Valle Varaita (Alpi Cozie). Boll R Comit Geol Ital 26:283–313
Udovkina NG, Muravitskaya GN, Laputina IP (1978) Phase equilibria in the talc-garnet-kyanite rocks of the Kokchetav block, Northern Kazakhstan. Isvestiya Akad Nauk SSSR Geol section 7:55–64 (in Russian)
Udovkina NG, Muravitskaya GN, Laputina IP (1980) Talc-garnet-kyanite rocks of the Kokchetav block, Northern Kazakhstan. Dokl Earth Sci section 237:202–205
Velde B (1967) Si content of natural phengites. Contrib Mineral Petrol 14:250–258
Vialon P (1966) Etude géologique du massif cristallin Dora-Maira, Alpes cottiennes internes, Italie. Thèse d'état Univ Grenoble

CHAPTER 15: KINETICS: FIELD

Austrheim, H. (1987) Eclogitization of lower crustal granulites by fluid migration through shear zones. *Earth and Planetary Science Letters*, **81**, 221–232.

This paper was one of the earliest by Håkon Austrheim to send a strong message to the metamorphic community not to proceed uncritically on the routine assumption that the P and T imposed on a metamorphic rock necessarily determines its mineralogy. Our senses appeared to have been dulled through the years by the apparent success of the metamorphic facies concept, which assumes the attainment of equilibrium under specific external conditions, followed by substantial preservation of the near-T_{max} parageneses as the rocks are later returned to the surface without regaining their lost volatiles. These conditions seem to be met in most of the packages of metamorphosed sediments + entrained volcanics now exposed worldwide. Were this not the case, we would not have the consistent development of metamorphic zones found in so many places.

Not surprisingly, kinetic difficulties arise in terranes composed of starting materials that consist predominantly of anhydrous plutonic igneous rocks or dry granulite facies rocks. Such rocks require the pervasive introduction of H_2O-rich fluid to catalyze reactions and produce mineral assemblages appropriate to the applied P-T conditions. Rubie (1986) was one of many writers to comment on this issue. Anhydrous protoliths may include the gabbro-ultramafic lower sections of ophiolites, layered igneous complexes emplaced in the middle or lower crust, silicic intrusions of batholith dimensions, and granite-depleted lower-crustal granulites. These rock masses may yield little or no evidence of their P-T history over long stretches of geological time. When the ultimate product of transformation should have been eclogite, the geodynamic consequences of zero or only partial reaction are serious.

The granulite-facies layered anorthositic rocks of the Bergen Arcs, western Norway (Austrheim and Griffin, 1985; Austrheim, 1987), have emerged in the last 20 y as a type area for the field study of kinetic problems and processes associated with the transformation of dry, deep-crustal – commonly igneous rocks – into eclogite. Austrheim and coworkers have carefully documented field evidence for the roles of introduced fluid and accompanying deformation in glacially polished outcrops over a substantial portion of the 1000 km^2 tectonic unit known as the Lindås Nappe in the Caledonian Bergen Arcs.

The field relations in places like Holsnøy Island are not particularly subtle. Typically, an outcrop of layered, garnet-bearing meta-anorthosite and corona gabbro-norite of Proterozoic age, with granulite facies recrystallization at 945 Ma (Sveconorwegian = Grenvillian), $P \leqslant 10$ kbar, $T = 800–850°C$, is seen to be cut by 460 Ma (Caledonian) shear zones up to 100 m wide in which kyanite-eclogite facies mineralogy ($P = 16–19$ kbar, $T = 650–700°C$) has developed. Some outcrops are free of eclogite overprint, others totally converted to eclogite, the majority showing partial conversion. Austrheim has estimated that granulite-facies mineralogy survived over ~50% of the terrane. The entire terrane, of course, was subjected to the pressure increase caused by ~20 km of crust superimposed on it by Caledonian subduction.

Eclogitization in the shear zones was not driven by the attainment of new P-T conditions, but instead proceeded irreversibly across all granulite lithotypes with the arrival of H_2O-rich fluid along cracks that developed in brittle dry granulite. In addition to omphacite, garnet, kyanite and quartz, the Holsnøy eclogites contain hydrate minerals such as clinozoisite, phengite, and calcic amphibole, and the carbonate calcite. There is no record of the P-T-t path over the 500 my interval separating the Grenville and Caledonian events. As suggested by Boundy et al. (1992), it is likely that the terrane cooled isobarically over this time to the 500–600°C characteristic of deep stable continental crust. Although the growth of hydrate minerals might suggest a retrograde change (Bjørnerud et al., 2002), it seems likely that the eclogite overprint was attended by an increase in both P and T. The straight-line path connecting the two events in Fig. 1 of Bjørnerud and Austrheim (2004) may be an oversimplification.

The clarity of the field relations in the Bergen arc exposures with respect to the outcrop-scale eclogitization of dry granulite crust was believed in 1987 to be unique (Austrheim, p. 229). Because of their simple but profound message, the Bergen rocks have since attracted much attention as a 'laboratory' for the study of reaction kinetics in metamorphism (Jamtveit et al., 1990; Klaper, 1991; Boundy et al., 1992; Erambert and Austrheim, 1993; Austrheim and Boundy, 1994; Matthey et al., 1994; Austrheim and Engvik, 1997; Boundy et al., 1997; Austrheim, 1998; Jamtveit et al., 2000; Bjørnerud et al., 2002; Boundy et al., 2002; Lund and Austrheim, 2003; Bjørnerud and Austrheim, 2004). These investigations have shown that incomplete eclogitization of the anorthositic rocks involved a broad temporal change from brittle to ductile behaviour as the amount of eclogite increased, with the following sequence of events: (1) seismic slip along cracks, creating friction-generated melt (pseudotachylyte) and quench-textured growth of skeletal eclogite minerals; (2) fracture-controlled eclogitization along cm- to m-wide ductile shear zones (± formation of anorthosite-block breccia); and (3) pervasive eclogitization. The introduction of fluid and alteration to hydrous eclogite selectively weakened the rocks, which in turn slowed the rate of fluid introduction and perhaps arrested it. Sluggish intergranular diffusion during the Proterozoic granulite metamorphism produced corona textures in the layered anorthosite-gabbro rocks of the Bergen Arcs. The eclogite reaction was initiated by frictional heat, followed by ingress of fluid that catalyzed and fed the growth of eclogite minerals. Nucleation was probably not rate-limiting at any time (cf. Rubie, 1998), because garnet and pyroxene were already present; the fluid promoted intergranular diffusion and allowed pervasive recrystallization to eclogite.

Many other field examples have come to light elsewhere with strong resemblance to the outcrop-scale relationships in the Bergen Arcs. Several examples of the partial transformation of magmatic and granulite-facies rocks have been described in the Western Gneiss region to the north of Bergen, where conditions locally reached those of the coesite-eclogite facies (Mørk, 1985; Krabbendam et al., 2000; Engvik et al., 2000, 2001; Wain et al., 2001; Lund and Austrheim, 2003). Alpine examples of the same process include the UHPM eclogitization of granitoids (Compagnoni and Maffeo, 1973; Oberhänsli et al., 1985; Biino and Compagnoni, 1992) and ophiolites (Wayte et al., 1989; Bucher and Fry, 1994, pp. 294–297). Outside Europe, notable examples have been described by Zhang and Liou (1997) in the UHPM Sulu terrane, China, by Molina et al. (2002) in HP rocks of the Polar Urals, Russia, and by John and Schenk (2003) in Precambrian gabbro in Zambia.

The broader implications of all these observations are numerous. (1) Although deep crust is generally assumed to deform by ductile flow, in the absence of fluid, brittle failure may occur instead, thereby promoting fluid infiltration and eventual transition to ductile behaviour. (2) Metamorphic reactions in deep dry crust up to 800°C may be inhibited and phase boundaries overstepped by pressure increments of as much as 1 Gpa (10 kbar). Large volumes of deep crust can exist metastably for extended periods of geological time. The assumption generally made in numerical geodynamic modelling that reactions proceed with gradual change in P and T under near-equilibrium conditions needs re-examination – a challenging problem for the future. (3) The geophysical signature of deep crust (density, seismic velocities and reflections) will be influenced by the extent of reaction allowed by the water budget (e.g. Hacker, 1996). The density (3.0 to 3.3 g/cm^3) and rheology changes induced by fluid-controlled eclogite formation and retrogression can have first-order effects on the subsequent geodynamic evolution of collision zones, influencing driving forces, buoyancy, exhumation, root formation, and elevations of mountains. (4) We should question the assumption that subducted lithosphere always contains enough volatiles in the right places to allow complete eclogitization at depth. The volume proportions of anhydrous magmatic rocks and dry granulites must be factored somehow into the geodynamic models.

The Holsnøy rocks have provided a good analogue for processes beneath the Himalaya in southern Tibet (Lund et al., 2004; Jackson et al., 2004). There, the underthrust lower crust of the Indian shield is believed to be dry anorthositic granulite. Earthquakes extend to depths of 70–90 km beneath the high Himalaya – sufficiently close to the Moho to suggest triggering by H_2O fluid derived from the mantle. The support of such unusually high mountains and the presence of deep roots are ascribed to the relative low density and great strength of metastable granulite at depth. Measurements of gravity, density and seismic velocities in the Himalaya region show that the high topography is incompatible with an eclogite-facies lower crust.

The cases of metastable persistence of dry rocks cited above are all from collisional terranes, but there is no reason not to expect to find some in type B orogenic belts. For example, aureole rocks on the east side of the 95 Ma Mt. Stuart quartz-diorite batholith, Washington State, show kyanite-staurolite overprint on andalusite-cordierite hornfels. Within the pluton, the only detectable manifestation of the doubling of the load on it (3 → 6 kbar) were rare microscopic cracks in metapelite enclaves with incipient growth of Barrovian minerals (Evans and Davidson, 1999).

We must accept the fact that, in the course of orogeny, large portions of deep crust, when composed of dry magmatic rocks or anhydrous granulite-facies rocks, can survive metastably in the field of eclogite as a result of the limited access of fluids. The onset of eclogitization resulting from collision seems to be soon followed by rapid exhumation into the upper crust, so that the eclogitization is incomplete.

References

Austrheim, H. (1987) Eclogitization of lower crustal granulites by fluid migration through shear zones. *Earth and Planetary Science Letters*, **81**, 221–232.

Austrheim, H, and Griffin, W.L. (1985) Shear deformation and eclogite formation within granulite-facies anorthosites of the Bergen Arcs, western Norway. *Chemical Geology*, **50**, 267–281.

Austrheim, H. (1998) Influence of fluid and deformation of the deep crust and consequences for the geodynamics of collision zones. Pp. 297–323 in: *When Continents Collide: Geodynamics and Geochemistry of Ultrahigh-pressure Rocks* (B.R. Hacker and J.G. Liou, editors). Kluwer Academic Publishers, Dordrecht, The Netherlands.

Austrheim, H. and Boundy, T.M. (1994) Pseudotachylytes generated during seismic faulting and eclogitization of the deep crust. *Science*, **265**, 82–83.

Austrheim, H. and Engvik, A.K. (1997) Fluid transport, deformation and metamorphism at depth in a collision zone. Pp. 123–138 in: *Fluid Flow and Transport in Rocks: Mechanisms and Effects* (B. Jamtveit and B. Yardley, editors). Chapman & Hall, New York.

Biino, G. and Compagnoni, R. (1992) Very-high pressure metamorphism of the Brossasco coronite metagranite, southern Dora-Maira massif, Western Alps. *Schweizerische Mineralogische und Petrographische Mitteilungen*, **72**, 347–363.

Bjørnerud, M.G. and Austrheim, H. (2004) Inhibited eclogite formation: the Key to rapid growth of strong and buoyant Archean continental crust. *Geology*, **32**, 765–768.

Bjørnerud, M.G., Austrheim, H. and Lund, M.D. (2002) Processes leading to eclogitization (densification) of subducted and tectonically buried crust. *Journal of Geophysical Research*, **107**, B10, 2252.

Boundy, T., Mezger, K. and Essene, E. (1997) Temporal and tectonic evolution of the granulite-eclogite association from the Bergen arcs, western Norway. *Lithos*, **39**, 159–178.

Boundy, T.M., Donohue, C.L., Essene, E.J., Mezger, K. and Austrheim, H. (2002) Discovery of eclogite-facies carbonate rocks from the Lindås Nappe, Caledonides, western Norway. *Journal of Metamorphic Geology*, **20**, 649–667.

Boundy, T.M., Fountain, D.M. and Austrheim, H. (1992) Structural development and petrofabrics of eclogite facies shear zones, Bergen Arcs, western Norway: implications for deep crustal deformation processes. *Journal of Metamorphic Geology*, **10**, 127–146.

Bucher, K. and Frey, M. (1994) *Petrogenesis of Metamorphic Rocks*. Springer-Verlag, Berlin, 318 pp.

Compagnoni, R. and Maffeo, B. (1973) Jadeite-bearing metagranites l.s. and related rocks in the Mount Mucrone area (Sesia-Lanzo zone, western Italian Alps). *Schweizerische Mineralogische und*

Petrographische Mitteilungen, **53**, 355–378.

Engvik, A.K., Austrheim, A. and Anderson, T.B. (2000) Structural, mineralogical and petrophysical effects on deep crustal rocks of fluid-limited polymetamorphism, Western Gneiss Region, Norway. *Journal of the Geological Society, London*, **157**, 121–134.

Engvik, A.K., Austrheim, A. and Erambert, M. (2001) Interaction between fluid flow, fracturing and mineral growth during eclogitization, an example from the Sunnfjord area, Western Gneiss Region, Norway. *Lithos*, **57**, 111–141.

Erambert, M. and Austrheim, H. (1993) The effect of fluid and deformation on zoning and inclusion patterns in polymetamorphic garnets. *Contributions to Mineralogy and Petrology*, **115**, 2304–214.

Evans, B.W. and Davidson, G.F. (1999). Kinetic control of metamorphic imprint during synplutonic loading of batholiths: An example from Mount Stuart, Washington. *Geology*, **27**, 415–418.

Hacker, B.R. (1996) Eclogite formation and rheology, buoyancy, seismicity and H_2O content of the oceanic crust. Pp. 337–346 in: *Subduction Top to Bottom* (G.E. Bebout, D.W. Scholl, S.H. Kirby and J.P. Platt, editors). Geophysical Monograph, **96**. American Geophysical Union.

Jackson, J.A., Austrheim, H., McKenzie, D. and Priestly, K. (2004) Metastability, mechanical strength, and the support of mountain belts. *Geology*, **32**, 625–628.

Jamtveit, B., Bucher-Nurminen, K. and Austrheim, H. (1990) Fluid controlled eclogitization of granulites in deep crustal shear zones, Bergen Arcs, western Norway. *Contributions to Mineralogy and Petrology*, **104**, 184–193.

Jamtveit, B.H., Austrheim, H. and Malte-Sørensen, A. (2000) Accelerated hydration of the Earth's deep crust induced by stress perturbations. *Nature*, **408**, 75–78.

John, T. and Schenk, V. (2003) Partial eclogitization of gabbroic rocks in a late Precambrian subduction zone (Zambia): prograde metamorphism triggered by fluid infiltration. *Contributions to Mineralogy and Petrology*, **146**, 174–191.

Klaper, E.M. (1991) Eclogite shear zones in a granulite-facies anorthosite comples – an example from the Bergen Arcs, Western Norway. *Schweizerische Mineralogische und Petrographische Mitteilungen*, **71**, 231–241.

Krabbendam, M., Wain, A. and Andersen, T.B. (2000) Pre-Caledonian granulite and gabbro enclaves in the Western Gneiss Region, Norway: indications of incomplete transition at high pressure. *Geological Magazine*, **137**, 235–255.

Lund, M.G. and Austrheim, H. (2003) High-pressure metamorphism and deep-crustal seismicity: evidence from contemporaneous formation of pseudotachylytes and eclogite facies coronas. *Tectonophysics*, **372**, 59–83.

Lund, M.G., Austrheim, H. and Erambert, M (2004) Earthquakes in the deep continental crust – insights from studies on exhumed high-pressure rocks. *Geophysical Journal International*, **158**, 569–576.

Mattey, D., Jackson, D.H., Harris, B.W. and Kelley, S. (1994) Isotopic constraints on fluid infiltration from an eclogite facies shear zone, Holsnøy, Norway. *Journal of Metamorphic Geology*, **12**, 311–325.

Molina, J.F., Austrheim, H., Glodny, J. and Rusin, A. (2002) The eclogites of the Marun-Key complex, Polar Urals (Russia): fluid control on reaction kinetics and metasomatism during high P metamorphism. *Lithos*, **61**, 55–78.

Mørk, M.B.E. (1985) A gabbro to eclogite transition on Flemsøy, Sunnemøre, western Norway. *Chemical Geology*, **50**, 283–310.

Oberhänsli, R., Hunziger, J.C., Martinotti, G. and Stern, W.B. (1985) Geochemistry, geochronology and petrology of Monte Mucrone: an example of eo-alpine eclogitization of Permian granitoids in the Sesia-Lanzo zone, western Alps, Italy. *Chemical Geology Isotope Geoscience Section*, **52**, 165–184

Rubie, D. (1986) The catalysis of mineral reactions by water and restrictions on the presence of aqueous fluid during metamorphism. *Mineralogical Magazine*, **50**, 399–415.

Wain, A., Waters, D.J. and Austrheim, H. (2001) Metastability of granulites and processes of eclogitisation in the UHP region of the Western Gneiss Region, Norway. *Journal of Metamorphic Geology*, **19**, 607–623.

Wayte, G.J., Worden, R.H., Rubie, D.C. and Droop, G.T.R. (1989) A TEM study of disequilibrium plagioclase breakdown at high pressure: the role of infiltrating fluid. *Contributions to Mineralogy and Petrology*, **101**, 426–437.

Zhang, R.Y. and Liou, J.G. (1997) Partial transformation of gabbro to coesite-bearing eclogite from Yangkou, the Sulu terrane, eastern China. *Journal of Metamorphic Geology*, **15**, 183–202.

Eclogitization of lower crustal granulites by fluid migration through shear zones

H. Austrheim

Mineralogisk-Geologisk Museum, Sars gate 1, N-0562 Oslo 5 (Norway)

Received July 21, 1986; revised version accepted October 20, 1986

The granulite facies assemblages of the anorthositic rocks of the Bergen Arcs (stable at 800–900°C and 10 kbar) have been transformed to eclogite facies assemblages (stable at 700–750°C and 16–19 kbar) in the vicinity of Caledonian shear zones. This section of the root zone of the Caledonian mountain chain reveals a deep polymetamorphic crust where Precambrian granulites (mean density 3.02 g/cm^3) and Caledonian eclogites (mean density 3.19 g/cm^3) alternate on a scale of meters over a minimum area of 3 × 12 km. Detailed mapping of three localities shows that eclogites account for up to 30–45% of the rock volume. The stabilitization of the eclogite mineralogy is controlled by fluids penetrating these deep crustal shear zones. The eclogitization is independent of preexisting compositional variation in this anorthosite-norite complex. The Bergen Arcs example suggests that the amount of eclogite versus granulites in the lowermost crust is a function of deformation and fluid access, rather than being controlled by T, P and rock composition alone. These relationships may explain the gradual increase in seismic velocity observed in some deep crustal sections and also the complex reflection pattern obtained from the lowermost crust in many areas.

1. Introduction

Information regarding the nature of the lower crust and the crust-mantle boundary is derived from three principal sources: (1) geophysical data, (2) xenolith material and (3) granulite facies terrains recognized as former deep crustal sections. In addition, theoretical and experimental petrology have had a great impact on our concepts of this part of the lithosphere. Recent research in these fields has abandoned the older model of a uniform lower crust of basaltic composition separated from the mantle by a sharp and continuous Moho. The emerging picture of the lower crust is one of great compositional diversity both vertically and laterally [1], with a predominance of mafic to intermediate granulites [2,3]. Anorthosites are encountered in xenolith populations [4] as well as in granulite facies terrains [5,6].

From investigations of exposed deep crustal sections, Fountain and Salisbury [2] find that the lower crust contains polymetamorphic and multiply deformed rocks and that the lower crust is a mixture of rocks which have evolved through deformation and metamorphism in a variety of tectonic environments.

Recent advances in geophysics have demonstrated that the Moho in many areas is not sharp but has a complex reflection pattern [7,8] with a gradual increase in seismic velocity in the lower crust [9]. O'Reilly and Griffin [10] explain this increase in seismic velocity below southeast Australia as due to an increasing amount of spinel lherzolite intercalated with gabbroic rocks. The deep crustal reflections have typically been interpreted in terms of rock boundaries, but it has also been suggested that they may represent mylonite zones [11] or layers rich in fluids [12].

The garnet granulite to eclogite phase transition has attracted attention in the discussion of the nature and evolution of the lower crust and the crust-mantle boundary [13]. Experimental results [14,15] have shown that the stable mineral assemblages in the lowermost crust in rocks of basaltic composition will be those of two pyroxene granulites, garnet granulites or eclogites and it is generally accepted today that granulites and eclogites are prominent rock types in this part of the lithosphere. The possibility that lower-grade rocks exist metastably in the lower crust has been mentioned by Kay and Kay [1] and has been argued for on kinetic grounds by Neugebauer and Spohn [16].

The present paper describes an area where arrested transformation from garnet granulites to eclogites in rocks of anorthositic-noritic composition can be observed directly and where it can be demonstrated that deformation and fluids are important parameters, in addition to pressure, temperature and rock composition.

2. Geological setting

The rocks dealt with in this account are from a major tectonic unit of high-grade rocks forming part of the Bergen Arcs (Fig. 1), a series of arcuate Caledonian thrust sheets centered around the town of Bergen, Western Norway [17]. This unit, which crops out over an area of more than 1000 km², contains abundant mafic material ranging from pure anorthosite to gabbroic anorthosite and norite [18]. In addition, various members of the charnokite suite are present, the most prominent of which is the type mangerite [19].

The anorthosites of the Bergen Arcs have traditionally been correlated with the anorthosites of the Jotun Nappe to the east (Fig. 1), based on lithological and metamorphic similarities [20]. The high-grade rocks of the Jotun Nappe have been proposed by Smithson et al. [21] to represent a thick slice of lower crust emplaced in the upper crust by Caledonian deformation. However, Loeschke and Nickelsen [22] pointed out that part of the Jotun Nappe must have reached the surface in pre-Caledonian time as it is overlain by Eocambrian sediments.

The geological history of the Bergen Arcs anorthosites can be divided into two major events [23]. A protracted Precambrian (Grenvillian) orogeny which included extensive magmatic activity, deformation and granulite facies metamorphism was followed by strong but localized Caledonian reworking. The majority of the Caledonian reworking zones contain hydrous amphibolite facies mineralogy [24], but in some areas Caledonian shear zones contain an eclogite facies mineralogy.

P, T estimates for the granulite facies mineral assemblages are 800–900°C and about 10 kbar pressure while a temperature of 700–750°C and a pressure of 16–19 kbar were obtained for the eclogite facies mineralogy [23]. The position of the terraine in the P, T plane in the interval between Grenvillian and Caledonian times is not known.

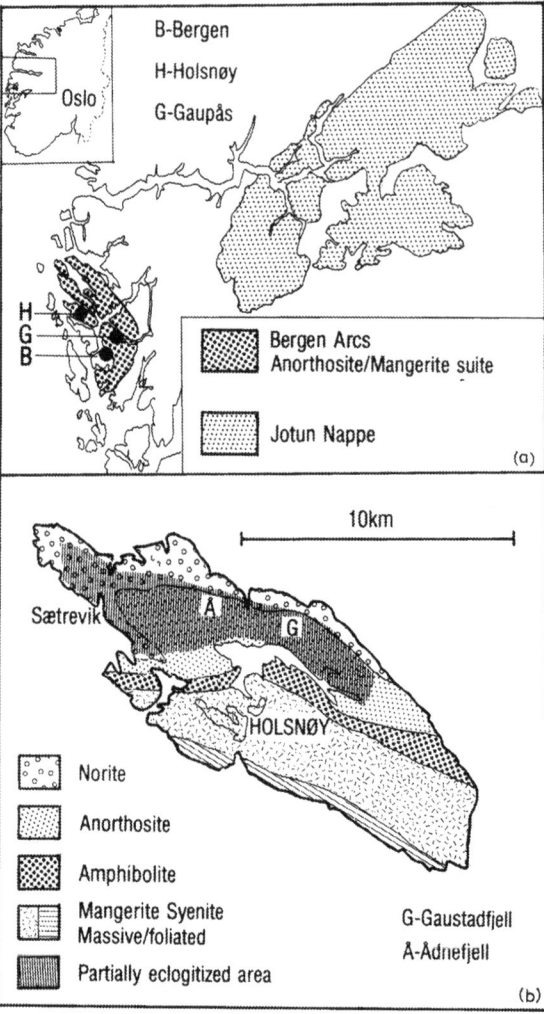

Fig. 1. (a) Location of the Bergen Arcs in relation to the Jotun Nappe. (b) Geological map of Holsnøy showing the approximate extent of the partly eclogitized area drawn on the map of Kolderup and Kolderup [18].

3. Field relations

Austrheim and Griffin [23] described formation of eclogites in a shear zone from Sætrevik on Holsnøy (Fig. 1), and indicated that similar shear zones are present in many parts of the Bergen Arcs.

Recent mapping in the area has revealed that eclogitization has affected a belt on northeastern Holsnøy which is 12 km long and 3–5 km broad (Fig. 1). In addition, eclogite shear zones have been found at Gaupås [25], some 40 km to the south. The eclogites at Gaupås are found in the same belt of anorthosites as those of Holsnøy, according to the map of Kolderup and Kolderup [18]. The extent of the partially eclogitized area around Gaupås is at present not known. It is clear, however, that the eclogitization is not confined to a few localities, but occurs on a regional scale. The possibility can not be ruled out that the larger area of anorthosites which shows reworking zones with an amphibolite facies mineralogy may have developed by hydration of a former eclogite mineralogy. In this case the partially eclogitized area would be 1-2 orders of magnitude greater than observed today.

3.1. Description of the eclogitized area

Outside the zones of Caledonian reworking the anorthosite complex alternates between well-banded parts, where the banding is defined by cm-thick mafic layers rich in diopside and garnet, and plagioclase-rich layers. This banding is referred to as the granulite facies foliation. In the coronitic parts the granulite facies foliation is defined by parallel-orientated disc-shaped coronas. These coronas, which typically have a core of orthopyroxene surrounded by clinopyroxene and garnet, were interpreted by Griffin [26] as formed during cooling of the anorthosite complex from magmatic temperature through a reaction between olivine and plagioclase. Griffin et al. [27] demonstrated that the corona forming reaction in this complex took place at temperatures $> 900°C$.

The eclogitization is typically confined to 0.2-5 m thick shear zones. In the majority of the shear zones the granulite facies foliation is dragged into the shear zone where it becomes parallel to the new eclogite facies foliation and eventually disappears in the central part of the shear zone (Fig. 2). Some shear zones have discontinuous margins (Figs. 3 and 4) where the granulite facies foliation is abruptly truncated. Within these shear zones, remnants of the granulite facies banding are seen in rotated blocks of competent mafic layers. The new foliation defined by the eclogite minerals flows around the mafic blocks in a complex flow pattern (Figs. 3 and 4).

The eclogitization can also be seen related to macroscopic kink bands in the anorthosite (Fig. 5). Here the granulite facies foliation can be traced right through the eclogitized area with a small offset of the bands along a mm-thick vein in the center of the kink band. The transition from granulite facies mineralogy to eclogite mineralogy, as represented by change in colour from white to dark green, takes place over a distance of less than 1 cm (Fig. 5). It is important to note that the

Fig. 2. Granulite facies anorthositic rock (lower part of photo) is transformed into eclogite (dark upper part). Note the bending of the granulite facies foliation and the development of the new eclogite facies foliation (upper part).

Fig. 3. **Discontinuous** shear zone (left-hand side) cuts granulite facies foliation in anorthosite gabbro. Note rotated block with relict granulite facies foliation (upper left corner). The light material in the eclogitized part consists of kyanite, phengite, clinozoisite, quartz and minor plagioclase. Note also the complex flow structures in the eclogitized part. Long dimension of sample is 30 cm.

margin of the eclogitizing front extends 5–10 cm outside the area actually kinked (Fig. 5). In this case the eclogitization bears a strong resemblance to hydrothermal veins [28].

In places, the eclogitization takes place in the neighbourhood of cracks (Fig. 6). Patches of eclogite which apparently bear no relation to shear zones can then be seen (Fig. 6). However, the distance to the nearest shear zone is always small, generally not more than 1 m, and unsheared eclogites often can be followed to a shear zone.

So far no structural analysis of the shear zones

Fig. 4. Discontinous shear zone (upper part of photo) cuts the banded anorthosite gabbro. Note blocks with relict granulite facies banding, surrounded by eclogite foliation.

Fig. 5. Eclogitization related to kinking of granulite facies foliation. Note that the granulite facies structure can be followed into the eclogitized zone. Note also the abrupt change from granulite facies anorthositic rocks (light colour) to the eclogite facies equivalent. The eclogitizing front extends ca. 5 cm behind the kinked part.

has been undertaken. In some localities the shear zones tend to be parallel. However, in most cases they curve around lenses of granulite facies anorthosites (Fig. 7) in a complex flow pattern.

3.2. Extent of eclogitization

The amount of eclogite formed in a given volume of rock varies from place to place. In the eclogitized belt on Holsnøy, localities of 1000 m² may be without or with only a few shear zones 1–2 m thick. In the most intensely eclogitized area, one can find localities of 1000 m² made up of completely eclogitized anorthositic material.

To quantify the extent of eclogitization, three

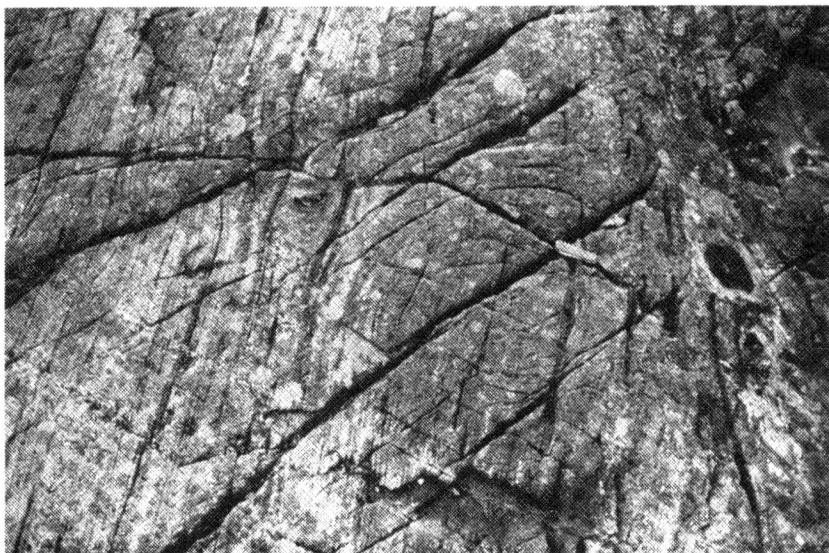

Fig. 6. Patch of eclogitized material outside a shear zone. Note the development of fracture systems/microshears.

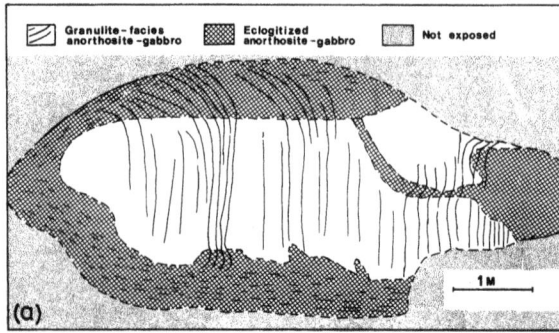
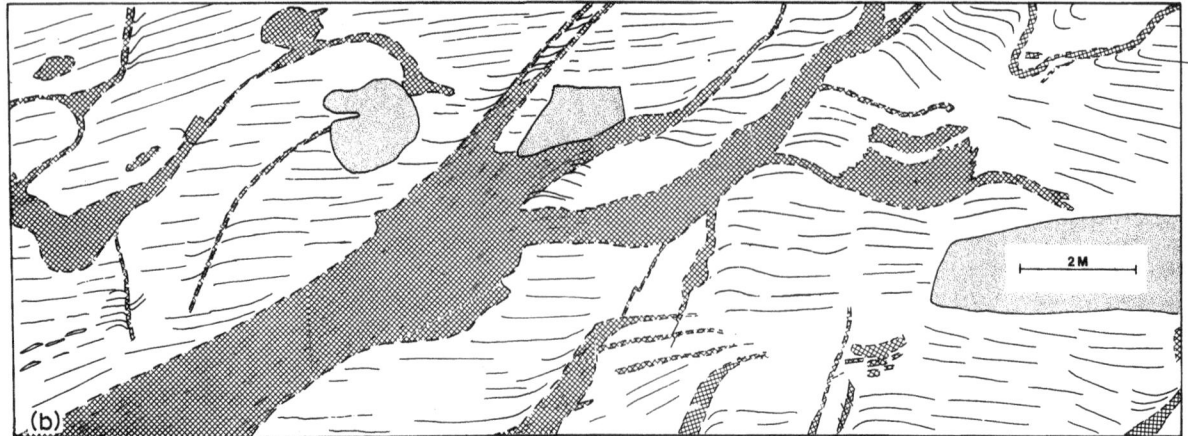

Fig. 7. Detailed maps of the three localities mentioned in the text. (a) Locality 1, southeast of Gaustadfjell. The eclogitized part amounts to 45% of the exposed surface. Note the lens-shaped form of the preserved granulite. (b) Locality 2, the southeastern top of Gaustadfjell. Eclogite is 30% of the exposure. (c) Locality 3, Ådnefjell. The right-hand side eclogite related to the major shear zone mentioned in the text. The eclogitized part of the mapped area amounts to 35%.

well-exposed localities of intermediate to high degree of eclogitization were chosen for detailed mapping.

Locality 1: Southeast of Gaustadfjell, Holsnøy. The mapped locality is a small lense-shaped exposure of 30 m² in the hillside up to Gaustadfjell (Fig. 1). The exposed surface, which is surrounded by vegetation on all sides, has a strike of 010 and dips 25° to the east. The eclogitized material is confined to the border of the outcrop (Fig. 7), indicating that the eclogite extends below the cover. From Fig. 7 the eclogite part was calculated to make up 45% of this outcrop.

Locality 2: the southeastern top of Gaustadfjell, Holsnøy. In a nearly horizontal outcrop of 3000 m² (Fig. 7) the amount of eclogite was calculated as 30%.

Locality 3: Ådnefjell, Holsnøy. A well-exposed horizontal area of 500 m² occurs just under the top of Ådnefjell. The area is bounded to the east by a major shear zone 5 m across, forming a small valley (Fig. 7). The amount of eclogite is approximately 35% of the mapped area.

4. Mineralogical and microstructural changes related to the granulite-eclogite phase transition

The granulite-eclogite phase transition as it occurs in relation to a shear zone at Sætrevik (Fig. 1) was described in detail by Austrheim and Griffin [23]. The main change is transformation of a plagioclase-rich rock into a rock rich in omphacite where plagioclase is absent or present in minor amounts. Austrheim and Griffin [23] list the following minerals as part of the granulite (I) and eclogite (II) mineral assemblages.

I. plagioclase (An_{35}–An_{65}), Al-rich diopside, garnet (I) ± orthopyroxene ± sulfur-rich scapolite ± dark brown hornblende.

II: omphacite (Jd_{50}), garnet (II), kyanite, clinozoisite, phengite, amphibole, quartz ± plagioclase.

A number of solid-solid and hydration reactions were listed to explain this transformation [23]. In the following a series of microstructural relationships, together with mineral compositions will be presented which stress the importance of the fluid phase during this transformation.

4.1. Calcite in eclogite

Calcite has not been found in the granulite facies parts of the anorthosite complex. In the eclogites, calcite can be present as separate grains in equilibrium with the eclogite minerals, and more importantly as inclusions in clinopyroxenes undergoing alteration from diopside to omphacite where Ca was displaced by Na in the clinopyroxene structure [23]. The calcite inclusions indicate that this change took place in the presence of CO_2-rich fluid.

4.2. Microcracks

Figs. 8 and 9 are microphotographs from the same cm-thick system of microfractures penetrating the anorthositic rock at Ådnefjell (Fig. 1). Where this crack system runs through plagioclase (Fig. 8), needles of clinozoisite arranged in a

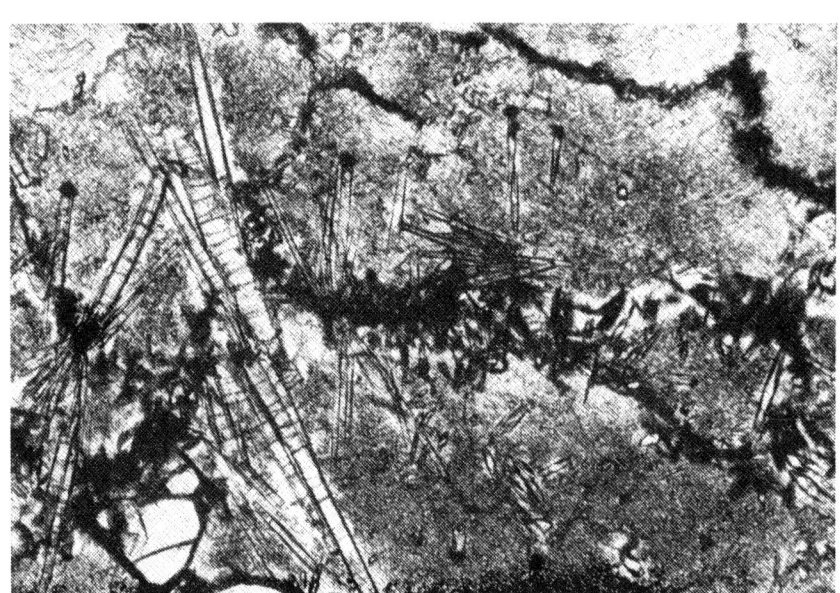

Fig. 8. Microfracture penetrating plagioclase in anorthositic rocks from Ådnefjell. Note (1) clouding of feldspar, (2) clinozoisite needles growing out from the main fracture. Plane-polarized light. Magnification 70×.

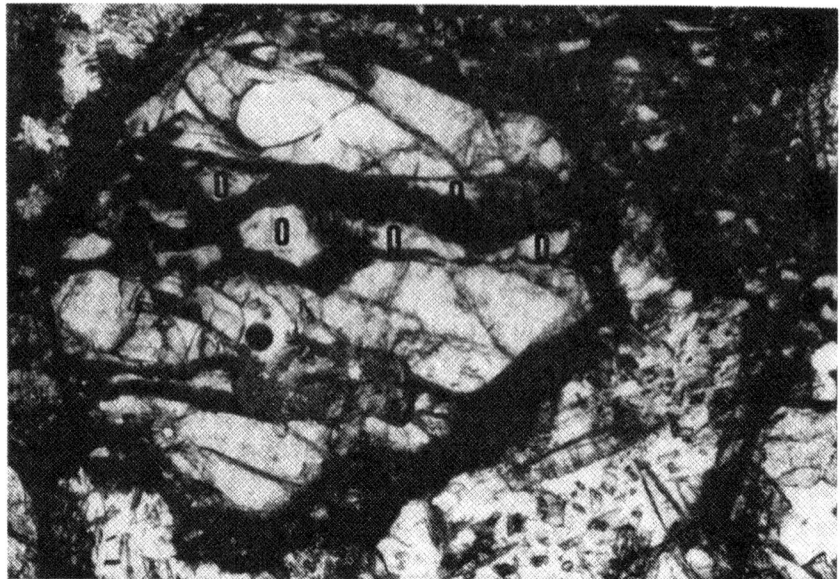

Fig. 9. Same fracture as shown in Fig. 8 penetrating garnet. Omphacite (O) is formed within garnet. Note clinozoisite inclusion in plagioclase. Plane-polarized light. Magnification 70×.

rosette-like texture branches out from the crack. The crack itself is locally found to be filled with quartz, but in most cases is seen as a dark mass. The crack leads to a strong clouding of plagioclase at a distance of 1–2 cm from the main fracture. Where the crack hits a granulite facies garnet (Fig. 9), the garnet is filled with omphacite (Jd_{50}).

In the main eclogite bodies the granulite facies garnets are recognised as cores to the more grossular-rich eclogite facies garnets [23]. In the granu-

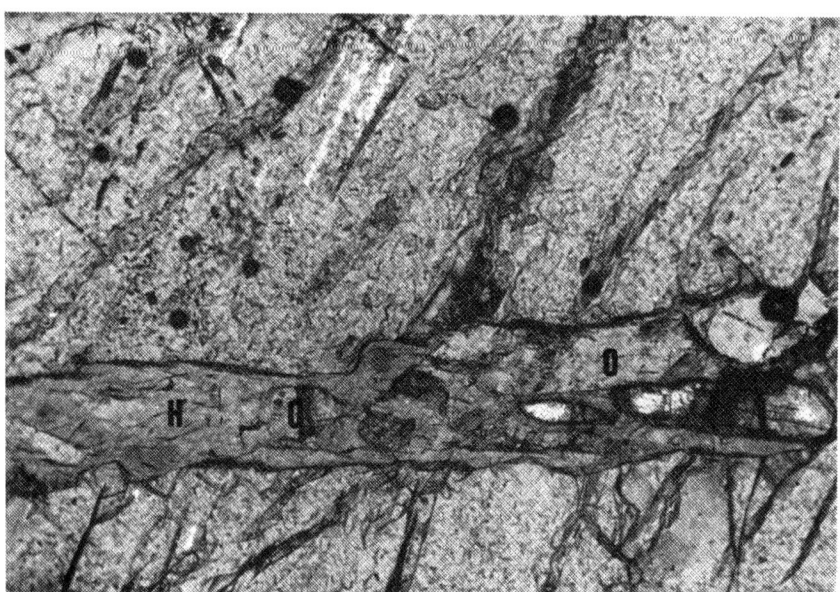

Fig. 10. Microcrack in granulite facies garnet filled with omphacite (O), Cl-rich hornblende (H), and kyanite (high relief). Plane-polarized light. Magnification 70×.

lite facies areas these garnets are inclusion-free or locally contain inclusions of diopside. In the eclogites the old granulite garnets locally contain cracks filled with omphacite, hornblende with up to 1 wt.% of Cl, and kyanite (Fig. 10).

These relationships suggest that the eclogitization was controlled by fluids penetrating the shear zones and that there was substantial element mobility in the neighbourhood of shear zones.

4.3. Density changes

Samples of granulites (13) and eclogites (21) weighing between 0.5 and 2.0 kg were collected and specific gravity was measured. The majority of the samples comes from the Ådnefjell locality and for each granulite a corresponding eclogite was collected from less than a meter away. Six additional samples of eclogitized anorthositic material, together with two mafic eclogites were also collected. The results are shown in Fig. 11. The density of the granulite facies anorthositic rocks ranges from 2.79 to 3.21 g/cm^3 with an average of 3.02 g/cm^3 and of the eclogitized anorthositic rocks from 3.06 to 3.33 g/cm^3 with an average of 3.19 g/cm^3. The spread within each group is mainly due to variation in the bulk composition. The dependence on bulk composition is also demonstrated by the mafic eclogites which have a density between 3.5 and 3.6 g/cm^3.

The eclogites of the Bergen Arcs show varying amounts of diopside/albite symplectite surrounding omphacite. Symplectite formation is related to decomposition of omphacite during uplift [29]. The density difference between eclogites and granulites as obtained by these measurements should therefore be regarded as a minimum value.

5. Discussion and conclusions

The Grenvillian garnet-granulites of the Bergen Arcs, with abundant anorthositic material, are a plausible lower crustal suite both compositionally and mineralogically. The Caledonian eclogites, however, were formed at depths in the crust only attained below young mountain chains, e.g. the Alps and the Himalayas. The situation corresponds to the Fountain and Salisbury [2] model where the lower crust is composed of rocks which are polymetamorphic and multiply deformed due to orogenic recycling. However, the alternation of anorthositic rocks on a meter scale between granulite and eclogite over an area of tens of km^2 introduces new concepts. The unique nature of the Bergen Arc exposures is more probably explained by somewhat unique tectonic processes of uplift and preservation than by rarity of the eclogite-forming process in mountain roots.

5.1. Kinetic and metasomatic aspects of the granulite eclogite transition

Green and Ringwood [30] showed experimentally that the P and T where eclogites form from granulites are strongly dependent on bulk rock composition. For example, the plagioclase-out boundary at 1100° lies 10–15 kbar higher for a gabbroic anorthosite than for an alkali-poor olivine tholeiite. This indicates that the boundary between granulites and eclogites should be lithologically dependent.

The present account, on the other hand, describes a transition from granulite to eclogite which would seem to be independent of whole rock composition within the range anorthosite–gabbroic anorthosite–norite. This follows from the observation that shear zones with eclogites crosscut the compositional layering defined by the granulite facies foliation. The regional extent of the eclogitization further demonstrates that we are not dealing with a simple P,T-isograd relation. Since the granulites and the eclogites of the Bergen Arcs must have followed the same P,T path we are left with the two following explanations:

(1) Fluids and deformation played an important role in the triggering metamorphic reac-

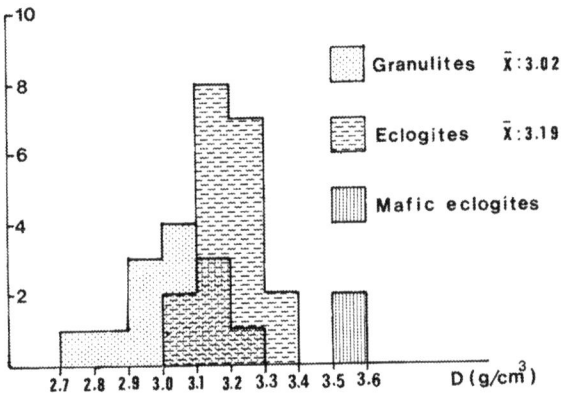

Fig. 11. Histogram showing density of granulite and eclogite facies anorthositic rocks. Two samples of mafic eclogites are also presented.

tion. This would require an extensive metastability in the lower crust.

(2) Small changes in fluid composition can lead to significant changes in mineral assemblages and mineral compositions [34]. Fluids can also facilitate element mobility and mass transport thereby changing the composition of a system.

The formation of eclogites with various hydrous minerals together with calcite and Cl-bearing hornblende from dry granulites strongly suggests that the fluid components were important in stabilizing the eclogite assemblages. The hornblende- and omphacite-filled veins in garnet further suggest element mobility at least on a local scale. It is therefore argued that the fluids penetrating these deep-crustal shear zones not only triggered metamorphic reactions but, more importantly, changed the thermodynamic properties and composition of the system to allow the metamorphic reactions to occur.

Fluid infiltration of the lower crust in the neighbourhood of shear zones has been reported from Southern India [31] where CO_2-rich fluids cause granulite facies metamorphism and also have a subtle metasomatic effect [32]. Schuiling and Kreulen [33] relate thermal domes to a CO_2-rich fluid stream from the mantle. Although the origin of the fluid penetrating the eclogite shear zones of the Bergen region is unknown, the described relationship suggests that the crust and the mantle undergo chemical exchanges via fluid motions. Regardless of this uncertainty, the described example from the Bergen Arcs shows that eclogitization of granulites can take place on a regional scale, unrelated to rock boundaries, and can produce a meter-scale alternation between granulites and eclogites.

5.2. The granulite-eclogite transition and the seismic signature of the lowermost crust

The transformation of granulites to eclogites is shown to have increased the rock density. This density change is likely to be reflected in a higher seismic velocity of eclogites as compared to granulites of the same composition and will give the partly eclogitized areas a seismic velocity intermediate between the eclogite and granulite facies anorthositic material. One can foresee from the Bergen Arcs example that the amount of eclogite versus granulite in the lower crust will vary not only with depth and temperature, but as a consequence of deformation and fluid migration. This must mean that the increasing seismic velocity with depth observed in some lower crustal sections can take place within the same rock type and reflect a higher proportion of eclogites versus granulite due to deformation and fluid transportation in the border region between crust and mantle.

The described relationships also give support to the Matthews and Cheadle [12] and Smithson et al.'s [11] suggestions that some of the complex deep crustal reflections may be related to fluids and mylonite zones in the lower crust. Phase transitions, new foliations, and the presence of fluids may enhance the seismic reflectivity. It is thus suggested that areas with high concentration of shear zones and eclogites, alternating with low strain areas of granulites, would result in a complex reflection pattern from a lower crust like the Bergen area. Combined with the observation that deformation in low crustal terrains tends to be concentrated in parallel shear belts, this may provide an adequate explanation for the fine scale layering observed in seismic profiling from the lower crust [35,36].

The quantification of these ideas in relation to the Bergen Arcs must await a detailed mapping and structural analysis of the eclogite shear zones combined with laboratory measurements of seismic velocities of the actual rocks and the shooting of seismic profiles in this area.

Acknowledgements

The author thanks Prof. W.L. Griffin, Prof. E.R. Neumann and Dr. M.B.E. Mørk for constructive criticism of the paper and B. Bjørnstad for assistance in the field. The careful reading and the suggestions made by an anonymous reviewer are gratefully acknowledged. The author acknowledges financial support from NAVF to the ILP project "Petrogenetic processes in the continental lithosphere". This is Contribution No. 14 from the Norwegian ILP project.

References

1 R.W. Kay and S.M. Kay, The nature of the lower continental crust: Interferences from geophysics, surface geology

and crustal xenoliths, Rev. Geophys. Space Phys. 19, 271–297, 1981.
2 D.M. Fountain and M.H. Salisbury, Exposed cross-sections through the continental crust: implications for crustal structure, petrology and evolution, Earth Planet. Sci. Lett. 56, 263–277, 1981.
3 W.L. Griffin and S.Y. O'Reilly, The composition of the lower crust and the nature of the continental Moho—xenolith evidence, in: Mantle Xenoliths, P.H. Nixon, ed., in press, Wiley, London, 1986.
4 E.R. Padovani and J.L. Carter, Aspect of the deep crustal evolution beneath south central New Mexico, in: The Earth's Crust, F. Heacock, ed., Am. Geophys. Union, Geophys. Monogr. 20, 19–56, 1977.
5 C. Pride and G.K. Muecke, Rare earth element geochemistry of the Scourian complex, NW Scotland—evidence for the granite-granulite link, Contrib. Mineral. Petrol. 73, 403–412, 1980.
6 I.F. Ermanovics and W.L. Davidson, The Pikwitonei granulites in relation to the north-western Superior Province of the Canadian Shield, in: The Early History of the Earth, B.F. Windley, ed., pp. 331–347, Wiley, New York, N.Y., 1976.
7 K. Fuchs, On the properties of deep crustal reflectors, Z. Geophys. 35, 133, 1969.
8 F.A. Cook, C.D. Brown, J.E. Oliver and S. Kaufman, The nature of the Moho on COCORP reflection data, EOS Trans. Am. Geophys. Union 59, 389, 1978.
9 D.M. Finlayson, C. Prodehl and C.D.N. Collins, Explosion seismic profiles, and implications for crustal evolution in southeastern Australia, Aust. Bur. Miner. Res., J. Aust. Geol. Geophys. 4, 243–252, 1979.
10 S.Y. O'Reilly and W.L. Griffin, A xenolith-derived geotherm for southeastern Australia and its geophysical implications, Tectonophysics, 111, 41–63, 1985.
11 S.B. Smithson, J.A. Brewer, S. Kaufman, J. Oliver and C. Hurich, Structure of the Laramide Wind River uplift, Wyoming, from COCORP deep reflection data and from gravity data, J. Geophys. Res. 84, 5955–5972, 1979.
12 D.M. Matthews and J. Cheadle, Deep reflection from the Caledonides and Variscides west of Britain and comparison of the Himalayas, in: Reflection Seismology: A Global Perspective, M. Barazangi and L. Brown, eds., Am. Geophys. Union, Geodyn. Ser. 13, 5–19, 1986.
13 P.J. Wyllie, The Dynamic Earth, 416 pp., John Wiley and Sons, Inc., 1971.
14 D.H. Green and A.E. Ringwood, An experimental investigation of the gabbro to eclogite transition and petrological implications, Geochim. Cosmochim. Acta 31, 767–833, 1967.
15 K. Ito and G.C. Kennedy, The fine structure of the basalt-eclogite transition, in: the Fiftieth Anniversary Symposia, B.A. Morgan, ed., Mineral. Soc. Am., Spec. Pap. 3, 1970.
16 H.J. Neugebauer and T. Spohn, Metastable phase transitions and progressive decline of gravitational energy: aspects of Atlantic type margin dynamics, in: Dynamics of Passive Margins, R.A. Scrutton, ed., Am. Geophys. Union, Geodyn. Ser., pp. 166–183, 1985.
17 B.A. Sturt, O. Skarpenes, A.T. Ohanian and I.R. Pringle, Reconnaissance Rb/Sr isochron study in the Bergen Arc system and regional implications, Nature (London) 253, 595–599, 1975.
18 C.F. Kolderup and N.H. Kolderup, Geology of the Bergen Arc System, Bergen Mus. Skrifter 20, 1–137, 1940.
19 C.F. Kolderup, Die Labradorfelsen des westlichen Norwegens, II. Die Labradorfelsen und die mit denselben verwandten Gesteine in dem Bergensgebiete, Bergen Mus. Årbog 12, 1–129, 1903.
20 V.M. Goldschmidt, Übersicht der Eruptivgesteine im Kaledonischen Gebirge zwischen Stavanger und Trondheim, Skr. Vidensks. Selsk.,Christallogr. 2, 1–140, 1916.
21 S. Smithson, P. Shive and S. Brown, Seismic velocity, reflections and structure of the crystalline crust, in: The Earth's Crust, J.F. Heacock, ed., Am. Geophys. Union, Geophys. Monogr. 20, 254–270, 1977.
22 J. Loeschke and R.P. Nickelsen, On the age and tectonic position of the Valdres Sparagmite in Slidre (Southern Norway), Neues Jahrb. Geol. Paläontol. 131, 337–367, 1968.
23 H. Austrheim and W.L. Griffin, Shear deformation and eclogite formation within granulite-facies anorthosites of the Bergen Arcs, western Norway, Chem. Geol. 50, 267–281, 1985.
24 H. Austrheim and B. Robins, Reaction involving hydration of orthopyroxene in anorthosite-gabbro, Lithos 14, 275–281, 1981.
25 H. Austrheim, Bergen Arcs, in: Excursion B1, Excursion in The Scandinavian Caledonides, W.L. Griffin and M.B.E. Mørk, eds., Uppsala Caledonide Symposium, 1981.
26 W.L. Griffin, Formation of eclogites and the coronas in anorthosites, Bergen Arcs, Norway, Geol. Soc. Am., Mem. 135, 37–63, 1972.
27 W.L. Griffin, M. Mellini, R. Oberti and G. Rossi, Evolution of coronas in Norwegian anorthosites: re-evaluation based on crystal-chemistry and microstructures, Contrib. Mineral. Petrol. 91, 330–339, 1985.
28 K. Bucher-Nurminen, The formation of metasomatic reaction veins in dolomitic marble roof pendants in the Bergell intrusion (Province Sondrio, Northern Italy), Am. J. Sci. 281, 1197–1222, 1981.
29 B.O. Mysen and W.L. Griffin, Pyroxene stoichiometry and breakdown of omphacite, Am. Mineral. 58, 60–63, 1972.
30 D.H. Green and A.E. Ringwood, A comparison of recent experimental data on the gabbro-garnet granulite-eclogite transition, J. Geol. 80, 277–288, 1972.
31 R.C. Newton and E.C. Hansen, The South Indian-Sri Lanka high-grade terrain as a possible deep-crust section, in: The Nature of the Lower Continental Crust, J.B. Dawson, ed., Geol. Soc. London Spec. Publ. 25, 297–307, 1986.
32 H.-J. Stähle, S. Hoernes and M. Raith, Granulitfazielle Umwandlung der Unterkruste: ein isochemischer Prozess?, Fortschr. Mineral. 63, 225, 1985.
33 R.D. Schuiling and Kreulen, Are thermal domes heated by CO_2-rich fluids from the mantle?, Earth Planet. Sci. Lett. 43, 298–302, 1979.
34 W.E. Glassley and D. Bridgewater, Fluid enhanced mass transport in deep crust and its influence on element abundances and isotope systems, in: The Deep Proterozoic Crust in the North Atlantic Provinces, A.C. Tobi and J.L.R. Touret, eds., NATO ASI Ser. 158, 105–119, 1985 (Reidel, Dordrecht).

35 S.B. Smithson, W.R. Pierson, W.L. Sharon and R.A. Johnson, Seismic reflection results from Precambrian crust, in: The Deep Proterozoic Crust in the North Atlantic Provinces, A.C. Tobi and J.L.R. Touret, eds., NATO ASI Ser. 158, 21–37, 1985 (Reidel, Dordrecht).

36 D.J.M. Oliver, S. Kaufman, R. Meyer and R. Pinney, Continuous seismic reflection profiling of the deep basement, Hardeman County, Texas, Geol. Soc. Am. Bull. 87, 1537–1546, 1976.

Others in the Landmark Series

LandmarkPapers I
Volcanic Petrology
IAN S.E. CARMICHAEL

THIS book summarizes the development of volcanic petrology from the 1920s to the present day and is a 'must-read' for all geologists interested in igneous rocks. (B.J. WOOD)

This book is dedicated to all those for whom volcanic rocks hold an abiding fascination. It has taken the efforts of geolgists (and let it be stressed, geochemists and experimentalists!) for more than a century to give volcanic petrology is depth, colour and intellectual ferment (I.S.E. CARMICHAEL).

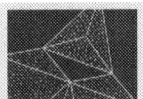

LandmarkPapers II
Structure Topology
FRANK C. HAWTHORNE

THE atomistic theory of the Greeks was a solely philosophical construct, and further development had to await a more sophisticated approach to Science. The first steps in this direction were taken by who else but Isaac Newton (1643–1727 AD). Although his ideas on action at a distance initially referred to planets, he also considered them as applying to atoms, and concluded from physical evidence involving surface tension and viscosity that there must be strong attractions between atoms. In what must be considered as insight of legendary proportions, Roger Joseph Boscovich (1711–1787), a Jesuit mathematician from Croatia (Malescio, 2003), proposed that at very short distances, atoms repulse each other, the repulsion increasing indefinitely as the particles become closer together, whereas at longer distances apart, atoms oscillate between attraction and repulsion.

In an editorial in *Nature*, Maddox (1988) noted that there is probably more information available on the spatial arrangements of atoms in crystals than on any other topic in Science, and considered it a "continuing scandal" that "it remains in general impossible to predict the structure of even the simplest crystalline solids from a knowledge of their chemical composition". (F.C. HAWTHORNE)

I think that the re-publication of these landmark papers, accompanied by the commentaries of Prof. Hawthorne, will be useful not only for undergraduate or PhD students, but for all structural mineralogists. This collection provides valuable insights into the evolution of structural mineralogy and its wider application to the petrology. I warmly recommend this volume to all mineralogists and to Earth sciences libraries. (G.D. GATTA, *MINERALOGICAL MAGAZINE*)

FRONT COVER – Granulite-facies metapelite from the Ivrea Zone, Italy; polarized-light photomicrograph taken with a 1/4 wavelength plate; showing sillimanite, biotite, cordierite (with pleochroic haloes), and garnet (courtesy of Peter Nievergelt, ETH Zuerich, Switzerland).

BACK COVER – Polarized-light photomicrograph of twinned coesite rimmed by polycrystalline and palisade quartz, enclosed in pyrope garnet; Parigi, Dora Maira, Italy (see Chapter 14) (courtesy of Hans-Peter Schertl, Ruhr University, Bochum, Germany).

www.ingramcontent.com/pod-product-compliance
Ingram Content Group UK Ltd.
Pitfield, Milton Keynes, MK11 3LW, UK
UKHW051525180426
11947UKWH00018B/1581